THEORY AND APPLICATIONS OF SPECIAL FUNCTIONS

A Volume Dedicated to Mizan Rahman

Developments in Mathematics

VOLUME 13

Series Editor:

Krishnaswami Alladi, *University of Florida, U.S.A.*

Aims and Scope

Developments in Mathematics is a book series publishing

(i) Proceedings of conferences dealing with the latest research advances,

(ii) Research monographs, and

(iii) Contributed volumes focusing on certain areas of special interest.

Editors of conference proceedings are urged to include a few survey papers for wider appeal. Research monographs, which could be used as texts or references for graduate level courses, would also be suitable for the series. Contributed volumes are those where various authors either write papers or chapters in an organized volume devoted to a topic of special/current interest or importance. A contributed volume could deal with a classical topic that is once again in the limelight owing to new developments.

THEORY AND APPLICATIONS OF SPECIAL FUNCTIONS

A Volume Dedicated to Mizan Rahman

Edited by

MOURAD E.H. ISMAIL
University of Central Florida, U.S.A.

ERIK KOELINK
Technische Universiteit Delft, The Netherlands

Springer

Library of Congress Cataloging-in-Publication Data

A C.I.P. record for this book is available from the Library of Congress.

ISBN 0-387-24231-7 e-ISBN 0-387-24233-3 Printed on acid-free paper.

Printed in the United States of America.

9 8 7 6 5 4 3 2 1 SPIN 11367123

springeronline.com

Contents

Preface

This volume, "Theory and Applications of Special Functions," is dedicated to Mizan Rahman in honoring him for the many important contributions to the theory of special functions that he has made over the years, and still continues to make. Some of the papers were presented at a special session of the American Mathematical Society Annual Meeting in Baltimore, Maryland, in January 2003 organized by Mourad Ismail.

Mizan Rahman's contributions are not only contained in his own papers, but also indirectly in other papers for which he supplied useful and often essential information. We refer to the paper on his mathematics in this volume for more information.

This paper contains some personal recollections and tries to describe Mizan Rahman's literary writings in his mother tongue, Bengali. An even more personal paper on Mizan Rahman is the letter by his sons, whom we thank for allowing us to reproduce it in this book.

The theory of special functions is very much an application driven field of mathematics. This is a very old field, dating back to the 18th century when physicists and mathematician were looking for solutions of the fundamental differential equations of mathematical physics. Since then the field has grown enormously, and this book reflects only part of the known applications.

About half of the mathematical papers in this volume deal with basic (or q-) hypergeometric series—in particular summation and transformation formulas—special functions of basic hypergeometric type, or multivariable analogs of basic hypergeometric series. This reflects the fact that basic hypergeometric series is one of the main subjects in the research of Mizan Rahman. The papers on these subjects in this volume are usually related to, or motivated by, different fields of mathematics, such as combinatorics, partition theory and representation theory. The other main subjects are hypergeometric series, and special functions of hypergeometric series, and generalities on special functions and orthogonal polynomials.

The papers in this volume on basic hypergeometric series can be subdivided into three groups: (1) papers on identities, such as integral representations, addition formulas, (bi-)orthogonality relations, for specific sets of special functions of basic hypergeometric type; (2) papers on summation and transformation formulas for single or multivariable basic hypergeometric series; and (3) papers related to combinatorics and Rogers-Ramanujan type identities. Some of the papers in this volume fall into more than just one class.

In the first group we find the two papers by Gasper and Rahman one on q-analogs of work of Tratnik on multivariable Wilson polynomials yielding multivariable orthogonal Askey-Wilson polynomials and its limit cases and the other paper on q-analogs of multivariable biorthogonal polynomials. The paper by Ismail and R. Zhang studies the q-analog $\mathcal{E}_q$ of the exponential function, giving, amongst other things, new proofs of the addition formula and its expression as a ${}_2\varphi_1$-series. They also present new derivations of the important Nassrallah-Rahman integral and connection coefficients for Askey-Wilson polynomials. Koornwinder gives an analytic proof of an addition formula for a three-parameter subclass of Askey-Wilson polynomials in the spirit of the Rahman-Verma addition formula for continuous q-ultraspherical polynomials. Stokman's paper simplifies previous work of Koelink and Stokman on the calculation of matrix elements of infinite dimensional quantum group representations as Askey-Wilson functions for which Rahman has supplied them with essential summation formulas. He uses integral representations for these matrix elements and shows how this can be extended to the case $|q| = 1$. Stokman's paper and Rahman's summation formulas are the motivation for Rosengren's paper using Stokman's method to extend Rahman's summation formulas. The paper by Abreu and Bustoz deals with completeness properties of Jackson's third (or the ${}_1\varphi_1$) q-Bessel function for its Fourier-Bessel expansion.

In the second group of papers on summation and transformation formulas for (multivariable) basic hypergeometric series we have the above mentioned short paper by Rosengren giving summation formulas involving bilateral sums of products of two basic hypergeometric series. Schlosser derives bilateral series from terminating ones both in the single and multivariable case. Multiple transformation formulas using the q-Pfaff-Saalscütz formula recursively are obtained by Chu. Kadell discusses various summation formulas as moments for little q-Jacobi polynomials, and extends this approach to non-terminating cases of these summation formulas.

The papers in the third group present some connections between basic hypergeometric series and, number theory and combinatorics, especially Rogers-Ramanujan type identities and the partition function. Andrews discusses two-variable analogs of the Gaussian polynomial and finite Rogers-Ramanujan type identities. In the paper by Berndt, Chan, Chan and Liaw the crank of partitions and its relation to entries in Ramanujan's notebooks are discussed. The paper by Chu also shows how the multiple transformation formulas yield multiple Rogers-Ramanujan type identities.

In this volume there are two papers that deal mainly with hypergeometric series. The paper by Stanton gives cubic and higher order transformation and summation formulas for hypergeometric series by splitting series up as a sum of r series, presenting q-analogs as well. The paper by Groenevelt, Koelink and Rosengren is solely devoted to hypergeometric series. The paper contains a summation formula where the summand involves a product of two Meixner-Pollaczek polynomials and a continuous dual Hahn polynomial. Then a Lie algebraic interpretation gives a transformation pair involving non-terminating ${}_3F_2$-series, which is proved analytically.

There are four papers that deal with aspects of the general theory for special functions and orthogonal polynomials and related subjects. Suslov's paper discusses a version of the Cauchy-Hadamard theorem giving the maximum domain of analyticity of expansions of functions into orthogonal polynomials of basic hypergeometric type. Szwarc discusses nonnegative linearization for both orthogonal polynomials and its associated polynomials, and shows the equivalence to a maximum principle for a canonically associated discrete boundary value problem. In Ruijsenaars's paper the L^2-asymptotics of orthogonal polynomials on $[-1, 1]$ having a c-function expansion as the degree tends to infinity is given explicitly with an exponentially decaying error term. The paper by C. Zhang deals with summation methods involving Jacobi's theta function, which gives a summation method for divergent series. This is applied to the confluence of ${}_2\varphi_1$-series to ${}_2\varphi_0$-series, and a new q-analog of the Γ-function.

The remaining papers do not fit into the scheme given above. Using basic hypergeometric series, Berg gives direct proofs of some results on distributions for exponential functionals of Lévy processes. In particular he obtains the corresponding Laplace and Mellin transforms, which have been previously obtained by different methods from stochastic processes. Clarkson discusses polynomials that occur in relation to rational solutions of the second, third and fourth Painlevé equations, in particular the Yablonskii-Vorob'ev, (generalized) Okamoto and generalized Hermite polynomials, and he demonstrates experimentally that the zeroes of these polynomials behave in a very regular fashion. DeDeo, Martínez,

Medrano, Minei, Stark and Terras study Ihara-Selberg zeta functions of Cayley graphs for the Heisenberg group over certain finite rings, and discuss a corresponding Artin L-function.

This book was prepared at the University of South Florida. Denise Marks put the book together and handled all correspondence and the galley proofs. We thank Denise for all she has done for this project. Working with Denise is always a pleasure.

We take this opportunity and thank all the speakers and participants in the American Mathematical Society Special Session, all the contributing authors, and the referees, in making this book a worthy tribute to Mizan Rahman.

Orlando, FL, and Delft,
July 2004.

Mourad E.H. Ismail
Erik Koelink

This volume is dedicated to Mizan Rahman in recognition of his contributions to special functions, q-series and orthogonal polynomials.

MIZAN RAHMAN, HIS MATHEMATICS AND LITERARY WRITINGS

Richard Askey
Department of Mathematics
University of Wisconsin
Madison, Wisconsin 53706

Mourad E.H. Ismail
Department of Mathematics
University of Central Florida
Orlando, FL 32618

Erik Koelink
Department of Mathematics
Technische Universiteit Delft
Delft, The Netherlands

Mizan studied at the University of Dhaka where he obtained his B.Sc. degree in mathematics and physics in 1953 and his M.Sc. in applied mathematics in 1954. He received a B.A. in mathematics from Cambridge University in 1958, and a M.A. in mathematics from Cambridge University in 1963. He was a senior lecturer at University of Dhaka from 1958 until 1962. Mizan decided to go abroad for his Ph.D. He went to the University of New Brunswick in 1962 and received his Ph.D. in 1965 with a thesis on Kinetic Theory of Plasma using singular integral equations techniques. After obtaining his Ph.D., Mizan became an assistant professor at Carleton University, where he spent the rest of his career. He is currently a distinguished professor emeritus there.

In this article we mainly discuss some of Mizan's mathematical results which are the most striking and influential, at least in our opinion. Needless to say, we cannot achieve completeness since Mizan has written so many interesting papers. The reference item preceded by **CV** refer to items under "Publications" on Mizan's CV while the ones without **CV** refer to references at the end of this article.

In the early part of his research career Mizan devoted some energy to questions involving statistical distributions resulting in the papers **CV**[5], **CV**[7] and **CV**[11]. Mizan spent the academic year 1972/73 at Bedford College, of the University of London on sabbatical and worked with Mike Hoare. In an e-mail to the editors, Hoare described how the liberal arts atmosphere of Bedford, set idyllically in Regent Park was well-suited for Mizan but the down side was that Physics at Bedford was a small department and "there was little resonance in the heavily algebracisized Mathematics Department under Paul Cohen." He added "This hardly seemed to matter, since we were both outsiders from what was most fashionable at the time."

Hoare's original plan was to study a one-dimensional gas model known as the Rayleigh piston, but his collaboration with Mizan went way beyond this goal. This resulted in **CV**[9] and **CV**[12]. Another problem suggested by Hoare involved urn models which made them soon realize that the urn models they were investigating were related to birth and death processes and Jacobi and Hahn polynomials. The result of their investigations are papers **CV**[13], **CV**[17], and **CV**[18]. Some probabilistic interpretations of identities for special functions were known, but it was not an active area of research. The Hoare-Rahman papers dealt with exactly solvable models where the eigenvalues and eigenfunctions have been found explicitly. Such questions led in a very natural way to certain kernels involving the Hahn and Krawtchouk polynomials. These kernels were reproducing kernels which take nonnegative values. More general bilinear forms involving orthogonal polynomials also appeared. The question of positivity of these kernels became important and Mizan started corresponding with R. Askey who, with G. Gasper, was working on positivity questions at the time and they were very knowledgeable about these questions. Through Askey and Gasper, Mizan Rahman was attracted to the theory of special functions and eventually to q-series. He mastered the subject very quickly and started contributing regularly to the subject. Within a few years, Mizan had become a world's expert in the theory of special functions in general and q-series in particular. It is appropriate here to quote from Mike Hoare's e-mail how he described the beginning of this activity. Mike wrote "After we had done some work on ... (Rayleigh Piston) ... I happened to mention a problem which I have been worrying away at for some years. This disarmingly simple notion arose from energy transfer in chemical kinetics (the Kassell model). Reformulated as a discrete 'urn model,' it corresponds to a Markoff chain for partitioning balls in boxes in which only a subset are randomized in each event. My eigenvalue solution for the simplest continuous case in Laguerre polynomials led to probability kernels (which are) effectively the same as those seen in the formulas of Erdélyi and Kogbetliantz in the 1930's special function theory." He then added "Once Mizan's interest was stimulated, he was off and running, with the early series of abstract papers you well know." Mike Hoare echoed the feelings of those of us who collaborated with Mizan when he wrote "To see Mizan at work was an amazing experience. He seldom had to cross anything out and [in] what seemed no time at all the sheets in his characteristically meticulous script would be delivered with a modest little gesture of triumph."

The Gegenbauer addition formula for the ultraspherical polynomials was found in 1875. It says

$$
\begin{aligned}
&C_n^{\nu}(\cos\theta\cos\varphi + z\sin\theta\sin\varphi) \\
&= \sum_{k=0}^{n} a_{k,n}(\nu)(\sin\theta)^k C_{n-k}^{\nu+k}(\cos\theta)(\sin\varphi)^k C_{n-k}^{\nu+k}(\cos\varphi) C_k^{\nu-1/2}(z),
\end{aligned}
\tag{1}
$$

where

$$
a_{k,n}(\nu) = \frac{\Gamma(2\nu-1)}{\Gamma^2(\nu)} \, \frac{\Gamma^2(k+\nu)\,(n-k)!(2k+2\nu-1)}{4^{-k}\,\Gamma(n+k+2\nu)}. \tag{2}
$$

The continuous q-ultraspherical polynomials first appeared in the work of L. J. Rogers from the 1890's on expansions of infinite products, which contained what later became known as the Rogers-Ramanujan identities. Their weight function and orthogonality relation were found in the late 1970's, (Askey and Ismail, 1983), (Askey and Wilson, 1985). Mizan recognized the importance of these polynomials and, in joint work with Verma, they extended the Gegenbauer addition formula to the continuous q-ultraspherical polynomials. In **CV**[48] they proved

$$
\begin{aligned}
C_n(z;\beta\,|\,q) = &\sum_{k=0}^{n} A_{k,n}(\beta) C_{n-k}\left(\cos\theta; \beta q^k\,|\,q\right) C_{n-k}\left(\cos\varphi; \beta q^k\right) \\
&\times p_k\left(z; \sqrt{\beta}\,e^{i(\theta+\varphi)}, \sqrt{\beta}\,e^{i(\theta-\varphi)}, \sqrt{\beta}\,e^{i(\varphi-\theta)}, \sqrt{\beta}\,e^{-i(\theta+\varphi)}\,|\,q\right),
\end{aligned}
\tag{3}
$$

where $A_{k,n}(\beta)$ are constants which are given in closed form, and the polynomial $p_n(x;a,b,c,d)$ is an Askey-Wilson polynomial. This result led to a product formula for the same polynomials. At the time the Rahman-Verma addition theorem was very surprising for two reasons. Firstly, the variables θ and φ appear in the parameters of the Askey-Wilson polynomial. Even more surprising is the fact that the terms in (3) factor in an appropriate symmetric way, since this factorization was not predicted by any structure known at the time. Only later partial explanations using representation theory of quantum groups have been given (Koelink, 1997).

Askey (Askey, 1970) raised the question of finding the domain of (α, β) within $\alpha > -1, \beta > -1$ which makes the linearization coefficients $a(m, n, k)$ in

$$P_n^{(\alpha,\beta)}(x)P_m^{(\alpha,\beta)}(x) = \sum_{k=|m-n|}^{m+n} a(m,n,k)P_k^{(\alpha,\beta)}(x). \tag{4}$$

nonnegative. E. Hylleraas (Hylleraas, 1962) showed that the coefficients $a(m, n, k)$ satisfy a three-term recurrence relation, and showed that the case $\alpha = \beta + 1$ leads to a closed form solution, as was the case when $\alpha = \beta$. For other (α, β) (except $\beta = -\frac{1}{2}$), the coefficients were represented as double sums, and this expression cannot be used for any of the applications the writers know except for computing a few of the coefficients. In (Gasper, 1970a) and (Gasper, 1970b), G. Gasper used the recurrence relation of Hylleraas to solve the problem of the positivity of these coefficients. Mizan started working on extending Gasper's results to the continuous q-Jacobi polynomials, where the problem is much more difficult. Rahman **CV**[27] identified $a(m, n, k)$ as a ${}_9F_8$ function and then used the ${}_9F_8$-representation to prove the nonnegativity of the linearization coefficients. Later Mizan **CV**[30] used the same technique to identify the q-analogue of $a(m, n, k)$ as a ${}_{10}\varphi_9$ and establish its nonnegativity for (α, β) in a certain subset of $(-1, \infty) \times (-1, \infty)$.

The linearization coefficients in (4) are integrals of products of three Jacobi polynomials. Din (Din, 1981) proved that

$$\int_{-1}^{1} P_m(x)P_n(x)Q_n(x)dx = 0, \quad \text{for } |m-n| < k < m+n, \tag{5}$$

where $\{P_n(x)\}$ are Legendre polynomials and $\{Q_n(x)\}$ are Legendre functions of the second kind. Askey, Koornwinder and Rahman **CV**[50] extended this to the ultraspherical polynomials. Rahman and Shah **CV**[39] summed the series, which is dual to (5), namely

$$F(\theta, \varphi, \psi) := \sum_{n=0}^{\infty} (n + 1/2)P_n(\cos\theta)P_n(\cos\varphi)Q_n(\cos\psi), \tag{6}$$

$0 < \theta, \varphi, \psi < \pi$. They proved that $F(\theta, \varphi, \psi) = 0$ for $|\theta - \varphi| < \psi < \theta + \varphi \leq \pi$, but $F(\theta, \varphi, \psi) = \Delta^{1/2}$ if $\psi < |\theta - \varphi|$, or $\pi < \theta + \varphi < 2\pi$ and $\theta + \varphi + \psi < 2\pi$. On the other hand, $F(\theta, \varphi, \psi) = -\Delta^{1/2}$, if $\pi \geq \psi > \theta + \varphi$. In the above

$$\begin{aligned}\Delta := {} & \sin((\theta + \varphi + \psi)/2)\sin((\theta + \varphi - \psi)/2) \\ & \times \sin((\theta - \varphi + \psi)/2)\sin((\varphi + \psi - \theta)/2).\end{aligned} \tag{7}$$

They also extended it to the ultraspherical polynomials. In **CV**[40], this was extended to the case where the sum involves a product of three ultraspherical polynomials and an ultraspherical function of the second kind. In **CV**[51], Rahman and Verma extended **CV**[39] in a different direction where it involved products of two q-ultraspherical polynomials and a q-ultraspherical function.

The Askey-Wilson integral (Askey and Wilson, 1985) is

$$\int_0^\pi \frac{\left(e^{2i\theta}, e^{-2i\theta}; q\right)_\infty}{\prod_{k=1}^{4} \left(a_k e^{i\theta}, a_k e^{-i\theta}; q\right)_\infty} d\theta = \frac{2\pi \left(a_1 a_2 a_3 a_4; q\right)_\infty}{(q;q)_\infty \prod_{1\le j<k\le 4} \left(a_j a_k; q\right)_\infty}. \tag{8}$$

This integral is an analogue of the beta integral and is the key ingredient in establishing the orthogonality of the Askey-Wilson polynomials. Nassrallah and Rahman **CV**[28] generalized this integral to what has become known as the Nassrallah-Rahman integral, namely

$$\begin{aligned} &\int_0^\pi \frac{\left(e^{2i\theta}, e^{-2i\theta}, be^{i\theta}, be^{-i\theta}; q\right)_\infty}{\prod_{k=1}^{5} (a_k e^{i\theta}, a_k e^{-i\theta}; q)_\infty} d\theta \\ &= \frac{2\pi \left(a_1a_2a_3a_4, a_1a_2a_3a_5, a_4a_5, ba_1, ba_2, ba_3; q\right)_\infty}{\left(q, a_1a_2a_3b; q\right)_\infty \prod_{1\le j<k\le 5} \left(a_j a_k; q\right)_\infty} \\ &\times {}_8W_7\left(a_1a_2a_3b/q; a_1a_2, a_1a_3, a_2a_3, b/a_4, b/a_5; a, a_4a_5\right). \end{aligned} \tag{9}$$

This is a very general extension of Euler's integral representation of the classical ${}_2F_1$ function. The sum (9) can be evaluated when $b = a_1a_2a_3a_4a_5$, and in this form it is an extension of (8). The integral evaluation (9) is precisely what is needed to introduce biorthogonal rational function generalizations of the Askey-Wilson polynomials, which was done by Mizan in **CV**[65]. In **CV**[75], **CV**[80], **CV**[90] and **CV**[91], Rahman and Suslov gave evaluations of several sums and integrals using the Pearson equation and quasi-periodicity of the integrand and summands. Earlier Ismail and Rahman **CV**[76] used the quasiperiodicity to evaluate, in closed form, certain series and integrals.

Ismail and Rahman **CV**[66] introduced and analyzed two families of orthogonal polynomials which arise as associated Askey-Wilson polynomials. This is the highest level in a hierarchy of associated polynomials of classical orthogonal polynomials starting from the Askey-Wimp and Ismail-Letessier-Valent associated Hermite and Laguerre polynomials from the mid-1980's. The Askey-Wilson polynomials are birth and death process polynomials, so their associated polynomials are also birth and death process polynomials. The death rate at zero population, μ_0, is either zero or follows the pattern of μ_n. Each definition of μ_0 leads to a family of orthogonal polynomials. Surprisingly, thanks to Mizan's insight and amazing computational power, one can get not only closed form representations for both families of polynomials, but also find the orthogonality measures of both families explicitly. The closed form expressions represent the polynomials in terms of the Askey-Wilson basis $\left\{\left(ae^{i\theta}, ae^{-i\theta}; q\right)_n\right\}$. The coefficients are ${}_{10}\varphi_9$. No other choice of μ_0 leads to orthogonal polynomials where the coefficients of their expansion in the Askey-Wilson basis is a single sum. The paper also gives a basis of solutions to the recurrence relation satisfied by the polynomials. Rahman and Tariq derived a Poisson and related kernels for associated continuous q-ultraspherical polynomials in **CV**[86] and reproducing kernels for associated Askey-Wilson polynomials in **CV**[88]. Earlier, Mizan **CV**[78] found generating functions for the Askey-Wilson polynomials. In **CV**[41] Mizan gave a q-analogue of Feldheim's kernel (Feldheim, 1941) which involves q-utraspherical polynomials. An integral representation analogous to the Weyl fractional integral in Fourier analysis is in **CV**[99].

Recall the notation $\breve{f}(z) = f(x)$, where $x = \left(z + z^{-1}\right)/2$, and the Askey-Wilson operator

$$(\mathcal{D}_q f)(x) = \frac{\breve{f}\left(q^{1/2}z\right) - \breve{f}\left(q^{-1/2}z\right)}{\breve{e}\left(q^{1/2}z\right) - \breve{e}\left(q^{-1/2}z\right)}, \tag{10}$$

where $e(x) = x$. Ismail raised the question of finding $\mathcal{D}_q^{-1}$, a right inverse to $\mathcal{D}_q$ on different L^2 spaces weighted by weight functions of different classical q-orthogonal polynomials. In **CV**[79] Ismail, Rahman and Zhang found an integral representation for $\mathcal{D}_q^{-1}$ on L^2 weighted by the weight function of the continuous q-Jacobi polynomials. They then proved that the spectrum of the compact operator $\mathcal{D}_q^{-1}$ is discrete and described completely the eigenvalues and eigenfunctions. This led to a generalization of the plane wave expansion in (Ismail and Zhang, 1994). Ismail and Rahman **CV**[100] found an integral representation of $\mathcal{D}_q^{-1}$ on L^2 weighted by the Askey-Wilson weight function. The kernel in this integral representation turned out to be very simple and all the complications are absorbed in a constant.

Mizan has served the mathematical community very well. He is a regular referee for many mathematics and physics journals. He co-organized a major meeting on special functions at the Fields Institute in Toronto in 1994 and co-edited its proceedings. One of the most important ways a mathematician can serve is in writing books which are needed. Gasper and Rahman (Gasper and Rahman, 1990) did this, and a second edition will appear shortly. The new edition will not only have more on q-series, but will contain a chapter on new work on elliptic hypergeometric series (Frenkel and Turaev, 1997), a very interesting new extension of hypergeometric and basic hypergeometric series. This extended the earlier trigonometric case in (Frenkel and Turaev, 1995). The Gasper-Rahman book also contains a treatment of the ${}_{10}\varphi_9$ biorthogonal rational functions **CV**[65] and its elliptic extensions. The book started because, according to George Gasper "Mizan was tired of having to repeatedly search papers for known formulas involving basic hypergeometric functions that were not contained in the books by Bailey or Slater." Mizan then suggested that he and Gasper should write an up-to-date book on basic hypergeometric functions. A first outline of this book dates back to 1982. Their book has become a much-cited classic, and Mizan and George have rendered the mathematical community a great service in writing this book.

Not only did Mizan co-author the definitive book on q-series, but he also wrote valuable review articles **CV**[64], **CV**[87], **CV**[97] and **CV**[98]. To the best of our knowledge, **CV**[98] is the first article which collects all the recent developments on associated orthogonal polynomials, which makes it a very valuable reference and teaching source.

Mizan's scientific contributions have been recognized and acknowledged. Part of his dissertation was included in a book on gases and plasmas by Wu (Wu, 1966). A special session was held in his honor at the the annual meeting of the American Mathematical Society held in Baltimore, Maryland. The session was well attended and highly successful. Several speakers expressed their mathematical debt to Mizan and noted his generosity with his ideas. He has helped younger mathematicians with suggestions and specific ideas on how to overcome certain hurdles and would not have his name as a joint author of the resulting paper(s). Mizan's contributions are well-appreciated by people working in special functions and related areas. R. W. (Bill) Gosper put it well when he wrote on April 7, 2004 "I can't begin to estimate Mizan Rahman's prowess as a q-slinger. All I know is that he alone could 'q' any hypergeometric identity that I could find. Sometimes the q-form was so unimaginable that I would have bet money there was none." He then added "And yet the memory that stands out was not a q. I exhibited to the usual gang of maniacs a really mysterious-looking infinite trig product identity, dug up with Macsyma. It wasn't even obvious that the nth term converged to 1. And that gentle man completely stung me with a reply that began, "Since this identity is rather elementary, let us prove the more general result .... That's when you know you're in the Big Leagues."

Mizan serves on the editorial board of an international journal *Integral Transforms and Special Functions.* He was been elected fellow of the Bangladesh Academy of Sciences in 2002. Since his retirement in 1996, Mizan has been a Distinguished Professor Emeritus at Carleton University.

Apart from papers in mathematics, Mizan has several publications in Bengali. He writes essays for several Bengali magazines, such as Parabaas, Dehes-Bideshe, Porshi, Natun Digonto, Probashi, Aamra, Obinashi Shobdorashi and Aakashleena on a regular basis. These essays are personal and dwell on the immigration experience, and are comparative studies of lifestyles, ethics and values in the societies of Bangladesh and India compared to the American and Canadian societies. Subjects that Mizan addresses cover raising children in a proper humanistic value system, and the problems that aging immigrants face. According to one of the editors, Samir Bhattacharya, his articles drew overwhelming response of appreciation from the readers, because "the language has an apparent simplicity, but is often lyrical and extremely touching, reasoning is clear—but above all, a deep humanism and his simplicity and integrity shine through," and, as Mr. Bhattacharya says: "I will publish any article from him any time."

Mizan Rahman has also published several books in Bengali; "Tirtho Aamar Gram" (My Village is My Pilgrimage), "Lal Nodi" (Red River) a collection of 25 of his essays, "Proshongo Nari" on women and "Album." He has been awarded for his contributions several times, including an Award for Excellence, a Best Writer Award, Award for Contributions to Bengali literature by several organizations in North America.

References

Andrews, G. E., Askey, R. A., and Roy, R. (1999). *Special Functions.* Cambridge University Press, Cambridge.

Askey, R. (1975). *Orthogonal Polynomials and Special Functions.* Regional Conference Series in Applied Mathematics. SIAM, Philadelphia, PA.

Askey, R. and Ismail, M. E. H. (1983). A generalization of ultraspherical polynomials. In Erdős, P., editor, *Studies in Pure Mathematics*, pages 55–78. Birkhäuser, Basel.

Askey, R. A. (1970). *Linearization of the product of orthogonal polynomials*, pages 131–138. Princeton Univ. Press, Princeton, N. J.

Askey, R. A. and Wilson, J. A. (1985). Some basic hypergeometric orthogonal polynomials that generalize Jacobi polynomials. *Memoirs Amer. Math. Soc.*, 54(319):iv+55.

Din, A. M. (1981). A simple sum formula for Clebsch-Gordan coefficients. *Lett. Math. Phys.*, 5:207–211.

Feldheim, E. (1941). Contributions à la théorie des polynomes de Jacobi. *Mat. Fiz. Lapok*, 48:453–504.

Frenkel, I. B. and Turaev, V. G. (1995). Trigonometric solutions of the Yang-Baxter equation, nets, and hypergeometric functions. In *Functional analysis on the eve of the 21st century, Vol. 1 (New Brunswick, NJ, 1993)*, volume 192 of *Progress in Mathematics*, pages 65–118. Birkhäuser Boston, Boston, MA.

Frenkel, I. B. and Turaev, V. G. (1997). *Elliptic solutions of the Yang-Baxter equation and modular hypergeometric functions*, pages 171–204. Birkhäuser Boston, Boston, MA.

Gasper, G. (1970a). Linearization of the product of Jacobi polynomials. I. *Canad. J. Math.*, 22:171–175.

Gasper, G. (1970b). Linearization of the product of Jacobi polynomials. II. *Canad. J. Math.*, 22:582–593.

Gasper, G. and Rahman, M. (1990). *Basic Hypergeometric Series.* Cambridge University Press, Cambridge.

Hylleraas, E. (1962). Linearization of products of Jacobi polynomials. *Math. Scand.*, 10:189–200.

Ismail, M. E. H. and Zhang, R. (1994). Diagonalization of certain integral operators. *Adv. Math.*, 109:1–33.

Koelink, E. (1997). Addition formulas for q-special functions. In Ismail, M. E. H., Masson, D. R., and Rahman, M., editors, *Special functions, q-series and related topics*, volume 14 of *Fields Institute Communications*, pages 109–129. American Mathematical Society, Providence, RI.

Wu, T.-Y. (1966). *Kinetic Theory of Gases and Plasma.* Addison-Wesley.

Ottawa
April 2004

CURRICULUM VITAE

A. NAME: Mizanur Rahman

B. DATE AND PLACE OF BIRTH: September 16, 1932, Dhaka, Bangladesh

C. EDUCATION:

Degree	Course	University	Year Received
B.Sc.	Math/Physics	Dhaka	1953
M.Sc.	Applied Math	Dhaka	1954
B.A.	Mathematics	Cambridge	1958
M.A.	Mathematics	Cambridge	1963
Ph.D.	Mathematics	New Brunswick	1965

D. EMPLOYMENT EXPERIENCE:

Date	Position	Institution
1958–62	Senior lecturer	Dhaka University
1962–65	Lecturer	University of New Brunswick
1965–69	Assistant Professor	Carleton University
1969–78	Associate Professor	Carleton University
1978–1998	Professor	Carleton University
1998-	Distinguished Research Professor (Retd.)	

E. ACADEMIC AND PROFESSIONAL RESPONSIBILITIES:

1. Teaching—Graduate Supervision

(i) Supervised one M.Sc. thesis in 1978.

(ii) Acted as a joint adviser with Professor M. R. Hoare for S. Raval, a Ph.D. student in Theoretical Physics at Bedford College, London University.

(iii) Supervised the Ph.D. theses of B. Nassrallah, M. J. Shah, and Q. Tariq.

2. Scholarly Studies & Research

a) **Research Interests**. Non-negative sums and integrals of special functions. Summation and transformation theorems of basic hypergeometric series. q-orthogonal polynomials and q-Fourier transforms.

b) **Publications**

1. "Collisional Relaxation in a Hard-sphere Gas" (with M. K. Sundaresan), *Phys. Lett.* **25A** (1967), 705–706.
2. "Continuum Eigenfunctions for a Hard-sphere Gas" (with M. K. Sundaresan), *Can. J. Phys.* **46** (1968), 2287–2290.
3. "Discrete Relaxation Modes for a Hard-sphere Gas" (with M. K. Sundaresan), *Can. J. Phys.* **46** (1968), 2463–2469.
4. "Continuum Eigenfunctions in the Neighbourhood of Singularities for a Uniform Hard-sphere Gas," *Can. J. Phys.* **48** (1970), 151–153.
5. "The Maximum Likelihood Estimate of the Non-Centrality Parameter of a Non-central F-variate" (with J. N. Pandey), *SIAM J. Math. Anal.* **2** (1971), 269–276. MR 46 # 10123.
6. "On the Integrability and Application of a Generalized Riccati Equation," *SIAM J. Appl. Math.* **21** (1971), 88–94. MR 44 # 5561.
7. "A Characterization of the Exponential Distribution" (with M. Ahsanullah), *J. Appl. Prob.* **9** (1972), 457–461. MR 49 # 8158.
8. "A Singular Inverse of a Matrix by Rank Annihilation" (with M. Ahsanullah), *Can. Math. Bull.* **15** (1973), 1–4. MR 48 # 306.
9. "On the Spectral Theory of Rayleigh's Piston I. The Discrete Spectrum" (with M. R. Hoare), *J. Phys. A* **6** (1973), 1461–1478. MR 50 # 15782.
10. "Explicit Form of the Distribution of the Behrens-Fisher d-Statistic" (with A. K. Md. Ehsanes Saleh), *J. Roy. Stat. Soc. B.* **36** (1974), 54–60. MR 50 # 8788.
11. "Bounds for Expected Values of Order Statistics" (with A. B. M. Lutful Kabir), *Comm. Stat.* **3** (6) (1974), 557–566. MR 50 # 1409.

12. "On the Spectral Theory of Rayleigh's Piston II. The Exact Singular Solution for Unit Mass Ratio" (with M. R. Hoare), *J. Phys. A* **7** (1974), 1070–1093. MR 50 # 15766.
13. "On the Spectral Theory of Rayleigh's Piston III. Exact Solution of the Absorbing Barrier Problem (γ = 1)" (with M. R. Hoare) *J. Phys. A* **9** (1976), 77–85.
14. "Construction of a Family of Positive Kernels from Jacobi Polynomials," *SIAM J. Anal.* **7** (1976), 92–116. MR 53 # 909.
15. "A Five-parameter Family of Positive Kernels from Jacobi Polynomials," *SIAM J. Math. Anal.* **7** (1976), 386–413. MR 53 # 11118.
16. "Some Positive Kernels and Bilinear Sums for Hahn Polynomials," *SIAM J. Math. Anal.* **7** (1976), 414–435. MR 53 # 11119.
17. "Exact Transform Solution of the One-Dimensional Special Rayleigh Problem" (with J. A. Barker, M. R. Hoare and S. Raval), *Can. J. Phys.* **55** (1977), 916–928.
18. "Stochastic Processes and Special Functions: On the Probabilistic Origin of some Positive Kernels Associated with Classical Orthogonal Polynomials" (with R. D. Cooper and M. R. Hoare), *J. Math. Anal. Appl.* **61** (1977), 262–291.
19. "On a Generalization of the Poisson Kernel for Jacobi Polynomials," *SIAM J. Math. Anal.* **8** (1977), 1014–1031. MR 56 # 16021.
20. "A Generalization of Gasper's Kernel for Hahn Polynomials: Application to Pollaczek Polynomials," *Can. J. Math.* **30** (1978) 133–46. MR 57 # 6544.
21. "A positive kernel for Hahn-Eberlein polynomials," *SIAM J. Math. Anal.* **9** (1978), 891–905. MR 80g # 33023.
22. "An elementary proof of Dunkl's Addition Theorem for Krawtchouk: polynomials," *SIAM J. Math. Anal.* **10** (1979), 438–445. MR 80j # 33022.
23. "Distributive Processes in Discrete Systems" (with M. R. Hoare), *Physica* **97 A** (1979), 1–41. MR 80i # 82018.

24. "A product formula and a non-negative Poisson kernel for Racah-Wilson polynomials," *Can. J. Math.* **30** (1980), 1501–1517. MR 82e # 33012.
25. "A stochastic matrix and bilinear sums for Racah-Wilson polynomials," *SIAM J. Math. Anal.* **12** (1981), 145–160. MR 82e # 33007.
26. "Families of biorthogonal rational functions in a discrete variable," *SIAM J. Math. Anal.* **12** (1981) 355–367.
27. "A non-negative representation of the linearization coefficients of the product of Jacobi polynomials," *Can. J. Math.* **33** (1981), 915–928.
28. "On the q-analogues of some transformations of nearly-poised hypergeometric series" (with B. Nassrallah), *Trans. Amer. Math. Soc.* **268** (1981), 211–229.
29. "Discrete orthogonal systems with respect to Dirichlet distribution," *Utilitas Mathematica* **20** (1981), 261–272.
30. "The linearization of the product of continuous q-Jacobi polynomials," *Can. J. Math.* **33** (1981), 961–987.
31. "Reproducing kernels and bilinear sums for q-Racah and q-Wilson polynomials," *Trans. Amer. Math. Soc.* **273** (1982), 483–508.
32. "The Rayleigh Model: Singular transport theory in one dimension" (with M. R. Hoare and S. Raval), *Phil. Trans. Roy. Soc. London* **A 305** (1982), 383–440.
33. "A Poisson kernel for continuous q-ultraspherical polynomials" (with G. Gasper), *SIAM J. Math. Anal.* **14** (1983), 409–420.
34. "Non-negative kernels in product formulas for q-Racah polynomials" (with G. Gasper), *J. Math. Anal. Appl.* **95** (1983), 304–318.
35. "Cumulative Bernoulli trials and Krawtchouk processes" (with M. R. Hoare), *Stochastic Processes and their Applications* **16** (1983), 113–139.
36. "Product formulas of Watson, Bailey and Bateman types and positivity of the Poisson kernel for q-Racah polynomials" (with G. Gasper), *SIAM J. Math. Anal.* **15** (1984), 768–789.

37. "A simple evaluation of Askey and Wilson's q-beta integral," *Proc. Amer. Math. Soc.* **92** (1984), 413–417.
38. "Projection formulas, a reproducing kernel and a generating function for q-Wilson polynomials" (with B. Nassrallah), *SIAM J. Math. Anal.* **16** (1985), 186–197.
39. "An infinite series with products of Jacobi polynomials and Jacobi functions of the second kind" (with M. J. Shah), *SIAM J. Math. Anal.* **16** (1985), 859–875.
40. "Sums of products of ultraspherical functions" (with M. J. Shah), *J. Math. Phys.* **26** (1985), 627–632.
41. "A q-extension of Feldheim's bilinear sum for Jacobi polynomials and some applications," *Can. J. Math.* **37** (1985), 551–576.
42. "A q-analogue of Appell's F_1 function and some quadratic transformation formulas for nonterminating basic hypergeometric series" (with B. Nassrallah), *Rocky Mtn. J. Math.* **16** (1986), 63–82.
43. "q-Wilson functions of the second kind," *SIAM J. Math. Anal.* **17** (1986), 1280–1286.
44. "Another conjectured q-Selberg integral," *SIAM J. Math. Anal.* **17** (1986), 1267–1279.
45. "A q-integral representation of Rogers' q-ultraspherical polynomials and some applications" (with A. Verma), *Constructive Approximation* **2** (1986), 1–10.
46. "A product formula for the continuous q-Jacobi polynomials," *J. Math. Anal. Appl.* **118** (1986), 309–322.
47. "Positivity of the Poisson kernel for the continuous q-Jacobi polynomials and some quadratic transformation formulas for basic hypergeometric series" (with G. Gasper), *SIAM J. Math. Anal.* **17** (1986), 970–999.
48. "Product and addition formulas for the continuous q-ultraspherical polynomials" (with A. Verma), *SIAM J. Math. Anal.* **17** (1986), 1461–1474.
49. "An integral representation of a ${}_{10}\varphi_9$ and continuous biorthogonal ${}_{10}\varphi_9$ rational functions," *Can. J. Math.* **38** (1986), 605–618.
50. "An integral of products of ultraspherical functions and a q-extension" (with R. Askey and T. H. Koorn-

winder), *J. London Math. Soc.* (2) **33** (1986), 133–148.

51. "Infinite sums and products of continuous q-ultraspherical functions" (with A. Verma), *Rocky Mount. J. Math.* **17** (1987), 371–384.
52. "An integral representation and some transformation properties of q-Bessel functions," *J. Math. Anal. Appl.* **125** (1987), 58–71.
53. "Solutions to Problems 86-3 and 86-4," *SIAM Review* **29** (1987), 130–131.
54. "Cummulative hypergeometric processes: a statistical role for the ${}_nF_{n-1}$ functions" (with M. R. Hoare), *J. Math. Anal. Appl.* **135** (1988), 615–626.
55. "A projection formula for the Askey-Wilson polynomials and an extension," *Proc. Amer. Math. Soc.* **103** (1988), 1099–1106.
56. "Some extensions of Askey-Wilson's q-beta integral and the corresponding orthogonal system," *Canad. Math. Bull.* **31** (1988), 467–476.
57. "An addition theorem and some product formulas for q- Bessel functions," *Canad. J. Math.* **40** (1988), 1203–1221.
58. "Some infinite integrals of q-Bessel functions," *Proceedings of the Ramanujan Birth Centennial Symposium on Classical Analysis*, December 26–28 (1987), N. K. Thakare (ed.), (1989), 119–137.
59. "A note on the orthogonality of Jackson's q-Bessel functions," *Canad. Math. Bull.* **32** (1989), 369–376.
60. "Some cubic summation formulas for basic hypergeometric series," *Utilitas Mathematica* **36** (1989), 161–172.
61. "A simple proof of Koornwinder's addition theorem for the little q-Legendre polynomials," *Proc. Amer. Math. Soc.* **107** (1989), 373–381.
62. "A nonterminating q-Clausen formula and some related product formulas" (with G. Gasper), *SIAM J. Math Anal.* **20** (1989), 1270–1282.
63. "An indefinite bibasic summation formula and some quadratic, cubic, and quartic summation and transformation formulas" (with G. Gasper), *Canad. J. Math.* **42** (1990), 1–27.

64. "Extensions of the beta integral and the hypergeometric function," *Proc. NATO – ASI* in "Orthogonal polynomials and their Applications," Paul Nevai (ed.), (1990), 319–344.
65. "Biorthogonality of a system of rational functions with respect to a positive measure on $[-1, 1]$," *SIAM J. Math. Anal.* **22** (1991), 1430–1441.
66. "The Associated Askey-Wilson polynomials" (with M. E. H. Ismail), *Trans. Amer. Math. Soc.* **328** (1991), 201–237.
67. "Complex weight functions for classical orthogonal polynomials" (with M. E. H. Ismail and D. R. Masson), *Can. J. Math.* **43** (1991), 1294–1308.
68. "Askey-Wilson functions of the first and second kinds: Series and integral representations of $C_n^2(x;\beta\,|\,q) + D_n^2(x;\beta\,|\,q)$," *J. Math. Anal. Appl.* **164** (1992), 263–284.
69. "Positivity of the Poisson kernel for the Askey-Wilson polynomials" (with Arun Verma), *Indian J. Math.* **33** (1991), 287–306.
70. "A cubic and a quintic summation formula," *Ganita* **43** (1992), 45–61.
71. "Some quadratic and cubic summation formulas for basic hypergeometric series," *Can. J. Math.* **45** #2 (1993), 394–411.
72. "Classical biorthogonal rational functions" (with S. K. Suslov), in Methods of Approximation Theory in Complex Analysis and Mathematical Physics, A. A. Gonchar and E. B. Saff (eds.), Lecture Notes in Math. **1550**, Springer-Verlag, pp. 131–146; Proceedings of International Seminars at Euler Mathematical Institute, Leningrad, May 1991.
73. "Quadratic transformation formulas for basic hypergeometric series" (with A. Verma), *Trans. Amer. Math. Soc.* **335** #1 (1993), 277–302.
74. "The Pearson equation and the beta integrals" (with S. K. Suslov), *SIAM J. Math. Anal.* **25** #2 (1994), 646–693.
75. "Barnes and Ramanujan-type integrals on the q-linear lattice" (with S. K. Suslov), *SIAM J. Math. Anal.* **25** #3 (1996), 1002–1022.

76. "Some basic bilateral sums and integrals" (with M. E. H. Ismail), *Par. J. Math.* **170** #2 (1995), 497–515.
77. "On the classical orthogonal polynomials" (with N. M. Atakishiyev and S. K. Suslov), *Constructive Approximation* **11** (1995), 181–226.
78. "Some generating functions for the associated Askey-Wilson polynomials," *J. Comp. Appl. Math.* **68** (1996), 287–296.
79. "Diagonalization of certain integral operators II" (with M. E. H. Ismail and R. Zhang), *J. Comp. Appl. Math.* **68** (1996), 163–196.
80. "A unified approach to the summation and integration formulas for basic hypergeometric series I" (with S. K. Suslov) *J. Stat. Planning and Inference* **54** (1996), 101–118.
81. "An integral representation of the very-well-poised ${}_8\psi_8$ series," *CRM Proceedings and Lecture Notes* **9** (1996), 281–288.
82. "Singular analogue of the Fourier transformation for the Askey-Wilson polynomials" (with S. K. Suslov), *CRM Proceedings and Lecture Notes* **9** (1996), 101–118.
83. "Some cubic summation and transformation formulas," *The Ramanujan Journal* **1** (1997), 299–308.
84. "Some summation theorems and transformation formulas for q-series" (with M. E. H. Ismail and S. K. Suslov), *Can. J. Math.* **49** (1997), 543–567.
85. "Enumeration of the k-poles" (with Z. Gao), *Annals of Combinatorics* **1** (1997), 55–66.
86. "Poisson kernel for the associated continuous q-ultraspherical polynomials" (with Q. Tariq), *Methods and Applications of Analysis*, **4** (1997), 77–90.
87. "The q-exponential functions, old and new," *Proceedings of the Dubna Conference on Integrable Systems*, (1994).
88. "A projection formula and a reproducing kernel for the associated Askey-Wilson polynomials" (with Q. M. Tariq), *Int. J. Math. and Stat. Sc.* **6** (1997), 141–160.

89. "The q-Laguerre polynomials and related moment problems" (with M. E. H. Ismail), *J. Math. Anal. Appl.* **218** (1998), 155–174.

90. "A unified approach to the summation and integration formulas for basic hypergeometric series II" (with S. K. Suslov), *Methods and Applications of Analysis* **5** (1998), 399–412.

91. "A unified approach to the summation and integration formulas for basic hypergeometric series III" (with S. K. Suslov), *Methods and Applications of Analysis* **5** (1998), 413–424.

92. "A q-extension of a product formula of Watson," *Questiones Mathematicae* **22**(1) (1999), 27–42.

93. "Addition formulas for q-Legendre type functions" (with Q. M. Tariq), *Methods and Applications of Analysis* **6** (1999), 3–20.

94. "Quadratic q-exponentials and connection coefficient problems" (with M. E. H. Ismail and D. Stanton), *Proc. Amer. Math. Soc.* **127** (1999), 2931–2941.

95. "A q-analogue of Weber-Schafheitlin integral of Bessel functions," *The Ramanujan Journal* **4** (2000), 251–265.

96. "A q-analogue of a product formula of Bailey and related results," in *Special Functions*, C. Dunkl, M. E. H. Ismail and R. Wong (eds.), World Scientific Publishing Co. (2000), pp. 262–281.

97. "The amazing first order linear equation," *Ganita* **51** (2000), 1–23.

98. "The associated classical orthogonal polynomials," in *Special Functions 2000*, J. Bustoz, M. E. H. Ismail and S. K. Suslov (eds), Kluwer Academic Publishers, (2001), pp. 255–280.

99. "Inverse operators, q-fractional integrals and q-Bernoulli polynomials" (with M. E. H. Ismail), *J. Approx. Theory* **114** (2002), 269–307.

100. "An inverse to the Askey-Wilson operator" (with M. E. H. Ismail), *Rocky Mount. J. Math.* **32** (2002), 657–678.

3. Professional activities

(i) Referee: I have refereed many papers for *SIAM J. Math. Anal.*, *Proc. Amer. Math. Soc.*, *J. Math. Phys.*, *Can. J. Math.*, *Rocky Mtn. J. Math.*, *Journal of Approximation Theory*, *Indian J. Math.* and *Ganita.*

(ii) Review: I have reviewed a fairly large number of papers for *Mathematical Reviews* since 1977.

(iii) Workshops: I was one of the three organizers of the 2-week workshop on "q-Series, Special Functions and Related Topics", June 12-23, 1994, in Toronto under the auspices of the Fields Institute.

(iv) Invited talks:

(1) Special session on orthogonal polynomials at the University of Michigan, Ann Arbor, Mich., Aug. 1980. (AMS summer meeting)

(2) Canadian Math. Soc. Winter meeting in Victoria, Dec. 10–12, 1981.

(3) Special meeting on Group Theory and Special functions at the Mathematical Institute at Oberwolfach, Germany, Mar. 13–19, 1983.

(4) Canadian Math. Soc. Summer meeting in Edmonton, June 21-23, 1984.

(5) International symposium on orthogonal polynomials and their applications, in Bar-le-Duc, France, Oct. 15–18, 1984.

(6) AMS meeting in Laramie, Wy., Aug. 11–15, 1985.

(7) AMS annual meeting at San Antonio, Texas in Jan. 1986.

(8) Gave a short course on Special Functions at the Research Institute in the University of Montreal in April–May 1986.

(9) Ramanujan centennial meeting at the University of Illinois, Urbana-Champaign, Aug. 1987.

(10) Ramanujan Birth Centennial Symposium on classical Analysis in Pune, India, Dec. 26–28, 1987.

(11) CMS summer meeting, June 1988.

(12) A 3-hour short course on q-series at the NATO Advanced Studies Institute on "Orthogonal Polynomials and their Applications," in Columbus, Ohio, May 22–June 3, 1989.

(13) SIAM 40th Anniversary Meeting in Los Angeles, July 20–29, 1992.

(14) Meeting on Difference equations in Integrable systems in Esterel, organized by Centre de Recherches Mathématiques of the University of Montreál, 1993.

(15) Meeting on Integrable Systems at Dubna, Russia, Summer 1994.

(16) International Workshop on Special Functions, Asymptotics, Harmonic Analysis and Mathematical Physics in Hong Kong, June 21–25, 1999.

(17) Centennial Mathematical Conference in Lucknow University, India, invited as the chief guest, December 31, 1999–January 4, 2000.

(18) Nato Advanced Study Institute Special Functions 2000: Current Perspective and Future Directions at Arizona State University, Tempe, AZ, May 29–June 9, 2000.

(19) AMS Annual Meeting at Baltimore, January 15–18, 2003: Special Session on Orthogonal Polynomials and Special Functions.

(v) Editorial Activities:

(a) Member of the Editorial Board of the journal: "Integral Transforms and Special Functions: An International Journal."

(b) From time to time I have lent a hand to Richard Bumby, the editor of the Problem Section of the AMS Math. Monthly.

(vi) Review of Grant Applications:

- I have reviewed some grant applications for NSERC, NSF and Austrian Math. Society.

4. Books:

- "Basic Hypergeometric Series" by G. Gasper and M. Rahman, published by Cambridge University Press in 1990.

- "Special Functions, q-Series and Related Topics," Fields Institute Communications, AMS (1997), edited by M. E. H. Ismail, D. Masson and M. Rahman.

5. Chapters in edited books:

- Part of Chapter 6 of "Kinetic Equations of Gases and Plasmas" by Ta-You Wu, (Addison-Wesley, 1966), specifically pp. 187–193, is based on my Ph.D. thesis.

6. Recent Grants:

- NSERC grant: $20,000 per year for three years, 1989–1992.
- NSERC grant: $20,000 per year for four years, 1992–1996.
- NSERC grant: $18,000 per year for five years, 1996–2000.
- NSERC grant: $12,000 per year for four years, 2000–2004.

7. Award and Honours:

- Scholarly Achievement Awards: 1980, 1983, 1986, 1988.
- Teaching Award: 1986.
- Election to a Fellowship of Bangladesh Academy of Science, 2003.

8. Bengali Literature:

i <u>Publications</u> (All in Bengali):

1. Tirtho Amar Gram (1994).
2. Lal Nodi (2001).
3. Proshongo Nari (2003).
4. Album (2003).

ii <u>Awards</u>:

1. Award of Excellence from Bangladesh Publications (Ottawa) in 1996.

2. Best Writer Award from Deshe Bideshe (by Readers' choice) in 1998.
3. Outstanding Achievement Award from Ottawa-Banglades Muslim Society in 2000.
4. Award for Contributions to Bengali Literature in North America from Bongo Shomyelon in 2002.

Dr. Mizanur Rahman — A personal anecdote ...

Dr. Mourad Ismail asked me to write a brief bio about my father, Dr. Mizanur Rahman. It will probably be more personal and emotional than factual. But what did you expect from a son? This bio will be for a collection of articles dedicated to him.

Apparently, the last book that Dad co-authored with Dr. George Gasper has been called a 'bible' in its field of Basic Hypergeometric Series. So, one would think that Dr. Rahman is a man of no small repute. So, why can I only think of him as my simple father?

Dad was born in Dhaka, Bangladesh. Most of the family originated from a small village called Hasnabad. He was the eldest of 5 boys and 4 girls. From the very beginning, he was responsible for taking care of most of his siblings, with some help from the two older sisters. Our grandmother was a homemaker, and our grandfather was a head clerk in the public service, working as the assistant to the District Magistrate. In spite of their humble status, my grandparents were firm believers in the power of education. So, they made sure that Dad went to school everyday (well, most days ...), did his homework, and studied for the tests. Passing with flying colours was his responsibility, and that he did.

Dad was one of the few who finished a double major in Math and Physics. His major was actually Physics, and the minor was supposed to be Biology or Chemistry. However, he disliked both. So, the university provided an option that they felt would be impossible one: if you don't want to minor in the other fields, then you would have to do a double major. Dad did, and earned the University medal for outstanding academic achievement! After Dhaka University, it was on to Cambridge in 1956. From what little we know of this time, it sounded like an extended field trip, with even a brief sojourn in Spain. If only grad school were this difficult all the time!

Dad married Parul Shamsun Nahar in July 1961. The marriage was partially arranged by a friend of his good friend, who also happened to be Mom's brother. After a wonderful boat trip from Karachi to London, they flew to Fredericton, New Brunswick in 1962. Dad was a grad student and a lecturer at the same time there. In 1965, he took a position in the Department of Mathematics and Statistics at Carleton University in Ottawa. This was the start of a long and successful career in teaching and research. Both of us brothers were born in Ottawa. Life had a comfortable and predictable rhythm to it. Dad left very early in the morning to go to work, regardless of the weather. And at night, we would run to greet him at the door. Every so often, Dad would take us to his office. This was a special treat for us two kids. We could see most of Ottawa from Dad's office — but the best part was eating those little sugar cubes that Dad used in his coffee. Especially fascinating was the prodigious amount of books, papers, reviews and miscellaneous stuff that Dad managed to cram into every available nook and cranny. Another 'perk' of working in academia was the sabbatical. To Dad, it meant an opportunity to do some intense research with various colleagues. For us, it usually meant an opportunity to travel. We went to Bangladesh one time. Another time, we spent a year in England. We came back with accents and a renewed appreciation for our food! I also remember that Mom was in the hospital a lot, dealing with progressive kidney failure. In spite of being so busy, Dad would take us fishing quite often. He cooked at home quite often. In the beginning, Mom would do all the cooking and housework. In spite of her failing health, she sacrificed everything to realize her dream: ensuring that her loved ones would be able to pursue successful careers. As Dad said, his 'modest successes were but a reflection of the sacrifices that Mom had to make throughout her painful life.' As we grew older, we would often wonder how someone so intelligent could be so 'detached' from normal things. Dad could be quite the 'absent-minded professor.' As I liked to say, he had a definite "Je ne sais pas" goofiness about him. Yet that trait was juxtaposed against a very deep and insightful wisdom. I know that it is Dad's teachings that have led to some degree of equanimity in my life. And now that he is a grandfather, I see him passing along those same kernels of wisdom to my children. If I had to use one word to summarize Dad's character, it would be honest — to a fault. Words can't really describe the myriad of situations that displayed his honesty. You really had to be there. As Dad's mathematical career advanced, it was amazing to see another side of him flourish. At heart, I suppose Dad was always an artist, a writer. So, over the last decade or so, he has been writing Bengali fiction more and more. Over that time, his following and stature has been quickly increasing. As a matter of fact, he has been awarded 2 national prizes in Canada, recognizing his contributions to Bengali literature. Additionally, Dad received an award at last year's international Bangladeshi conference, held in Atlanta. It too recognized his talents in Bengali literature. Now that Dad is "retired," we would have expected him to have time for us. No such luck. He is 200% busy with continuing mathematical research, writing math books and his Bengali writing. He has been an inspiration to his two sons in so many ways. He is a man of honesty, integrity, curiosity, and has a wonderful sense of humour. He has lived his life with a pure heart and single-minded devotion to family and work. He is a hard act to follow.

To Dad,

From your sons,

Babu S. Rahman
Raja A. Rahman

Mizanur Rahman by Michael Hoare

This is an outline of how we first met and how Mizan came to branch out on his extraordinary mathematical career. We first corresponded and eventually met at Bedford College, London University in 1971/72. He was then working on Neutron Transport Theory (singular integral equations), his Ph.D. subject, in which I had an interest from a more general statistical physics viewpoint. With a sabbatical coming up he had written to the outstanding English expert on neutron transport, Mike Williams at Queen Mary College Nuclear Engineering Department, with a view to spending a year there. For some reason Mike couldn't take him in and suggested that he come to me at Bedford instead. This worked splendidly, and I dare say the liberal arts atmosphere of the college, set idyllically in Regent's Park, was a distinct improvement on that of Nuclear Engineering in the East End. The only downside was that we were a very small Physics department and there was little resonance with the heavily algebraicized Math department under Paul Cohn. This hardly seemed to matter, since we were both to an extent outsiders from what were the fashionable subjects at the time.

We started out on his home ground, the singular integral equation arising from the form of one-dimensional gas model known as 'Rayleigh's piston' and this led to the first calculations of its eigenvalue spectrum, with characteristic mixed discrete and continuum sets. In the course of this he admitted me to the faith, convincing me that Cauchy Principal Values and Hadamard finite parts could be made tangible and did not need to handled as though matters of higher metaphysics. About this time I happened to mention a problem in combinatorics that had been fascinating me for some time, in fact since my post-doc days at the University of Washington. This arose from a disarmingly simple model in chemical kinetics which involved the partitioning of energy quanta between different vibrational degrees of freedom in colliding polyatomic molecules. Reformulated as a 'urn model,' its iterations corresponded to a Markov chain for partitioning balls in boxes, with only a subset randomized at each event. My earlier eigenvalue solution for a continuum version of this had come out in Laguerre polynomials and led to probability transition kernels with action very similar to formulae of Erdélyi and Kogbetlianz in 1930s special function theory, though at the time without any probabilistic interpretation. (I was happily able to meet Erdélyi in Edinburgh shortly before his death and he was delighted to know that the formulae had a 'practical' side.) Mizan seized on the implications of this problem and its generalizations and before long was off and running with his first series of papers on special function theory, while my 'statistical physics' by-products followed at more leisurely intervals. I tended to keep cautiously within the bounds of 'physical' models, while Mizan was soon off into the never never land of q-theory. After he returned to Carleton we managed to strike lucky with grants from the NRC of Canada and the SRC in London, which kept us in funds for several years of visits to and fro as the work progressed. (Happy days of the '70s and beneficent Research Councils). Mizan even managed to come to Stuttgart in '77 when my turn for a sabbatical came round and it was here that we sorted out what I have ever since felt was the real 'gem' among our various generalizations. This was the discovery of a new take on Bernoulli Trials — the idea of 'cumulative trials' in which one has the right to 'throw again' on the subset of trials that fail, in order to achieve complete success. That such a simple idea could have lain dormant for over 200 years still amazes me, but no trace of our results in the earlier literature has ever turned up.

Mizan Rahman by Steve Milne

The well-known classic book "Basic Hypergeometric Series" by George Gasper and Mizan Rahman has been immensely helpful to me, both in my research and teaching. My work in multiple basic hypergeometric series, especially that on multivariable ${}_{10}\varphi_9$ transformations, was facilitated by the one-variable treatment in this book. Furthermore, my Ph.D. students all learned q-series from my graduate special topics courses based on this wonderful book or its notes. It will continue to be essential to my program for many years to come.

ON THE COMPLETENESS OF SETS OF q-BESSEL FUNCTIONS $J_\nu^{(3)}(x;q)$

L. D. Abreu*
Department of Mathematics
Universidade de Coimbra
Coimbra, PORTUGAL
daniel@mat.uc.pt

J. Bustoz†
Department of Mathematics
Arizona State University
Tempe, Arizona 85287-1804
bustoz@asu.edu

Abstract We study completeness of systems of third Jackson q-Bessel functions by two quite different methods. The first uses a Dalzell-type criterion and relies on orthogonality and the evaluation of certain q-integrals. The second uses classical entire function theory.

1. Introduction

For $0 < q < 1$ define the q-integral on the interval $(0, a)$ by

$$\int_0^a f(x)d_qx = (1-q)\sum_{n=0}^{\infty} f\left(aq^n\right)aq^n. \tag{1.1}$$

*Research partially supported by Fundacaõ para a Ciência e Tecnologia and Centro de Mátemática da Universidade de Coimbra.

†Joaquín Bustoz (1939–2003) passed away in August 2003 as a consequence of a car accident. He will be missed both as a mathematician and for his work on teaching mathematics, in particular on getting students from minorities into higher education.

$L_q^2(0,1)$ will denote the Hilbert space associated with the inner product

$$\langle f,g\rangle = \int_0^1 f(x)g(x)d_qx.$$

It is a well known fact that the third Jackson q-Bessel function $J_\nu^{(3)}(z;q)$, defined as

$$J_\nu^{(3)}(z;q) = z^\nu \frac{(q^{\nu+1};q)_\infty}{(q;q)_\infty} \sum_{k=0}^{\infty} (-1)^k \frac{q^{\frac{k(k+1)}{2}}}{(q^{\nu+1};q)_k\,(q;q)_k} z^{2k} \tag{1.2}$$

satisfies the orthogonality relation

$$\begin{aligned} &\int_0^1 xJ_\nu\left(j_{n\nu}qx;q^2\right)J_\nu\left(j_{m\nu}qx;q^2\right)d_qx \\ &= \frac{q-1}{2q^2}\,J_{\nu+1}\left(j_{n\nu};q^2\right)J_\nu\left(j_{n\nu};q^2\right)\delta_{n,m} \end{aligned} \tag{1.3}$$

where $j_{1\nu} < j_{2\nu} < \cdots$ are the zeros of $J_\nu^{(3)}\left(z;q^2\right)$ arranged in ascending order. Important information on the zeros of $J_\nu^{(3)}\left(z;q^2\right)$ has been given recently (Ismail, 2003; Koelink and Swarttouw, 1994; Koelink, 1999; Abreu et al., 2003). The orthogonality relation (1.3) is a consequence of the second order difference equation of Sturm-Liouville type satisfied by the functions $J_\nu^{(3)}\left(z;q^2\right)$ (Swartouw, 1992; Koelink and Swarttouw, 1994). In this paper we consider completeness properties of the third q-Bessel function in the spaces $L_q(0,1)$ and $L_q^2(0,1)$. We will approach the problem from two substantially different directions. In one case we will apply a q-version of the Dalzell Criterion (Higgins, 1977) to prove completeness of the system $\left\{J_\nu^3\left(j_{n\nu}qx;q^2\right)\right\}$ in $L_q^2(0,1)$. In another case we will use the machinery of entire functions and the Phragmén-Lindelöf principle to prove completeness of the system $\left\{J_\mu^3\left(j_{n\nu}qx;q^2\right)\right\}$, $\mu,\nu>0$ in $L_q^1(0,1)$. This theorem is in the spirit of classical results on Bessel functions (Boas and Pollard, 1947) that state the completeness of systems $\{J_\nu(\lambda_n(z))\}$ where the numbers λ_n are allowed a certain freedom. Although the entire function argument is more general, there is reason to present the Dalzell Criterion approach as well because it relies solely on techniques of q-integration and on properties of orthogonal expansions in a Hilbert space. Also, this approach requires the calculation of some q-integrals of q-Bessel functions that parallel results for classical Bessel functions. Thus this method of proof extends the q-theory of orthogonal functions.

The third Jackson q-Bessel function was also studied by Exton (Exton, 1983) and sometimes appears in the literature as *The Hahn-Exton q-Bessel Function.* There are other two analogues of the Bessel function introduced by Jackson (Jackson, 1904). The notation of Ismail (Ismail, 1982; Ismail, 2003), denoting all three analogues by $J_\nu^{(k)}(z;q)$, $k = 1,2,3$ has become common and we adhere to it here. However, because the present work will deal exclusively with $J_\nu^{(3)}\left(z;q^2\right)$, to simplify notation we write from now on

$$J_\nu(z) = J_\nu^{(3)}\left(z;q^2\right).$$

It is critical to keep in mind that in definition (1.2) the q-Bessel function is defined with base q, whereas in defining $J_\nu(z)$ we have changed the base to q^2. Thus the series definition for $J_\nu(z)$ is

$$J_\nu(z) = z^\nu \frac{\left(q^{2\nu+2};q^2\right)_\infty}{\left(q^2;q^2\right)_\infty} \sum_{k=0}^{\infty} (-1)^k \frac{q^{k(k+1)}}{\left(q^{2\nu+2};q^2\right)_\infty \left(q^2;q^2\right)_\infty} z^{2k}.$$

Let $z_{n\nu}$, $n = 1,2,\ldots$ denote the positive roots of $J_\nu^{(3)}(z;q)$ arranged in increasing order. From (Kvitsinsky, 1995) we have that

$$\sum_{n=1}^{\infty} \left(z_{n\nu}\right)^{-2} = \frac{q}{(1-q)\left(1-q^{\nu+1}\right)}. \tag{1.4}$$

Replacing q by q^2, we find for the roots $j_{n\nu}$ of $J_\nu(z)$ that

$$\sum_{n=1}^{\infty} \left(j_{n\nu}\right)^{-2} = \frac{q}{\left(1-q^2\right)\left(1-q^{2\nu+2}\right)}. \tag{1.5}$$

Expression (1.5) will be used in Section 2.

2. Completeness: A Dalzell type criterion

It is easy to verify (Higgins, 1977) that if $\{\Phi_n\}$ and $\{\Psi_n\}$ are two sequences in a Hilbert space H, with Ψ_n complete in H and Φ_n complete in Ψ_n and orthogonal in H, then Φ_n is also complete in H. Then, if Ψ_n is complete in H, a necessary and sufficient condition for the orthogonal sequence Φ_n to be complete in H is that it satisfies the Parseval relation

$$\sum_n \left|\langle \Phi_n, \Psi_k\rangle\right|^2 = \left\|\Psi_k\right\|, \text{ for every } \Psi_k,\ k = 0,1,\ldots. \tag{2.1}$$

This fact was used by Dalzell to derive a completeness criterion and apply it to several sequences of special functions (Dalzell, 1945). In this section we will derive a similar criterion suitable to be used in $L_q^2(0,1)$. Then, we use it to prove completeness in $L_q^2(0,1)$ of the orthonomal set of functions

$$\Phi_n(x) = \frac{x^{\frac{1}{2}} J_\nu\left(j_{n\nu} q x\right)}{\left\| x^{\frac{1}{2}} J_\nu\left(j_{n\nu} q x\right) \right\|}.$$

To do so, we will evaluate explicitly some q-integrals using the results from the preceding section. We start by stating and proving the following lemma:

Lemma 2.1. *Let $g \in L_q^2(0,1)$ such that $g(q^n) > 0$, $n = 0,1,2\ldots$. Define $\chi_n(x) = 1$ if $x \in [0,q^n]$ and $\chi_n(x) = 0$ otherwise. Then $\{g\chi_n\}$ is complete in $L_q^2(0,1)$.*

Proof. Let $f \in L_q^2(0,1)$ be such that

$$\int_0^1 f(x)g(x)\chi_n(x)d_qx = 0,\ n = 0,1,2,\ldots.$$

Now, by (1.1) and using the fact that $\chi_n\left(q^k\right) = 0$ if $k < n$, we get:

$$A_n = \sum_{k=n}^{\infty} f\left(q^k\right) g\left(q^k\right) q^k = 0.$$

Then,

$$0 = A_n - A_{n+1} = f\left(q^n\right) g\left(q^n\right) q^n$$

because $g\left(q^n\right) > 0$ it follows that

$$f\left(q^n\right) = 0,\ n = 0,1,2,\ldots.$$

□

Theorem 2.2. *Let $g \in L_q^2(0,1)$ such that $g(q^n) > 0$, $n = 1,2\ldots$ and let $w(x)$ be such that $\int_0^1 w(x)d_qx$ exists and $w(q^n) > 0$, $n = 1,2\ldots$. Then an orthonormal sequence $\{\Phi_n\} \subset L_q^2(0,1)$ is complete in $L_q^2(0,1)$ if and only if*

$$\sum_n^{\infty} \int_0^1 \left| \int_0^r \Phi_n(x)g(x)d_qx \right|^2 w(r)d_qr = \int_0^1 \left[\int_0^r |g(x)|^2 d_qx \right] w(r)d_qr. \quad (2.2)$$

Proof. Writing $\Psi_k = g\chi_k$ in (2.1), by the preceding lemma, the sequence $\{\Phi_n\}$ is complete in $L_q^2(0,1)$ if and only if

$$\sum_n^\infty \left| \int_0^1 \Phi_n(x)g(x)\chi_k(x)d_qx \right|^2 = \int_0^1 |g(x)\chi_k(x)|^2\, d_qx,\ k = 0, 1, \ldots .$$

that is

$$\sum_n^\infty \left| \int_0^r \Phi_n(x)g(x)d_qx \right|^2$$

$$= \int_0^r |g(x)|^2\, d_qx \text{ for every } r \in \left\{q^k, k = 0, 1, \ldots\right\}.$$

Integrating both sides of this relation after multiplying by $w(x)$, one gets the relation (2.2). On the other hand, if (2.2) holds, then define

$$F(r) = \int_0^r |g(x)|^2\, d_qx - \sum_n^\infty \left| \int_0^r \Phi_n(x)g(x)d_qx \right|^2 .$$

From the hypothesis,

$$\int_0^1 F(r)w(r)d_qr = 0.$$

Observing that by the Bessel inequality, $F(r)$ is non-negative, we get

$$F\left(q^k\right) = 0,\ k = 1, 2, \ldots .$$

□

We proceed to evaluate two important q-integrals.

Lemma 2.3. *For every real number* r,

$$\int_0^r x^{\nu+1} J_\nu\left(j_{n\nu}qx\right) d_qx = \frac{1-q}{qj_{nv}}\, r^{\nu+1} J_{\nu+1}\left(j_{n\nu}qr\right).$$

Proof. Express

$$\int_0^r x^{\nu+1} J_\nu\left(j_{n\nu}qx\right) d_qx$$

using the power series expansion (1.2). Then interchange the q-integral with the sum and use the following fact:

$$\int_0^r x^{2\nu+2k+1} d_q x = (1-q) r^{2\nu+2k+2} \sum_{n=0}^{\infty} q^{n(2\nu+2k+2)}$$
$$= \frac{1-q}{1-q^{2\nu+2k+2}} r^{2\nu+2k+2}.$$

Rearranging terms the result follows in a straightforward manner. □

Lemma 2.4.

$$\int_0^1 x J_{\nu+1}^2 (j_{n\nu} q x)\, d_q x = \int_0^1 x J_\nu^2 (j_{n\nu} q x)\, d_q x.$$

Proof. Consider the following formula from (Koelink and Swarttouw, 1994):

$$\int_0^1 x J_{\nu+1}^2 (aqz) d_q x = \frac{(1-q)q^\nu}{-2a} \{a J_{\nu+2}(aq) J'_{\nu+1}(a)$$
$$-J_{\nu+2}(aq) J_{\nu+1}(a) - aq J'_{\nu+2}(aq) J_{\nu+1}(a)\} \tag{2.3}$$

and

$$J_{\nu+1}(xq) = q^{-\nu-1} \left(\frac{1-q^{2\nu}}{x} J_\nu(x) - J_{\nu-1}(x) \right). \tag{2.4}$$

Shift $\nu \to \nu + 1$ in (2.4) and set $x = j_{n\nu}$. This yields

$$J_{\nu+2}(j_{n\nu} q) = \frac{1-q^{2\nu+2}}{q^{\nu+2} j_{n\nu}} J_{\nu+1}(j_{n\nu}).$$

Taking derivatives in both members of (2.4), changing $\nu \to \nu + 1$ and again setting $x = j_{n\nu}$ the result is

$$J'_{\nu+2}(j_{n\nu} q) = q^{-\nu-3} \Big\{ \left(1 - q^{2\nu+2}\right) \Big[\frac{1}{j_{n\nu}} J'_{\nu+1}(j_{n\nu})$$
$$- \frac{1}{j_{n\nu}^2} J_{\nu+1}(j_{n\nu}) \Big] - J_\nu(j_{n\nu}) \Big\}.$$

Substituting this in (2.3) we get the simplification:

$$\int_0^1 x J_{\nu+1}^2(j_{n\nu}qx)\,d_qx = \frac{(1-q)}{-2q^2} J_\nu'(j_{n\nu}) J_{\nu+1}(j_{n\nu})$$

$$= -\frac{1}{2}(1-q)q^{\nu-2} J_{\nu+1}(j_{n\nu}q) J_\nu(j_{n\nu}) = \int_0^1 x J_\nu^2(j_{n\nu}qx)\,d_qx$$

where (1.3) was used in the last identity. □

Theorem 2.5. *The orthonormal sequence* $\{\Phi_n\}$ *defined by*

$$\Phi_n(x) = \frac{x^{\frac{1}{2}} J_\nu(j_{n\nu}qx)}{\left\| x^{\frac{1}{2}} J_\nu(j_{n\nu}qx) \right\|}$$

is complete in $L_q^2(0,1)$.

Proof. In (2.2) take $\{\Phi_n\}$ defined as above, $g(x) = x^{\nu+\frac{1}{2}}$ and $w(r) = r^{-2\nu-1}$. We need thus to prove the identity

$$\sum_{n=1}^{\infty} \int_0^1 \left| \int_0^r \frac{x^{\frac{1}{2}} J_\nu(j_{n\nu}qx)}{\left\| x^{\frac{1}{2}} J_\nu(j_{n\nu}qx) \right\|} x^{\nu+1} d_qx \right|^2 r^{-2\nu-1} d_qr$$

$$= \int_0^1 \left[\int_0^r |g(x)|^2 \, d_qx \right] w(r) d_qr.$$

Lemma 1 and Lemma 2 allow us to reduce the left hand member of above to:

$$\frac{(1-q)^2}{q^2} \sum_{n=1}^{\infty} \frac{1}{j_{n\nu}^2},$$

that is,

$$\frac{1-q}{(1+q)(1-q^{2\nu+2})}$$

by (1.5). It is straightforward to compute

$$\int_0^1 \left[\int_0^r |g(x)|^2 \, d_qx \right] w(r) d_qr = \frac{1-q}{(1+q)(1-q^{2\nu+2})},$$

and the Theorem is proved. □

3. Completeness: An entire function approach

From (1.2) we can write

$$J_\nu(w) = \frac{(q^{2\nu+2};q^2)_\infty}{(q^2;q^2)_\infty} z^\nu F_\nu(w),$$

where

$$F_\nu(w) = \sum_{k=0}^{\infty} \frac{(-1)^k q^{k(k+1)} z^{2k}}{(q^{2\nu+2};q^2)_k \, (q^2;q^2)_k}.$$

The function $F_\nu(w)$ is entire and it is directly shown that $F_\nu(w)$ has order zero.

Set $G(w) = \int_0^1 g(x) F_\mu(qwx) d_q x$, and $h(w) = \frac{G(w)}{F_\nu(w)}$.

Lemma 3.1. *If $\mu > 0$, $\nu > 0$ and $g(x) \in L_q^1(0,1)$ then $h(w)$ is entire of order* 0.

Proof. We first show that $G(w)$ is entire of order 0. From the definition of the q-integral we have

$$G(w) = (1-q) \sum_{k=0}^{\infty} g\left(q^k\right) F_\mu\left(wq^{k+1}q^k\right). \tag{3.1}$$

The series in (3.1) converges uniformly in any disk $|w| \le R$. Hence $G(w)$ is entire. Recall that the order $\rho(f)$ of an entire function $f(w)$ is given by

$$\rho(f) = \limsup_{r \to \infty} \frac{\ln \ln M(r;f)}{\ln r}$$

where

$$M(r,f) = \max_{|w| \le r} |f(w)|.$$

From (3.1)

$$M(r;G) \le M(r;F_\mu) \int_0^1 |g(x)| \, d_q x.$$

Since $\rho(F_\mu) = 0$ we have that $\rho(G) = 0$.

Both the numerator and the denominator of $h(w)$ are entire functions of order 0. If we write $G(w)$ and $F_\nu(w)$ as canonical products, each factor of $F_\nu(w)$ divides out with a factor of $G(w)$ by the hypothesis of Theorem 3.3. $h(w)$ is thus entire of order 0. □

Lemma 3.2. *If $\mu > 0$, $\nu > 0$, and $0 < q < 1$ then the quotient $\frac{F_\mu(q^m w)}{F_\nu(w)}$ is bounded on the imaginary w axis.*

Proof. We will make use of the simple inequality

$$(q^\alpha; q)_\infty < (q^\alpha; q)_k < 1, \quad \alpha > 0, \quad 0 < q < 1.$$

Using this inequality we get for $w = iy$, y real,

$$F_\mu(q^m iy) = \sum_{n=0}^{\infty} \frac{q^{n(n+1)} q^{2mn} y^{2n}}{(q^{2\mu+2}; q^2)_n (q^2; q^2)_n} < \frac{1}{(q^{2\mu+2}; q^2)_\infty} \sum_{n=0}^{\infty} \frac{q^{n(n+1)} y^{2n}}{(q^2; q^2)_n},$$

$$F_\nu(iy) = \sum_{n=0}^{\infty} \frac{q^{n(n+1)} y^{2n}}{(q^{2\nu+2}; q^2)_n (q^2; q^2)_n} > \sum_{n=0}^{\infty} \frac{q^{n(n+1)} y^{2n}}{(q^2; q^2)_n}.$$

Thus we have

$$0 \le \frac{F_\mu(q^m; y)}{F_\nu(iy)} < \frac{1}{(q^{2\mu+2}; q^2)_\infty}.$$

□

Theorem 3.3. *Let $\mu > 0$, $\nu > 0$ and $g(x) \in L_q^1(0,1)$. If*

$$\int_0^1 g(x) J_\mu(q j_{n\nu} x)\, d_q x = 0,$$

$n = 1, 2, \ldots$ *then $g(x) = 0$ for $x = q^m$, $m = 0, 1, \ldots$.*

Proof. Lemma 3.2 implies that $h(iy)$ is bounded. Since $h(w)$ is entire of order 0, we can apply one of the versions of the Phragmén-Lindelöf theorem (Levin, 1980, p. 49) and Lemma 3.2 and conclude that $h(w)$ is bounded in the entire w-plane. Next by Liouville's theorem we conclude that $h(w)$ is constant. Say that $h(w) \equiv C$. We will prove that $C = 0$. We have

$$G(w) - CF_\nu(w) \equiv 0.$$

In infinite series form this equality produces an identity of the form

$$\sum_{k=0}^{\infty} A_k w^{2k} \equiv 0.$$

From the identity theorem for analytic functions we conclude that $A_k = 0$. Calculating A_k we find

$$\frac{q^{k(k+1)+2k}(1-q)(-1)^k}{(q^{2\mu+2}; q^2)_k (q^2; q^2)_k} \sum_{j=0}^{\infty} g(q^j)\, q^{(2k+1)j}$$

$$- \frac{q^{k(k+1)}(-1)^k}{(q^{2\nu+2}; q^2)_k (q^2; q^2)_k} C = 0, \quad k = 0, 1, 2, \ldots.$$

Dividing out common factors and letting $k \to \infty$ gives $C = 0$. We can now conclude that $G(w) \equiv 0$, or that is,

$$\int_0^1 g(x) J_\mu(wqx) d_q x \equiv 0.$$

We complete the proof with a simple argument that gives $g(q^m) = 0$, $m = 0, 1, \ldots$. If $G(w) \equiv 0$ then

$$\sum_{j=0}^{\infty} g\left(q^j\right) q^{(2k+1)j} = 0.$$

Letting $k \to 0$ gives $g(1) = 0$. Then dividing by q^{2k} and again letting $k \to \infty$ gives $g(q) = 0$. Continuing this process we have $g(q^m) \equiv 0$ and the proof of the theorem is complete. □

References

Abreu, L. D., Bustoz, J., and Cardoso, J. L. (2003). The roots of the third Jackson q-Bessel function. *Internat. J. Math. Math. Sci.*, 67:4241–4248.

Boas, R. P. and Pollard, H. (1947). Complete sets of Bessel and Legendre functions. *Ann. of Math.*, 48:366–384.

Dalzell, D. P. (1945). On the completeness of a series of normal orthogonal functions. *J. Lond. Math. Soc.*, 20:87–93.

Exton, H. (1983). *q-Hypergeometric Functions and Applications.* Ellis Horwood, Chichester.

Higgins, J. R. (1977). *Completeness and basis properties of sets of special functions.* Cambridge University Press, London, New York, Melbourne.

Ismail, M. E. H. (1982). The zeros of basic Bessel functions, the functions $J_{v+ax}(x)$, and associated orthogonal polynomials. *J. Math. Anal. Appl.*, 86:1–19.

Ismail, M. E. H. (2003). Some properties of jackson's third q-bessel function. Preprint.

Jackson, F. H. (1904). On generalized functions of Legendre and Bessel. *Transactions of the Royal Society of Edinburgh*, 41:1–28.

Koelink, H. T. (1999). Some basic Lommel polynomials. *Journal of Approximation Theory*, 96:345–365.

Koelink, H. T. and Swarttouw, R. F. (1994). On the zeros of the Hahn-Exton q-Bessel function and associated q-Lommel polynomials. *J. Math. Anal. Appl.*, 186:690–710.

Kvitsinsky, A. A. (1995). Spectral zeta functions for q-Bessel equations. *J. Phys. A: Math. Gen.*, 28:1753–1764.

Levin, B. Y. (1980). *Distribution of zeros of Entire Functions.* American Mathematical Society, Providence, RI.

Swartouw, R. F. (1992). *The Hahn-Exton q-Bessel Function.* PhD thesis, Technische Universiteit Delft.

a-GAUSSIAN POLYNOMIALS AND FINITE ROGERS-RAMANUJAN IDENTITIES

George E. Andrews*
Department of Mathematics
The Pennsylvania State University
University Park, PA 16802
andrews@math.psu.edu

Abstract Classical Gaussian polynomials are generalized to two variable polynomials. The first half of the paper is devoted to a full account of this extension and its inherent properties. The final part of the paper considers the role of these polynomials in finite identities of the Rogers-Ramanujan type.

Keywords: Rogers-Ramanujan identities, Gaussian polynomials

1. Introduction

Our object in this paper is to better understand certain classical generalizations of the Rogers-Ramanujan identities (Andrews, 1976, p. 104):

$$\sum_{n=0}^{\infty} \frac{q^{n^2}}{(q;q)_n} = \frac{1}{(q;q^5)_\infty (q^4;q^5)_\infty}, \tag{1.1}$$

and

$$\sum_{n=0}^{\infty} \frac{q^{n^2+n}}{(q;q)_n} = \frac{1}{(q^2;q^5)_\infty (q^3;q^5)_\infty}, \tag{1.2}$$

where $|q| < 1$ and

$$(A;q)_n = (A;q)_\infty / (Aq^n;q)_\infty, \tag{1.3}$$

and

$$(A;q)_\infty = \prod_{j=0}^{\infty} (1 - Aq^j). \tag{1.4}$$

*Partially supported by National Science Foundation Grant DMS9206993.

The majority of early proofs of (1.2) and (1.3) were based on the following theorem which W. N. Bailey (Bailey, 1972, p. 8, line 4) called an "a-generalization."

$$\sum_{n=0}^{\infty} \frac{q^{n^2} a^n}{(q;q)_n} = \frac{1}{(aq;q)_\infty} \left\{ 1 + \sum_{n=1}^{\infty} \frac{(-1)^n a^{2n} q^{n(5n-1)/2} \left(1 - aq^{2n}\right) (aq;q)_{n-1}}{(q;q)_n} \right\}. \tag{1.5}$$

As is well-known, Watson proved this identity as a limiting case of his q-analog of Whipple's theorem (Watson, 1929).

There occur in the literature two refinements of (1.5) in which the series on the left of the identity is replaced by a polynomial. Namely (Andrews, 1974; Bressoud, 1981a; Paule, 1994; Zeilberger, 1990).

$$\sum_{n=0}^{N} a^n q^{n^2} \begin{bmatrix} N \\ n \end{bmatrix}; q \Bigg] = \sum_{n=0}^{N} (-1)^n q^{n(5n-1)/2} \left(1 - aq^{2n}\right) \begin{bmatrix} N \\ n \end{bmatrix}; q \Bigg] \frac{1}{(aq^n;q)_{N+1}} \tag{1.6}$$

and (Bressoud, 1981b, eq. (3.5)).

$$\sum_{n=0}^{N} a^n q^{n^2} \left[\begin{matrix} N \\ n \end{matrix} ; q \right] = \sum_{N \geqq 2n \geqq 0} (-1)^n a^{2n} q^{n(5n-1)/2} \left(1 - aq^{2n}\right) \left[\begin{matrix} N \\ n \end{matrix} ; q \right] \left[\begin{matrix} N-n \\ n \end{matrix} ; q \right] (q;q)_n \frac{\left(a^2 q^{N+2n+1}; q\right)_{N-2n}}{(aq^n;q)_{N+1-n}}, \tag{1.7}$$

where

$$\left[\begin{matrix} N \\ n \end{matrix} ; q \right] = \begin{cases} 0 & \text{if } n < 0 \text{ or } n > N \\ \frac{(q;q)_N}{(q;q)_n (q;q)_{N-n}} & \text{otherwise} \end{cases} \tag{1.8}$$

is the Gaussian polynomial or q-binomial coefficient.

Now there is something rather surprising about (1.6) and (1.7) that is readily observed upon examination. The left sides of both (1.6) and (1.7) are polynomials term by term and consequently the sums are polynomials. However it is not the case that the right-hand side of either (1.6) or (1.7) is obviously a polynomial in that the terms of the sums are mostly rational functions with non-trivial denominators.

For example, when $N = 2$, (1.6) asserts

$$1 + aq(1+q) + a^2 q^4 = \frac{1}{(1-aq)(1-aq^2)} - \frac{a^2 q^2 (1+q)}{(1-aq)(1-aq^3)} + \frac{a^4 q^9}{(1-aq^2)(1-aq^2)} \tag{1.9}$$

and (1.7) asserts (after cancelling common factors)

$$1 + aq(1+q) + a^2q^4 = \frac{(1-a^2q^3)(1+aq^2)}{(1-aq)} - \frac{a^2q^2(1-q^2)}{(1-aq)} \tag{1.10}$$

One of the objects of this paper is to present a new representation for the polynomial on the left of (1.6) or (1.7) that converges to the right-hand side of (1.5) and is a polynomial term by term. To accomplish this we shall require the development of an "a-generalization" of Gaussian polynomials.

Our new identity asserts

$$\begin{aligned}
&\sum_{n=0}^{N} a^n q^{n^2} \begin{bmatrix} N \\ n \end{bmatrix}_{;q,q} \\
&= \sum_{0 \leqq 2n \leqq N} (-1)^n a^{2n} q^{n(5n-1)/2} \begin{bmatrix} N \\ n \end{bmatrix}_{;q,q} \begin{bmatrix} 2N+1-2n \\ N-2n \end{bmatrix}_{;q,aq^n} \\
&- \sum_{0 \leqq 2n \leqq N-1} (-1)^n a^{2n+1} q^{n(5n+3)/2} \begin{bmatrix} N \\ n \end{bmatrix}_{;q,q} \begin{bmatrix} 2N-2n \\ N-2n-1 \end{bmatrix}_{;q,aq^n}.
\end{aligned} \tag{1.11}$$

The a-Gaussian polynomial $\begin{bmatrix} N \\ n \end{bmatrix}_{;q,a}$ will be defined and studied in Sections 2 and 3. Propositions 3.1 and 3.2 show that (1.11) converges directly to (1.5). Now for $N = 2$, (1.11) asserts

$$\begin{aligned}
&1 + aq(1+q) + a^2q^4 \\
&\quad = (1 + a + aq + aq^2 + a^2 + a^2q + 2a^2q^2 + a^2q^3 + a^2q^4) \\
&\quad - a^2q^2(1+q) - a(1 + a + aq + aq^2)
\end{aligned} \tag{1.12}$$

As we shall see in Sections 2 and 3, the a-Gaussian polynomials have their own intrinsic surprises and appeal. However, it is natural to ask why one would want (1.11) when it would seem that (1.6) and (1.7) would suffice as finitized versions of (1.5). We shall discuss this question further in Section 6. For now, we merely note that the long standing Borwein conjectures (Andrews, 1995) are merely assertions about polynomials that are, in fact, finitizations of classical Rogers-Ramanujan type identities (Andrews, 1995, Sec. 4). Consequently, in depth studies of such polynomials is clearly in order, and it is to be hoped that a-Gaussian polynomials may add some insight in this area.

In addition, our work here contributes to further elucidation of truncated Rogers-Ramanujan series, a topic suggested by Ramanujan and studied from the point of view of Bailey Chains in (Andrews, 1993).

2. a-Gaussian polynomials

The definition for a-Gaussian polynomials is, at first glance, rather unilluminating. So we preface it with a discussion of what we are striving for.

To begin with, it is well-known (Andrews, 1976, Ch. 3, Th. 3.1) that the Gaussian polynomial

$$\begin{bmatrix} N+M \\ M \end{bmatrix}; q$$

is the generating function for partitions with largest part $\leqq N$ and number of parts $\leqq M$. So, for example

$$\begin{aligned}\begin{bmatrix} 5 \\ 2 \end{bmatrix}; q &= 1 + q + 2q^2 + 2q^3 + 2q^4 + q^5 + q^6 \\ &= 1 + q + q^2 + q^{1+1} + q^3 + q^{2+1} + q^{3+1} + q^{2+2} + q^{3+2} + q^{3+3}.\end{aligned}$$

Now as is noted in (Andrews, 1976, Ch. 2) often one needs a two variable generating function in which a second variable records the number of parts of the partition being generated. Thus one would like to generalize the above polynomial to

$$\begin{aligned}1 + aq + aq^2 + a^2q^{1+1} + aq^3 + a^2q^{2+1} + a^2q^{3+1} \\ + a^2q^{2+2} + a^2q^{3+2} + a^2q^{3+3} \\ = 1 + aq\begin{bmatrix} 3 \\ 1 \end{bmatrix}; q + a^2q^2\begin{bmatrix} 4 \\ 2 \end{bmatrix}; q.\end{aligned}$$

Proposition 5 below makes clear that our a-Gaussian polynomials achieve this initial objective.

Definition 2.1. *For integers N and j with $N \geqq 0$*

$$\begin{bmatrix} N \\ j \end{bmatrix}; q, a = \begin{cases} 0 & \text{if } j < 0 \\ 1 & \text{if } j = 0 \text{ or } N \\ \sum_{h=0}^{j} a^h \begin{bmatrix} N-j+h-1 \\ h \end{bmatrix}; q & \text{if } 0 < j < N \\ (aq^{N-j}; q)_{j-N} & \text{if } j > N. \end{cases} \tag{2.1}$$

Remark 2.2. *The cases $j \leq N$ and $j \geq N$ actually coincide if one interprets $\begin{bmatrix} -A \\ m \end{bmatrix}; q$ in the standard way. We have chosen to use the several separate lines to emphasize that the polynomial is a finite product* *when $j > N$. The more succinct representation would have sacrificed clarity.*

We shall now prove seven propositions about a-Gaussian polynomials. The first one establishes that we have truly generalized the classical Gaussian polynomials. Propositions 2.4–2.6 are the natural extensions of the Pascal triangle recurrences for Gaussian polynomials. Proposition 2.7 establishes the connection with partitions that we described at the beginning of this section. Proposition 2.8 is a naturally terminating representation of a-Gaussian polynomials. Proposition 2.9 is the natural extension of the finite geometric series summation to a-Gaussian polynomials.

Proposition 2.3. *For integers N and j with $N \geqq 0$,* $\begin{bmatrix} N \\ j \end{bmatrix} ; q, q \Big] = \begin{bmatrix} N \\ j \end{bmatrix} ; q \Big]$.

Proof. Clearly both sides are identically 0 if $j < 0$ or $j > N$. Also both sides equal 1 when $j = 0$ or N. Finally, for $0 < j < N$

$$\left[\begin{matrix} N \\ j \end{matrix} ; q, q\right] = \sum_{h=0}^{j} q^h \left[\begin{matrix} N-j+h-1 \\ h \end{matrix} ; q\right] = \left[\begin{matrix} N \\ j \end{matrix} ; q\right] \tag{2.2}$$

by (Andrews, 1976, p. 37, eq. (3.3.9)). □

Proposition 2.4. *For integers N and j with $N \geqq 1$,*

$$\left[\begin{matrix} N \\ j \end{matrix} ; q, a\right] = \left[\begin{matrix} N-1 \\ j \end{matrix} ; q, a\right] + aq^{N-j-1} \left[\begin{matrix} N-1 \\ j-1 \end{matrix} ; q, a\right]. \tag{2.3}$$

Proof. If $j < 0$, then both sides are 0. If $j = 0$, then both sides equal 1. If $0 < j < N - 1$, we see that

$$\begin{aligned}
\begin{bmatrix} N \\ j \end{bmatrix} ;q,a \Big] &= \sum_{h=0}^{j} a^h \begin{bmatrix} N-j+h-1 \\ h \end{bmatrix} ;q \Big] \\
&= \sum_{h=0}^{j} a^h \left(\begin{bmatrix} N-1-j+h-1 \\ h \end{bmatrix} ;q \Big] \right. \\
&\qquad \left. + q^{N-j-1} \begin{bmatrix} N-1-j+h-1 \\ h-1 \end{bmatrix} ;q,q \Big] \right) \\
&\qquad \text{(by (Andrews, 1976, p. 35, eq. (3.3.3)))} \\
&= \begin{bmatrix} N-1 \\ j \end{bmatrix} ;q,a \Big] + aq^{N-j-1} \sum_{h=0}^{j-1} a^h \begin{bmatrix} N-1-j+h \\ h \end{bmatrix} \\
&= \begin{bmatrix} N-1 \\ j \end{bmatrix} ;q,a \Big] + aq^{N-j-1} \begin{bmatrix} N-1 \\ j-1 \end{bmatrix} ;q,a \Big] .
\end{aligned}$$

Noting that

$$\begin{bmatrix} N \\ N-1 \end{bmatrix} ;q,a \Big] = \sum_{h=0}^{j} a^h = \frac{1-a^{j+1}}{1-a}, \tag{2.4}$$

we see that the case $j = N - 1$ also falls into place because

$$\begin{aligned}
\begin{bmatrix} N \\ N-1 \end{bmatrix} ;q,a \Big] = 1 + a \sum_{h=0}^{j-1} a^h &= \begin{bmatrix} N-1 \\ N-1 \end{bmatrix} ;q,a \Big] \\
&\quad + aq^{N-(N-1)-1} \begin{bmatrix} N-1 \\ N-2 \end{bmatrix} ;q,a \Big] .
\end{aligned} \tag{2.5}$$

The case $j = N$ asserts

$$1 = \left(1 - aq^{-1}\right) + aq^{-1} \cdot 1$$

which is obvious.

Finally, if $j > N$

$$\begin{aligned}
&\left[\begin{matrix} N-1 \\ j \end{matrix} ;q,a\right] + aq^{N-j-1}\left[\begin{matrix} N-1 \\ j-1 \end{matrix} ;q,a\right] \\
&= \left(aq^{N-1-j};q\right)_{j-(N-1)} + aq^{N-j-1}\left(aq^{(N-1)-(j-1)};q\right)_{(j-1)-(N-1)} \\
&= \left(aq^{N-j};q\right)_{j-N}\left(\left(1-aq^{N-1-j}\right)+aq^{N-j-1}\right) \\
&= \left(aq^{N-j};q\right)_{j-N} \\
&= \left[\begin{matrix} N \\ j \end{matrix} ;q,a\right].
\end{aligned}$$

Thus Proposition 2.4 is established. □

Proposition 2.5. *For integers N and j with $N \geqq 1$,*

$$\left[\begin{matrix} N \\ j \end{matrix} ;q,a\right] = \left[\begin{matrix} N-1 \\ j \end{matrix} ;q,aq\right] + a\left[\begin{matrix} N-1 \\ j-1 \end{matrix} ;q,a\right]$$

Proof. If $j < 0$, then both sides of this equation are identically 0. If $j = 0$, then both sides equal 1. If $0 < j < N-1$, then

$$\begin{aligned}
&\left[\begin{matrix} N \\ j \end{matrix} ;q,a\right] = \sum_{h=0}^{j} a^h \left[\begin{matrix} N-j+h-1 \\ h \end{matrix} ;q\right] \\
&= \sum_{h=0}^{j} a^h \left(\left[\begin{matrix} N-j+h-2 \\ h-1 \end{matrix} ;q\right] + q^h\left[\begin{matrix} N-j+h-2 \\ h \end{matrix} ;q\right]\right) \\
&= \sum_{h=0}^{j-1} a^{h+1}\left[\begin{matrix} N-j+h-1 \\ h \end{matrix} ;q\right] + \sum_{h=0}^{j}(aq)^h\left[\begin{matrix} N-1-j+h-1 \\ h \end{matrix} ;q\right] \\
&= a\left[\begin{matrix} N-1 \\ j-1 \end{matrix} ;q,a\right] + \left[\begin{matrix} N-1 \\ j \end{matrix} ;q,aq\right].
\end{aligned}$$

If $j = N-1$, the assertion is

$$1 + a + \cdots a^{N-1} = a\left(1 + a + \cdots + a^{N-2}\right) + 1,$$

which is immediate.

If $j = N$, the assertion is

$$1 = (1-a) + a$$

which is obvious.

Finally if $j > N$,

$$\begin{aligned}
&\left[\begin{matrix} N-1 \\ j \end{matrix} ;q,aq\right] + a\left[\begin{matrix} N-1 \\ j-1 \end{matrix} ;q,a\right] \\
&= a\left(qq^{(N-1)-j};q\right)_{j-n(N-1)} + a\left(aq^{(N-1)-(j-1)};q\right)_{(j-1)-(N-1)} \\
&= \left(aq^{N-j};q\right)_{j-N}((1-a)+a) \\
&= \left(aq^{N-j};q\right)_{j-n} \\
&= \left[\begin{matrix} N \\ j \end{matrix} ;q,a\right].
\end{aligned}$$

This proves Proposition 2.5. □

Proposition 2.6. *For integers N and j with $N \geqq 1$,*

$$\left[\begin{matrix} N \\ j \end{matrix} ;q,a\right] = \left[\begin{matrix} N-1 \\ j-1 \end{matrix} ;q,a\right] + a^j\left[\begin{matrix} N-1 \\ j \end{matrix} ;q\right].$$

Proof. If $j < 0$, then both sides are 0. If $j = 0$, then both sides equal 1.
If $0 < j < N-1$, then

$$\begin{aligned}
&\left[\begin{matrix} N \\ j \end{matrix} ;q,a\right] - \left[\begin{matrix} N-1 \\ j-1 \end{matrix} ;q,a\right] \\
&= \sum_{h=0}^{j} a^h \left[\begin{matrix} N-j+h-1 \\ h \end{matrix} ;q\right] - \sum_{h=0}^{j-1} a^h \left[\begin{matrix} (N-1)-(j-1)+h-1 \\ h \end{matrix} ;q\right] \\
&= a^j\left[\begin{matrix} N-1 \\ j \end{matrix} ;q\right].
\end{aligned}$$

If $j = N-1$, the assertion is

$$1 + a + a^2 + \cdots + a^{N-1} = \left(1 + a + a^2 + \cdots + a^{N-2}\right) + a^{N-1}.$$

If $j = N$, the assertion is $1 = 1$.
If $j > N$, then

$$\begin{aligned}
&\left[\begin{matrix} N \\ j \end{matrix} ;q,a\right] - \left[\begin{matrix} N-1 \\ j-1 \end{matrix} ;q,a\right] \\
&= \left(aq^{N-j};q\right)_{j-N} - \left(aq^{(N-1)-(j-1)};q\right)_{(j-1)-(N-1)} \\
&= 0 = a^j\left[\begin{matrix} N-1 \\ j \end{matrix} ;q\right].
\end{aligned}$$

Thus Proposition 2.6 is proved. □

Proposition 2.7. *For nonnegative integers*

$$\begin{bmatrix} N+M \\ N \end{bmatrix}; q, aq \Bigg] = \sum_{n,m \geqq 0} p(N,M,n,m) a^m q^n,$$

where $p(N,M,n,m)$ is the number of partitions of n into m parts with $m \leqq M$ and each part $\leqq N$.

Proof. It is well known (Andrews, 1976, Th. 3.1, p. 33) that

$$\left[\begin{matrix} N+M \\ M \end{matrix} ; q\right]$$

is the generating function for partitions with $\leqq M$ parts each $\leqq N$. Hence

$$q^h \begin{bmatrix} N+h-1 \\ h \end{bmatrix}$$

is the generating function for partitions with exactly h parts each $\leqq N$. Consequently

$$\begin{aligned}
&\sum_{n,m \geqq 0} p(N,M,n,m) a^m q^n \\
&= \sum_{h=0}^{M} a^h q^h \begin{bmatrix} N+h-1 \\ h \end{bmatrix} \\
&= \left[\begin{matrix} N+M \\ M \end{matrix} ; q, aq\right],
\end{aligned}$$

as desired □

Proposition 2.8. *For nonnegative integers N and j,*

$$\left[\begin{matrix} N \\ j \end{matrix} ; q, a\right] = \frac{1}{(q;q)_j} \sum_{i=0}^{j} \left[\begin{matrix} j \\ i \end{matrix} ; q\right] a^{j-i} \left(aq^{N-j}; q\right)_i (q/a;q)_{j-i}.$$

Proof. If $N = 0$, the sum on the right is

$$\begin{aligned}
&\frac{1}{(q;q)_j}\sum_{i=0}^{j}\begin{bmatrix} j \\ i \end{bmatrix}_q a^{j-i}(-1)^i a^i q^{-ji+\binom{i}{2}}(q/a;q)_j \\
&= \frac{a^j(q/a;q)_j}{(q;q)_j}\sum_{i=0}^{j}\begin{bmatrix} j \\ i \end{bmatrix}_q (-1)^i q^{\binom{i}{2}-ji} \\
&= \frac{a^j(q/a;q)_j}{(q;q)_j}\left(q^{-j};q\right)_j \qquad \text{(by (Andrews, 1976, p. 35, eq. (3.3.6))} \\
&= (-1)^j q^{-\binom{j+1}{2}} a^j (q/a;q)_j \\
&= \left(aq^{-j};q\right)_j \\
&= \begin{bmatrix} 0 \\ j \end{bmatrix}_{q,a}
\end{aligned}$$

for all $j \geqq 0$.

If, on the other hand, $j = 0$, the sum on the right is equal to 1 which is $\begin{bmatrix} N \\ 0 \end{bmatrix}_{q,a}$.

We can conclude the proof of the proposition by showing that the right-hand side of the asserted identity satisfies the recurrence given in Proposition 2.4 thus permitting a double induction on N and j to conclude matters.

We denote by $R(N, j)$ the right-hand side of the equation asserted in the proposition.

$$\begin{aligned}
&R(N,j) - R(N-1,j) \\
&= \frac{1}{(q;q)_j}\sum_{i=0}^{j}\begin{bmatrix} j \\ i \end{bmatrix}_q a^{j-i}(q/a;q)_{j-i}\left(\left(aq^{N-j};q\right)_i - \left(aq^{N-1-j};q\right)_i\right) \\
&= \frac{1}{(q;q)_j}\sum_{i=0}^{j}\begin{bmatrix} j \\ i \end{bmatrix}_q a^{j-i}(q/a;q)_{j-i}\left(aq^{N-j};q\right)_i \left(\left(1 - aq^{N-j+i-1}\right)\right. \\
&\quad \left.- \left(1 - aq^{N-1-j}\right)\right) \\
&= \frac{1}{(q;q)_j}\sum_{i=0}^{j}\begin{bmatrix} j \\ i \end{bmatrix}_q a^{j-i}(q/a;q)_{j-i}\left(aq^{N-j};q\right)_{i-1} aq^{N-1-j}\left(1 - q^i\right) \\
&= \frac{aq^{N-1-j}}{(q;q)_{j-1}}\sum_{i=0}^{j-1}\begin{bmatrix} j-1 \\ i \end{bmatrix}_q a^{j-1-i}(q/a;q)_{j-1-i}\left(aq^{(N-1)-(j-1)};q\right)_i \\
&= aq^{N-1-j}R(N-1,j-1),
\end{aligned}$$

and Proposition 2.8 is proved. □

Proposition 2.9. *For N and j nonnegative integers*

$$\begin{bmatrix} N+j+1 \\ j \end{bmatrix}_{;q,a} = \frac{1}{(q;q)_N} \sum_{i=0}^{N} \begin{bmatrix} N \\ i \end{bmatrix}_{;q} \frac{(-1)^i q^{\binom{i+1}{2}} \left(1 - a^{j+1} q^{i(j+1)}\right)}{(1-aq^i)}$$

Proof.

$$\begin{aligned}
\begin{bmatrix} N+j+1 \\ j \end{bmatrix}_{;q,a} &= \sum_{h=0}^{j} \begin{bmatrix} N+h \\ h \end{bmatrix}_{;q} a^h \\
&= \sum_{h=0}^{j} \frac{(q^{h+1};q)_N}{(q;q)_N} a^h \\
&= \frac{1}{(q;q)_N} \sum_{h=0}^{j} \sum_{i=0}^{N} \begin{bmatrix} N \\ i \end{bmatrix}_{;q} (-1)^i q^{\binom{i+1}{2}+hi} a^h \\
&\qquad \text{(by (Andrews, 1976, p. 36, eq. (3.3.6)))} \\
&= \frac{1}{(q;q)_N} \sum_{i=0}^{N} \begin{bmatrix} N \\ i \end{bmatrix}_{;q} (-1)^i q^{\binom{i+1}{2}} \frac{\left(1 - a^{j+1} q^{i(j+1)}\right)}{(1-aq^i)},
\end{aligned}$$

by the finite geometric series summation. □

3. Limiting Cases and Identities

The previous section described fundamental formulas and recurrences for the a-Gaussian polynomials. In this section, we examine the limiting values of these polynomials (Propositions 3.1 and 3.2), and we show how they fit into a generalized Chu-Vandermonde summation (Proposition 3.3). Proposition 3.4 provides a useful reduction formula.

Proposition 3.1. *For $|a| < 1$, $|q| < 1$,*

$$\lim_{N\to\infty} \begin{bmatrix} N+M \\ N \end{bmatrix}_{;q,a} = \frac{1}{(a;q)_M}.$$

Proof.

$$\begin{aligned}
\lim_{N\to\infty} \begin{bmatrix} N+M \\ N \end{bmatrix}_{;q,a} &= \lim_{N\to\infty} \sum_{h=0}^{N} \begin{bmatrix} M+h-1 \\ h \end{bmatrix}_{;q} a^h \\
&= \frac{1}{(a;q)_M} \qquad \text{(by (Andrews, 1976, p. 36, eq. (3.3.7)))}.
\end{aligned}$$

□

Proposition 3.2. *If $|a| < 1$, $|q| < 1$, and A, B, C and D are integers with $A > C > 0$, then*

$$\lim_{N\to\infty} \begin{bmatrix} AN+B \\ CN+D \end{bmatrix}_{;q,a} = \frac{1}{(a;q)_\infty}.$$

Proof.

$$\lim_{N\to\infty} \begin{bmatrix} AN+B \\ CN+D \end{bmatrix}_{;q,a} = \lim_{N\to\infty} \sum_{h=0}^{CN+D} \begin{bmatrix} (A-C)N+B-D+h-1 \\ h \end{bmatrix}_{;q} a^h$$

$$= \sum_{h=0}^{\infty} \frac{a^h}{(q;q)_h} = \frac{1}{(a;q)_\infty}$$

by (Andrews, 1976, p. 19, eq. (2.2.5)). □

Proposition 3.3. *If R, N, and j are non-negative integers then*

$$\begin{bmatrix} n \\ j \end{bmatrix}_{;q,a} = \sum_{i=0}^{R} a^i q^{i(n+i-j-R-1)} \begin{bmatrix} R \\ i \end{bmatrix}_{;q} \begin{bmatrix} n-R \\ j-i \end{bmatrix}_{;q,a}.$$

Proof. We shall prove this result by showing that the right-hand side does not depend on R and is equal to the left-hand side when $R = 0$ (the latter is immediately obvious).

$$\sum_{i=0}^{R} a^i q^{i(n+i-j-R-1)} \begin{bmatrix} R \\ i \end{bmatrix}_{;q} \begin{bmatrix} n-R \\ j-i \end{bmatrix}_{;q,a}$$

$$= \sum_{i=0}^{R} a^i q^{i(n+i-j-R-1)} \begin{bmatrix} R \\ i \end{bmatrix}_{;q} \left(\begin{bmatrix} n-R \\ j-i \end{bmatrix}_{;q,a} + aq^{n-R-j+i-1} \begin{bmatrix} n-R-1 \\ j-i-1 \end{bmatrix}_{;q,a} \right)$$

$$= \sum_{i=0}^{R} a^i q^{i(n+i-j-R-1)} \begin{bmatrix} R \\ i \end{bmatrix}_{;q} \begin{bmatrix} n-(R+1) \\ j-i \end{bmatrix}_{;q,a}$$

$$+\sum_{i=0}^{R+1} a^i q^{(i-1)(n+i-1-j-R-1)} \begin{bmatrix} R \\ i-1 \end{bmatrix}; q \Big] q^{n-R-j+i-2} \begin{bmatrix} n-(R+1) \\ j-i \end{bmatrix}; q, a \Big]$$

$$= \sum_{i=0}^{R+1} a^i q^{i(n+i-j-(R+1)-1)} \left(q^i \begin{bmatrix} R \\ i \end{bmatrix}; q \Big] + \begin{bmatrix} R \\ i-1 \end{bmatrix}; q \Big] \right) \begin{bmatrix} n-(R+1) \\ j-i \end{bmatrix}; q, a \Big]$$

$$= \sum_{i=0}^{R+1} a^i q^{i(n+i-j-(R+1)-1)} \begin{bmatrix} R+1 \\ i \end{bmatrix}; q \Big] \begin{bmatrix} n-(R+1) \\ j-i \end{bmatrix}; q, a \Big].$$

Thus the sum on the right-hand side of the asserted identity is unaltered when R is replaced by $R+1$. Consequently it is equal to its value at $R = 0$ which is $\left[\begin{matrix} n \\ j \end{matrix} ; q, a \right]$ as asserted. □

Proposition 3.4. *For nonnegative integers r, n, m,*

$$\left[\begin{matrix} n+m \\ n \end{matrix} ; q, aq^r \right] = \sum_{j=0}^{r} a^j (-1)^j q^{\binom{j}{2}} (q^m; q)_j \left[\begin{matrix} r \\ j \end{matrix} ; q \right] \left[\begin{matrix} n+m \\ n-j \end{matrix} ; q, a \right].$$

Proof. We proceed by induction on r. When $r = 0$, the assertion is a tautology. At $r+1$,

$$\left[\begin{matrix} n+m \\ n \end{matrix} ; q, aq^{r+1} \right]$$

$$= \sum_{j=0}^{n} \left[\begin{matrix} m+j-1 \\ j \end{matrix} ; q \right] a^j q^{(r+1)j}$$

$$= \sum_{j=0}^{n} \left[\begin{matrix} m+j-1 \\ j \end{matrix} ; q \right] a^j \left(q^{rj} - q^{rj} (1-q^j) \right)$$

$$= \sum_{j=0}^{n} \left[\begin{matrix} m+j-1 \\ j \end{matrix} ; q \right] a^j q^{rj} - \sum_{j=0}^{n} (1-q^m) \left[\begin{matrix} m+j-1 \\ j-1 \end{matrix} ; q \right] a^j q^{rj}$$

$$= \sum_{j=0}^{n} \left[\begin{matrix} m+j-1 \\ j \end{matrix} ; q \right] a^j q^{rj} - (1-q^m) \sum_{j=0}^{n-1} \left[\begin{matrix} m+j \\ j \end{matrix} ; q \right] a^{j+1} q^{r(j+1)}$$

$$= \left[\begin{matrix} n+m \\ n \end{matrix} ; q, aq^r \right] - aq^r (1-q^m) \left[\begin{matrix} n+m \\ n \end{matrix} ; q, aq^r \right]$$

$$= \sum_{j=0}^{r} a^j (-1)^j q^{\binom{j}{2}} (q^m; q)_j \left[\begin{matrix} r \\ j \end{matrix} ; q \right] \left[\begin{matrix} n+m \\ n-j \end{matrix} ; q, a \right]$$

$$+\sum_{j=0}^{r} a^{j+1}(-1)^{j+1}q^{\binom{j}{2}}q^{r}\,(q^m)_{j+1}\begin{bmatrix} r \\ j \end{bmatrix}_{q}\begin{bmatrix} n+m \\ n-1-j \end{bmatrix}_{q,a}$$

$$=\sum_{j=0}^{r+1} a^{j}(-1)^{j}q^{\binom{j}{2}}\,(q^m;q)_j\begin{bmatrix} r \\ j \end{bmatrix}_{q}\begin{bmatrix} n+m \\ n-j \end{bmatrix}_{q,a}$$

$$\sum_{j=0}^{r+1} a^{j}(-1)^{j}q^{\binom{j-1}{2}+r}\,(q^m;q)_j\begin{bmatrix} r \\ j-1 \end{bmatrix}_{q}\begin{bmatrix} n+m \\ n-j \end{bmatrix}_{q,a}$$

$$=\sum_{j=0}^{r+1} a^{j}(-1)^{j}q^{\binom{j}{2}}\,(q^m;q)_j\begin{bmatrix} n+m \\ n-j \end{bmatrix}_{q,a}\left(\begin{bmatrix} r \\ j \end{bmatrix}_{q}+q^{r-j+1}\begin{bmatrix} r \\ j-1 \end{bmatrix}_{q}\right)$$

$$=\sum_{j=0}^{r+1} a^{j}(-1)^{j}q^{\binom{j}{2}}\,(q^m;q)_j\begin{bmatrix} r+1 \\ j \end{bmatrix}_{q}\begin{bmatrix} n+m \\ n-j \end{bmatrix}_{q,a}$$

(by (Andrews, 1976, p. 35, eq. (3.3.3))).

Hence Proposition 3.4 follows by induction on r. □

4. a-Generalizations of Finite Rogers-Ramanujan Type Identities

In Section 1, equation (1.11) is the special case $k=2$, $m=N$ of the following result:

Theorem 4.1. *For m, N, k nonnegative integers with $k>0$*

$$\sum_{s\geqq 0}(-1)^s a^{ks} q^{s((2k+1)s-1)/2}\begin{bmatrix} N \\ s \end{bmatrix}_{q}\begin{bmatrix} N+m+1-ks \\ m-ks \end{bmatrix}_{q,aq^s}$$

$$-\sum_{s\geqq 0}(-1)^s a^{ks+1} q^{s((2k+1)s+3)/2}\begin{bmatrix} N \\ s \end{bmatrix}_{q}\begin{bmatrix} N+m-ks \\ m-ks \end{bmatrix}_{q,aq^s}$$

$$=\sum_{\substack{n_1\geqq n_2\geqq\cdots\geqq n_{k-1}\geqq 0\\ n_1+n_2+\cdots n_{k-1}\leqq m}}\frac{q^{n_1^2+n_2^2+\cdots+n_{k-1}^2}a^{n_1+n_2+\cdots+n_{k-1}}(q;q)_N}{(q;q)_{N-n_1}(q;q)_{n_1-n_2}(q;q)_{n_2-n_3}\cdots(q;q)_{n_{k-2}-n_{k-1}}(q;q)_{n_{k-1}}}$$

Remark 4.2. *If we let $m,N\to\infty$ take $k=1$ and invoke Propositions 3.1 and 3.2, we retrieve* (1.5) *term by term.*

Proof. Call the left side of this identity $L(m)$ and the right side $R(m)$. We proceed by induction on m.

Clearly $L(0)=R(0)=1$.

Furthermore, it is immediate that

$$
\begin{aligned}
&R(m) - R(m-1) \\
&= a^m \left[a^m\right] \sum_{n_1 \geqq \cdots \geqq n_{k-1} \geqq 0} \frac{a^{n_1^2+n_2^2+\cdots+n_{k-1}^2} a^{n_1+n_2+\cdots+n_{k-1}} (q;q)_N}{(q;q)_{N-n_1}(q;q)_{n_1-n_2}\cdots(q;q)_{n_{k-2}-n_{k-1}}(q;q)_{n_k-1}}
\end{aligned}
$$

where $[a^m] \sum_{j=0}^{\infty} c_j a^j = c_m$.

On the other hand,

$$
\begin{aligned}
&L(m) - L(m-1) \\
&= \sum_{s \geqq 0} (-1)^s a^{ks} q^{s((2k+1)s-1)/2} \begin{bmatrix} N \\ s \end{bmatrix}; q \Bigg] \\
&\qquad \times \left(\begin{bmatrix} N+m+1-ks \\ m-ks \end{bmatrix}; q, aq^s \Bigg] - \begin{bmatrix} N+m-ks \\ m-ks-1 \end{bmatrix}; q, aq^s \Bigg] \right) \\
&- \sum_{s \geqq 0} (-1)^s a^{ks+1} q^{s((2k=1)s+3)/2} \begin{bmatrix} N \\ s \end{bmatrix}; q \Bigg] \\
&\qquad \left(\begin{bmatrix} N+m-ks \\ m-ks-1 \end{bmatrix}; q, aq^s \Bigg] - \begin{bmatrix} N+m-ks-1 \\ m-ks-2 \end{bmatrix}; q, aq^s \Bigg] \right) \\
&= \sum_{s \geqq 0} (-1)^s a^{ks} q^{s((2k+1)s-1)/2} \begin{bmatrix} N \\ s \end{bmatrix}; q \Bigg] (aq^s)^{m-ks-1} \begin{bmatrix} N+m-ks \\ m-ks \end{bmatrix}; q \Bigg] \\
&- \sum_{s \geqq 0} (-1)^s a^{ks+1} q^{s((2k+1)s+3)/2} \begin{bmatrix} N \\ s \end{bmatrix}; q \Bigg] (aq^s)^{m-ks-1} \begin{bmatrix} N+m-ks-1 \\ m-ks-1 \end{bmatrix}; q \Bigg] \\
&\qquad\qquad \text{(by Proposition 2.6)} \\
&= a^m \left\{ \sum_{s \geqq 0} (-1)^s q^{\binom{s}{2}+ms} \begin{bmatrix} N \\ s \end{bmatrix}; q \Bigg] \begin{bmatrix} N+m-ks \\ m-ks \end{bmatrix}; q \Bigg] \right. \\
&\qquad \left. - \sum_{s \geqq 0} (-1)^s q^{\binom{s+1}{2}+ms} \begin{bmatrix} N \\ s \end{bmatrix}; q \Bigg] \begin{bmatrix} N+m-ks-1 \\ m-ks-1 \end{bmatrix}; q \Bigg] \right\}.
\end{aligned}
$$

Hence the object of proving

$$L(m) - L(m-1) = R(m) - R(m-1)$$

reduces to proving

$$\sum_{s\geqq 0}(-1)^s q^{\binom{s}{2}+ms}\begin{bmatrix} N \\ s \end{bmatrix}; q\Bigg]\begin{bmatrix} N+m-ks \\ m-ks \end{bmatrix}; q\Bigg]$$

$$-\sum_{s\geqq 0}(-1)^s q^{\binom{s+1}{2}+ms}\begin{bmatrix} N \\ s \end{bmatrix}; q\Bigg]\begin{bmatrix} N+m-ks-1 \\ m-ks-1 \end{bmatrix}; q\Bigg]$$

$$=[a^m]\sum_{n_1\geqq\cdots\geqq n_{k-1}\geqq 0}\frac{q^{n_1^2+n_2^2+\cdots+n_{k-1}^2}a^{n_1+n_2+\cdots+n_{k-1}}(q;q)_N}{(q;q)_{N-n_1}(q;q)_{n_1-n_2}\cdots(q;q)_{n_{k-2}-n_{k-1}}(q;q)_{n_{k-1}}}.$$

This latter result is provable using an identity of J. Stembridge (Stembridge, 1990, Theorem 1.3 (b) with k replaced by $k-1$ and z replaced by aq). Namely

$$[a^m]\sum_{n_1\geqq\cdots\geqq n_{k-1}\geqq 0}\frac{q^{n_1^2+n_2^2+\cdots+n_{k-1}^2}a^{n_1+\cdots+n_{k-1}}(q;q)_N}{(q;q)_{N-n_1}(q;q)_{n_1-n_2}\cdots(q;q)_{n_{k-2}-n_{k-1}}(q;q)_{n_{k-1}}}$$

$$=[a^m]\sum_{n=0}^{N}(-1)^n a^{kn}q^{kn^2+\binom{n}{2}}\begin{bmatrix} N \\ n \end{bmatrix}; q\Bigg]\frac{(1-aq^{2n})}{(aq^n;q)_{N+1}}$$

$$=[a^m]\left\{\sum_{n=0}^{N}(-1)^n a^{kn}q^{kn^2+\binom{n}{2}}\begin{bmatrix} N \\ n \end{bmatrix}; q\Bigg](1-aq^{2n})\sum_{h=0}^{\infty}\begin{bmatrix} N+h \\ h \end{bmatrix}; q\Bigg]a^h q^{nh}\right\}$$

$$=\sum_{n=0}^{N}(-1)^n q^{\binom{n}{2}+mn}\begin{bmatrix} N \\ n \end{bmatrix}; q\Bigg]\begin{bmatrix} N+m-kn \\ m-kn \end{bmatrix}; q\Bigg];$$

$$-\sum_{n=0}^{N}(-1)^n q^{\binom{n+1}{2}+mn}\begin{bmatrix} N \\ n \end{bmatrix}; q\Bigg]\begin{bmatrix} N+m-kn-1 \\ m-kn-1 \end{bmatrix}; q\Bigg];$$

thus the induction step is established, and Theorem 4.1 is proved. □

We conclude this section with two reductions of Theorem 4.1 using Proposition 3.4. These results will allow us to obtain the single variable identities of the next section.

Corollary 4.3. *For m, N, k nonnegative integers with $k > 0$*

$$\sum_{\substack{n_1 \geqq \cdots \geqq n_{k-1} \\ n_1 + \cdots + n_{k-1} \leq m}} \frac{q^{n_1^2+n_2^2+\cdots+n_{k-1}^2} a^{n_1+n_2+\cdots+n_{k-1}} (q;q)_N}{(q;q)_{N-n_1}(q;q)_{n_1-n_2}\cdots(q;q)_{n_{k-2}-n_{k-1}}(q)_{n_{k-1}}}$$

$$= \begin{bmatrix} N+m \\ m \end{bmatrix}; q, aq \Bigg] + \sum_{s=1}^{N} (-1)^s a^{ks} q^{s((2k+1)s-1)/2} \begin{bmatrix} N \\ s \end{bmatrix}; q \Bigg]$$

$$\sum_{j=0}^{s-1} a^j (-1)^j q^{\binom{j+1}{2}} \left(q^{n+1};q\right)_j \begin{bmatrix} s-1 \\ j \end{bmatrix}; q \Bigg]$$

$$\times \left(\begin{bmatrix} N+m+1-ks \\ m-ks-j \end{bmatrix}; q, aq \Bigg] - aq^{2s} \begin{bmatrix} N+m-ks \\ m-ks-j-1 \end{bmatrix}; q, aq \Bigg] \right).$$

Proof. Apply Proposition 3.4 (with $r = s-1$ and a replaced aq) to each of the a-Gaussian polynomials in Theorem 4.1. The terms with $s = 0$ are instead combined using Proposition 2.5. □

Corollary 4.4. *For m, N, k non-negative with $k > 0$*

$$\sum_{\substack{n_1 \geqq \cdots \geqq n_{k-1} \geqq 0 \\ n_1 + \cdots + n_{k-1} \leq m}} \frac{q^{n_1^2+n_2^2+\cdots+n_{k-1}^2} a^{n_1+n_2+\cdots+n_{k-1}} (q;q)_N/(q;q)_{N-n_1}}{(q;q)_{n_1-n_2}\cdots(q;q)_{n_1-n_2}\cdots(q;q)_{n_{k-2}-n_{k-1}}(q;q)_{n_{k-1}}}$$

$$= \sum_{s=0}^{N} (-1)^s a^{ks} q^{s((2k+1)s-1)/2} \begin{bmatrix} N \\ s \end{bmatrix}; q \Bigg] \sum_{j=0}^{s} a^j (-1)^j q^{\binom{j}{2}} (q^{N+1};q)_j \begin{bmatrix} s \\ j \end{bmatrix}; q \Bigg]$$

$$\times \left(\begin{bmatrix} N+m+1-ks \\ m-ks-j \end{bmatrix}; q, a \Bigg] - aq^{2s} \begin{bmatrix} N+m-ks \\ m-ks-j-1 \end{bmatrix}; q, a \Bigg] \right).$$

Proof. Apply Proposition 3.4 (with $r = s$) to each of the a-Gaussian polynomials in Theorem 4.1. □

5. Single Variable Polynomial Rogers-Ramanujan Generalizations

Schur (Schmüdgen, 1990) was the first to prove the Rogers-Ramanujan identities as a limiting case of polynomial identities. Namely, he proved

$$\sum_{0 \leqq 2j \leqq n} q^{j^2} \begin{bmatrix} n-j \\ j \end{bmatrix}; q \Bigg] = \sum_{j=-\infty}^{\infty} (-1)^j q^{j(5j+1)/2} \begin{bmatrix} n \\ \left\lfloor \frac{n-5j}{2} \right\rfloor \end{bmatrix}; q \Bigg], \tag{5.1}$$

and

$$\sum_{0 \leqq 2j \leqq n-1} q^{j^2+j} \begin{bmatrix} n-1-j \\ j \end{bmatrix} ; q \Bigg] \tag{5.2}$$
$$= \sum_{j=-\infty}^{\infty} (-1)^j q^{j(5j-3)/2} \begin{bmatrix} n \\ \left\lfloor \frac{n-5j}{2} \right\rfloor + 1 \end{bmatrix} ; q \Bigg].$$

The relationship of (5.1) and (5.2) to more general results is discussed extensively in (Andrews, 1989, esp. §9).

To everyone's surprise, David Bressoud (Bressoud, 1981b) found a completely different polynomial refinement:

$$\sum_{j=0}^{n} q^{j^2} \left[\begin{matrix} n \\ j \end{matrix} ; q \right] = \sum_{j=-\infty}^{\infty} (-1)^j q^{j(5j+1)/2} \left[\begin{matrix} 2n \\ n+2j \end{matrix} ; q \right], \tag{5.3}$$

and

$$(1-q^{n+1}) \sum_{j=0}^{n} q^{j^2+j} \left[\begin{matrix} n \\ j \end{matrix} ; q \right] \tag{5.4}$$
$$= \sum_{j=-\infty}^{\infty} (-1)^j q^{j(5j+3)/2} \left[\begin{matrix} 2n+2 \\ n+2j+2 \end{matrix} ; q \right].$$

Identities (5.3) and (5.4) have been placed in the context of more general q-hypergeometric identities (Andrews and Berkovich, 2002).

The list does not stop here. At least two further polynomial refinements of the Rogers-Ramanujan identities have been found (Andrews, 1974), (Andrews, 1990, p. 3, eqs. (1.11) and (1.12)). Most recently, S. O. Warnaar (Warnaar, 2002) has found extensive partial sum Rogers-Ramanujan identities.

It should be noted that in each of the examples given above (and in those alluded to in (Andrews, 1974) and (Andrews, 1989)) all the sums terminate naturally. In other words, the index of summation is extended over all values that produce non-zero summands.

As we shall see, we may set $a = 1$ in Corollary 4.3 and $a = q$ in Corollary 4.4 in order to obtain *partial sums* of the Rogers-Ramanujan polynomial. Our results are quite different from those of Warnaar in (Warnaar, 2002).

To this end we require a definition and a lemma.

Definition 5.1. $E_n(x,q) = \lim_{N\to\infty} \left[\begin{matrix} N \\ n \end{matrix} ; q, x \right] = \sum_{j=0}^{n} \frac{x^j}{(q;q)_j}$.

We remark in passing that Euler proved (Andrews, 1976, p. 19, eq. (2.2.5)) $E_\infty(x,q) = \dfrac{1}{(x;q)_\infty}$.

Lemma 5.2. *For non-negative integer N, M and t*

$$\sum_{j=0}^{t}(-1)^j q^{\binom{j+1}{2}} \left(q^{N+1};q\right)_j \begin{bmatrix} t \\ j \end{bmatrix}_q \begin{bmatrix} N+M \\ M-j \end{bmatrix}_q = \begin{bmatrix} N+M \\ M \end{bmatrix}_q (q;q)_t E_t\left(q^{M+1},q\right).$$

Proof.

$$\begin{aligned} &\sum_{j=0}^{t}(-1)^j q^{\binom{j+1}{2}} \left(q^{N+1};q\right)_j \begin{bmatrix} t \\ j \end{bmatrix}_q \begin{bmatrix} N+M \\ M-j \end{bmatrix}_q \\ &= \begin{bmatrix} N+M \\ M \end{bmatrix}_q \lim_{c\to\infty} \sum_{j=0}^{t} \frac{\left(q^{-t};q\right)_j \left(q^{-M};q\right)_j c^j q^{(M+t+1)j}}{(q;q)_j (c;q)_j} \\ &= \begin{bmatrix} N+M \\ M \end{bmatrix}_q (q;q)_t \sum_{j=0}^{t} \frac{q^{j(M+1)}}{(q)_j} \\ &= \begin{bmatrix} N+M \\ M \end{bmatrix}_q (q;q)_t E_t\left(q^{M+1},q\right), \end{aligned}$$

where the penultimate assertion follows from the last line on page 38 of (Andrews, 1976) with $b \to q^{-t}$, $a \to q^{-M}$, and $t \to q^{M+t+1}$. □

Theorem 5.3.

$$\begin{aligned} &\sum_{\substack{n_1 \geqq \cdots \geqq n_{k-1} \geqq 0 \\ n_1+\cdots+n_{k-1} \leqq m}} \frac{q^{n_1^2+n_2^2+\cdots+n_{k-1}^2}(q;q)_N}{(q;q)_{N-n_1}(q;q)_{n_1-n_2}\cdots(q;q)_{n_{k-2}-n_{k-1}}(q;q)_{n_{k-1}}} \\ &= \begin{bmatrix} N+M \\ m \end{bmatrix}_q + \sum_{s=1}^{N}(-1)^s q^{s((2k+1)s-1)/2} \begin{bmatrix} N \\ s \end{bmatrix}_q (q;q)_{s-1} \\ &\quad \times \left(\begin{bmatrix} N+M+1-ks \\ m-ks \end{bmatrix}_q E_{s-1}\left(q^{m-ks+1},q\right) \right. \\ &\quad \left. -q^{2s} \begin{bmatrix} N+M-ks \\ m-ks-1 \end{bmatrix}_q E_{s-1}\left(q^{m-ks},q\right) \right) \end{aligned}$$

Proof. Set $a = 1$ in Corollary 4.3 and invoke Lemma 5.2 for the inner sum with $t = s - 1$. □

Theorem 5.4.

$$\sum_{\substack{n_1 \geqq \cdots \geqq n_{k-1} \\ n_1+\cdots+n_{k-1} \leqq m}} \frac{q^{n_1^2+\cdots+n_{k-1}^2+n_1+\cdots+n_{k-1}}(q;q)_N}{(q;q)_{N-n_1}(q;q)_{n_1-n_2}\cdots(q;q)_{n_1}}$$

$$= \sum_{s=0}^{N}(-1)^s q^{s((2h+1)s+(2k-1))/2} \begin{bmatrix} N \\ s \end{bmatrix}; q \Big] (q;q)_s$$

$$\times \left(\begin{bmatrix} N+m+1-ks \\ m-ks \end{bmatrix}; q \Big] E_s\left(q^{m-ks+1}, q\right) \right.$$

$$\left. -q^{2s+1} \begin{bmatrix} N+m-ks \\ m-ks-1 \end{bmatrix}; q \Big] E_s\left(q^{m-k2}; q\right) \right)$$

Proof. Set $a = q$ in Corollary 4.3 and invoke Lemma 5.2 for the inner sum with $t = s$. □

6. Conclusion

The primary object of this paper has been the development of a-Gaussian polynomials. In light of their natural partition-theoretic interpretation (Proposition 2.7), it is surprising that they have not been studied previously. It seems extremely likely that Proposition 2.7 has already suggested itself to many workers. The first thing one notices is that for a-Gaussian polynomials there is no lovely product formula like (1.8) only a less satisfying sum (Proposition 2.8) which reduces to (1.8) when $a = q$. In addition, the symmetry identity (Andrews, 1976, p. 35, eq. (3.3.2))

$$\begin{bmatrix} N \\ m \end{bmatrix}; q \Big] = \begin{bmatrix} N \\ N-m \end{bmatrix}; q \Big]$$

has no simple analog for a-Gaussian polynomials. It may well be that these two deficits discouraged further investigation especially in light of the fact that the definition of a-Gaussian polynomials contains a sum that is not naturally terminating.

A secondary object of this paper has been the study of the polynomial refinements of "a-generalizations" of Rogers-Ramanujan type identities. Such studies almost always have in mind (or, at least, in the back of their mind) the famous Borwein conjecture (Andrews, 1995). Namely, if

$$\left(q;q^3\right)_n \left(q^2;q^3\right)_n = A_n\left(q^3\right) - qB_n\left(q^3\right) - q^2 C_n\left(q^3\right), \tag{6.1}$$

then each of $A_n(q)$, $B_n(q)$ and $C_n(q)$ has non-negative coefficients.

It is not hard to show (Andrews, 1995, p. 491) that

$$A_n(q) = \sum_{j=-\infty}^{\infty} (-1)^j q^{j(9j+1)/2} \begin{bmatrix} 2n \\ n+3j \end{bmatrix} ;q \Bigg], \tag{6.2}$$

$$B_n(q) = \sum_{j=-\infty}^{\infty} (-1)^j q^{j(9j-5)/2} \left[\begin{matrix} 2n \\ n+3j-1 \end{matrix} ;q \right], \tag{6.3}$$

$$C_n(q) = \sum_{j=-\infty}^{\infty} (-1)^j q^{j(9j+7)/2} \left[\begin{matrix} 2n \\ n+3j+1 \end{matrix} ;q \right], \tag{6.4}$$

Much is known about polynomials of this general nature. Indeed the main theorem in (Andrews et al., 1987) shows that many such polynomials must have non-negative coefficients.

However, the right-hand side of (5.3) is *not* covered by (Andrews et al., 1987), but nonetheless, we see easily that it has non-negative coefficient by inspection of the left-hand side of (5.3).

While the investigation of polynomial "a-generalizations" has not here led to further information on the Borwein conjecture, it should be pointed out that it has provided new insights on truncated Rogers-Ramanujan identities, a topic treated from a wholly different viewpoint in (Andrews, 1993).

References

Andrews, G. E. (1974). Problem 74–12. *SIAM Review*, 16.

Andrews, G. E. (1976). *The Theory of Partitions*. Addison-Wesley, Reading. (Reissued: Cambridge University Press, Cambridge, 1984, 1998).

Andrews, G. E. (1989). On the proofs of the Rogers-Ramanujan identities. In Stanton, D., editor, *q-Series and Partitions*, volume 18 of *IMA Volume in Math. and Its Appl.*, pages 1–14. Springer, New York.

Andrews, G. E. (1990). *q*-trinominial coefficients and Rogers-Ramanujan type identities. In Berndt et al., B., editor, *Analytic Number Theory*, Progr. Math., pages 1–11. Birkhauser, Boston.

Andrews, G. E. (1993). On Ramanujan's empirical calculation for the Rogers-Ramanujan identities. In *A tribute to Emil Grosswald: number theory and related analysis*, volume 143 of *Contemp. Math.*, pages 27–35. American Mathematical Society, Providence, RI.

Andrews, G. E. (1995). On a conjecture of Peter Borwein. *J. Symbolic Computation*, 20:487–501.

Andrews, G. E. and Berkovich, A. (2002). The WP-Bailey tree and its implications. *J. London Math. Soc. (2)*, 66(3):529–549.

Andrews, G. E., Bressoud, D. M., Baxter, R. J., Burge, W., Forrester, P. J., and Viennot, G. (1987). Partitions with prescribed hook differences. *Europ. J. Math.*, 8:341–350.

Bailey, W. N. (1949). Identities of the Rogers-Ramanujan type. *Proc. London Math. Soc. (2)*, 50:1–10.

Bressoud, D. M. (1981a). Solution to problem 74–12. *SIAM Review*, 23:101–104.

Bressoud, D. M. (1981b). Some identities for terminating q-series. *Math. Proc. Camb. Phil. Soc.*, 89:211–223.

Paule, P. (1994). Short and easy computer proofs of the Rogers-Ramanujan identities and identities of similar type. *Electronic J. Combin.*, 1:Research Paper 10, approx. 9 pp. (electronic).

Schur, I. (1917). Ein beitrag zur additiven zahlentheorie. In *S.-B. Preuss. Akad. Wiss.*, Phys.-Math. Klasse, pages 302–321. (Reprinted: Gesamm, Abhand., Vol. 2, Springer, Berlin, 1973, pp. 117–136).

Stembridge, J. (1990). Hall-Littlewood functions, plane partitions, and the Rogers-Ramanujan identities. *Trans. Amer. Math. Soc.*, 319:469–498.

Warnaar, S. O. (2002). Partial sum analogs of the Rogers-Ramanujan identities. *J. Comb. Th. (A)*, 99:143–161.

Watson, G. N. (1929). A new proof of the Rogers-Ramanujan identities. *J. London Math. Soc.*, 4:4–9.

Zeilberger (a.k.a. S. B. Ekhad and S. Tre), D. (1990). A purely verification proof of the first Rogers-Ramanujan identity. *J. Comb. Th. (A)*, 54(2):309–311.

ON A GENERALIZED GAMMA CONVOLUTION RELATED TO THE q-CALCULUS

Christian Berg
Department of Mathematics
University of Copenhagen
Universitetsparken 5, DK-2100
DENMARK
berg@math.ku.dk

Abstract We discuss a probability distribution I_q depending on a parameter $0 < q < 1$ and determined by its moments $n!/(q;q)_n$. The treatment is purely analytical. The distribution has been discussed recently by Bertoin, Biane and Yor in connection with a study of exponential functionals of Lévy processes.

Keywords: q-calculus, infinitely divisible distribution

1. Introduction

In (Bertoin et al., 2004) Bertoin et al. studied the distribution I_q of the exponential functional

$$\mathcal{I}_q = \int_0^\infty q^{N_t}\, dt, \tag{1.1}$$

where $0 < q < 1$ is fixed and $(N_t, t \geq 0)$ is a standard Poisson process. They found the density $i_q(x)$, $x > 0$ and its Laplace and Mellin transforms. They also showed that a simple construction from I_q leads to the density

$$\lambda_q(x) = \frac{1}{\log(1/q)(q, -x, -q/x; q)_\infty}, \tag{1.2}$$

found by Askey, cf. (Askey, 1989), and having log-normal moments. The notation in (1.2) is the standard notation from (Gasper and Rahman, 1990), see below.

The distribution I_q has also appeared in recent work of Cowan and Chiu (Cowan and Chiu, 1994), Dumas et al. (Dumas et al., 2002) and Pakes (Pakes, 1996).

The proofs in (Bertoin et al., 2004) rely on earlier work on exponential functionals which use quite involved notions from the theory of stochastic processes, see (Carmona et al., 1994; Carmona et al., 1997).

The purpose of this note is to give a self-contained analytic treatment of the distribution I_q and its properties.

In Section 2 we define a convolution semigroup $(I_{q,t})_{t>0}$ of probabilities supported by $[0,\infty[$, and it is given in terms of the corresponding Bernstein function $f(s) = \log(-s;q)_\infty$ with Lévy measure ν on $]0,\infty[$ having the density

$$\frac{d\nu}{dx} = \frac{1}{x}\sum_{n=0}^{\infty} \exp\left(-xq^{-n}\right). \tag{1.3}$$

The function $1/\log(-s;q)_\infty$ is a Stieltjes transform of a positive measure which is given explicitly, and this permits us to determine the potential kernel of $(I_{q,t})_{t>0}$.

The measure $I_q := I_{q,1}$ is a *generalized Gamma convolution* in the sense of Thorin, cf. (Thorin, 1977b; Thorin, 1977a). The moment sequence of I_q is shown to be $n!/(q;q)_n$, and the nth moment of $I_{q,t}$ is a polynomial of degree n in t. We give a recursion formula for the coefficients of these polynomials. We establish that I_q has the density

$$i_q(x) = \sum_{n=0}^{\infty} \exp(-xq^{-n})\frac{(-1)^n q^{n(n-1)/2}}{(q;q)_n(q;q)_\infty}.$$

A treatment of the theory of generalized Gamma convolutions can be found in Bondesson's monograph (Bondesson, 1992). The recent paper (Biane et al., 2001) contains several examples of generalized Gamma convolutions which are also distributions of exponential functionals of Lévy processes.

We shall use the notation and terminology from the theory of basic hypergeometric functions for which we refer the reader to the monograph by Gasper and Rahman (Gasper and Rahman, 1990). We recall the q-shifted factorials

$$(z;q)_n = \prod_{k=0}^{n-1}\left(1 - zq^k\right), z \in \mathbb{C}, 0 < q < 1, n = 1, 2, \ldots, \infty$$

and $(z;q)_0 = 1$. Note that $(z;q)_\infty$ is an entire function of z.

For finitely many complex numbers $z_1, z_2, \ldots, z_p$ we use the abbreviation

$$(z_1, z_2, \ldots, z_p; q)_n = (z_1; q)_n (z_2; q)_n \cdots (z_p; q)_n.$$

The q-shifted factorial is defined for arbitrary complex index λ by

$$(z; q)_\lambda = \frac{(z; q)_\infty}{(zq^\lambda; q)_\infty},$$

and this is related to Jackson's function Γ_q defined by

$$\Gamma_q(z) = \frac{(q; q)_{z-1}}{(1-q)^{z-1}} = \frac{(q; q)_\infty}{(q^z; q)_\infty}(1-q)^{1-z}. \tag{1.4}$$

In Section 3 we introduce the entire function $h(z) = \Gamma(z)\,(q^z; q)_\infty$ and use it to express the Mellin transform of I_q. We finally show that the density λ_q given in (1.2) can be written as the product convolution of I_q and another related distribution, see Theorem 3.2 below. The Mellin transform of the density λ_q can be evaluated as a special case of the Askey-Roy beta-integral given in (Askey and Roy, 1986) and in particular we have, see also (Askey, 1989):

$$\int_0^\infty \frac{t^c}{(-t, -q/t; q)_\infty}\frac{dt}{t} = (q; q)_\infty \frac{\Gamma(c)\Gamma(1-c)}{\Gamma_q(c)\Gamma_q(1-c)}(1-q), \quad c \in \mathbb{C} \setminus \mathbb{Z}. \tag{1.5}$$

The value of (1.5) is an entire function of c and equals $h(c)h(1-c)/(q;q)_\infty$.

The following formulas about the q-exponential functions, cf. (Gasper and Rahman, 1990), are important in the following:

$$e_q(z) = \sum_{n=0}^\infty \frac{z^n}{(q; q)_n} = \frac{1}{(z; q)_\infty}, \quad |z| < 1, \tag{1.6}$$

$$E_q(z) = \sum_{n=0}^\infty \frac{q^{n(n-1)/2} z^n}{(q; q)_n} = (-z; q)_\infty, \quad z \in \mathbb{C}. \tag{1.7}$$

2. The analytic method

We recall that a function $\varphi : \,]0, \infty[\, \mapsto [0, \infty[$ is called *completely monotonic*, if it is C^∞ and $(-1)^k \varphi^{(k)}(s) \geq 0$ for $s > 0$, $k = 0, 1, \ldots$. By the Theorem of Bernstein completely monotonic functions have the form

$$\varphi(s) = \int_0^\infty e^{-sx}\, d\alpha(x), \tag{2.1}$$

where α a non-negative measure on $[0, \infty[$. Clearly $\varphi(0+) = \alpha([0, \infty[)$. The equation (2.1) expresses that φ is the Laplace transform of the measure α.

To establish that a probability η on $[0, \infty[$ is infinitely divisible, one shall prove that its Laplace transform can be written

$$\int_0^\infty e^{-sx}\, d\eta(x) = \exp(-f(s)), \quad s \geq 0,$$

where the non-negative function f has a completely monotonic derivative. If η is infinitely divisible, there exists a convolution semigroup $(\eta_t)_{t>0}$ of probabilities on $[0, \infty[$ such that $\eta_1 = \eta$ and it is uniquely determined by

$$\int_0^\infty e^{-sx}\, d\eta_t(x) = e^{-tf(s)}, \quad s > 0,$$

cf. (Berg and Forst, 1975), (Bertoin, 1996). The function f is called the *Laplace exponent* or *Bernstein function* of the semigroup. It has the integral representation

$$f(s) = as + \int_0^\infty \left(1 - e^{-sx}\right) d\nu(x), \tag{2.2}$$

where $a \geq 0$ and the Lévy measure ν on $]0, \infty[$ satisfies the integrability condition $\int x/(1+x)\, d\nu(x) < \infty$. If f is not identically zero the convolution semigroup is transient with potential kernel $\kappa = \int_0^\infty \eta_t\, dt$, and the Laplace transform of κ is $1/f$ since

$$\int_0^\infty e^{-sx}\, d\kappa(x) = \int_0^\infty \left(\int_0^\infty e^{-sx}\, d\eta_t(x)\right) dt = \int_0^\infty e^{-tf(s)}\, dt = \frac{1}{f(s)}. \tag{2.3}$$

The generalized Gamma convolutions η are characterized among the infinitely divisible distributions by the following property of the corresponding Bernstein function f, namely by f' being a Stieltjes transform, i.e. of the form

$$f'(s) = a + \int_0^\infty \frac{d\mu(x)}{s+x}, \quad s > 0,$$

where $a \geq 0$ and μ is a non-negative measure on $[0, \infty[$. The relation between μ and ν is that

$$\frac{d\nu}{dx} = \frac{1}{x} \int_0^\infty e^{-xy}\, d\mu(y).$$

This result was used in (Berg, 1981) to simplify the proof of a theorem of Thorin (Thorin, 1977b), stating that the Pareto distribution is a generalized Gamma convolution.

Theorem 2.1. *Let $0 < q < 1$ be fixed. The function*

$$f(s) = \log(-s;q)_\infty = \sum_{n=0}^\infty \log(1 + sq^n), \quad s \geq 0 \tag{2.4}$$

is a Bernstein function. The corresponding convolution semigroup $((I_{q,t})_{t>0})$ consists of generalized Gamma convolutions and we have

$$\int_0^\infty e^{-sx}\, dI_{q,t}(x) = e^{-tf(s)} = \frac{1}{(-s;q)_\infty^t}, \quad s > 0. \tag{2.5}$$

The potential kernel $\kappa_q = \int_0^\infty I_{q,t}\, dt$ has the following completely monotonic density

$$k_q(x) = 1 - q + \int_1^\infty e^{-xy} \varphi_q(y)\, dy, \tag{2.6}$$

where φ is the continuous function

$$\varphi_q(x) = \begin{cases} n\left(\log^2 |(x;q)_\infty| + n^2\pi^2\right)^{-1} & \text{if } q^{-(n-1)} < x < q^{-n}, \\ 0 & \text{if } x = q^{-(n-1)}, \end{cases} \tag{2.7}$$

$n = 1, 2, \ldots.$

Proof. The function f defined by (2.4) has the derivative

$$f'(s) = \sum_{n=0}^\infty \frac{1}{s + q^{-n}},$$

showing that f' is a Stieltjes transform with $a = 0$, and μ is the discrete measure with mass 1 in each of the points q^{-n}, $n \geq 0$. In particular f is a Bernstein function with $a = 0$ and Lévy measure given by (1.3).

Since

$$\frac{\log(1+s)}{s} = \int_1^\infty \frac{1}{x+s}\frac{dx}{x}$$

we get ($[x]$ denoting the integer part of x)

$$\frac{f(s)}{s} = \int_1^\infty \frac{[\log x/\log(1/q)]+1}{(x+s)x}\,dx$$

showing that $f(s)/s$ is a Stieltjes transform. It follows by the Reuter-Itô Theorem, cf. (Itô, 1974), (Reuter, 1956), (Berg, 1980), that $1/f(s)$ is a Stieltjes transform. Since f is an increasing function mapping $]-1,\infty[$ onto the real line with $f(0)=0$ and $f'(0)=1/(1-q)$ we get

$$\frac{1}{f(s)} = \frac{1-q}{s} + \int_1^\infty \frac{d\mu(x)}{x+s},$$

where

$$d\mu(x) = \lim_{y\to 0+} \frac{-1}{\pi}\,\mathrm{Im}\{\frac{1}{f(-x+iy)}\}\,dx$$

in the vague topology.

For $x \in \left]q^{-(n-1)}, q^{-n}\right[$, $n = 1, 2, \dots$ we find

$$\begin{aligned}\lim_{y\to 0+} \frac{1}{f(-x+iy)} &= \left(\sum_{k=0}^{n-1} \log|1-q^k x| + in\pi + \sum_{k=n}^{\infty} \log(1-q^k x)\right)^{-1} \\ &= \left(\log|(x;q)_\infty| + in\pi\right)^{-1}.\end{aligned}$$

These expressions define in fact a continuous function on $[1,\infty[$, vanishing at the points q^{-n}, $n \geq 0$, so the measure μ has the density φ given by (2.7). Using that the Stieltjes transformation is the second iteration of the Laplace transformation, the assertion about the potential kernel κ_q follows. □

Denoting by $\mathcal{E}_a, a > 0$ the exponential distribution with density $a\exp(-ax)$ on the positive half-line, we have

$$\int_0^\infty e^{-sx}\,d\mathcal{E}_a(x) = e^{-\log(1+s/a)}, \quad s \geq 0,$$

so we can write $I_q := I_{q,1}$ as the infinite convolution

$$I_q = *_{n=0}^{\infty}\, \mathcal{E}_{q^{-n}}.$$

If we let $\Gamma_{a,t}$ denote the Gamma distribution with density

$$x \mapsto \frac{a^t}{\Gamma(t)} x^{t-1} e^{-ax}, \quad x > 0,$$

then we similarly have

$$I_{q,t} = *_{n=0}^{\infty} \Gamma_{q^{-n},t}.$$

Specializing (2.5) to $t = 1$ we have

$$\int_0^\infty e^{-sx}\, dI_q(x) = \frac{1}{(-s;q)_\infty}, \quad s > 0, \tag{2.8}$$

and since the right-hand side of (2.8) is meromorphic in $\mathbb{C}$ with poles at $s = -q^{-n}$, $n \geq 0$ and in particular holomorphic for $|s| < 1$, we know that I_q has moments of any order with

$$s_n(I_q) = (-1)^n D^n \left\{ \frac{1}{(-s;q)_\infty} \right\}_{s=0}, \quad n = 0, 1, \ldots,$$

cf. (Lukacs, 1960, p. 136). Here and in the following we denote by $s_n(\mu)$ the nth moment of the measure μ. However by (1.6) we have

$$\frac{1}{(-s;q)_\infty} = \sum_{n=0}^{\infty} \frac{(-s)^n}{(q;q)_n}, \tag{2.9}$$

hence

$$s_n(I_q) = \frac{n!}{(q;q)_n}. \tag{2.10}$$

Since I_q has an analytic characteristic function, the corresponding Hamburger moment problem is determinate. By Stirling's formula we have

$$\sum_{n=0}^{\infty} \frac{1}{\sqrt[2n]{s_{2n}(I_q)}} = \infty,$$

so also Carleman's criterion shows the determinacy, cf. (Akhiezer, 1965). By (Berg, 1985, Cor. 3.3) follows that $I_{q,t}$ is determinate for all $t > 0$ and by (Berg, 2000) the n'th moment $s_n(I_{q,t})$ is a polynomial of degree n in t given by

$$s_n(I_{q,t}) = \sum_{k=1}^{n} c_{n,k} t^k, \quad n \geq 1, \tag{2.11}$$

where the coefficients $c_{n,k}$ satisfy the recurrence equation

$$c_{n+1,l+1} = \sum_{k=l}^{n} c_{k,l} \binom{n}{k} \sigma_{n-k}.$$

Here $\sigma_n = (-1)^n f^{(n+1)}(0)$, where f is given by (2.4), cf. (Berg, 2000, Prop. 2.4), so σ_n is easily calculated to be

$$\sigma_n = \frac{n!}{1-q^{n+1}}, \quad n \geq 0.$$

It follows also by (Berg, 2000) that

$$c_{n,n} = \sigma_0^n = (1-q)^{-n}, \quad c_{n,1} = \sigma_{n-1} = (n-1)!/(1-q^n).$$

Defining $d_{n,k} = (1-q)^k c_{n,k}$ we have

$$s_n\left(I_{q,t}\right) = \sum_{k=1}^{n} d_{n,k} \left(\frac{t}{1-q}\right)^k, \quad n \geq 1, \tag{2.12}$$

and

$$d_{n+1,l+1} = n! \sum_{k=l}^{n} \frac{d_{k,l}}{k!} \left(\sum_{j=0}^{n-k} q^j\right)^{-1}, \quad l = 0, 1, \ldots, n. \tag{2.13}$$

In particular

$$d_{n,n} = 1, \quad d_{n,n-1} = \frac{\binom{n}{2}}{1+q}, \quad d_{n,1} = (n-1)! \left(\sum_{j=0}^{n-1} q^j\right)^{-1}.$$

We give the first coefficients

$$d_{1,1} = 1$$

$$d_{2,2} = 1, \quad d_{2,1} = \frac{1}{1+q}$$

$$d_{3,3} = 1, \quad d_{3,2} = \frac{3}{1+q}, \quad d_{3,1} = \frac{2}{1+q+q^2}.$$

It follows by induction using (2.13) that $d_{n,k}$ as a function of q has a finite limit for $q \to 1^-$.

The image measures $\mu_t = \tau_{1-q}\left(I_{q,t}\right)$ under $\tau_{1-q}(x) = x(1-q)$ form a convolution semigroup $(\mu_t)_{t>0}$ with

$$\int_0^\infty e^{-sx}\, d\mu_t(x) = \frac{1}{(-s(1-q);q)_\infty^t}, \quad s > 0,$$

and

$$s_n(\mu_t) = \sum_{k=1}^{n} d_{n,k}(1-q)^{n-k}t^k.$$

It follows that $s_n(\mu_t) \to t^n$ for $q \to 1^-$, so $\lim_{q\to 1^-} \mu_t = \delta_t$ weakly by the method of moments. This is also in accordance with

$$\lim_{q\to 1^-} \frac{1}{(-s(1-q);q)_\infty^t} = e^{-st},$$

because the q-exponential function E_q given in (1.7) converges to the exponential function in the following sense

$$\lim_{q\to 1^-} E_q(z(1-q)) = \exp(z),$$

cf. (Gasper and Rahman, 1990).

Remark 2.2. *Consider a non-zero Bernstein function f. In (Carmona et al., 1994; Carmona et al., 1997) it was proved by probabilistic methods that the sequence*

$$s_n = \frac{n!}{f(1)\cdot\ldots\cdot f(n)}$$

is a determinate Stieltjes moment sequence, meaning that it is the moment sequence of a unique probability on $[0,\infty[$. The special case $f(s) = 1-q^s$ gives the moment sequence (2.10). *In (Berg and Duran, 2004) the above result of (Carmona et al., 1994; Carmona et al., 1997) is obtained as a special case of the following result:*

Let (a_n) be a non-vanishing Hausdorff moment sequence. Then (s_n) defined by $s_0 = 1$ and $s_n = 1/(a_1\cdot\ldots\cdot a_n)$ for $n \geq 1$ is a normalized Stieltjes moment sequence.

In order to find an expression for I_q we consider the discrete signed measure

$$\mu_q = \sum_{k=0}^{\infty} \frac{(-1)^k q^{k(k+1)/2}}{(q;q)_k(q;q)_\infty}\delta_{q^k} \tag{2.14}$$

with moments

$$s_n(\mu_q) = \sum_{n=0}^{\infty} k\frac{(-1)^k q^{nk} q^{k(k+1)/2}}{(q;q)_k(q;q)_\infty} = \frac{(q^{n+1};q)_\infty}{(q;q)_\infty} = \frac{1}{(q;q)_n},$$

where we have used (1.7). In particular, the signed measure μ_q has mass 1.

For measures ν, τ on $]0,\infty[$ we define the product convolution $\nu \diamond \tau$ as the image of the product measure $\nu \otimes \tau$ under $x, y \mapsto xy$. The product convolution is the ordinary convolution of measures on the locally compact abelian group $]0,\infty[$ with multiplication as group operation. In particular we have

$$\int f \, d\nu \diamond \tau = \iint f(xy) \, d\nu(x) \, d\tau(y).$$

From this equation we get the moment equation

$$s_n(\nu \diamond \tau) = s_n(\nu) s_n(\tau),$$

hence $s_n(\mu_q \diamond \mathcal{E}_1) = n!/(q;q)_n$, which shows that $\mu_q \diamond \mathcal{E}_1$ has the same moments as I_q. Since the first measure is not known to be non-negative, we cannot conclude right-away that the two measures are equal, although I_q is Stieltjes determinate. We shall show that $\mu_q \diamond \mathcal{E}_1$ has a density $i_q(x)$, which is non-negative. Since $\delta_a \diamond \mathcal{E}_1 = \mathcal{E}_{1/a}$ for $a > 0$, it is easy to see that

$$i_q(x) = \sum_{n=0}^{\infty} \exp\left(-xq^{-n}\right) \frac{(-1)^n q^{n(n-1)/2}}{(q;q)_n (q;q)_\infty}, \tag{2.15}$$

but it is not obvious that $i_q(x) \geq 0$.

Proposition 2.3. *The function $i_q(x)$ given by (2.15) is non-negative for $x > 0$. Therefore $I_q = \mu_q \diamond \mathcal{E}_1 = i_q(x) 1_{]0,\infty[}(x) \, dx$.*

Proof. The Laplace transform of the function i_q is

$$\sum_{n=0}^{\infty} \frac{(-1)^n q^{n(n-1)/2}}{(s + q^{-n})(q;q)_n (q;q)_\infty}, \tag{2.16}$$

which is the partial fraction expansion of $1/(-s;q)_\infty$, since the residue of this function at the pole $s = q^{-n}$ is

$$\frac{(-1)^n q^{n(n-1)/2}}{(q;q)_n (q;q)_\infty}.$$

We claim that

$$\frac{1}{(-s;q)_\infty} = \sum_{n=0}^{\infty} \frac{(-1)^n q^{n(n-1)/2}}{(s + q^{-n})(q;q)_n (q;q)_\infty}, \quad s \neq q^{-n}, n = 0, 1, \ldots, \tag{2.17}$$

which shows that I_q and i_q have the same Laplace transform, so i_q is the density of I_q and hence non-negative.

To see the equation (2.17) we note that the left-hand side minus the right-hand side of the equation is an entire function φ, and by (2.9) we get

$$\frac{\varphi^{(n)}(0)}{n!} = \frac{(-1)^n}{(q;q)_n} - \frac{(-1)^n}{(q;q)_\infty} \sum_{k=0}^{\infty} q^{(n+1)k} \frac{(-1)^k q^{k(k-1)/2}}{(q;q)_k},$$

but by (1.7) the sum above equals $\left(q^{n+1};q\right)_\infty$, and we get $\varphi^{(n)}(0)/n! = 0$, which shows that φ is identically zero.

We call the attention to the fact that the identity (2.17) was also used in the work (Dumas et al., 2002) of Dumas et al., but it is in fact a special case of Jackson's transformations, see (III 4) in (Gasper and Rahman, 1990) with $b = 1$, $a = -s$, $z = q$. □

Let R_q denote the following positive discrete measure

$$R_q = (q;q)_\infty \sum_{k=0}^{\infty} \frac{q^k}{(q;q)_k} \delta_{q^k} \tag{2.18}$$

with moments

$$s_n(R_q) = (q;q)_\infty \sum_{k=0}^{\infty} \frac{\left(q^{n+1}\right)^k}{(q;q)_k} = (q;q)_n, \tag{2.19}$$

by (1.6). We claim that μ_q given by (2.14) and R_q are the inverse of each other under the product convolution, i.e.

$$\delta_1 = \mu_q \diamond R_q. \tag{2.20}$$

This amounts to proving that

$$\sum_{k=0}^{n} \frac{(-1)^k q^{k(k+1)/2} q^{n-k}}{(q;q)_k (q;q)_{n-k}} = \delta_{n0},\ n \geq 0,$$

but this follows by Cauchy multiplication of the power series (1.6), (1.7). Combining Proposition 2.3 with (2.20) we get:

Corollary 2.4. *The following factorization hold*

$$\mathcal{E}_1 = I_q \diamond R_q,$$

which corresponds to the factorization of the moments of $\mathcal{E}_1$ *as*

$$n! = \frac{n!}{(q;q)_n} \cdot (q;q)_n.$$

Remark 2.5. *The factorization of Corollary 2.4 is a special case of a general factorization in (Bertoin and Yor, 2001):*

$$\mathcal{E}_1 = I_f \diamond R_f, \quad n! = \frac{n!}{f(1)\cdot\ldots\cdot f(n)} \cdot (f(1)\cdot\ldots\cdot f(n)),$$

where f is a non-zero Bernstein function (2.2), and I_f, R_f are determined by their moments

$$s_n(I_f) = \frac{n!}{f(1)\cdot\ldots\cdot f(n)}, \quad s_n(R_f) = f(1)\cdot\ldots\cdot f(n).$$

3. The entire function $h(z) := \Gamma(z)(q^z;q)_\infty$

Since the Gamma function has simple poles at $z=-n$, $n=0,1,\ldots$, where $(q^z;q)_\infty$ has simple zeros, it is clear that the product $h(z) := \Gamma(z)\,(q^z;q)_\infty$ is entire. We have

$$\begin{aligned} h(0) &= \lim_{z\to 0}\Gamma(z)\,(q^z;q)_\infty = \lim_{z\to 0}\Gamma(z+1)\,(q^{z+1};q)_\infty\,\frac{1-q^z}{z} \\ &= (q;q)_\infty \log(1/q), \end{aligned}$$

and from this it is easy to see that

$$h(-n) = \frac{(q;q)_n}{n!} q^{-n(n+1)/2} (q;q)_\infty \log(1/q). \tag{3.1}$$

Proposition 3.1. *For $z \in \mathbb{C}$ we have*

$$\int_0^\infty x^z \, dI_q(x) = \frac{h(z+1)}{(q;q)_\infty}. \tag{3.2}$$

Proof. For $\operatorname{Re} z > -1$ the following calculation holds by (2.15) and (1.7):

$$\begin{aligned} \int_0^\infty x^z \, dI_q(x) &= \sum_{n=0}^\infty \frac{(-1)^n q^{n(n-1)/2}}{(q;q)_n (q;q)_\infty} \int_0^\infty x^z e^{-xq^{-n}}\,dx \\ &= (q;q)_\infty^{-1}\Gamma(z+1)(q^{z+1};q)_\infty. \end{aligned}$$

Since the right-hand side is entire and I_q is a positive measure, we get by a classical result (going back to Landau for Dirichlet series, see (Widder, 1941, p. 58)) that the integral in (3.2) must converge for all $z \in \mathbb{C}$, and therefore the equation holds for all $z \in \mathbb{C}$. □

When discussing measures on $]0,\infty[$ it is useful to consider this set as a locally compact group under multiplication. The Haar measure is then

$dm(x) = (1/x)\,dx$, and it is useful to consider the density of a measure with respect to the Haar measure m. The Mellin transformation is the Fourier tranformation of the locally compact abelian group $(]0,\infty[,\cdot)$, and when the dual group is realized as the additive group $\mathbb{R}$, the Mellin transformation of a finite measure μ on $]0,\infty[$ is defined as

$$\mathcal{M}(\mu)(\xi) = \int_0^\infty x^{-i\xi}\,d\mu(x), \quad \xi \in \mathbb{R}.$$

We get from (3.2) that

$$\mathcal{M}(I_q)(\xi) = \frac{h(1-i\xi)}{(q;q)_\infty}. \tag{3.3}$$

From Proposition 3.1 it follows that I_q has negative moments of any order, and from (3.1) we get in particular that

$$J_q := \frac{1}{x\log(1/q)}\,dI_q(x) \tag{3.4}$$

is a probability.

The image of J_q under the reflection $x \mapsto 1/x$ is denoted $\check{J}_q$.

Theorem 3.2. *The product convolution $L_q := I_q \diamond \check{J}_q$ has the density* (1.2)

$$\lambda_q(x) = \frac{1}{\log(1/q)(q,-x,-q/x;q)_\infty}$$

with respect to Lebesgue measure on the half-line.

Proof. For $z \in \mathbb{C}$ we clearly have

$$\int_0^\infty x^z\,dL_q(x) = \int_0^\infty x^z\,dI_q(x) \int_0^\infty x^{-z}\,dJ_q(x),$$

and by (3.2) we get

$$\int_0^\infty x^z\,dL_q(x) = \frac{h(z+1)h(-z)}{\log(1/q)(q;q)_\infty^2}.$$

By (1.5) it follows that for $z \in \mathbb{C}$

$$\int_0^\infty x^z\,dL_q(x) = \int_0^\infty x^z\lambda_q(x)\,dx,$$

so $L_q = \lambda_q(x)\,dx$. □

Remark 3.3. *In (Bertoin et al., 2004) the authors prove Theorem 3.2 by showing that*

$$\int_0^\infty x^z\,dL_q(x) = \frac{1}{(q;q)_\infty^3 \log(1/q)} \int_0^\infty x^z \left(\sum_{n=-\infty}^{\infty} \frac{(-1)^n q^{n(n+1)/2}}{x+q^n} \right) dx$$

for $-1 < \operatorname{Re} z < 0$, *and then they prove the partial fraction expansion of the meromorphic density* $\lambda_q(x)$

$$\lambda_q(x) = \frac{1}{(q;q)_\infty^3 \log(1/q)} \sum_{n=-\infty}^{\infty} \frac{(-1)^n q^{n(n+1)/2}}{x+q^n}.$$

Remark 3.4. *The moments of* $\check{J}_q$ *are given by*

$$s_n(\check{J}_q) = \frac{(q;q)_n}{n!} q^{-(n+1)n/2}, \tag{3.5}$$

so

$$\sum_{n=0}^{\infty} \frac{1}{\sqrt[2n]{s_n(\check{J}_q)}} < \infty.$$

Therefore Carleman's criterion gives no information about determinacy of $\check{J}_q$. *By the Krein criterion, cf. (Berg, 1995), (Stoyanov, 2000), we can conclude that*

$$\int_0^\infty \frac{\log i_q(x)}{\sqrt{x}(1+x)}\,dx = -\infty,$$

because I_q *is determinate. The substitution* $x = 1/y$ *in this integral leads to*

$$\int_0^\infty \frac{\log i_q(1/y)}{\sqrt{y}(1+y)}\,dy = -\infty, \tag{3.6}$$

but since

$$\check{J}_q = \frac{i_q(1/y)\,dy}{y\log(1/q)},$$

we see that (3.6) gives no information about indeterminacy of $\check{J}_q$. *We do not know if* $\check{J}_q$ *is determinate or indeterminate, and as a factor of an indeterminate distribution* L_q *none of these possibilities can be excluded.*

References

Akhiezer, N. I. (1965). *The Classical Moment Problem and Some Related Questions in Analysis.* Oliver and Boyd, Edinburgh.

Askey, R. (1986). Limits of some q-laguerre polynomials. *J. Approx. Theory*, 46:213–216.

Askey, R. (1989). Orthogonal polynomials and theta functions. In Ehrenpreis, L. and Gunning, R. C., editors, *Theta functions—Bowdoin 1987, Part 2 (Brunswick, ME, 1987)*, volume 49 of *Proc. Symp. Pure Math.*, pages 299–321. Amer. Math. Soc., Providence, RI.

Askey, R. and Roy, R. (1986). More q-beta integrals. *Rocky Mountain J. Math.*, 16:365–372.

Berg, C. (1980). Quelques remarques sur le cône de Stieltjes. In *Seminar on Potential Theory, Paris, No. 5. (French)*, volume 814 of *Lecture Notes in Mathematics*, pages 70–79. Springer, Berlin.

Berg, C. (1981). The Pareto distribution is a generalized Γ-convolution. A new proof. *Scand. Actuarial J.*, pages 117–119.

Berg, C. (1985). On the preservation of determinacy under convolution. *Proc. Amer. Math. Soc.*, 93:351–357.

Berg, C. (1995). Indeterminate moment problems and the theory of entire functions. *J. Comp. Appl. Math.*, 65:27–55.

Berg, C. (2000). On infinitely divisible solutions to indeterminate moment problems. In Dunkl, C., Ismail, M., and Wong, R., editors, *Special Functions (Hong Kong, 1999)*, pages 31–41. World Scientific, Singapore.

Berg, C. and Duran, A. J. (2004). A transformation from Hausdorff to Stieltjes moment sequences. *Ark. Mat.* To appear.

Berg, C. and Forst, G. (1975). *Potential theory on locally compact abelian groups.* Springer-Verlag, New York.

Bertoin, J. (1996). *Lévy processes.* Cambridge University Press, Cambridge.

Bertoin, J., Biane, P., and Yor, M. (2004). *Poissonian exponential functionals, q-series, q-integrals, and the moment problem for the log-normal distribution.* Birkhäuser. To appear.

Bertoin, J. and Yor, M. (2001). On subordinators, self-similar Markov processes and some factorizations of the exponential variable. *Elect. Comm. in Probab.*, 6:95–106.

Biane, P., Pitman, J., and Yor, M. (2001). Probability laws related to the Jacobi theta and Riemann zeta functions, and Brownian excursions. *Bull. Amer. Math. Soc.*, 38:435–465.

Bondesson, L. (1992). *Generalized gamma convolutions and related classes of distributions and densities*, volume 76 of *Lecture Notes in Statistics.* Springer, New York.

Carmona, P., Petit, F., and Yor, M. (1994). Sur les fonctionelles exponentielles de certains processus de Lévy. *Stochastics and Stochastics Reports*, 47:71–101.

Carmona, P., Petit, F., and Yor, M. (1997). On the distribution and asymptotic results for exponential functionals of Lévy processes. In Yor, M., editor, *Exponential functionals and principal values related to Brownian motion*, Biblioteca de la Revista Matemática Iberoamericana, pages 73–121. Revista Matemática Iberoamericana, Madrid.

Cowan, R. and Chiu, S.-N. (1994). A stochastic model of fragment formation when DNA replicates. *J. Appl. Probab.*, 31:301–308.

Dumas, V., Guillemin, F., and Robert, P. (2002). A Markovian analysis of additive-increase multiplicative-decrease algorithms. *Adv. in Appl. Probab.*, 34:85–111.

Gasper, G. and Rahman, M. (1990). *Basic Hypergeometric Series*. Cambridge University Press, Cambridge.

Itô, M. (1974). Sur les cônes convexes de Riesz et les noyaux complètement sous-harmoniques. *Nagoya Math. J.*, 55:111–144.

Lukacs, E. (1960). *Characteristic functions*. Griffin, London.

Pakes, A. G. (1996). Length biasing and laws equivalent to the log-normal. *J. Math. Anal. Appl.*, 197:825–854.

Reuter, G. E. H. (1956). Über eine Volterrasche Integralgleichung mit totalmonotonem Kern. *Arch. Math.*, 7:59–66.

Stoyanov, J. (2000). Krein condition in probabilistic moment problems. *Bernoulli*, 6:939–949.

Thorin, O. (1977a). On the infinite divisibility of the lognormal distribution. *Scand. Actuarial. J.*, (3):121–148.

Thorin, O. (1977b). On the infinite divisibility of the Pareto distribution. *Scand. Actuarial. J.*, (1):31–40.

Widder, D. V. (1941). *The Laplace Transform*, volume 6 of *Princeton Mathematical Series*. Princeton University Press, Princeton, N. J.

RAMANUJAN AND CRANKS

Bruce C. Berndt*
Department of Mathematics
University of Illinois
1409 West Green Street
Urbana, IL 61801
berndt@math.uiuc.edu

Heng Huat Chan†
Department of Mathematics
National University of Singapore
2 Science Drive 2
Singapore 117543
REPUBLIC OF SINGAPORE
chanhh@math.nus.edu.sg

Song Heng Chan
Department of Mathematics
University of Illinois
1409 West Green Street
Urbana, IL 61801
songchan@math.uiuc.edu

Wen-Chin Liaw‡
Department of Mathematics
National Chung Cheng University
Min-Hsiung, Chia-Yi 62101, Taiwan
REPUBLIC OF CHINA
wcliaw@math.ccu.edu.tw

*Research partially supported by grant MDA904-00-1-0015 from the National Security Agency.
†Research partially supported by National University of Singapore Academic Research Fund, Project Number R146000027112.
‡Research partially supported by grant NSC91-2115-M-194-016 from the National Science Council of the Republic of China.

Abstract The existence of the crank was first conjectured by F. J. Dyson in 1944 and was later established by G. E. Andrews and F. G. Garvan in 1987. However, much earlier, in his lost notebook, Ramanujan studied the generating function $F_a(q)$ for the crank and offered several elegant claims about it, although it seems unlikely that he was familiar with all the combinatorial implications of the crank. In particular, Ramanujan found several congruences for $F_a(q)$ in the ring of formal power series in the two variables a and q. An obscure identity found on page 59 of the lost notebook leads to uniform proofs of these congruences. He also studied divisibility properties for the coefficients of $F_a(q)$ as a power series in q. In particular, he provided ten lists of coefficients which he evidently thought exhausted these divisibility properties. None of the conjectures implied by Ramanujan's tables have been proved.

1. Introduction

In attempting to find a combinatorial interpretation for Ramanujan's famous congruences for the partition function $p(n)$, the number of ways of representing the positive integer n as a sum of positive integers, in 1944, F. J. Dyson [7] defined the *rank of a partition* to be the largest part minus the number of parts. Let $N(m,n)$ denote the number of partitions of n with rank m, and let $N(m,t,n)$ denote the number of partitions of n with rank congruent to m modulo t. Then Dyson conjectured that

$$N(k,5,5n+4) = \frac{p(5n+4)}{5}, \qquad 0 \le k \le 4, \tag{1.1}$$

and

$$N(k,7,7n+5) = \frac{p(7n+5)}{7}, \qquad 0 \le k \le 6, \tag{1.2}$$

which yield combinatorial interpretations of Ramanujan's famous congruences $p(5n+4) \equiv 0 \pmod 5$ and $p(7n+5) \equiv 0 \pmod 7$, respectively. These conjectures, as well as further conjectures of Dyson, were first proved by A. O. L. Atkin and H. P. F. Swinnerton-Dyer [4] in 1954. The generating function for $N(m,n)$ is given by

$$\sum_{m=-\infty}^{\infty}\sum_{n=0}^{\infty} N(m,n)a^m q^n = \sum_{n=0}^{\infty} \frac{q^{n^2}}{(aq;q)_n(q/a;q)_n}, \tag{1.3}$$

where $|q| < 1$ and $|q| < |a| < 1/|q|$. Although, to the best of our knowledge, Ramanujan was unaware of the concept of the rank of a partition, he recorded theorems on its generating function in his lost notebook; in particular, see [20, p. 20].

The corresponding analogue does not hold for $p(11n+6) \equiv 0 \pmod{11}$, and so Dyson conjectured the existence of a *crank*. In his doctoral dissertation [11], F. G. Garvan defined vector partitions which became the forerunners of the crank. The *true crank* was discovered by G. E. Andrews and Garvan on June 6, 1987, at a student dormitory at the University of Illinois.

Definition 1.1. *For a partition π, let $\lambda(n)$ denote the largest part of π, let $\mu(\pi)$ denote the number of ones in π, and let $\nu(\pi)$ denote the number of parts of π larger than $\mu(\pi)$. The crank $c(\pi)$ is then defined to be*

$$c(\pi) = \begin{cases} \lambda(\pi), & \text{if } \mu(\pi) = 0, \\ \nu(\pi) - \mu(\pi), & \text{if } \mu(\pi) > 0. \end{cases} \tag{1.4}$$

For $n \neq 1$, let $M(m,n)$ denote the number of partitions of n with crank m, while for $n = 1$, we set

$$M(0,1) = -1, M(-1,1) = M(1,1) = 1, \text{and } M(m,1) = 0 \text{ otherwise}.$$

Let $M(m,t,n)$ denote the number of partitions of n with crank congruent to m modulo t. The main theorem of Andrews and Garvan [2] relates $M(m,n)$ with vector partitions. In particular, the generating function for $M(m,n)$ is given by

$$\sum_{m=-\infty}^{\infty} \sum_{n=0}^{\infty} M(m,n) a^m q^n = \frac{(q;q)_\infty}{(aq;q)_\infty (q/a;q)_\infty}. \tag{1.5}$$

The crank not only leads to a combinatorial interpretation of $p(11n + 6) \equiv 0 \pmod{11}$, as predicted by Dyson, but also to similar interpretations for $p(5n+4) \equiv 0 \pmod 5$ and $p(7n+5) \equiv 0 \pmod 7$.

Theorem 1.2. *With $M(m,t,n)$ defined above,*

$$M(k,5,5n+4) = \frac{p(5n+4)}{5}, \qquad 0 \le k \le 4,$$
$$M(k,7,7n+5) = \frac{p(7n+5)}{7}, \qquad 0 \le k \le 6,$$
$$M(k,11,11n+6) = \frac{p(11n+6)}{11}, \qquad 0 \le k \le 10.$$

An excellent introduction to cranks can be found in Garvan's survey paper [12]. Also, see [3] for an interesting article on relations between the ranks and cranks of partitions.

2. Entries on Pages 179 and 180

At the top of page 179 in his lost notebook [20], Ramanujan defines a function $F(q)$ and coefficients λ_n, $n \ge 0$, by

$$F(q) := F_a(q) := \frac{(q;q)_\infty}{(aq;q)_\infty (q/a;q)_\infty} =: \sum_{n=0}^{\infty} \lambda_n q^n. \tag{2.1}$$

Thus, $F_a(q)$ is the generating function for cranks, and by (1.5), for $n > 1$,

$$\lambda_n = \sum_{m=-\infty}^{\infty} M(m,n)a^m.$$

He then offers two congruences for $F_a(q)$. These congruences, like others in the sequel, are to be regarded as congruences in the ring of formal power series in the two variables a and q. First, however, we need to define Ramanujan's theta function $f(a,b)$ by

$$f(a,b) := \sum_{n=-\infty}^{\infty} a^{n(n+1)/2} b^{n(n-1)/2}, \qquad |ab| < 1, \tag{2.2}$$

which satisfies the Jacobi triple product identity [5, p. 35, Entry 19]

$$f(a,b) = (-a;ab)_\infty(-b;ab)_\infty(ab;ab)_\infty. \tag{2.3}$$

The two congruences are then given by the following two theorems.

Theorem 2.1.

$$F_a(\sqrt{q}) \equiv \frac{f(-q^3,-q^5)}{(-q^2;q^2)_\infty} + \left(a - 1 + \frac{1}{a}\right)\sqrt{q}\,\frac{f(-q,-q^7)}{(-q^2;q^2)_\infty} \pmod{a^2 + \frac{1}{a^2}}. \tag{2.4}$$

Theorem 2.2.

$$\begin{aligned} F_a(q^{1/3}) \equiv{}& \frac{f(-q^2,-q^7)f(-q^4,-q^5)}{(q^9;q^9)_\infty} \\ &+ \left(a - 1 + \frac{1}{a}\right) q^{1/3}\frac{f(-q,-q^8)f(-q^4,-q^5)}{(q^9;q^9)_\infty} \\ &+ \left(a^2 + \frac{1}{a^2}\right) q^{2/3}\frac{f(-q,-q^8)f(-q^2,-q^7)}{(q^9;q^9)_\infty} \pmod{a^3 + 1 + \frac{1}{a^3}}. \end{aligned} \tag{2.5}$$

Note that $\lambda_2 = a^2 + a^{-2}$, which trivially implies that $a^4 \equiv -1 \pmod{\lambda_2}$ and $a^8 \equiv 1 \pmod{\lambda_2}$. Thus, in (2.4), a behaves like a primitive 8th root of unity modulo λ_2. On the other hand, $\lambda_3 = a^3 + 1 + a^{-3}$, from which it follows that $a^9 \equiv -a^6 - a^3 \equiv 1 \pmod{\lambda_3}$. So in (2.5), a behaves like a primitive 9th root of unity modulo λ_3.

This now leads us to the following definition.

Definition 2.3. *Let $P(q)$ denote any power series in q. Then the t-dissection of P is given by*

$$P(q) =: \sum_{k=0}^{t-1} q^k P_k(q^t). \tag{2.6}$$

Thus, if we let $a = \exp(2\pi i/8)$ and replace q by q^2, (2.4) implies the 2-dissection of $F_a(q)$, while if we let $a = \exp(2\pi i/9)$ and replace q by q^3, (2.5) implies the 3-dissection of $F_a(q)$. The first proofs of (2.4) and (2.5) in the forms where a is replaced by the respective primitive root of unity were given by Garvan [14]; his proof of (2.5) uses a Macdonald identity for the root system A_2.

3. Entries on Pages 18 and 20

Ramanujan gives the 5-dissection of $F_a(q)$ on pages 18 and 20 of his lost notebook [20], with the better formulation on page 20. It is interesting that Ramanujan does not give the two variable form, analogous to those in (2.4) and (2.5), from which the 5-dissection would follow by setting a to be a primitive fifth root of unity. Proofs of the 5-dissection have been given by Garvan [13] and A. B. Ekin [9]. To describe this dissection, we first set

$$f(-q) := f(-q, -q^2) = (q;q)_\infty, \tag{3.1}$$

by (2.3).

Theorem 3.1. *If ζ is a primitive fifth root of unity and $f(-q)$ is defined by* (3.1), *then*

$$\begin{aligned} F_\zeta(q) = {} & \frac{f(-q^{10}, -q^{15})}{f^2(-q^5, -q^{20})} f^2(-q^{25}) \\ & + (\zeta - 1 + \zeta^{-1}) q \frac{f^2(-q^{25})}{f(-q^5, -q^{20})} \\ & + (\zeta^2 + \zeta^{-2}) q^2 \frac{f^2(-q^{25})}{f(-q^{10}, -q^{15})} \\ & + (\zeta^3 + 1 + \zeta^{-3}) q^3 \frac{f(-q^5, -q^{20})}{f^2(-q^{10}, -q^{15})} f^2(-q^{25}). \end{aligned}$$

For completeness, we state Theorem 3.1 in the two variable form as a congruence. But first, for brevity, it will be convenient to define

$$A_n := a^n + a^{-n} \tag{3.2}$$

and

$$S_n := \sum_{k=-n}^{n} a^k. \tag{3.3}$$

Theorem 3.2. *With $f(-q)$, A_n, and S_n defined by* (3.1)–(3.3), *respectively,*

$$\begin{aligned}F_a(q) \equiv &\frac{f(-q^{10},-q^{15})}{f^2(-q^5,-q^{20})}f^2(-q^{25})\\ &+(A_1-1)q\frac{f^2(-q^{25})}{f(-q^5,-q^{20})}+A_2q^2\frac{f^2(-q^{25})}{f(-q^{10},-q^{15})}\\ &+(A_3+1)q^3\frac{f(-q^5,-q^{20})}{f^2(-q^{10},-q^{15})}f^2(-q^{25}) \pmod{S_2}.\end{aligned}$$

As we have seen, by letting a be a root of unity, we can derive a dissection from a congruence in the ring of formal power series in two variables. In fact, the converse is true, and this is proved in [6].

4. Entries on Pages 70 and 71

The first explicit statement and proof of the 7-dissection of $F_a(q)$ was given by Garvan [13, Thm. 5.1]; another proof was later found by Ekin [9]. Although Ramanujan did not state the 7-dissection of $F_a(q)$, he clearly knew it, because the six quotients of theta functions that appear in the 7-dissection are found on the bottom of page 71 (written upside down) in his lost notebook. We record the two variable form here.

Theorem 4.1. *With $f(a,b)$ defined by* (2.2), *$f(-q)$ defined by* (3.1), *and A_n and S_n defined by* (3.2) *and* (3.3), *respectively,*

$$\begin{aligned}F_a(q) \equiv &\frac{1}{f(-q^7)}\{f^2(-q^{21},-q^{28})+(A_1-1)qf(-q^{14},-q^{35})f(-q^{21},-q^{28})\\ &+A_2q^2f^2(-q^{14},-q^{35})+(A_3+1)q^3f(-q^7,-q^{42})f(-q^{21},-q^{28})\\ &-A_1q^4f(-q^7,-q^{42})f(-q^{14},-q^{35})-(A_2+1)q^6f^2(-q^7,-q^{42})\}\\ &\pmod{S_3}.\end{aligned}$$

The first appearance of the 11-dissection of $F_a(q)$ in the literature also can be found in Garvan's paper [13, Thm. 6.7]. However, again, it is very likely that Ramanujan knew the 11-dissection, since he offers the quotients of theta functions which appear in the 11-dissection on page 70 of his lost notebook [20]. Further proofs were found by Ekin [8], [9], and a reformulation of Garvan's result was given by M. D. Hirschhorn [15]. We state the 11-dissection in the two variable form as a congruence.

Theorem 4.2. *With A_n and S_n defined by (3.2) and (3.3), respectively,*

$$F_a(q) \equiv \frac{1}{(q^{11};q^{11})_\infty (q^{121};q^{121})_\infty^2} \Big\{ ABCD + (A_1 - 1)\, qA^2BE + A_2 q^2 AC^2 D + (A_3+1)\, q^3 ABD^2 + (A_2 + A_4 + 1)\, q^4 ABCE - (A_2 + A_4)\, q^5 B^2 CE + (A_1 + A_4)\, q^7 ABDE - (A_2 + A_5 + 1)\, q^{19} CDE^2 - (A_4 + 1)\, q^9 ACDE - A_3 q^{10} BCDE \Big\} \pmod{S_5},$$

where $A = f(-q^{55}, -q^{66})$, $B = f(-q^{77}, -q^{44})$, $C = f(-q^{88}, -q^{33})$, $D = f(-q^{99}, -q^{22})$, and $E = f(-q^{110}, -q^{11})$.

The present authors have recently given two proofs of each of Theorems 2.1, 2.2, 3.2, 4.1, and 4.2 in [6]. Our first proofs of each theorem use a method of "rationalization" which is like the method employed by Garvan [13], [14] in proving the dissections where a is replaced with a primitive root of unity. Our second method employs a formula found on page 59 in Ramanujan's lost notebook [20]. In fact, as we shall see in the next section, Ramanujan actually does not record a formula, but instead records "each side" without stating an equality.

5. Entries on Pages 58 and 59

On page 58 in his lost notebook [20], Ramanujan recorded the following power series:

$$\begin{aligned}
&1 + q(a_1 - 1) + q^2 a_2 + q^3(a_3+1) + q^4(a_4 + a_2 + 1) \\
&+ q^5(a_5 + a_3 + a_1 + 1) + q^6(a_6 + a_4 + a_3 + a_2 + a_1 + 1) \\
&+ q^7(a_3+1)(a_4+a_2+1) + q^8 a_2(a_6 + a_4 + a_3 + a_2 + a_1 + 1) \\
&+ q^9 a_2(a_3+1)(a_4+a_2+1) + q^{10} a_2(a_3+1)(a_5+a_3+a_1+1) \\
&+ q^{11} a_1 a_2(a_8 + a_5 + a_4 + a_3 + a_2 + a_1 + 2) \\
&+ q^{12}(a_3 + a_2 + a_1 + 1)(a_5 + a_4 + a_3 + a_2 + a_1 + 1) \\
&\quad \times (a_4 - 2a_3 + 2a_2 - a_1 + 1) \\
&+ q^{13}(a_1 - 1)(a_2 - a_1 + 1) \\
&\quad \times (a_{10} + 2a_9 + 2a_8 + 2a_7 + 2a_6 + 4a_5 + 6a_4 + 8a_3 + 9a_2 + 9a_1 + 9) \\
&+ q^{14}(a_2+1)(a_3+1)(a_4+a_2+1)(a_5 - a_3 + a_1 + 1) \\
&+ q^{15} a_1 a_2 (a_5 + a_4 + a_3 + a_2 + a_1 + 1)(a_7 - a_6 + a_4 + a_1) \\
&+ q^{16}(a_3+1)(a_3+a_2+a_1+1)(a_5+a_4+a_3+a_2+a_1+1) \\
&\quad \times (a_5 - 2a_4 + 2a_3 - 2a_2 + 3a_1 - 3)
\end{aligned}$$

$$
\begin{aligned}
&+ q^{17}(a_2+1)(a_3+1)(a_5+a_4+a_3+a_2+a_1+1)(a_7-a_6+a_3+a_1-1)\\
&+ q^{18}(a_4+a_2+1)(a_3+a_2+a_1+1)(a_5+a_4+a_3+a_2+a_1+1)\\
&\quad \times (a_6-2a_5+a_4+a_3-a_2+1)\\
&+ q^{19}a_2(a_1-1)(a_4+a_2+1)(a_3+a_2+a_1+1)\\
&\quad \times (a_9-a_7+a_4+2a_3+a_2-1)\\
&+ q^{20}(a_2-a_1+1)(a_3+1)(a_5+a_4+a_3+a_2+a_1+1)\\
&\quad \times (a_{10}+a_6+a_4+a_3+2a_2+2a_1+3)\\
&+ q^{21}a_1a_2(a_3+1)(a_2-a_1+1)(a_5+a_4+a_3+a_2+a_1+1)\\
&\quad \times (a_8-a_6+a_4+a_1+2)\\
&+\cdots . \qquad (5.1)
\end{aligned}
$$

Although Ramanujan did not indicate the meaning of his notation a_n, in fact,

$$a_n := a^n + a^{-n}, \qquad (5.2)$$

and indeed Ramanujan has written out the first 21 coefficients in the power series representation of the crank $F_a(q)$. (We have corrected a misprint in the coefficient of q^{21}.)

On the following page, beginning with the coefficient of q^{13}, Ramanujan listed some (but not necessarily all) of the factors of the coefficients up to q^{26}. The factors he recorded are

$$
\begin{aligned}
&13. \quad (a_1-1)(a_2-a_1+1) \qquad (5.3)\\
&14. \quad (a_2+1)(a_3+1)(a_4+a_2+1)\\
&15. \quad a_1a_2(a_5+a_4+a_3+a_2+a_1+1)\\
&16. \quad (a_3+1)(a_3+a_2+a_1+1)(a_5+a_4+a_3+a_2+a_1+1)\\
&17. \quad (a_2+1)(a_3+1)(a_5+a_4+a_3+a_2+a_1+1)\\
&18. \quad (a_4+a_2+1)(a_3+a_2+a_1+1)(a_5+a_4+a_3+a_2+a_1+1)\\
&19. \quad a_2(a_1-1)(a_4+a_2+1)(a_3+a_2+a_1+1)\\
&20. \quad (a_3+1)(a_2-a_1+1)(a_5+a_4+a_3+a_2+a_1+1)\\
&21. \quad a_1a_2(a_3+1)(a_2-a_1+1)(a_5+a_4+a_3+a_2+a_1+1)\\
&22. \quad a_2(a_3+1)(a_1-1)\\
&23. \quad (a_1-1)(a_4+a_2+1)\\
&24. \quad (a_3+1)(a_4+a_2+1)(a_3+a_2+a_1+1)\\
&25. \quad a_2(a_1-1)(a_5+a_4+a_3+a_2+a_1+1)\\
&26. \quad a_2(a_3+1)(a_3+a_2+a_1+1).
\end{aligned}
$$

Ramanujan did not indicate why he recorded only these factors. However, it can be noted that in each case he recorded linear factors only when the leading index is ≤ 5. To the left of each n, $15 \leq n \leq 26$, are the unexplained numbers 16×16, undecipherable, $27 \times 27, -25, 49, -7 \cdot 19, 9, -7, -9, -11 \cdot 15, -11$, and -4, respectively.

6. Congruences for the Coefficients λ_n on Pages 179 and 180

On pages 179 and 180 in his lost notebook [20], Ramanujan offers ten tables of indices of coefficients λ_n satisfying certain congruences. On page 61 in [20], he offers rougher drafts of nine of the ten tables; Table 6 is missing on page 61. Unlike the tables on pages 179 and 180, no explanations are given on page 61. Clearly, Ramanujan calculated factors well beyond the factors recorded on pages 58 and 59 of his lost notebook given in Section 5.5. To verify Ramanujan's claims, we calculated λ_n up to $n = 500$ with the use of Maple V. Ramanujan evidently thought that each table is complete in that there are no further values of n for which the prescribed divisibility property holds. However, we are unable to prove any of these assertions.

Table 1. $\lambda_n \equiv 0 \pmod{a^2 + \frac{1}{a^2}}$

Thus, Ramanujan indicates which coefficients λ_n have a_2 as a factor. The 47 values of n with a_2 as a factor of λ_n are

$$\begin{gathered}2, 8, 9, 10, 11, 15, 19, 21, 22, 25, 26, 27, 28, 30, 31, 34, 40, 42, 45,\\ 46, 47, 50, 55, 57, 58, 59, 62, 66, 70, 74, 75, 78, 79, 86, 94, 98,\\ 106, 110, 122, 126, 130, 142, 154, 158, 170, 174, 206.\end{gathered}$$

Replacing q by q^2 in (2.4), we see that Table 1 contains the degree of q for those terms with zero coefficients for both

$$\frac{f(-q^6, -q^{10})}{(-q^4; q^4)_\infty} \quad \text{and} \quad q\frac{f(-q^2, -q^{14})}{(-q^4; q^4)_\infty}. \tag{6.1}$$

Table 2. $\lambda_n \equiv 1 \pmod{a^2 + \frac{1}{a^2}}$

To interpret this table properly, we return to the congruence given in (2.4). Replacing q by q^2, we see that Ramanujan has recorded all the degrees of q of the terms (except for the constant term) with coefficients equal to 1 in the power series expansion of

$$\frac{f(-q^6, -q^{10})}{(-q^4; q^4)_\infty}. \tag{6.2}$$

The 27 values of n given by Ramanujan are

$$14, 16, 18, 24, 32, 48, 56, 72, 82, 88, 90, 104, 114, 138, 146,$$
$$162, 178, 186, 194, 202, 210, 218, 226, 234, 242, 250, 266.$$

Table 3. $\lambda_n \equiv -1 \pmod{a^2 + \frac{1}{a^2}}$

This table is to be understood in the same way as the previous table, except that now Ramanujan is recording the indices of those terms with coefficients equal to -1 in the power series expansion of (6.2). Here Ramanujan missed one value, namely, $n = 214$. The 27 (not 26) values of n are then given by

$$4, 6, 12, 20, 36, 38, 44, 52, 54, 60, 68, 76, 92, 102, 118,$$
$$134, 150, 166, 182, 190, 214, 222, 238, 254, 270, 286, 302.$$

Table 4. $\lambda_n \equiv a - 1 + \frac{1}{a} \pmod{a^2 + \frac{1}{a^2}}$

We again return to the congruence given in (2.4). Note that $a-1+1/a$ occurs as a factor of the second expression on the right side. Thus, replacing q by q^2, Ramanujan records the indices of all terms of

$$q\frac{f(-q^2, -q^{14})}{(-q^4; q^4)_\infty} \tag{6.3}$$

with coefficients that are equal to 1. The 22 values of n which give the coefficient 1 are equal to

$$1, 7, 17, 23, 33, 39, 41, 49, 63, 71, 73, 81,$$
$$87, 89, 95, 105, 111, 119, 121, 127, 143, 159.$$

Table 5. $\lambda_n \equiv -\left(a - 1 + \frac{1}{a}\right) \pmod{a^2 + \frac{1}{a^2}}$

The interpretation of this table is analogous to the preceding one. Now Ramanujan determines those coefficients in the expansion of (6.3) which are equal to -1. His table of 23 values of n includes

$$3, 5, 13, 29, 35, 37, 43, 51, 53, 61, 67, 69, 77,$$
$$83, 85, 91, 93, 99, 107, 115, 123, 139, 155.$$

Table 6. $\lambda_n \equiv 0 \pmod{a + \frac{1}{a}}$

Ramanujan thus gives here those coefficients which have a_1 as a factor. There are only three values, namely, when n equals

$$11, 15, 21.$$

These three values can be discerned from the table on page 59 of the lost notebook.

From the calculation

$$\frac{(q;q)_\infty}{(aq;q)_\infty(q/a;q)_\infty} \equiv \frac{(q;q)_\infty}{(-q^2;q^2)_\infty} = \frac{f(-q)f(-q^2)}{f(-q^4)} \pmod{a+\frac{1}{a}},$$

where $f(-q)$ is defined by (3.1), we see that in Table 6 Ramanujan recorded the degree of q for the terms with zero coefficients in the power series expansion of

$$\frac{f(-q)f(-q^2)}{f(-q^4)}. \tag{6.4}$$

For the next three tables, it is clear from the calculation

$$\frac{(q;q)_\infty}{(aq;q)_\infty(q/a;q)_\infty} \equiv \frac{(q^2;q^2)_\infty}{(-q^3;q^3)_\infty} = \frac{f(-q^2)f(-q^3)}{f(-q^6)} \pmod{a-1+\frac{1}{a}},$$

that Ramanujan recorded the degree of q for the terms with coefficients 0, 1, and -1, respectively, in the power series expansion of

$$\frac{f(-q^2)f(-q^3)}{f(-q^6)}. \tag{6.5}$$

Table 7. $\lambda_n \equiv 0 \pmod{a-1+\frac{1}{a}}$

The 19 values satisfying the congruence above are, according to Ramanujan,

$$1, 6, 8, 13, 14, 17, 19, 22, 23, 25,$$
$$33, 34, 37, 44, 46, 55, 58, 61, 82.$$

Table 8. $\lambda_n \equiv 1 \pmod{a-1+\frac{1}{a}}$

The 26 values of n found by Ramanujan are

$$5, 7, 10, 11, 12, 18, 24, 29, 30, 31, 35, 41, 42, 43,$$
$$47, 49, 53, 54, 59, 67, 71, 73, 85, 91, 97, 109.$$

As in Table 2, Ramanujan ignored the value $n = 0$.

Table 9. $\lambda_n \equiv -1 \pmod{a-1+\frac{1}{a}}$

The 26 values of n found by Ramanujan are

$$2, 3, 4, 9, 15, 16, 20, 21, 26, 27, 28, 32, 38, 39,$$
$$40, 52, 56, 62, 64, 68, 70, 76, 94, 106, 118, 130.$$

Table 10. $\lambda_n \equiv 0 \pmod{a+1+\frac{1}{a}}$

Ramanujan has but two values of n such that λ_n satisfies the congruence above, and they are when n equals

$$14, 17.$$

From the calculation

$$\frac{(q;q)_\infty}{(aq;q)_\infty(q/a;q)_\infty} \equiv \frac{(q;q)_\infty^2}{(q^3;q^3)_\infty} = \frac{f^2(-q)}{f(-q^3)} \pmod{a+1+\frac{1}{a}},$$

it is clear that Ramanujan recorded the degree of q for the terms with zero coefficients in the power series expansion of

$$\frac{f^2(-q)}{f(-q^3)}. \tag{6.6}$$

The infinite products in (6.2)–(6.6) do not appear to have monotonic coefficients for sufficiently large n. However, if these infinite products are dissected properly, then we conjecture that the coefficients in the dissections are indeed monotonic. Hence, for (6.2), (6.3), (6.4), (6.5), and (6.6), we must study, respectively, the dissections of

$$\frac{f(-q^6,-q^{10})}{(-q^4;q^4)_\infty}, \qquad \frac{f(-q^2,-q^{14})}{(-q^4;q^4)_\infty},$$

$$\frac{f(-q)f(-q^2)}{f(-q^4)}, \qquad \frac{f(-q^2)f(-q^3)}{f(-q^6)}, \qquad \frac{f^2(-q)}{f(-q^3)}.$$

For each of the five products given above, we have determined certain dissections.

We require an addition theorem for theta functions found in Chapter 16 of Ramanujan's second notebook [19], [5, p. 48, Entry 31]. Our applications of this lemma lead to the desired dissections.

Lemma 6.1. *If* $U_n = \alpha^{n(n+1)/2}\beta^{n(n-1)/2}$ *and* $V_n = \alpha^{n(n-1)/2}\beta^{n(n+1)/2}$ *for each integer* n, *then*

$$f(U_1,V_1) = \sum_{k=0}^{N-1} U_k f\left(\frac{U_{N+k}}{U_k}, \frac{V_{N-k}}{U_k}\right). \tag{6.7}$$

Setting $(\alpha,\beta,N) = (-q^6,-q^{10},4)$ and $(-q^4,-q^{12},2)$ in (6.7), we obtain, respectively,

$$f(-q^6,-q^{10}) = A - q^6B - q^{10}C + q^{28}D, \tag{6.8}$$

$$f(-q^4,-q^{12}) = f(q^{24},q^{40}) - q^4 f(q^8,q^{56}), \tag{6.9}$$

where $A := f(q^{120},q^{136})$, $B := f(q^{72},q^{184})$, $C := f(q^{56},q^{200})$, and $D := f(q^8,q^{248})$.

Setting $(\alpha,\beta,N) = (-q,-q^2,3)$ in (6.7), we obtain

$$f(-q) = f(-q^{12},-q^{15}) - qf(-q^6,-q^{21}) - q^2 f(-q^3,-q^{24}). \tag{6.10}$$

For (6.2), the 8-dissection (with, of course, the odd powers missing) is given by

$$\begin{aligned}\frac{f(-q^6,-q^{10})}{(-q^4;q^4)_\infty} &= \frac{f(-q^6,-q^{10})f(-q^4,-q^{12})}{f(-q^{16})}\\ &= \frac{1}{f(-q^{16})}\{A - q^6B - q^{10}C + q^{28}D\}\\ &\quad\times\{f(q^{24},q^{40}) - q^4 f(q^8,q^{56})\}\\ &= \frac{1}{f(-q^{16})}\Big\{Af(q^{24},q^{40}) - q^{32}Df(q^8,q^{56})\\ &\quad + q^2\left[q^8Bf(q^8,q^{56}) - q^8Cf(q^{24},q^{40})\right]\\ &\quad + q^4\left[-Af(q^{24},q^{40}) + q^{24}Df(q^8,q^{56})\right]\\ &\quad + q^6\left[-Bf(q^{24},q^{40}) + q^8Cf(q^8,q^{56})\right]\Big\},\end{aligned}$$

where we have applied (6.8) and (6.9) in the penultimate equality.

For (6.6), we have the 3-dissection,

$$\begin{aligned}\frac{f^2(-q)}{f(-q^3)} &= \frac{1}{(q^3;q^3)_\infty}\{f(-q^{12},-q^{15}) - qf(-q^6,-q^{21}) - q^2f(-q^3,-q^{24})\}^2\\ &= \frac{1}{(q^3;q^3)_\infty}\Big\{f^2(-q^{12},-q^{15}) + 2q^3f(-q^6,-q^{21})f(-q^3,-q^{24})\\ &\quad - q\left[2f(-q^{12},-q^{15})f(-q^6,-q^{21}) - q^3f^2(-q^3,-q^{24})\right]\\ &\quad + q^2\left[f^2(-q^6,-q^{21}) - 2f(-q^{12},-q^{15})f(-q^3,-q^{24})\right]\Big\},\end{aligned}$$

where we have applied (6.10) in the first equality. For (6.3), (6.4), and (6.5), we have derived an 8-dissection, a 4-dissection, and a 6-dissection, respectively. Furthermore, we make the following conjecture.

Conjecture 6.2. *Each component of each of the dissections for the five products given above has monotonic coefficients for powers of q above* 1400.

We have checked the coefficients for each of the five products up to $n = 2000$. For each product, we give below the values of n after which their dissections appear to be monotonic and strictly monotonic, respectively.

(6.2)	1262	1374
(6.3)	719	759
(6.4)	149	169
(6.5)	550	580
(6.6)	95	95

Our conjectures on the dissections of (6.4), (6.5), and (6.6) have motivated the following stronger conjecture.

Conjecture 6.3. *For any positive integers α and β, each component of the $(\alpha+\beta+1)$-dissection of the product*

$$\frac{f(-q^{\alpha})f(-q^{\beta})}{f(-q^{\alpha+\beta+1})}$$

has monotonic coefficients for sufficiently large powers of q.

We remark that our conjectures for (6.4), (6.5), and (6.6) are then the special cases of Conjecture 6.3 when we set $(\alpha,\beta)=(1,2),(2,3)$, and $(1,1)$, respectively.

Setting $(\alpha,\beta,N)=(-q^6,-q^{10},2)$ and $(-q^2,-q^{14},2)$ in (6.7), we obtain, respectively,

$$f(-q^6,-q^{10})=f(q^{28},q^{36})-q^6 f(q^4,q^{60}) \tag{6.11}$$

and

$$f(-q^2,-q^{14})=f(q^{20},q^{44})-q^2 f(q^{12},q^{52}). \tag{6.12}$$

After reading our conjectures for (6.2) and (6.3), Garvan made the following stronger conjecture.

Conjecture 6.4. *Define b_n by*

$$\begin{aligned}\sum_{n=0}^{\infty} b_n q^n &= \frac{f(-q^6,-q^{10})}{(-q^4;q^4)_\infty}+q\frac{f(-q^2,-q^{14})}{(-q^4;q^4)_\infty}\\ &= \frac{f(q^{28},q^{36})}{(-q^4;q^4)_\infty}+q\frac{f(q^{20},q^{44})}{(-q^4;q^4)_\infty}-q^6\frac{f(q^4,q^{60})}{(-q^4;q^4)_\infty}\\ &\quad -q^3\frac{f(q^{12},q^{52})}{(-q^4;q^4)_\infty},\end{aligned}$$

where we have applied (6.11) *and* (6.12) *in the last equality. Then*

$$\begin{aligned} (-1)^n b_{4n} &\geq 0, && \textit{for all } n \geq 0, \\ (-1)^n b_{4n+1} &\geq 0, && \textit{for all } n \geq 0, \\ (-1)^n b_{4n+2} &\geq 0, && \textit{for all } n \geq 0, n \neq 3, \\ (-1)^{n+1} b_{4n+3} &\geq 0, && \textit{for all } n \geq 0. \end{aligned}$$

Furthermore, each of these subsequences are eventually monotonic.

It is clear that the monotonicity of the subsequences in Conjecture 6.4 implies the monotonicity of the dissections of (6.2) and (6.3) as stated in Conjecture 6.2.

In [1], Andrews and R. Lewis made three conjectures on the inequalities between the rank counts $N(m,t,n)$ and between the crank counts $M(m,t,n)$. Two of them, [1, Conj. 2 and Conj. 3] directly imply that Tables 10 and 6, respectively, are complete. Recently, using the circle method, D. M. Kane [16] proved the former conjecture. It follows immediately from [16, Cor. 2] that Table 10 is complete.

7. Page 182: Partitions and Factorizations of Crank Coefficients

On page 182 in his lost notebook [20], Ramanujan returns to the coefficients λ_n in the generating function (2.1) of the crank. He factors λ_n, $1 \leq n \leq 21$, as before, but singles out nine particular factors by giving them special notation. The criterion that Ramanujan apparently uses is that of multiple occurrence, i.e., each of these nine factors appears more than once in the 21 factorizations, while other factors not favorably designated appear only once. Ramanujan uses these factorizations to compute $p(n)$, which, of course, arises from the special case $a = 1$ in (2.1), i.e.,

$$\frac{1}{(q;q)_\infty} = \sum_{n=0}^{\infty} p(n) q^n, \qquad |q| < 1.$$

Ramanujan evidently was searching for some general principles or theorems on the factorization of λ_n so that he could not only compute $p(n)$ but say something about the divisibility of $p(n)$. No theorems are stated by Ramanujan. Is it possible to determine that certain factors appear in some precisely described infinite family of values of λ_n? It would be interesting to speculate on the motivations which led Ramanujan to make these factorizations.

The factors designated by Ramanujan are

$$\rho_1 = a_1 - 1,$$

$$\begin{aligned}
\rho &= a_2 - a_1 + 1, \\
\rho_2 &= a_2, \\
\rho_3 &= a_3 + 1, \\
\rho_4 &= a_1 a_2, \\
\rho_5 &= a_4 + a_2 + 1, \\
\rho_7 &= a_3 + a_2 + a_1 + 1, \\
\rho_9 &= (a_2 + 1)(a_3 + 1), \\
\rho_{11} &= a_5 + a_4 + a_3 + a_2 + a_1 + 1.
\end{aligned}$$

At first glance, there does not appear to be any reasoning behind the choice of subscripts; note that there is no subscript for the second value. However, observe that in each case, the subscript

$n =$ (as a sum of powers of a) the number of terms with positive coefficients minus the number of terms with negative coefficients in the representation of ρ_n, when all expressions are expanded out, or if $\rho_n = \rho_n(a)$, we see that $\rho_n(1) = n$.

The reason ρ does not have a subscript is that the value of n in this case would be $3 - 2 = 1$, which has been reserved for the first factor. These factors then lead to rapid calculations of values for $p(n)$. For example, since $\lambda_{10} = \rho\rho_2\rho_3\rho_7$, then

$$p(10) = 1 \cdot 2 \cdot 3 \cdot 7 = 42.$$

In the table below, we provide the content of this page.

$$\begin{aligned}
p(1) &= 1, & \lambda_1 &= \rho_1, \\
p(2) &= 2, & \lambda_2 &= \rho_2, \\
p(3) &= 3, & \lambda_3 &= \rho_3, \\
p(4) &= 5, & \lambda_4 &= \rho_5, \\
p(5) &= 7, & \lambda_5 &= \rho_7\rho, \\
p(6) &= 11, & \lambda_6 &= \rho_1\rho_{11}, \\
p(7) &= 15, & \lambda_7 &= \rho_3\rho_5, \\
p(8) &= 22, & \lambda_8 &= \rho_1\rho_2\rho_{11}, \\
p(9) &= 30, & \lambda_9 &= \rho_2\rho_3\rho_5, \\
p(10) &= 42, & \lambda_{10} &= \rho\rho_2\rho_3\rho_7, \\
p(11) &= 56, & \lambda_{11} &= \rho_4\rho_7(a_5 - a_4 + a_2),
\end{aligned}$$

$$\begin{aligned}
p(12) &= 77, & \lambda_{12} &= \rho_7\rho_{11}(a_4 - 2a_3 + 2a_2 - a_1 + 1),\\
p(13) &= 101, & \lambda_{13} &= \rho\rho_1\,(a_{10} + 2a_9 + 2a_8 + 2a_7 + 3a_6\\
&&&\qquad +4a_5 + 6a_4 + 8a_3 + 9a_2 + 9a_1 + 9)\,,\\
p(14) &= 135, & \lambda_{14} &= \rho_5\rho_9(a_5 - a_3 + a_1 + 1),\\
p(15) &= 176, & \lambda_{15} &= \rho_4\rho_{11}(a_7 - a_6 + a_4 + a_1),\\
p(16) &= 231, & \lambda_{16} &= \rho_3\rho_7\rho_{11}(a_5 - 2a_4 + 2a_3 - 2a_2 + 3a_1 - 3),\\
p(17) &= 297, & \lambda_{17} &= \rho_9\rho_{11}(a_7 - a_6 + a_3 + a_1 - 1),\\
p(18) &= 385, & \lambda_{18} &= \rho_5\rho_7\rho_{11}(a_6 - 2a_5 + a_4 + a_3 - a_2 + 1),\\
p(19) &= 490, & \lambda_{19} &= \rho_1\rho_2\rho_5\rho_7(a_9 - a_7 + a_4 + 2a_3 + a_2 - 1),\\
p(20) &= 627, & \lambda_{20} &= \rho\rho_3\rho_{11}(a_{10} + a_6 + a_4 + a_3 + 2a_2 + 2a_1 + 3),\\
p(21) &= 792, & \lambda_{21} &= \rho\rho_3\rho_4\rho_{11}(a_8 - a_6 + a_4 + a_1 + 2).
\end{aligned}$$

8. Further Entries on Page 59

Further down page 59, Ramanujan offers the quotient (with one misprint corrected)

$$\begin{aligned}
\Big(1 &+ q(a_1-2) + q^2(a_2-a_1) + q^3(a_3-a_2) + q^4(a_4-a_3) + \cdots\\
&- \left(q^3(a_1-2) + q^5(a_2-a_1) + q^7(a_3-a_2) + q^9(a_4-a_3) + \cdots\right)\\
&+ \left(q^6(a_1-2) + q^9(a_2-a_1) + q^{12}(a_3-a_2) + q^{15}(a_4-a_3) + \cdots\right)\\
&- \left(q^{10}(a_1-2) + q^{14}(a_2-a_1) + q^{18}(a_3-a_2) + q^{22}(a_4-a_3) + \cdots\right)\\
&+ \left(q^{15}(a_1-2) + q^{20}(a_2-a_1) + q^{25}(a_3-a_2) + \cdots\right)\\
&- \left(q^{21}(a_1-2) + \cdots\right)\Big)\Big/\\
&\quad (1 - q - q^2 + q^5 + q^7 - q^{12} - q^{15} + q^{22} + \cdots).
\end{aligned} \tag{8.1}$$

In more succinct notation, (8.1) can be rewritten as

$$\frac{1 - \sum\limits_{m=1,n=0}^{\infty} (-1)^m q^{m(m+1)/2+mn}(a_{n+1} - a_n)}{(q;q)_\infty}, \tag{8.2}$$

where now $a_0 := 2$. Scribbled underneath (8.1) are the first few terms of (5.1) through q^5. Thus, although not claimed by Ramanujan, (8.1) is, in fact, equal to $F_a(q)$. We state this in the next theorem, with a_n replaced by A_n.

Theorem 8.1. *If A_n is given by (3.2), then, if $|q| < \min(|a|, 1/|a|)$,*

$$\frac{(q;q)_\infty^2}{(aq;q)_\infty(q/a;q)_\infty} = 1 - \sum_{m=1,n=0}^{\infty} (-1)^m q^{m(m+1)/2+mn}(A_{n+1} - A_n). \quad (8.3)$$

It is easily seen that Ramanujan's Theorem 8.1, which we prove in the next section, is equivalent to a theorem discovered independently by R. J. Evans [10, eq. (3.1)], V. G. Kač and D. H. Peterson [17, eq. (5.26)], and Kač and M. Wakimoto [18, middle of p. 438]. As remarked in [17], the identity, in fact, appears in the classic text of J. Tannery and J. Molk [21, Sect. 486].

Theorem 8.2. *Let*

$$r_k = (-1)^k q^{k(k+1)/2}. \quad (8.4)$$

Then

$$\frac{(q;q)_\infty^2}{(aq;q)_\infty(q/a;q)_\infty} = \sum_{k=-\infty}^{\infty} \frac{r_k(1-a)}{1-aq^k}. \quad (8.5)$$

A notable feature of the authors' [6] second method, based on Theorem 8.1 or Theorem 8.2, for establishing Ramanujan's five congruences is that elegant identities arise in the proofs. For example, in the proof of Theorem 2.1, we need to prove that

$$\sum_{k=-\infty}^{\infty} r_k \frac{q^k - 1}{1+q^{4k}} = q\frac{(q;q)_\infty}{(-q^4;q^4)_\infty} f(-q^2, -q^{14})$$

and

$$\sum_{k=-\infty}^{\infty} r_k \frac{q^k + 1}{1+q^{4k}} = \frac{(q;q)_\infty}{(-q^4;q^4)_\infty} f(-q^6, -q^{10}),$$

where r_k is defined by (8.4). To prove Theorem 2.2, we need to prove

$$\sum_{k=-\infty}^{\infty} r_k \frac{q^k - 1}{1+q^{3k}+q^{6k}} = q(q;q)_\infty \frac{f(-q^3, -q^{24})f(-q^{12}, -q^{15})}{(q^{27};q^{27})_\infty}$$

and two similar identities.

On page 59, below the list of factors and above the two foregoing series, Ramanujan records two further series, namely,

$$S_1(a,q) := \frac{1}{1+a} + \sum_{n=1}^{\infty} \left(\frac{(-1)^n q^{n(n+1)/2}}{1+aq^n} + \frac{(-1)^n q^{n(n+1)/2}}{a+q^n} \right) \quad (8.6)$$

and

$$S_2(a,q) := 1 + \sum_{m=1,n=0}^{\infty} (-1)^{m+n} q^{m(m+1)/2+mn}(a_{n+1}+a_n), \tag{8.7}$$

where here $a_0 := 1$. No theorem is claimed by Ramanujan, but the following theorem, to be proved in the next section, holds.

Theorem 8.3. *With* $S_1(a,q)$ *and* $S_2(a,q)$ *defined by* (8.6) *and* (8.7), *respectively,*

$$(1+a)S_1(a,q) = S_2(a,q) = F_{-a}(q).$$

9. Proofs of Theorems 8.1 and 2.1

Proof of Theorem 8.1. Our proof is different from that of Evans [10], Kač and Peterson [17], and Kač and Wakimoto [18]. We employ the partial fraction decomposition

$$\begin{aligned}\frac{(q;q)_\infty^2}{(aq;q)_\infty(q/a;q)_\infty} &= 1 + \sum_{n=1}^{\infty}(-1)^n q^{n(n-1)/2}(1+q^n)\\ &\times \left\{1 - \frac{1-q^n}{1+q^n}\sum_{m=0}^{\infty} a^m q^{mn} - \frac{1-q^n}{1+q^n}\sum_{m=1}^{\infty} a^{-m}q^{mn}\right\}, \end{aligned}\tag{9.1}$$

found in Garvan's paper [13, eq. (7.16)]. From (9.1), we find that

$$\begin{aligned}&\frac{(q;q)_\infty^2}{(aq;q)_\infty(q/a;q)_\infty}\\ &= 1 + \sum_{n=1}^{\infty}(-1)^n q^{n(n-1)/2}\Bigg((1+q^n)\\ &\qquad - (1-q^n)\sum_{m=0}^{\infty} a^m q^{mn} - (1-q^n)\sum_{m=1}^{\infty} a^{-m}q^{mn}\Bigg)\\ &= 1 + \sum_{n=1}^{\infty}(-1)^n q^{n(n-1)/2}\left(2 - (1-q^n)\sum_{m=0}^{\infty} q^{mn}(a^m + a^{-m})\right)\\ &= 1 + \sum_{n=1}^{\infty}(-1)^n q^{n(n-1)/2}\left(2 - \sum_{m=0}^{\infty} q^{mn}A_m + \sum_{m=0}^{\infty} q^{(m+1)n}A_m\right)\\ &= 1 + \sum_{n=1}^{\infty}(-1)^n q^{n(n-1)/2}\left(-\sum_{m=1}^{\infty} q^{mn}A_m + \sum_{m=1}^{\infty} q^{mn}A_{m-1}\right)\\ &= 1 + \sum_{m,n=1}^{\infty}(-1)^n q^{n(n-1)/2+mn}(A_{m-1}-A_m)\end{aligned}$$

$$= 1 - \sum_{m=0,n=1}^{\infty} (-1)^n q^{n(n+1)/2+mn} \left(A_{m+1} - A_m\right),$$

which is (8.3), but with the roles of m and n reversed. □

Proof of Theorem 2.1. Multiply (8.6) throughout by $(1+a)$ to deduce that

$$\begin{aligned}
(1+a)S_1(a,q) &= 1 + (1+a)\sum_{n=1}^{\infty}\left(\frac{(-1)^n q^{n(n+1)/2}}{1+aq^n} + \frac{(-1)^n q^{n(n+1)/2}}{a+q^n}\right) \\
&= 1 + (1+a)\sum_{n=1}^{\infty}\left(\frac{(-1)^n q^{n(n+1)/2}}{1+aq^n} + \frac{(-1)^{-n} q^{n(n-1)/2}}{1+aq^{-n}}\right) \\
&= 1 + (1+a)\sum_{n\neq 0}\frac{(-1)^n q^{n(n+1)/2}}{1+aq^n} \\
&= \sum_{n=-\infty}^{\infty}\frac{(-1)^n q^{n(n+1)/2}(1+a)}{1+aq^n} \\
&= \frac{(q;q)_\infty^2}{(-aq;q)_\infty(-q/a;q)_\infty},
\end{aligned} \tag{9.2}$$

by an application of (8.5).

Secondly,

$$\begin{aligned}
S_2(a,q) &= 1 + \sum_{m=1,n=0}^{\infty} (-1)^m q^{m(m+1)/2+mn} \\
&\quad \times \left(-(-a)^{n+1} - (-a)^{-n-1} + (-a)^n + (-a)^{-n}\right) \\
&= \frac{(q;q)_\infty^2}{(-aq;q)_\infty(-q/a;q)_\infty},
\end{aligned} \tag{9.3}$$

by Theorem 8.1. Thus, (9.2) and (9.3) yield Theorem 2.1. □

10. Conclusion

From the abundance of material in the lost notebook on factors of the coefficients λ_n of the generating function (2.1) for cranks, $F_a(q)$, Ramanujan clearly was eager to find some general theorems with the likely intention of applying them in the special case of $a = 1$ to determine arithmetical properties of the partition function $p(n)$. Although he was able to derive five beautiful congruences for $F_a(q)$, the kind of arithmetical theorem that he was seeking evidently eluded him. Indeed, general theorems on the divisibility of λ_n by sums of powers of a appear extremely difficult, if not impossible, to obtain. Moreover, demonstrating that the tables in Section 5.6 are complete seems to be a formidable challenge.

Garvan discovered a 5-dissection of $F_a(q)$, where a is any primitive 10th root of unity, in [14, eq. (2.16)]. This is, to date, the only dissection identity for the generating function of cranks that does not appear in Ramanujan's lost notebook. It would also be interesting to uncover new dissection identities of $F_a(q)$ when a is a primitive root of unity of order greater than 11.

Acknowledgments

We are grateful to Frank Garvan for several corrections and useful suggestions.

References

[1] G. A. Andrews and R. Lewis, *The rank and cranks of partitions moduli* 2, 3, *and* 4, J. Number Theory **85** (2000), no. 1, 74–84.

[2] G. E. Andrews and F. G. Garvan, *Dyson's crank of a partition*, Bull. Amer. Math. Soc. (N.S.) **18** (1988), 167–171.

[3] A. O. L. Atkin and F. G. Garvan, *Relations between the ranks and cranks of partitions*, Ramanujan J. **7** (2003), no. 1-3, 343–366.

[4] A. O. L. Atkin and H. P. F. Swinnerton-Dyer, *Some properties of partitions*, Proc. London Math. Soc. (3) **4** (1954), 84–106.

[5] B. C. Berndt, *Ramanujan's notebooks, part iii*, Springer-Verlag, New York, 1991.

[6] B. C. Berndt, H. H. Chan, S. H. Chan, and W.-C. Liaw, *Cranks and dissections in Ramanujan's lost notebook*, Submitted for publication, 2003.

[7] F. J. Dyson, *Some guesses in the theory of partitions*, Eureka (Cambridge) **8** (1944), 10–15.

[8] A. B. Ekin, *Inequalities for the crank*, J. Combin. Thy. Ser. A **83** (1998), 283–289.

[9] ______, *Some properties of partitions in terms of crank*, Trans. Amer. Math. Soc. **352** (2000), 2145–2156.

[10] R. J. Evans, *Generalized Lambert series*, Analytic Number Theory (Allerton Park, IL, 1995) (B. C. Berndt, H. G. Diamond, and A. J. Hildebrand, eds.), Progr. Math., vol. 1, Birkhäuser Boston, Boston, MA, 1996, pp. 357–370.

[11] F. G. Garvan, *Generalizations of Dyson's rank*, Ph.D. thesis, Pennsylvania State University, University Park, PA, 1986.

[12] ______, *Combinatorial interpretations of Ramanujan's partition congruences*, Ramanujan Revisited (Urbana-Champaign, Ill., 1987) (G. E. Andrews, R. A. Askey, B. C. Berndt, K. G. Ramanathan, and R. A. Rankin, eds.), Academic Press, Boston, MA, 1988, pp. 29–45.

[13] ———, *New combinatorial interpretations of Ramanujan's partition congruences mod* 5, 7 *and* 11, Trans. Amer. Math. Soc. **305** (1988), 47–77.

[14] ———, *The crank of partitions mod* 8, 9 *and* 10, Trans. Amer. Math. Soc. **322** (1990), 79–94.

[15] M. D. Hirschhorn, *A birthday present for Ramanujan*, Amer. Math. Monthly **97** (1990), 398–400.

[16] D. M. Kane, *Resolution of a conjecture of Andrews and Lewis involving cranks of partitions*, Proc. Amer. Math. Soc. **132** (2004), 2247–2256.

[17] V. G. Kač and D. H. Peterson, *Infinite-dimensional Lie algebras, theta functions and modular forms*, Adv. in Math. **53** (1984), 125–264.

[18] V. G. Kač and M. Wakimoto, *Integrable highest weight modules over affine superalgebras and number theory*, Lie Theory and Geometry, Progr. Math., vol. 123, Birkhäuser Boston, Boston, MA, 1994, pp. 415–456.

[19] S. Ramanujan, *Notebooks, vols. 1, 2*, Tata Institute of Fundamental Research, Bombay, 1957.

[20] ———, *The lost notebook and other unpublished papers*, Narosa, New Delhi, 1988.

[21] J. Tannery and J. Molk, *éléments de la Théorie des Fonctions Elliptiques*, Paris, 1898.

THE SAALSCHÜTZ CHAIN REACTIONS AND MULTIPLE q-SERIES TRANSFORMATIONS

CHU Wenchang
Department of Applied Mathematics
Dalian University of Technology
Dalian 116024, P. R. China

Current Address:
Dipartimento di Matematica
Università degli Studi di Lecce
Lecce-Arnesano P. O. Box 193
73100 Lecce, ITALIA
chu.wenchang@unile.it

Abstract By recursive use of the q-Saalschütz summation formula, we investigate further the Saalschütz chain reactions introduced by the author in (Chu, 2002). Some general series transformations which express basic terminating series in terms of finite multiple sums will be established. As applications, we derive by means of Jackson's ${}_6\varphi_5$-series identity three transformations including one due to Andrews (1975). These transformations yield further a number of multiple Rogers-Ramanujan identities, whose research was initiated and developed mainly by Andrews and Bressoud from the middle of seventieth up to now.

1. Introduction and notation

For two complex numbers q and x, the shifted-factorial of order n with base q is defined by

$$(x;q)_0 \equiv 1 \text{ and } (x;q)_n = (1-x)(1-xq)\cdots(1-xq^{n-1}) \text{ for } n = 1, 2, \ldots . \tag{1.1a}$$

When $|q| < 1$, the infinite product

$$(x;q)_\infty = \prod_{k=0}^{\infty}(1 - xq^k) \tag{1.1b}$$

allows us consequently to express

$$(x;q)_n = (x;q)_\infty / (xq^n;q)_\infty \tag{1.1c}$$

where n can be an arbitrary real number.

The product and fraction forms of the shifted factorials are abbreviated throughout the paper respectively to

$$[a,\, b, \cdots c; q]_n = (a;q)_n\,(b;q)_n \cdots (c;q)_n \tag{1.1d}$$

$$\left[\begin{matrix} a, & b, & \cdots, & c \\ A, & B, & \cdots, & C \end{matrix} \middle| \, q\right]_n = \frac{(a;q)_n\,(b;q)_n \cdots (c;q)_n}{(A;q)_n\,(B;q)_n \cdots (C;q)_n}. \tag{1.1e}$$

Following Bailey (Bailey, 1935) and Slater (Slater, 1966), the basic hypergeometric series is defined by

$${}_{1+r}\varphi_s\left[\begin{matrix} a_0, & a_1, & \cdots, & a_r \\ & b_1, & \cdots, & b_s \end{matrix} \middle| \, q;\, z\right] = \sum_{n=0}^{\infty} z^n \left[\begin{matrix} a_0, & a_1, & \cdots, & a_r \\ q, & b_1, & \cdots, & b_s \end{matrix} \middle| \, q\right]_n \tag{1.2}$$

where the base q will be restricted to $|q| < 1$ for non-terminating q-series.

Among the basic hypergeometric formulas, we reproduce three of them for our subsequent references. The first is the q-Saalschütz theorem (cf. (Bailey, 1935, Chapter 8) and (Slater, 1966, §3.3)):

$${}_3\varphi_2\left[\begin{matrix} q^{-n}, & a, & b \\ & c, & q^{1-n}ab/c \end{matrix} \middle| \, q;\, q\right] = \left[\begin{matrix} c/a, & c/b \\ c, & c/ab \end{matrix} \middle| \, q\right]_n. \tag{1.3}$$

The second is the very well-poised formula due to Jackson (cf. (Bailey, 1935, Chapter 8) and (Slater, 1966, §3.3)):

$${}_6\varphi_5\left[\begin{matrix} a, & q\sqrt{a}, & -q\sqrt{a}, & b, & c, & d \\ & \sqrt{a}, & -\sqrt{a}, & qa/b, & qa/c, & qa/d \end{matrix} \middle| \, q;\, \frac{qa}{bcd}\right] \tag{1.4a}$$

$$= \left[\begin{matrix} qa, qa/bc, qa/bd, qa/cd \\ qa/b, qa/c, qa/d, qa/bcd \end{matrix} \middle| \, q\right]_\infty, \qquad \big(|qa/bcd| < 1\big). \tag{1.4b}$$

The third and the last one is Watson's q-analogue of the Whipple transformation (cf. (Bailey, 1935, Chapter 8)):

$${}_8\varphi_7\left[\begin{matrix} a, & q\sqrt{a}, & -q\sqrt{a}, & b, & c, & d, & e, & q^{-m} \\ & \sqrt{a}, & -\sqrt{a}, & qa/b, & qa/c, & qa/d, & qa/e, & aq^{m+1} \end{matrix} \middle| \, q;\, \frac{q^{2+m}a^2}{bcde}\right]$$

$$= \left[\begin{matrix} qa, qa/bc \\ qa/b, qa/c \end{matrix} \middle| \, q\right]_m \times {}_4\varphi_3\left[\begin{matrix} q^{-m}, & b, & c, & qa/de \\ & qa/d, & qa/e, & q^{-m}bc/a \end{matrix} \middle| \, q;\, q\right].$$

The celebrated Rogers-Ramanujan identities (cf. (Slater, 1966, §3.5)) read as:

$$1+\sum_{m=1}^{\infty}\frac{q^{m^2}}{(1-q)(1-q^2)\cdots(1-q^m)}=\prod_{n=0}^{\infty}\frac{1}{(1-q^{1+5n})(1-q^{4+5n})} \tag{1.5a}$$

$$1+\sum_{m=1}^{\infty}\frac{q^{m+m^2}}{(1-q)(1-q^2)\cdots(1-q^m)}=\prod_{n=0}^{\infty}\frac{1}{(1-q^{2+5n})(1-q^{3+5n})}. \tag{1.5b}$$

Bailey (Bailey, 1947; Bailey, 1948) discovered numerous identities of such kind. A systematic collection was done by Slater (Slater, 1951; Slater, 1952). Some more recent results may be found in Gessel-Stanton (Gessel and Stanton, 1983).

In their work on multiple Rogers-Ramanujan identities, Andrews and Bressoud et al. (Agarwal et al., 1987; Andrews, 1984; Andrews, 1986; Bressoud, 1980a; Bressoud, 1988) introduced the powerful Bailey chains and Bailey lattice. They found general transformations which express multiple unilateral sums into a single (unilateral) basic hypergeometric series involving two sequences (Bailey pair) connected by an inverse series relation. The latter can be reformulated, in particular settings, as a bilateral basic hypergeometric series. By evaluating the bilateral sum with the Jacobi triple or the quintuple product formulas, they derived with great success many multiple Rogers-Ramanujan identities.

By iterating the q-Saalschütz formula (1.3), *the Saalschütz chain reactions* under "finite condition" has been introduced by the author in (Chu, 2002) to study the ordinary and basic hypergeometric series with integer differences between numerator and denominator parameters. We will investigate further in the next section *the Saalschütz chain reactions* without finite condition and establish transformation theorems (from 2.4 to 2.7) of the same nature as Bailey chains due to Andrews and Bressoud but with one (or two) independent arbitrary sequence(s). Then we proceed in Section 3 and 4 to derive explicitly several specific transformation formulas (without indeterminate sequence). Finally in the last section, we conclude with thirty multiple Rogers-Ramanujan identities which are simply limiting cases of the transformations presented in this paper combined with the Jacobi triple product identity.

The purpose of this paper is not to present a general cover of the Rogers-Ramanujan identities and their multiple counterparts through *the Saalschütz chain reactions.* Instead, it will be limited to illustrate how to explore this method potentially to generate multiple basic hypergeometric transformations and produce multiple Rogers-Ramanujan identities.

2. The Saalschütz chain reactions

For nonnegative integers k, M with $k \geq M$ and three indeterminates a, x, y, the q-Saalschütz formula (1.3) tells us that

$$
\begin{aligned}
{}_3\varphi_2\left[\begin{matrix} q^{M-k}, & q^{M+k}a, & qa/xy \\ & q^{1+M}a/x, & q^{1+M}a/y \end{matrix} \,\middle|\, q;\, q\right] &= \left[\begin{matrix} q^M y, & q^{1-k}/x \\ q^{1+M}a/x, & q^{-k}y/a \end{matrix} \,\middle|\, q\right]_{k-M} \\
= \left[\begin{matrix} x, & y \\ qa/x, & qa/y \end{matrix} \,\middle|\, q\right]_k \left(\frac{qa}{xy}\right)^k &\Big/ \left[\begin{matrix} x, & y \\ qa/x, & qa/y \end{matrix} \,\middle|\, q\right]_M \left(\frac{qa}{xy}\right)^M
\end{aligned}
$$

which may be restated explicitly as follows:

$$
\left(\frac{qa}{xy}\right)^k \left[\begin{matrix} x, & y \\ qa/x, & qa/y \end{matrix} \,\middle|\, q\right]_k = \sum_{m \geq 0} q^m \frac{(qa/xy;q)_m}{(q;q)_m} [x,y;q]_M \tag{2.1a}
$$

$$
\times \frac{[q^{M-k},\, q^{M+k}a;q]_m}{[qa/x, qa/y;q]_{M+m}} \left(\frac{qa}{xy}\right)^M. \tag{2.1b}
$$

Denote the multiple summation index and its partial sums respectively by

$$
\tilde{m} = (m_1, m_2, \ldots, m_n) \tag{2.2a}
$$

$$
M_k = \sum_{i=1}^{k} m_i, \; 0 \leq k \leq n. \tag{2.2b}
$$

With $1 \leq \iota \leq n$, we may rewrite (2.1) with subscripts as

$$
\begin{aligned}
\left(\frac{qa}{x_\iota y_\iota}\right)^k \left[\begin{matrix} x_\iota, & y_\iota \\ qa/x_\iota, & qa/y_\iota \end{matrix} \,\middle|\, q\right]_k &= \sum_{m_\iota \geq 0} q^{m_\iota} \frac{(qa/x_\iota y_\iota;q)_{m_\iota}}{(q;q)_{m_\iota}} [x_\iota, y_\iota;q]_{M_{\iota-1}} \\
&\times \frac{[q^{M_{\iota-1}-k},\, q^{M_{\iota-1}+k}a;q]_{m_\iota}}{[qa/x_\iota, qa/y_\iota;q]_{M_\iota}} \left(\frac{qa}{x_\iota y_\iota}\right)^{M_{\iota-1}}.
\end{aligned}
$$

Then the recursive product (*the Saalschütz chain reactions*) of the expression just displayed for $\iota = 1, 2, \ldots, n$ reasults compactly in the following

Lemma 2.1 (Multiple sum: Andrews (Andrews, 1979a, Eq. 5.2)). *With k being a nonnegative integer, there holds*

$$\prod_{\iota=1}^{n}\left[\begin{matrix}x_\iota,\ y_\iota\\ qa/x_\iota,\ qa/y_\iota\end{matrix}\middle|\, q\right]_k\left(\frac{qa}{x_\iota y_\iota}\right)^k=\sum_{\tilde{m}\geq 0}\frac{(qa/x_ny_n;q)_{m_n}}{(q;q)_{m_n}}\left[\begin{matrix}q^{-k},\ q^ka\\ qa/x_n,qa/y_n\end{matrix}\middle|\, q\right]_{M_n}$$
$$\times\ q^{M_n}\prod_{i=1}^{n-1}\frac{(qa/x_iy_i;q)_{m_i}}{(q;q)_{m_i}}\left[\begin{matrix}x_{i+1},\ y_{i+1}\\ qa/x_i,qa/y_i\end{matrix}\middle|\, q\right]_{M_i}\left(\frac{qa}{x_{i+1}y_{i+1}}\right)^{M_i}$$

where the multiple summation index $\tilde{m}$ runs over all $m_i \geq 0$ for $i = 1, 2, \ldots, n$.

In this lemma, replacing x_i by qa/x_i and then a by aq^{-k}, we may state the limiting case $k \to \infty$, (which did not appear in Andrews (Andrews, 1979a) explicitly), as follows:

Corollary 2.2 (Multiple sum).

$$\prod_{\iota=1}^{n}\left[\begin{matrix}x_\iota/a,\ y_\iota\\ x_\iota,\ y_\iota/a\end{matrix}\middle|\, q\right]_\infty=\sum_{\tilde{m}\geq 0}(a;q)_{M_n}\prod_{i=1}^{n}\frac{(x_i/y_i;q)_{m_i}}{(q;q)_{m_i}}\frac{(y_i;q)_{M_{i-1}}}{(x_i;q)_{M_i}}\left(\frac{y_i}{a}\right)^{m_i}.$$

When $a \to \infty$ and $y_i \to 0$, it reduces to the following

Corollary 2.3 (Andrews (Andrews, 1979a, §6)).

$$\prod_{\iota=1}^{n}\frac{1}{(x_\iota;q)_\infty}=\sum_{\tilde{m}\geq 0}q^{\binom{M_n}{2}}\prod_{i=1}^{n}\frac{x_i^{m_i}\,q^{\binom{m_i}{2}}}{(q;q)_{m_i}(x_i;q)_{M_i}}.$$

For this identity, Milne (Milne, 1980, Thm. 3.1) has given an alternate derivation.

For two natural numbers u and v with $1 \leq u \leq v \leq n$, and complex indeterminates c and $\{x_i, y_i\}$, denote factorial fractions by

$$\Lambda_k^{(c)}\left[\begin{matrix}\{x_\iota,y_\iota\}\\ [u,v]\end{matrix}\right]=\prod_{i=u}^{v}\left[\begin{matrix}x_i,\ y_i\\ c/x_i,c/y_i\end{matrix}\middle|\, q\right]_k\left(\frac{c}{x_iy_i}\right)^k$$
$$\Omega_{\tilde{m}}^{(c)}\left[\begin{matrix}\{x_\iota,y_\iota\}\\ [u,v]\end{matrix}\right]=\prod_{i=u}^{v}\frac{(c/x_iy_i;q)_{m_i}}{(q;q)_{m_i}}\left[\begin{matrix}x_{i+1},y_{i+1}\\ c/x_i,c/y_i\end{matrix}\middle|\, q\right]_{M_i}\left(\frac{c}{x_{i+1}y_{i+1}}\right)^{M_i}.$$

It is trivial to note that

$$\Lambda_k^{(c)}\left[\begin{matrix}\{x_\iota,y_\iota\}\\ [u,\ v]\end{matrix}\right]=\prod_{i=u}^{v}\Lambda_k^{(c)}\left[\begin{matrix}\{x_\iota,y_\iota\}\\ [i,\ i]\end{matrix}\right].$$

Then the multiple sum in the lemma may be expressed as

$$\Lambda_{k}^{(qa)}\begin{bmatrix}\{x_i,y_i\}\\ [1,n]\end{bmatrix}=\sum_{\tilde{m}\geq 0}\frac{(qa/x_ny_n;q)_{m_n}}{(q;q)_{m_n}}\Omega_{\tilde{m}}^{(qa)}\begin{bmatrix}\{x_i,y_i\}\\ [1,n-1]\end{bmatrix}$$
$$\times\, q^{M_n}\begin{bmatrix}q^{-k},q^ka\\ qa/x_n,qa/y_n\end{bmatrix}q\Bigg]_{M_n}.$$

Multiplying both sides by $\frac{(a;q)_k}{(q;q)_k}W_k$ for suitable W_k and then summing over all $k\geq 0$, we establish the following general transformation theorem, which may be considered as a counterpart of the main result obtained in (Chu, 2002, Thm. 2).

Theorem 2.4 (Well-poised transformation). *For a complex sequence $\{W_k\}$, there holds the following multiple basic hypergeometric transformation*

$$\sum_{k\geq 0}\frac{(a;q)_k}{(q;q)_k}\Lambda_{k}^{(qa)}\begin{bmatrix}\{x_i,y_i\}\\ [1,n]\end{bmatrix}W_k=\sum_{\tilde{m}\geq 0}(-1)^{M_n}\frac{(qa/x_ny_n;q)_{m_n}}{[qa/x_n,qa/y_n;q]_{M_n}}q^{-\binom{M_n}{2}}$$
$$\times\frac{(a;q)_{2M_n}}{(q;q)_{m_n}}\Omega_{\tilde{m}}^{(qa)}\begin{bmatrix}\{x_i,y_i\}\\ [1,n-1]\end{bmatrix}$$
$$\times\sum_{k\geq 0}q^{-kM_n}\frac{(aq^{2M_n};q)_k}{(q;q)_k}W_{k+M_n}$$

provided that both series are well-defined and convergent.

Rewriting the well-poised transformation in Theorem 2.4 as

$$\sum_{k\geq 0}\frac{(a;q)_k}{(q;q)_k}\Lambda_{k}^{(qa)}\begin{bmatrix}\{x_i,y_i\}\\ [1,\ell]\end{bmatrix}W_k=\sum_{m_1,m_2,\cdots,m_\ell\geq 0}(-1)^{M_\ell}\frac{(qa/x_\ell y_\ell;q)_{m_\ell}}{[qa/x_\ell,qa/y_\ell;q]_{M_\ell}}\tag{2.3a}$$

$$\times\,\Omega_{\tilde{m}}^{(qa)}\begin{bmatrix}\{x_i,\ y_i\}\\ [1,\ell-1]\end{bmatrix}\frac{(a;q)_{2M_\ell}}{(q;q)_{m_\ell}}q^{-\binom{M_\ell}{2}}\tag{2.3b}$$

$$\times\sum_{k\geq 0}q^{-kM_\ell}\frac{(aq^{2M_\ell};q)_k}{(q;q)_k}W_{k+M_\ell}\tag{2.3c}$$

and then performing the substitution

$$W_k=\alpha_k\left\{\gamma_k\Lambda_{k}^{(a)}\begin{bmatrix}\{x_j,y_j\}\\ [1+\ell,n]\end{bmatrix}-\gamma_{k+1}\Lambda_{k+1}^{(a)}\begin{bmatrix}\{x_j,y_j\}\\ [1+\ell,n]\end{bmatrix}\right\}$$

we can manipulate the last sum (2.3c), shortly as S(2.3c) with respect to k according to the formal series rearrangement

$$\sum_{k=0}^{\infty} \mathcal{U}_k \, \nabla \mathcal{V}_k = \mathcal{V}_0 \mathcal{U}_{-1} + \sum_{k=0}^{\infty} \mathcal{V}_k \, \Delta \mathcal{U}_k \tag{2.4}$$

where ∇ and Δ are forward and backward difference operators:

$$\nabla F(n) = F(n) - F(n+1)$$
$$\Delta F(n) = F(n) - F(n-1).$$

This can be proceeded as follows:

$$\begin{aligned}
\mathrm{S}(2.3c) &= \sum_{k\geq 0} \frac{(aq^{2M_\ell};q)_k}{(q;q)_k} \alpha_{k+M_\ell} \nabla \left\{ \gamma_{k+M_\ell} \Lambda^{(a)}_{k+M_\ell} \begin{bmatrix} \{x_j, y_j\} \\ [1+\ell, n] \end{bmatrix} \right\} q^{-kM_\ell} \\
&= \sum_{k\geq 0} \Delta \left\{ q^{-kM_\ell} \frac{(aq^{2M_\ell};q)_k}{(q;q)_k} \alpha_{k+M_\ell} \right\} \gamma_{k+M_\ell} \Lambda^{(a)}_{k+M_\ell} \begin{bmatrix} \{x_j, y_j\} \\ [1+\ell, n] \end{bmatrix} \\
&= \sum_{k\geq 0} \frac{(aq^{2M_\ell-1};q)_k}{(q;q)_k} \Lambda^{(aq^{2M_\ell})}_k \begin{bmatrix} \{x_j q^{M_\ell}, y_j q^{M_\ell}\} \\ [1+\ell, n] \end{bmatrix} \Lambda^{(a)}_{M_\ell} \begin{bmatrix} \{x_j, y_j\} \\ [1+\ell, n] \end{bmatrix} \\
&\times \frac{q^{-kM_\ell} \gamma_{k+M_\ell}}{1 - aq^{2M_\ell-1}} \left\{ (1 - aq^{k-1+2M_\ell}) \alpha_{k+M_\ell} - q^{M_\ell}(1-q^k) \alpha_{k-1+M_\ell} \right\}.
\end{aligned}$$

Now applying the transformation in Theorem 2.4 to the penultimate line, we obtain with some simplification, the following expression

$$\begin{aligned}
\mathrm{S}(2.3c) = &\sum_{m_{1+\ell}, m_{2+\ell}, \ldots, m_n \geq 0} q^{\binom{M_\ell}{2} - \binom{M_n}{2}} \frac{(a/x_n y_n;q)_{m_n}}{(q;q)_{m_n}} \frac{(aq^{2M_\ell-1};q)_{2M_n-2M_\ell}}{[a/x_n, a/y_n;q]_{M_n}} \\
&\times \frac{(-1)^{M_n-M_\ell}}{1-aq^{2M_\ell-1}} \Omega^{(a)}_{\tilde{m}} \begin{bmatrix} \{x_j \; y_j\} \\ [1+\ell, n-1] \end{bmatrix} \left(\frac{a}{x_{1+\ell} y_{1+\ell}} \right)^{M_\ell} [x_{1+\ell}, y_{1+\ell};q]_{M_\ell} \\
&\times \sum_{k\geq 0} q^{-kM_n} \frac{(aq^{2M_n-1};q)_k}{(q;q)_k} \left\{ \begin{matrix} (1-aq^{k-1+M_n+M_\ell}) \, \alpha_{k+M_n} \\ -q^{M_\ell}(1-q^{k+M_n-M_\ell}) \alpha_{k-1+M_n} \end{matrix} \right\} \gamma_{k+M_n}.
\end{aligned}$$

Replacing (2.3c) with this result leads (2.3) to the following almost-poised series transformation.

Theorem 2.5 (Almost-poised transformation). *For two complex sequences $\{\alpha_k, \gamma_k\}$, there holds the following multiple basic hypergeomet-*

ric transformation

$$\sum_{k\geq 0} \alpha_k \frac{(a;\,q)_k}{(q;\,q)_k} \Lambda_k^{(qa)}\begin{bmatrix}\{x_i,y_i\}\\ [1,\ell]\end{bmatrix} \nabla\left\{\gamma_k \Lambda_k^{(a)}\begin{bmatrix}\{x_j,y_j\}\\ [1+\ell,n]\end{bmatrix}\right\} \tag{2.5a}$$

$$= \sum_{\tilde{m}\geq 0} \frac{(a/x_n y_n;q)_{m_n}}{(q;q)_{m_n}} \Omega_{\tilde{m}}^{(qa)}\begin{bmatrix}\{x_i,y_i\}\\ [1,\ell]\end{bmatrix} \Omega_{\tilde{m}}^{(a)}\begin{bmatrix}\{x_j,y_j\}\\ [1+\ell,n-1]\end{bmatrix} q^{-\binom{M_n}{2}} \tag{2.5b}$$

$$\times \frac{(-1)^{M_n}}{1-a/q} \frac{(a/q;q)_{2M_n}}{[a/x_n, a/y_n;q]_{M_n}} \sum_{k\geq 0} q^{-kM_n} \frac{\left(aq^{2M_n-1};q\right)_k}{(q;q)_k} \bar{\alpha}_{k+M_n}\gamma_{k+M_n} \tag{2.5c}$$

provided that both series are well-defined and convergent with

$$\bar{\alpha}_{k+M_n} = \left(q^{-M_\ell} - aq^{k-1+M_n}\right)\alpha_{k+M_n} - \left(1-q^{k+M_n-M_\ell}\right)\alpha_{k-1+M_n}. \tag{2.6}$$

According to (2.4), the k-sum in (2.5c) may be reformulated as

$$\sum_{k\geq 0} \frac{\alpha_{k-1+M_n}\bar{\gamma}_{k+M_n}}{1-aq^{2M_n-2}} \frac{\left(aq^{2M_n-2};q\right)_k}{(q;q)_k} q^{-kM_n} \tag{2.7}$$

where we have defined dually

$$\bar{\gamma}_{k+M_n} = (1-q^k)\left(q^{M_n-M_\ell} - aq^{k-2+2M_n}\right)\gamma_{k-1+M_n} \tag{2.8a}$$

$$- (1-q^{k+M_n-M_\ell})(1-aq^{k-2+2M_n})\,\gamma_{k+M_n}. \tag{2.8b}$$

In Theorem 2.5, take

$$\gamma_k \equiv 1.$$

Then it is easy to check the factorization

$$\bar{\gamma}_{k+M_n} = -(1-q^{M_n-M_\ell})(1-aq^{2k-2+2M_n})$$

which leads us to the following transformation:

Theorem 2.6 (Almost-poised transformation). *For a complex sequence $\{\alpha_k\}$, there holds the following multiple basic hypergeometric* *transformation*

$$\sum_{k\geq 0}\frac{(a;q)_k}{(q;q)_k}\alpha_k\,\Lambda_k^{(qa)}\begin{bmatrix}\{x_i,y_i\}\\ [1,\ell]\end{bmatrix}\nabla\left\{\Lambda_k^{(a)}\begin{bmatrix}\{x_j,y_j\}\\ [1+\ell,n]\end{bmatrix}\right\} \tag{2.9a}$$

$$=\sum_{\tilde{m}\geq 0}(-1)^{1+M_n}\frac{1-q^{M_n-M_\ell}}{1-a/q}\frac{(a/q;q)_{2M_n}}{[a/x_n,a/y_n;q]_{M_n}}q^{-\binom{M_n}{2}} \tag{2.9b}$$

$$\times\frac{(a/x_ny_n;q)_{m_n}}{(q;q)_{m_n}}\Omega_{\tilde{m}}^{(qa)}\begin{bmatrix}\{x_i,y_i\}\\ [1,\ell]\end{bmatrix}\Omega_{\tilde{m}}^{(a)}\begin{bmatrix}\{x_j,y_j\}\\ [1+\ell,n-1]\end{bmatrix} \tag{2.9c}$$

$$\times\sum_{k\geq 0}q^{-kM_n}\frac{1-aq^{2M_n-2+2k}}{1-aq^{2M_n-2}}\frac{(aq^{2M_n-2};q)_k}{(q;q)_k}\alpha_{k-1+M_n} \tag{2.9d}$$

provided that both series are well-defined and convergent.

Taking instead in Theorem 2.5

$$\alpha_k=\frac{(qa/e;q)_k}{(e;q)_k}\left(\frac{e}{a}\right)^k$$

we can compute without difficulty that

$$\bar{\alpha}_{k+M_n}=(1-aq^{2k-1+2M_n})\frac{1-q^{-M_\ell}e/a}{1-e/a}\frac{(a/e;q)_{k+M_n}}{(e;q)_{k+M_n}}\left(\frac{e}{a}\right)^{k+M_n}$$

which leads us to the following transformation:

Theorem 2.7 (Almost-poised transformation). *For a complex sequence $\{\gamma_k\}$, there holds the following multiple basic hypergeometric transformation*

$$\sum_{k\geq 0}\begin{bmatrix}a,qa/e\\ q,e\end{bmatrix}q\bigg]_k\left(\frac{e}{a}\right)^k\Lambda_k^{(qa)}\begin{bmatrix}\{x_i,y_i\}\\ [1,\ell]\end{bmatrix}\nabla\left\{\gamma_k\Lambda_k^{(a)}\begin{bmatrix}\{x_j,y_j\}\\ [1+\ell,n]\end{bmatrix}\right\} \tag{2.10a}$$

$$=\sum_{\tilde{m}\geq 0}(-1)^{M_n}\frac{(a/x_ny_n;q)_{m_n}}{(q;q)_{m_n}}\frac{1-q^{-M_\ell}e/a}{1-e/a}\frac{(a/e;q)_{M_n}}{(e;q)_{M_n}}\left(\frac{e}{a}\right)^{M_n} \tag{2.10b}$$

$$\times\frac{(a;q)_{2M_n}}{[a/x_n,a/y_n;q]_{M_n}}\Omega_{\tilde{m}}^{(qa)}\begin{bmatrix}\{x_i,y_i\}\\ [1,\ell]\end{bmatrix}\Omega_{\tilde{m}}^{(a)}\begin{bmatrix}\{x_j,y_j\}\\ [1+\ell,n-1]\end{bmatrix}q^{-\binom{M_n}{2}} \tag{2.10c}$$

$$\times\sum_{k\geq 0}\frac{1-aq^{2M_n-1+2k}}{1-aq^{2M_n-1}}\begin{bmatrix}aq^{2M_n-1},q^{M_n}a/e\\ q,\quad q^{M_n}e\end{bmatrix}q\bigg]_k\left(\frac{e}{a}q^{-M_n}\right)^k\gamma_{k+M_n} \tag{2.10d}$$

provided that both series are well-defined and convergent.

3. Basic Almost-poised Transformations

By specifying the W and α,γ-sequences in the transformation theorems established in the last section, we will derive three explicit multiple series transformations which exemplify a larger class of such relations.

In Theorem 2.4, take

$$W_k = \frac{1-q^{2k}a}{1-a}\left[\begin{matrix} b, & c, & d \\ qa/b, & qa/c, & qa/d \end{matrix}\middle| q\right]_k \left(\frac{qa}{bcd}\right)^k.$$

Then the inner sum with respect to k becomes

$$ {}_6\varphi_5\left[\begin{matrix} aq^{2M}, q^{1+M}\sqrt{a}, -q^{1+M}\sqrt{a}, bq^M, cq^M, dq^M \\ q^M\sqrt{a}, -q^M\sqrt{a}, q^{1+M}a/b, q^{1+M}a/c, q^{1+M}a/d \end{matrix}\middle| q; \frac{q^{1-M}a}{bcd}\right]$$
$$\times \frac{1-aq^{2M_n}}{1-a}\left[\begin{matrix} b, & c, & d \\ qa/b, qa/c, qa/d \end{matrix}\middle| q\right]_{M_n} \left(\frac{qa}{bcd}\right)^{M_n}$$

with the ${}_6\varphi_5$-series evaluated by Jackson's formula as

$$\left(\frac{-a}{bcd}\right)^{-M} \frac{[qa/b,\, qa/c,\, qa/d; q]_M}{(qa;q)_{2M_n}\,(bcd/a;q)_M} q^{\binom{M_n}{2}} \left[\begin{matrix} qa,\, qa/bc,\, qa/bd,\, qa/cd \\ qa/b,\, qa/c,\, qa/d,\, qa/bcd \end{matrix}\middle| q\right]_\infty$$

This leads us to the following

Theorem 3.1 (Andrews (Andrews, 1975, Theorem 4) and (Andrews, 1979a, §5)). *For complex parameters $\{a,b,c,d\}$, and indeterminates $\{x_k, y_k\}_{k=1}^n$ with $X = x_1x_2\cdots x_n$ and $Y = y_1y_2\cdots y_n$, there holds the multiple basic hypergeometric series transformation*

$$ {}_{2n+6}\varphi_{5+2n}\left[\begin{matrix} a,\, q\sqrt{a},\, -q\sqrt{a},\, & b, & c, & d, & \{x_k,\, y_k\} \\ \sqrt{a},\, -\sqrt{a}, & qa/b, & qa/c, & qa/d, & \{qa/x_k, qa/y_k\} \end{matrix}\middle| q; \frac{(qa)^{1+n}}{bcdXY}\right] \tag{3.1a}$$

$$= \left[\begin{matrix} qa,\, qa/bc,\, qa/bd,\, qa/cd \\ qa/b,\, qa/c,\, qa/d,\, qa/bcd \end{matrix}\middle| q\right]_\infty \sum_{\tilde{m}\geq 0} \Omega_{\tilde{m}}^{(qa)}\left[\begin{matrix}\{x_i,\, y_i\} \\ [1, n-1]\end{matrix}\right] q^{M_n} \tag{3.1b}$$

$$\times \frac{(qa/x_ny_n;q)_{m_n}}{(q;q)_{m_n}} \left[\begin{matrix} b, \quad c, \quad d \\ bcd/a,\, qa/x_n,\, qa/y_n \end{matrix}\middle| q\right]_{M_n} \tag{3.1c}$$

provided that both series are well-defined and terminated by one of b, c or d.

In Theorem 2.6, take

$$\alpha_k = \left[\begin{matrix} b, & c, & d \\ qa/b, qa/c, qa/d \end{matrix}\middle| q\right]_k \left(\frac{q^2a}{bcd}\right)^k.$$

Then the sum with respect to k in (2.9d) becomes

$$
{}_6\varphi_5\left[\begin{matrix} aq^{2M_n-2}, \sqrt{a}q^{M_n}, -\sqrt{a}q^{M_n}, bq^{M_n-1}, cq^{M_n-1}, dq^{M_n-1} \\ \sqrt{a}q^{M_n-1}, -\sqrt{a}q^{M_n-1}, q^{M_n}a/b, q^{M_n}a/c, q^{M_n}a/d \end{matrix} \,\middle|\, q; \frac{aq^{2-M_n}}{bcd}\right]
$$
$$
\times \left[\begin{matrix} b, & c, & d \\ qa/b, & qa/c, & qa/d \end{matrix} \,\middle|\, q\right]_{M_n-1} \left(\frac{q^2a}{bcd}\right)^{M_n-1}
$$

with the ${}_6\varphi_5$-series under terminating condition evaluated as

$$
q^{\binom{M_n}{2}} \left[\begin{matrix} qa, qa/bc, qa/bd, qa/cd \\ qa/b, qa/c, qa/d, qa/bcd \end{matrix} \,\middle|\, q\right]_\infty
$$
$$
\times \frac{[qa/b, qa/c, qa/d; q]_{M_n-1}}{(qa;q)_{2M_n-2}(bcd/a;q)_{M_n-1}} \left(\frac{-qa}{bcd}\right)^{1-M_n}.
$$

Substituting these into Theorem 2.6, we get the following transformation:

Theorem 3.2 (Almost-poised transformation). *There holds the multiple basic hypergeometric transformation*

$$
\sum_{k\geq 0} \left[\begin{matrix} a, & b, & c, & d \\ q, & qa/b, & qa/c, & qa/d \end{matrix} \,\middle|\, q\right]_k \left(\frac{q^2a}{bcd}\right)^k \Lambda_k^{(qa)}\left[\begin{matrix} \{x_i, y_i\} \\ [1, \ell] \end{matrix}\right] \nabla \left\{ \Lambda_k^{(a)}\left[\begin{matrix} \{x_j, y_j\} \\ [1+\ell, n] \end{matrix}\right]\right\} \tag{3.2a}
$$

$$
= \frac{q/a}{(1-q/b)(1-q/c)(1-q/d)} \left[\begin{matrix} a, & qa/bc, & qa/bd, & qa/cd \\ qa/b, & qa/c, & qa/d, & q^2a/bcd \end{matrix} \,\middle|\, q\right]_\infty \tag{3.2b}
$$

$$
\times \sum_{\tilde{m}\geq 0} \frac{(a/x_ny_n;q)_{m_n}}{(q;q)_{m_n}} \left[\begin{matrix} b/q, & c/q, & d/q \\ bcd/aq, & a/x_n, & a/y_n \end{matrix} \,\middle|\, q\right]_{M_n} q^{M_n} \tag{3.2c}
$$

$$
\times\, \Omega_{\tilde{m}}^{(qa)}\left[\begin{matrix} \{x_i, y_i\} \\ [1, \ell] \end{matrix}\right] \Omega_{\tilde{m}}^{(a)}\left[\begin{matrix} \{x_j, y_j\} \\ [1+\ell, n-1] \end{matrix}\right] (1 - q^{M_n - M_\ell}) \tag{3.2d}
$$

provided that both series are well-defined and terminated by one of b, c or d.

In Theorem 2.7, take

$$
\gamma_k = \left[\begin{matrix} b, & c, & d, & e \\ a/b, & a/c, & a/d, & a/e \end{matrix} \,\middle|\, q\right]_k \left(\frac{a^2}{bcde}\right)^k.
$$

Then the sum with respect to k in (2.10d) becomes

$$
{}_6\varphi_5\left[\begin{matrix} aq^{2M_n-1}, q^{M_n}\sqrt{qa}, -q^{M_n}\sqrt{qa}, bq^{M_n}, cq^{M_n}, dq^{M_n} \\ q^{M_n}\sqrt{a/q}, -q^{M_n}\sqrt{a/q}, q^{M_n}a/b, q^{M_n}a/c, q^{M_n}a/d \end{matrix} \middle| q; \frac{aq^{-M_n}}{bcd}\right]
\times \left[\begin{matrix} b, & c, & d, & e \\ a/b, & a/c, & a/d, & a/e \end{matrix} \middle| q\right]_{M_n} \left(\frac{a^2}{bcde}\right)^{M_n}
$$

with the ${}_6\varphi_5$-series under terminating condition evaluated as

$$
(-1)^{M_n} q^{\binom{M_n}{2}} \left[\begin{matrix} a, a/bc, a/bd, a/cd \\ a/b, a/c, a/d, a/bcd \end{matrix} \middle| q\right]_\infty
\times \frac{[a/b, a/c, a/d; q]_{M_n}}{(a;q)_{2M_n}(qbcd/a;q)_{M_n}} \left(\frac{qbcd}{a}\right)^{M_n}.
$$

Substituting these into Theorem 2.7, we get the following transformation:

Theorem 3.3 (Almost-poised transformation). *There holds the multiple basic hypergeometric transformation*

$$
\sum_{k\geq 0} \frac{(a;q)_k}{(q;q)_k} \nabla \left\{ \left[\begin{matrix} b, & c, & d, & e \\ a/b, & a/c, & a/d, & a/e \end{matrix} \middle| q\right]_k \left(\frac{a^2}{bcde}\right)^k \Lambda_k^{(a)}\left[\begin{matrix} \{x_j, y_j\} \\ [1+\ell, n] \end{matrix}\right] \right\} \tag{3.3a}
$$

$$
\times \Lambda_k^{(qa)}\left[\begin{matrix} \{x_i, y_i\} \\ [1, \ell] \end{matrix}\right] \frac{(qa/e;q)_k}{(e;q)_k}\left(\frac{e}{a}\right)^k = \left[\begin{matrix} a, a/bc, a/bd, a/cd \\ a/b, a/c, a/d, a/bcd \end{matrix} \middle| q\right]_\infty \tag{3.3b}
$$

$$
\times \sum_{\tilde{m}\geq 0} \frac{(a/x_n y_n; q)_{m_n}}{(q;q)_{m_n}} \left[\begin{matrix} b, \quad c, \quad d \\ qbcd/a, a/x_n, a/y_n \end{matrix} \middle| q\right]_{M_n} q^{M_n} \tag{3.3c}
$$

$$
\times \Omega_{\tilde{m}}^{(qa)}\left[\begin{matrix} \{x_i, y_i\} \\ [1, \ell] \end{matrix}\right] \Omega_{\tilde{m}}^{(a)}\left[\begin{matrix} \{x_j, y_j\} \\ [1+\ell, n-1] \end{matrix}\right] \frac{1-q^{-M_\ell}e/a}{1-e/a} \tag{3.3d}
$$

provided that both series are well-defined and terminated by one of b, c or d.

4. Reductions and Consequences

The transformations displayed in the last section may be reduced by limiting process to several known and unknown results as consequences.

4.1 George Andrews

In Theorem 3.1, let $c \to \infty$ and $x_i, y_i \to \infty$ for $i = 2, 3, \ldots, n$. The result may be stated as

Proposition 4.1. *There holds the multiple basic hypergeometric transformation*

$$\sum_{\tilde{m}\geq 0} q^{M_n-M_n^2}\left[\begin{matrix} qa/xy \\ qa/x, qa/y \end{matrix}\Big| q\right]_{m_1} \frac{[b,d;q]_{M_n}}{(bd)^{M_n}} \prod_{i=1}^{n} \frac{a^{M_i}q^{M_i^2}}{(q;q)_{m_i}} = \left[\begin{matrix} qa/b, qa/d \\ qa,\ a/bd \end{matrix}\Big| q\right]_\infty$$
$$\times \sum_{k\geq 0} q^{(2n-1)\binom{k}{2}} \frac{1-aq^{2k}}{1-a} \left[\begin{matrix} a, & b, & d, & x, & y \\ q, & qa/b, & qa/d, & qa/x, & qa/y \end{matrix}\Big| q\right]_k \left\{\frac{-(qa)^{n+1}}{bdxy}\right\}^k.$$

This may be considered as an extension of Andrews (Andrews, 1975, Corollary 4.1).

If taking further $b = q^{-M}$, $d \to \infty$ and $x = 1$, $y \to \infty$, then Proposition 4.1 reduces to the following:

Corollary 4.2 (Andrews (Andrews, 1979b, Eq. 4.1) and (Andrews, 1981, Eqs. 1.5 & 3.1)).

$$\sum_{\tilde{m}\geq 0} \frac{(q^{1+M-M_n};q)_{M_n}}{(q\xi;\,q)_{m_1}} \prod_{k=1}^{n} \frac{q^{M_k^2}\xi^{M_k}}{(q;\,q)_{m_k}} = \frac{1}{(q\xi;\,q)_\infty}$$

whose limiting version reads as

$$\sum_{\tilde{m}\geq 0} \frac{1}{(q\xi;\,q)_{m_1}} \prod_{k=1}^{n} \frac{q^{M_k^2}\xi^{M_k}}{(q;\,q)_{m_k}} = \frac{1}{(q\xi;\,q)_\infty}.$$

Again in Theorem 3.1, let $c \to \infty$, and $x_i = -\sqrt{qa}$, $y_i \to \infty$ for $i = 2, 3, \ldots, n$. The result may be stated as

Proposition 4.3. *There holds the multiple basic hypergeometric transformation*

$$\sum_{\tilde{m}\geq 0} q^{-\binom{M_n}{2}} \left[\begin{matrix} qa/xy, -\sqrt{qa} \\ qa/x, qa/y \end{matrix}\Big| q\right]_{m_1} \left(\frac{\sqrt{qa}}{bd}\right)^{M_n}$$
$$\times \left[\begin{matrix} b, d \\ -\sqrt{qa} \end{matrix}\Big| q\right]_{M_n} \prod_{i=1}^{n} \frac{a^{M_i/2}q^{M_i^2/2}}{(q;q)_{m_i}} = \left[\begin{matrix} qa/b, qa/d \\ qa,\ a/bd \end{matrix}\Big| q\right]_\infty$$
$$\times \sum_{k\geq 0} q^{n\binom{k}{2}} \frac{1-aq^{2k}}{1-a} \left[\begin{matrix} a, & b, & d, & x, & y \\ q, qa/b, qa/d, qa/x, qa/y \end{matrix}\Big| q\right]_k \left\{\frac{-(qa)^{\frac{n+3}{2}}}{bdxy}\right\}^k.$$

4.2 David Bressoud

Letting $e \to 0$ and $c \to \infty$ in Theorem 3.3, we may state the result as

Theorem 4.4 (Almost-poised transformation). *There holds the multiple basic hypergeometric transformation*

$$\sum_{\tilde{m}\geq 0}\frac{(a/x_ny_n;q)_{m_n}}{(q;q)_{m_n}}\left[\begin{matrix}b,\quad d\\ a/x_n,a/y_n\end{matrix}\Bigg|q\right]_{M_n}\left(\frac{a}{bd}\right)^{M_n}$$
$$\times\,\Omega^{(qa)}_{\tilde{m}}\left[\begin{matrix}\{x_i,y_i\}\\ [1,\ell]\end{matrix}\right]\Omega^{(a)}_{\tilde{m}}\left[\begin{matrix}\{x_j,y_j\}\\ [1+\ell,n-1]\end{matrix}\right]$$
$$=\left[\begin{matrix}a/b,a/d\\ a,\ a/bd\end{matrix}\Bigg|q\right]_\infty\sum_{k\geq 0}(-1)^k\frac{(a;q)_k}{(q;q)_k}\Lambda^{(qa)}_k\left[\begin{matrix}\{x_i,y_i\}\\ [1,\ell]\end{matrix}\right]q^{\binom{1+k}{2}}$$
$$\times\,\nabla\left\{\Lambda^{(a)}_k\left[\begin{matrix}\{x_j,y_j\}\\ [1+\ell,n]\end{matrix}\right]\left[\begin{matrix}b,\quad d\\ a/b,\ a/d\end{matrix}\Bigg|q\right]_k\left(\frac{a}{bd}\right)^k\right\}$$

The central theorem of Bressoud (Bressoud, 1980a, Thm. 1) is equivalent to the limiting case $b\to\infty$ and $x_i,y_i\to\infty$ for $1<i\leq\ell$. In Theorem 4.4, taking $x_i,y_i\to\infty$ further for $i=2,3,\ldots,n$, we obtain the following

Proposition 4.5. *There holds the multiple basic hypergeometric transformation*

$$\sum_{\tilde{m}\geq 0}\frac{[b,d;q]_{M_n}}{(bd)^{M_n}}\left[\begin{matrix}qa/xy\\ qa/x,qa/y\end{matrix}\Bigg|q\right]_{m_1}q^{-M_n^2-\sum_{\iota=\ell+1}^{n-1}M_\iota}\prod_{i=1}^{n}\frac{a^{M_i}q^{M_i^2}}{(q;q)_{m_i}}\tag{4.1a}$$
$$=\left[\begin{matrix}a/b,a/d\\ a,\ a/bd\end{matrix}\Bigg|q\right]_\infty\sum_{k\geq 0}\left[\begin{matrix}a,\ b,\quad d,\quad x,\quad y\\ q,\ a/b,\ a/d,\ qa/x,\ qa/y\end{matrix}\Bigg|q\right]_k\left(\frac{-qa^{n+1}}{bdxy}\right)^k\tag{4.1b}$$
$$\times\,q^{k\ell+(2n-1)\binom{k}{2}}\left\{1-\frac{a}{bd}\frac{(1-bq^k)(1-dq^k)}{(1-q^ka/b)(1-q^ka/d)}\left(q^{2k}a\right)^{n-\ell}\right\}.\tag{4.1c}$$

The limiting case $b,\ d,\ x,\ y\to\infty$ of Proposition 4.5 reads, with replacement $a\to q\xi$ as a formula on Alder polynomials

Corollary 4.6 (Andrews (Andrews, 1974, Eq. 2.5)).

$$\sum_{\tilde{m}\geq 0}q^{\sum_{\iota=1}^{\ell}M_\iota}\prod_{k=1}^{n}\frac{q^{M_k^2}\xi^{M_k}}{(q;\,q)_{m_k}}=\frac{1}{(q\xi;q)_\infty}\sum_{k=0}^{\infty}\xi^{k(n+1)}\,q^{\binom{k+1}{2}+k^2(n+1)+k\ell}$$
$$\times\,(-1)^k\frac{(q\xi;\,q)_k}{(q;\,q)_k}\left\{1-\left(q^{1+2k}\xi\right)^{1+n-\ell}\right\}.$$

4.3 Other transformations

For $c\to\infty$, Theorem 3.2 may be restated as

Theorem 4.7 (Almost-poised transformation). *There holds the multiple basic hypergeometric transformation*

$$\sum_{\tilde{m}\geq 0}\left(1-q^{M_n-M_\ell}\right)\frac{(a/x_ny_n;q)_{m_n}}{(q;q)_{m_n}}\left[\begin{matrix}b/q,\ d/q\\ a/x_n,\ a/y_n\end{matrix}\Big| q\right]_{M_n}\left(\frac{qa}{bd}\right)^{M_n}$$

$$\times\Omega_{\tilde{m}}^{(qa)}\left[\begin{matrix}\{x_i,y_i\}\\ [1,\ell]\end{matrix}\right]\Omega_{\tilde{m}}^{(a)}\left[\begin{matrix}\{x_j,y_j\}\\ [1+\ell,n-1]\end{matrix}\right]=\frac{(1-q/b)(1-q/d)}{q/a}$$

$$\times\left[\begin{matrix}qa/b,qa/d\\ a,\ qa/bd\end{matrix}\Big| q\right]_\infty\sum_{k\geq 0}(-1)^k\left[\begin{matrix}a,\quad b,\quad d\\ q,qa/b,qa/d\end{matrix}\Big| q\right]_k\left(\frac{qa}{bd}\right)^k$$

$$\times\Lambda_k^{(qa)}\left[\begin{matrix}\{x_i,y_i\}\\ [1,\ell]\end{matrix}\right]\nabla\left\{\Lambda_k^{(a)}\left[\begin{matrix}\{x_j,y_j\}\\ [1+\ell,n]\end{matrix}\right]\right\}q^{\binom{1+k}{2}}.$$

Its further limiting case $x_i,y_i\to\infty$ for $i=2,3,\ldots,n$ reads as

Proposition 4.8. *There holds the multiple basic hypergeometric transformation*

$$\sum_{\tilde{m}\geq 0}\left(1-q^{M_n-M_\ell}\right)\frac{[b/q,d/q;q]_{M_n}}{(bd)^{M_n}}\left[\begin{matrix}qa/xy\\ qa/x,qa/y\end{matrix}\Big| q\right]_{m_1}$$

$$\times q^{M_n-M_n^2-\sum_{\iota=\ell+1}^{n-1}M_\iota}\prod_{i=1}^{n}\frac{a^{M_i}q^{M_i^2}}{(q;q)_{m_i}}$$

$$=\left[\begin{matrix}qa/b,qa/d\\ a,\ qa/bd\end{matrix}\Big| q\right]_\infty\sum_{k\geq 0}q^{k\ell+(2n-1)\binom{k}{2}}\left\{1-\left(q^{2k}a\right)^{n-\ell}\right\}$$

$$\times\frac{(1-q/b)(1-q/d)}{q/a}\left[\begin{matrix}a,\quad b,\quad d,\quad x,\quad y\\ q,\ qa/b,\ qa/d,\ qa/x,\ qa/y\end{matrix}\Big| q\right]_k\left(\frac{-q^2a^{n+1}}{bdxy}\right)^k.$$

Letting c, $e\to\infty$ in Theorem 3.3, we may state the result as

Theorem 4.9 (Almost-poised transformation). *There holds the multiple basic hypergeometric transformation*

$$\sum_{\tilde{m}\geq 0}\frac{(a/x_ny_n;q)_{m_n}}{(q;q)_{m_n}}\left[\begin{matrix}b,\quad d\\ a/x_n,a/y_n\end{matrix}\Big| q\right]_{M_n}\left(\frac{a}{bd}\right)^{M_n}$$

$$\times\Omega_{\tilde{m}}^{(qa)}\left[\begin{matrix}\{x_i,y_i\}\\ [1,\ell]\end{matrix}\right]\Omega_{\tilde{m}}^{(a)}\left[\begin{matrix}\{x_j,y_j\}\\ [1+\ell,n-1]\end{matrix}\right]q^{-M_\ell}$$

$$=\left[\begin{matrix}a/b,a/d\\ a,\ a/bd\end{matrix}\Big| q\right]_\infty\sum_{k\geq 0}\left(\frac{-1}{a}\right)^k\frac{(a;q)_k}{(q;q)_k}\Lambda_k^{(qa)}\left[\begin{matrix}\{x_i,y_i\}\\ [1,\ell]\end{matrix}\right]q^{-\binom{k}{2}}$$

$$\times\nabla\left\{\Lambda_k^{(a)}\left[\begin{matrix}\{x_j,y_j\}\\ [1+\ell,n]\end{matrix}\right]\left[\begin{matrix}b,\quad d\\ a/b,\ a/d\end{matrix}\Big| q\right]_k\left(\frac{q^{k-1}a^2}{bd}\right)^k\right\}$$

Its further limiting case $x_i, y_i \to \infty$ for $i = 2, 3, \ldots, n$ of this theorem reads as

Proposition 4.10. *There holds the multiple basic hypergeometric transformation*

$$\sum_{\tilde{m}\geq 0} \frac{[b,d;q]_{M_n}}{(bd)^{M_n}} \left[\begin{matrix} qa/xy \\ qa/x, qa/y \end{matrix} \middle| q\right]_{m_1} q^{-M_n^2 - \sum_{\iota=\ell}^{n-1} M_\iota} \prod_{i=1}^{n} \frac{a^{M_i} q^{M_i^2}}{(q;q)_{m_i}}$$
$$= \left[\begin{matrix} a/b, a/d \\ a, a/bd \end{matrix} \middle| q\right]_\infty \sum_{k\geq 0} (-1)^k \left\{1 - \frac{a}{bd} \frac{(1-q^k b)(1-q^k d)}{(1-q^k a/b)(1-q^k a/d)} (q^{2k} a)^{1+n-\ell}\right\}$$
$$\times\ q^{\binom{k}{2}(2n-1)} \left[\begin{matrix} a, & b, & d, & x, & y \\ q, & a/b, & a/d, & qa/x, & qa/y \end{matrix} \middle| q\right]_k \left(\frac{a^{1+n} q^\ell}{bdxy}\right)^k.$$

5. Multiple Rogers-Ramanujan Identities

Under various specifications, the k-sums appeared in the propositions demonstrated in last section may be evaluated in closed forms by the Jacobi triple product identity (cf. Bailey (Bailey, 1935, Chapter 8))

$$[q, x, q/x; q]_\infty = \sum_{m=-\infty}^{+\infty} (-1)^m q^{\binom{m}{2}} x^m$$
$$= \sum_{n=0}^{\infty} (-1)^n \left\{1 - (q/x)^{1+2n}\right\} q^{\binom{n}{2}} x^n$$

which allows us to derive the following multiple Rogers-Ramanujan identities:

Example 1 ($a = 1$: $b, d \to \infty$ and $x, y = \pm q^{1/2}$ in Proposition 4.1).

$$\sum_{\tilde{m}\geq 0} \frac{(-1;q)_{m_1}}{(q;q^2)_{m_1}} \prod_{k=1}^{n} \frac{q^{M_k^2}}{(q;q)_{m_k}} = \frac{[q^{1+2n}, -q^n, -q^{1+n}; q^{1+2n}]_\infty}{(q;q)_\infty}.$$

Example 2 ($a = q$: $b, d \to \infty$ and $x, y = \pm q^{1/2}$ in Proposition 4.1).

$$\sum_{\tilde{m}\geq 0} \frac{(-q;q)_{m_1}}{(q;q^2)_{1+m_1}} \prod_{k=1}^{n} \frac{q^{M_k(1+M_k)}}{(q;q)_{m_k}} = \frac{[q^{1+2n}, -q^{1+2n}, -q^{1+2n}; q^{1+2n}]_\infty}{(q;q)_\infty}.$$

Example 3 ($a = q$: $x = -q$ and $b, d, y \to \infty$ in Proposition 4.5).

$$\sum_{\tilde{m}\geq 0} \frac{q^{\sum_{\iota=1}^{\ell} M_\iota}}{(-q;q)_{m_1}} \prod_{k=1}^{n} \frac{q^{M_k^2}}{(q;q)_{m_k}} = \frac{[q^{2+2n}, q^{n+\ell+1}, q^{1+n-\ell}; q^{2+2n}]_\infty}{(q;q)_\infty}$$

It has first been discovered in (Agarwal and Bressoud, 1989, Eq. 1.1), (Bressoud, 1980a, Eq. 3.4), (Bressoud, 1980b, Thm. 1: $\delta = 0$), (Bressoud, 1981, Eq. 5.6), (Bressoud, 1989, Eq. 1.1) and (Bressoud et al., 2000, p. 7-2) by Bressoud et al.

Example 4 ($a = q$: $x = -q$ **and** b, d, $y \to \infty$ **in Proposition 4.8; cf. (Stembridge, 1990, Eq. I-11)).**

$$\sum_{\tilde{m}\geq 0} \frac{q^{-M_n} - q^{-M_\ell}}{(-q;q)_{m_1}} q^{\sum_{\iota=1}^{\ell} M_\iota} \prod_{k=1}^{n} \frac{q^{M_k^2}}{(q;q)_{m_k}} = \frac{\left[q^{2+2n},\, q^{n-\ell},\, q^{n+\ell+2}; q^{2+2n}\right]_\infty}{(q;q)_\infty}.$$

Example 5 ($a = q$: b, d, x, $y \to \infty$ **in Proposition 4.5).**

$$\sum_{\tilde{m}\geq 0} q^{\sum_{\iota=1}^{\ell} M_\iota} \prod_{k=1}^{n} \frac{q^{M_k^2}}{(q;q)_{m_k}} = \frac{\left[q^{3+2n},\, q^{n+\ell+2},\, q^{1+n-\ell}; q^{3+2n}\right]_\infty}{(q;q)_\infty}$$

As a common generalization of the Rogers-Ramanujan identities (1.5a)-(1.5b), this one has been found in (Agarwal et al., 1987, Eq. 1.5), (Agarwal and Bressoud, 1989, Eq. 0.3), (Andrews, 1974, Eq. 1.7), (Andrews, 1976, Eq. 7.3.7), (Andrews, 1979b, Eq. 1.3), (Andrews, 1981, Eq. 1.1), (Andrews, 1984, Eq. 1.3), (Andrews, 1986, Eqs. 3.45–46), (Bressoud, 1980a, Eq. 3.2), (Bressoud, 1980b, Thm. 1: $\delta = 1$), (Bressoud, 1981, Eqs. 5.3 & 6.1), (Bressoud, 1989, Eq. 0.3), (Bressoud et al., 2000, pp. 4-1 & 7-1), (Garrett et al., 1999, Eq. 4.1) and (Stembridge, 1990, Eqs. c–d) mainly by Andrews and Bressoud et al.

Example 6 ($a = q$: b, d, x, $y \to \infty$ **in Proposition 4.8).**

$$\sum_{\tilde{m}\geq 0} \left(q^{-M_n} - q^{-M_\ell}\right) q^{\sum_{\iota=1}^{\ell} M_\iota} \prod_{k=1}^{n} \frac{q^{M_k^2}}{(q;q)_{m_k}} = \frac{\left[q^{3+2n}, q^{n-\ell}, q^{n+\ell+3}; q^{3+2n}\right]_\infty}{(q;q)_\infty}.$$

Example 7 ($a = q$: $b = -q$, $d \to \infty$ **and** x, $y = \pm q^{1/2}$ **in Proposition 4.3).**

$$\sum_{\tilde{m}\geq 0} \frac{(-q;q)_{m_1}^2}{(q;q^2)_{1+m_1}} \prod_{k=1}^{n} \frac{q^{\binom{1+M_k}{2}}}{(q;q)_{m_k}} = \frac{(-q;q)_\infty}{(q;q)_\infty} \left[q^{1+n}, -q^{1+n}, -q^{1+n}; q^{1+n}\right]_\infty.$$

Example 8 ($a = q$: $b = -q$, $d \to \infty$ **and** x, $y = \pm q^{1/2}$ **in Proposition 4.1).**

$$\sum_{\tilde{m}\geq 0} \frac{(-q;q)_{m_1}(-q;q)_{M_n}}{(q;q^2)_{1+m_1}} q^{-\binom{1+M_n}{2}} \prod_{k=1}^{n} \frac{q^{M_k(1+M_k)}}{(q;q)_{m_k}}$$
$$= \frac{(-q;q)_\infty}{(q;q)_\infty} \left[q^{2n}, -q^{2n}, -q^{2n}; q^{2n}\right]_\infty.$$

Example 9 ($a = 1$: $b = -q$, $d \to \infty$ and $x = y = -1$ in Proposition 4.1).

$$\sum_{\tilde{m}\geq 0} \frac{(q;\,q)_{m_1}(-q;\,q)_{M_n}}{(-q;\,q)^2_{m_1}} q^{-\binom{1+M_n}{2}} \prod_{k=1}^{n} \frac{q^{M_k^2}}{(q;\,q)_{m_k}}$$
$$= 2\frac{(-q;\,q)_\infty}{(q;\,q)_\infty} \left[q^{2n},\, q^n,\, q^n; q^{2n}\right]_\infty.$$

Example 10 ($a = q$: $b = x = -q$ and $d, y \to \infty$ in Proposition 4.8).

$$\sum_{\tilde{m}\geq 0} \left(q^{-M_n} - q^{-M_\ell}\right) \frac{(-1;q)_{M_n}}{(-q;q)_{m_1}} q^{-\binom{M_n}{2}+\sum_{\iota=1}^{\ell} M_\iota} \prod_{k=1}^{n} \frac{q^{M_k^2}}{(q;q)_{m_k}}$$
$$= 2\frac{(-q;q)_\infty}{(q;q)_\infty} \left[q^{1+2n},\, q^{n-\ell},\, q^{n+\ell+1}; q^{1+2n}\right]_\infty.$$

Example 11 ($a = q$: ${}^{b=-q,\, d=-q^{1/2}}_{x=q^{1/2},\, y\to\infty} \mid q \to q^2$ in Proposition 4.3).

$$\sum_{\tilde{m}\geq 0} \frac{(-q^2;\,q^2)_{m_1}(-q;\,q^2)_{M_n}}{(q;\,q^2)_{1+m_1}} q^{-M_n^2} \prod_{k=1}^{n} \frac{q^{M_k(1+M_k)}}{(q^2;\,q^2)_{m_k}}$$
$$= \frac{(-q;\,q)_\infty}{(q;\,q)_\infty} \left[q^{2+2n},\, -q^{2+2n},\, -q^{2+2n}; q^{2+2n}\right]_\infty.$$

Example 12 ($a = q$: $b = -q$ and $d, x, y \to \infty$ in Proposition 4.8).

$$\sum_{\tilde{m}\geq 0} \left(q^{-M_n} - q^{-M_\ell}\right) (-1;q)_{M_n}\, q^{-\binom{M_n}{2}+\sum_{\iota=1}^{\ell} M_\iota} \prod_{k=1}^{n} \frac{q^{M_k^2}}{(q;q)_{m_k}}$$
$$= 2\frac{(-q;q)_\infty}{(q;q)_\infty} \left[q^{2+2n},\, q^{n-\ell},\, q^{n+\ell+2}; q^{2+2n}\right]_\infty.$$

Example 13 ($a = q$: ${}^{b=-q,\, d=-q^{1/2}}_{x=q^{1/2},\, y\to\infty} \mid q \to q^2$ in Proposition 4.1).

$$\sum_{\tilde{m}\geq 0} \frac{(-q;q)_{2M_n}}{(q;q^2)_{1+m_1}} q^{-M_n-2M_n^2} \prod_{k=1}^{n} \frac{q^{2M_k(1+M_k)}}{(q^2;q^2)_{m_k}}$$
$$= \frac{(-q;q)_\infty}{(q;q)_\infty} \left[q^{4n}, -q^{4n}, -q^{4n}; q^{4n}\right]_\infty.$$

Example 14 ($a=1$: $\left.{}^{b=-1,\,d=-q^{1/2}}_{x,\,y\to\infty}\right|\, q\to q^2$ **in Proposition 4.1).**

$$\sum_{\tilde m\geq 0}(-1;q)_{2M_n}\, q^{M_n-2M_n^2}\prod_{k=1}^{n}\frac{q^{2M_k^2}}{(q^2;q^2)_{m_k}}$$
$$=\frac{(-q;q)_\infty}{(q;q)_\infty}\left[q^{2+4n},\, q^{1+2n},\, q^{1+2n};q^{2+4n}\right]_\infty.$$

Example 15 ($a=1$: $\left.{}^{b,\,d\to\infty}_{x=-1,\,y=-q^{1/2}}\right|\, q\to q^2$ **in Proposition 4.1).**

$$\sum_{\tilde m\geq 0}\frac{(q;q^2)_{m_1}}{(-q;q)_{2m_1}}\prod_{k=1}^{n}\frac{q^{2M_k^2}}{(q^2;q^2)_{m_k}}=\frac{\left[q^{2+4n},q^{1+2n},q^{1+2n};q^{2+4n}\right]_\infty}{(q^2;q^2)_\infty}.$$

Example 16 ($a=1$: $\left.{}^{b,\,d\to\infty}_{x=q^{1/2},\,y\to\infty}\right|\, q\to q^2$ **in Proposition 4.1; cf. (Bressoud et al., 2000, p. 8-2)).**

$$\sum_{\tilde m\geq 0}\frac{1}{(q;q^2)_{m_1}}\prod_{k=1}^{n}\frac{q^{2M_k^2}}{(q^2;q^2)_{m_k}}=\frac{\left[q^{4+4n},-q^{1+2n},-q^{3+2n};q^{4+4n}\right]_\infty}{(q^2;q^2)_\infty}.$$

Example 17 ($a=1$: $\left.{}^{b=-q^{1/2},\,d\to\infty}_{x=-1,\,y=-\varepsilon q^{1/2}}\right|\, q\to q^2$ **in Proposition 4.3).**

$$\sum_{\tilde m\geq 0}\frac{(-q;q^2)_{m_1}(\varepsilon q;q^2)_{m_1}}{(-q^2;q^2)_{m_1}(-\varepsilon q;q^2)_{m_1}}\prod_{k=1}^{n}\frac{q^{M_k^2}}{(q^2;q^2)_{m_k}}$$
$$\overset{\varepsilon=\pm 1}{=\!=\!=}\frac{(-q;q^2)_\infty}{(q^2;q^2)_\infty}\left[q^{2+2n},\varepsilon q^{1+n},\varepsilon q^{1+n};q^{2+2n}\right]_\infty.$$

Example 18 ($a=1$: $\left.{}^{b=-q^{1/2},\,d\to\infty}_{x=-1,\,y\to\infty}\right|\, q\to q^2$ **in Proposition 4.3).**

$$\sum_{\tilde m\geq 0}\frac{(-q;q^2)_{m_1}}{(-q^2;q^2)_{m_1}}\prod_{k=1}^{n}\frac{q^{M_k^2}}{(q^2;q^2)_{m_k}}=\frac{(-q;q^2)_\infty}{(q^2;q^2)_\infty}\left[q^{4+2n},q^{2+n},q^{2+n};q^{4+2n}\right]_\infty.$$

Example 19 ($a=q$: $\left.{}^{-b=q^{1/2},\,d\to\infty}_{x=-q,\,y=q^{1/2}}\right|\, q\to q^2$ **in Proposition 4.1).**

$$\sum_{\tilde m\geq 0}\frac{(-q;q^2)_{m_1}(-q;q^2)_{M_n}}{(q;q^2)_{1+m_1}(-q^2;q^2)_{m_1}}q^{-M_n^2}\prod_{k=1}^{n}\frac{q^{2M_k(1+M_k)}}{(q^2;q^2)_{m_k}}$$
$$=\frac{(-q;q^2)_\infty}{(q^2;q^2)_\infty}\left[q^{4n},\,-q^{4n},\,-q^{4n};q^{4n}\right]_\infty.$$

Example 20 ($a=q$: $\substack{b=-q^{3/2},\, d\to\infty \\ x=-q,\, y=-q^{1/2}} \Big|\, q\to q^2$ **in Proposition 4.8).**

$$q\sum_{\tilde{m}\geq 0}\left(q^{-2M_n}-q^{-2M_\ell}\right)\frac{\left(-q;q^2\right)_{M_n}\left(q;q^2\right)_{m_1}}{(-q;q)_{1+2m_1}}q^{-M_n^2+2\sum_{\iota=1}^{\ell}M_\iota}\prod_{k=1}^{n}\frac{q^{2M_k^2}}{\left(q^2;q^2\right)_{m_k}}$$
$$=\frac{\left(-q;q^2\right)_\infty}{\left(q^2;q^2\right)_\infty}\left[q^{4n},\ q^{2n-2\ell},\ q^{2n+2\ell};q^{4n}\right]_\infty.$$

Example 21 ($a=q$: $\substack{b=-q^{1/2},\, d\to\infty \\ x=q^{1/2},\, y\to\infty} \Big|\, q\to q^2$ **in Proposition 4.1).**

$$\sum_{\tilde{m}\geq 0}\frac{(-q;\,q^2)_{M_n}}{(q;\,q^2)_{1+m_1}}q^{-M_n^2}\prod_{k=1}^{n}\frac{q^{2M_k(1+M_k)}}{(q^2;\,q^2)_{m_k}}$$
$$=\frac{(-q;\,q^2)_\infty}{(q^2;\,q^2)_\infty}\left[q^{2+4n},\ -q^{2+4n},\ -q^{2+4n};q^{2+4n}\right]_\infty.$$

Example 22 ($a=q$: $\substack{b=-q^{1/2},\, d\to\infty \\ x=-q,\, y\to\infty} \Big|\, q\to q^2$ **in Proposition 4.5; cf. (Bressoud, 1980a, Eq. 3.9)).**

$$\sum_{\tilde{m}\geq 0}\frac{\left(-q;q^2\right)_{M_n}}{\left(-q^2;q^2\right)_{m_1}}q^{-M_n^2+2\sum_{\iota=1}^{\ell}M_\iota}\prod_{k=1}^{n}\frac{q^{2M_k^2}}{\left(q^2;q^2\right)_{m_k}}$$
$$=\frac{\left(-q;q^2\right)_\infty}{\left(q^2;q^2\right)_\infty}\left[q^{2+4n},\ q^{2n-2\ell+1},\ q^{1+2n+2\ell};q^{2+4n}\right]_\infty.$$

Example 23 ($a=q$: $\substack{b=-q^{1/2},\, d\to\infty \\ x=-q^{3/2},\, y\to\infty} \Big|\, q\to q^2$ **in Proposition 4.8).**

$$\sum_{\tilde{m}\geq 0}\left(1-q^{2M_n-2M_\ell}\right)\frac{\left(-q^{-1};q^2\right)_{M_n}}{\left(-q;q^2\right)_{m_1}}q^{-M_n^2+2\sum_{\iota=1}^{\ell}M_\iota}\prod_{k=1}^{n}\frac{q^{2M_k^2}}{\left(q^2;q^2\right)_{m_k}}$$
$$=\frac{\left(-q;q^2\right)_\infty}{\left(q^2;q^2\right)_\infty}\left[q^{2+4n},\ q^{2n-2\ell},\ q^{2n+2\ell+2};q^{2+4n}\right]_\infty.$$

Example 24 ($a=q$: $\substack{b=-q^{1/2},\, d\to\infty \\ x,\, y\to\infty} \Big|\, q\to q^2$ **in Proposition 4.5).**

$$\sum_{\tilde{m}\geq 0}\left(-q;q^2\right)_{M_n}q^{-M_n^2+2\sum_{\iota=1}^{\ell}M_\iota}\prod_{k=1}^{n}\frac{q^{2M_k^2}}{\left(q^2;q^2\right)_{m_k}}$$
$$=\frac{\left(-q;q^2\right)_\infty}{\left(q^2;q^2\right)_\infty}\left[q^{4+4n},\ q^{2n-2\ell+1},\ q^{3+2n+2\ell};q^{4+4n}\right]_\infty$$

This identity appeared in (Agarwal and Bressoud, 1989, Eq. 1.2), (Andrews, 1975, Corollary 4.3), (Andrews, 1976, Eq. 7.4.4), (Bressoud, 1980a, Eq. 3.8) and (Bressoud, 1989, Eq. 1.2) due to Andrews and Bressoud et al.

Example 25 ($a = 1$: $\substack{b=-1,\, d\to\infty \\ x\to 0,\, y=-\varepsilon q^{1/2}} \big|\, q \to q^2$ **in Proposition 4.3**).

$$\sum_{\tilde{m}\geq 0}(-\varepsilon)^{m_1}\frac{(-q;\,q^2)_{m_1}(-1;\,q^2)_{M_n}}{(-q\varepsilon;\,q^2)_{m_1}(-q;\,q^2)_{M_n}}q^{M_n-m_1}\prod_{k=1}^{n}\frac{q^{M_k^2}}{(q^2;\,q^2)_{m_k}}$$
$$\overset{\varepsilon=\pm 1}{=\!=\!=}\frac{(-q^2;\,q^2)_\infty}{(q^2;\,q^2)_\infty}\left[q^{2n},\,\varepsilon q^n,\,\varepsilon q^n;q^{2n}\right]_\infty.$$

Example 26 ($a = 1$: $\substack{b=-1,\, d\to\infty \\ x,\, y=\pm q^{1/2}} \big|\, q \to q^2$ **in Proposition 4.3**).

$$\sum_{\tilde{m}\geq 0}\frac{(-1;\,q^2)_{M_n}(-1;\,q^2)_{m_1}}{(-q;\,q^2)_{M_n}(q;\,q^2)_{m_1}}q^{M_n}\prod_{k=1}^{n}\frac{q^{M_k^2}}{(q^2;\,q^2)_{m_k}}$$
$$=\frac{(-q^2;\,q^2)_\infty}{(q^2;\,q^2)_\infty}\left[q^{2+2n},\,-q^{1+n},\,-q^{1+n};q^{2+2n}\right]_\infty.$$

Example 27 ($a = q$: $\substack{b=-q,\, d\to\infty \\ x=-q^{1/2},\, y\to\infty} \big|\, q \to q^2$ **in Proposition 4.3**).

$$\sum_{\tilde{m}\geq 0}\frac{(-q^2;\,q^2)_{m_1}}{(-q;\,q^2)_{1+m_1}}\prod_{k=1}^{n}\frac{q^{M_k(1+M_k)}}{(q^2;\,q^2)_{m_k}}=\frac{(-q^2;\,q^2)_\infty}{(q^2;\,q^2)_\infty}\left[q^{4+2n},\,q,\,q^{3+2n};q^{4+2n}\right]_\infty.$$

Example 28 ($a = 1$: $\substack{b=-1,\, d\to\infty \\ x,\, y\to\infty} \big|\, q \to q^2$ **in Proposition 4.3**).

$$\sum_{\tilde{m}\geq 0}\frac{(-1;\,q^2)_{M_n}(-q;\,q^2)_{m_1}}{(-q;\,q^2)_{M_n}}q^{M_n}\prod_{k=1}^{n}\frac{q^{M_k^2}}{(q^2;\,q^2)_{m_k}}$$
$$=\frac{(-q^2;\,q^2)_\infty}{(q^2;\,q^2)_\infty}\left[q^{6+2n},\,q^{3+n},\,q^{3+n};q^{6+2n}\right]_\infty.$$

Example 29 ($a = 1$: $b = d = -1$ **and** $x = -q$, $y \to \infty$ **in Proposition 4.1**).

$$\sum_{\tilde{m}\geq 0}\frac{(-1;\,q)_{M_n}^2}{(-1;\,q)_{m_1}}q^{M_n-M_n^2}\prod_{k=1}^{n}\frac{q^{M_k^2}}{(q;\,q)_{m_k}}=\frac{(-q;\,q)_\infty^2}{(q;\,q)_\infty^2}\left[q^{2n},\,q^n,\,q^n;q^{2n}\right]_\infty.$$

Example 30 ($a = 1$: $\left.{}^{b=d=-1}_{x=-q,\, y\to\infty}\right| q \to q^2$ **in Proposition 4.3).**

$$\sum_{\tilde{m}\geq 0} \frac{(-1;\, q^2)^2_{M_n}(-q;\, q^2)_{m_1}}{(-q;\, q^2)_{M_n}(-1;\, q^2)_{m_1}} q^{2M_n - M_1^n} \prod_{k=1}^{n} \frac{q^{M_k^2}}{(q^2;\, q^2)_{m_k}}$$
$$= \frac{(-q^2;\, q^2)^2_\infty}{(q^2;\, q^2)^2_\infty} \left[q^{2+2n},\, q^{1+n},\, q^{1+n}; q^{2+2n}\right]_\infty .$$

These examples are selected from about two hundreds multiple Rogers-Ramanujan identities derived from the propositions displayed in the last section. More identities of such kind may be found in (Agarwal and Bressoud, 1989), (Andrews, 1984), (Bressoud, 1980a; Bressoud, 1989) and (Stembridge, 1990), mainly due to Andrews and Bressoud. For the most recent development, we refer to (Bressoud et al., 2000), (Garrett et al., 1999), (Schilling and Warnaar, 1997), (Stanton, 2001) and (Warnaar, 2001).

Acknowledgments

The author thanks to the anonymous referee for the careful corrections and useful suggestions.

References

Agarwal, A. K., Andrews, G. E., and Bressoud, D. M. (1987). The Bailey lattice. *J. Indian Math. Soc. (N.S.)*, 51:57–73.

Agarwal, A. K. and Bressoud, D. M. (1989). Lattice paths and multiple basic hypergeometric series. *Pacific J. Math.*, 136:209–228.

Andrews, G. E. (1974). An analytic generalization of the Rogers-Ramanujan identities for odd moduli. *Proc. Nat. Acad. Sci. U.S.A.*, 71:4082–4085.

Andrews, G. E. (1975). Problems and prospects for basic hypergeometric functions. In Askey, R., editor, *The Theory and Applications of Special Functions*, pages 191–224. Academic Press, New York.

Andrews, G. E. (1976). *The Theory of Partitions*, volume 2 of *Encyclopedia of Math. and Appl.* Cambridge University Press, Cambridge.

Andrews, G. E. (1979a). Connection coefficient problems and partitions. In *Relations between combinatorics and other parts of mathematics (Proc. Sympos. Pure Math., Ohio State Univ., Columbus, Ohio, 1978)*, Proc. Sympos. Pure Math., XXXIV, pages 1–24. Amer. Math. Soc., Providence, RI.

Andrews, G. E. (1979b). Partitions and Durfee dissection. *Amer. J. Math.*, 101(3):735–742.

Andrews, G. E. (1981). Multiple q-series identities. *Houston J. Math.*, 7(1):11–22.

Andrews, G. E. (1984). Multiple series Rogers-Ramanujan type identities. *Pacific J. Math.*, 114(2):267–283.

Andrews, G. E. (1986). *q-Series: Their Development and Application in Analysis, Number Theory, Combinatorics, Physics, and Computer Algebra*, volume 66 of

CBMS Regional Conference Series in Mathematics. Published for the Conference Board of the Mathematical Sciences, Washington, DC.

Bailey, W. N. (1935). *Generalized Hypergeometric Series.* Cambridge University Press, Cambridge.

Bailey, W. N. (1947). Some identities in combinatory analysis. *Proc. London Math. Soc. (2)*, 49:421–435.

Bailey, W. N. (1948). Identities of the Rogers-Ramanujan type. *Proc. London Math. Soc. (2)*, 50:1–10.

Bressoud, D. M. (1980a). Analytic and combinatorial generalizations of the Rogers-Ramanujan identities. *Mem. Amer. Math. Soc.*, 24(227):54.

Bressoud, D. M. (1980b). An analytic generalization of the Rogers-Ramanujan identities with interpretation. *Quart. J. Math. Oxford Ser. 2*, 31:385–399.

Bressoud, D. M. (1981). On partitions, orthogonal polynomials and the expansion of certain infinite products. *Proc. London Math. Soc. (3)*, 42:478–500.

Bressoud, D. M. (1988). The Bailey lattice: an introduction. In *Ramanujan Revisited (Urbana-Champaign, Ill., 1987)*, pages 57–67. Academic Press, Boston, MA.

Bressoud, D. M. (1989). Lattice paths and the Rogers-Ramanujan identities. In *Number theory, Madras 1987*, volume 1395 of *Lecture Notes in Mathematics*, pages 140–172. Springer-Verlag, Berlin-New York.

Bressoud, D. M., Ismail, M., and Stanton, D. (2000). Change of base in Bailey pairs. *Ramanujan J.*, 4(4):435–453.

Chu, W. (2002). The Saalschütz chain reactions and bilateral basic hypergeometric series. *Constructive Approx.*, 18:579–598.

Garrett, T., Ismail, M., and Stanton, D. (1999). Variants of the Rogers-Ramanujan identities. *Adv. Appl. Math.*, 23(3):274–299.

Gessel, I. and Stanton, D. (1983). Applications of q-Lagrange inversion to basic hypergeometric series. *Trans. Amer. Math. Soc.*, 277:173–201.

Milne, S. C. (1980). A multiple series transformation of the very well poised ${}_{4+2k}\Psi_{2k+4}$. *Pacific J. Math.*, 91:419–430.

Schilling, A. and Warnaar, S. O. (1997). A higher-level Bailey lemma: Proof and application. Preprint.

Slater, L. J. (1951). A new proof of Rogers' transformations of infinite series. *Proc. London Math. Soc. (2)*, 53:460–475.

Slater, L. J. (1952). Further identities of the Rogers-Ramanujan type. *Proc. London Math. Soc. (2)*, 54:147–167.

Slater, L. J. (1966). *Generalized Hypergeometric Functions.* Cambridge University Press, Cambridge.

Stanton, D. (2001). *The Bailey-Rogers-Ramanujan group*, volume 291 of *Contemporary Mathematics*, pages 55–70. Amer. Math. Soc., Providence, RI.

Stembridge, J. R. (1990). Hall-Littlewood functions, plane partitions, and the Rogers-Ramanujan identities. *Trans. Amer. Math. Soc.*, 319:469–498.

Warnaar, S. O. (2001). *The generalized Borwein conjecture I: The Burge transform*, volume 291 of *Contemporary Mathematics*, pages 243–267. Amer. Math. Soc., Providence, RI.

PAINLEVÉ EQUATIONS AND ASSOCIATED POLYNOMIALS

Peter A. Clarkson
Institute of Mathematics, Statistics & Actuarial Science
University of Kent
Canterbury, CT2 7NF
UNITED KINGDOM
P.A.Clarkson@kent.ac.uk

Abstract In this paper we are concerned with rational solutions and associated polynomials for the second, third and fourth Painlevé equations. These rational solutions are expressible as in terms of special polynomials. The structure of the roots of these polynomials is studied and it is shown that these have a highly regular structure.

1. Introduction

In this paper we discuss hierarchies of rational solutions and associated polynomials for the second, third and fourth Painlevé equations ($\mathrm{P_{II}}$–$\mathrm{P_{IV}}$)

$$w'' = 2w^3 + zw + \alpha, \tag{1.1}$$

$$w'' = \frac{(w')^2}{w} - \frac{w'}{z} + \frac{\alpha w^2 + \beta}{z} + \gamma w^3 + \frac{\delta}{w}, \tag{1.2}$$

$$w'' = \frac{(w')^2}{2w} + \frac{3}{2}w^3 + 4zw^2 + 2\left(z^2 - \alpha\right)w + \frac{\beta}{w}, \tag{1.3}$$

where $' = \mathrm{d}/\mathrm{d}z$ and α, β, γ and δ are arbitrary constants.

The six Painlevé equations ($\mathrm{P_I}$–$\mathrm{P_{VI}}$), were discovered by Painlevé, Gambier and their colleagues whilst studying second order ordinary differential equations of the form

$$w'' = F\left(z, w, w'\right), \tag{1.4}$$

where F is rational in w' and w and analytic in z. The Painlevé equations can be thought of as nonlinear analogues of the classical special functions. Indeed Iwasaki, Kimura, Shimomura and Yoshida (Iwasaki et al., 1991) characterize the six Painlevé equations as "the most important nonlinear ordinary differential equations" and state that "many specialists believe that during the twenty-first century the Painlevé functions will become new members of the community of special functions." The general solutions of the Painlevé equations are transcendental in the sense that they cannot be expressed in terms of (known) classical functions and so require the introduction of a new transcendental function to describe their solution. However it is well-known that P_{II}–P_{VI}, possess hierarchies of rational solutions for special values of the parameters (see, for example, (Airault, 1979; Albrecht et al., 1996; Bassom et al., 1995; Fokas and Ablowitz, 1982; Fukutani et al., 2000; Gromak, 1999; Gromak et al., 2002; Okamoto, 1987a; Okamoto, 1987b; Okamoto, 1986; Okamoto, 1987c; Umemura and Watanabe, 1997; Umemura and Watanabe, 1998; Vorob'ev, 1965; Watanabe, 1995; Yablonskii, 1959; Yuan and Li, 2002) and the references therein). These hierarchies are usually generated from "seed solutions" using the associated Bäcklund transformations and frequently can be expressed in the form of determinants through "τ-functions".

Vorob'ev (Vorob'ev, 1965) and Yablonskii (Yablonskii, 1959) expressed the rational solutions of P_{II} in terms of the logarithmic derivative of certain polynomials which are now known as the *Yablonskii-Vorob'ev polynomials.* Okamoto (Okamoto, 1986) obtained analogous polynomials related to some of the rational solutions of P_{IV}, these polynomials are now known as the *Okamoto polynomials.* Further Okamoto noted that they arise from special points in parameter space from the point-of-view of symmetry, which is associated to the affine Weyl group of type $A_2^{(2)}$. Umemura (Umemura, 2003) associated analogous special polynomials with certain rational and algebraic solutions of P_{III}, P_V and P_{VI} which have similar properties to the Yablonskii-Vorob'ev polynomials and the Okamoto polynomials; see also (Noumi M. and H., 1998; Umemura, 1998; Umemura, 2001; Yamada, 2000). Subsequently there have been several studies of special polynomials associated with the rational solutions of P_{II} (Fukutani et al., 2000; Kajiwara and Masuda, 1999a; Kajiwara and Ohta, 1996; Taneda, 2000), the rational and algebraic solutions of P_{III} (Kajiwara and Masuda, 1999b; Ohyama, 2001), the rational solutions of P_{IV} (Fukutani et al., 2000; Kajiwara and Ohta, 1998; Noumi and Yamada, 1999), the rational solutions of P_V (Masuda et al., 2002; Noumi and Yamada, 1998b) and the algebraic solutions of P_{VI} (Kirillov and Taneda, 2002b; Kirillov and Taneda, 2002a; Masuda, 2002; Taneda, 2001a; Taneda, 2001b). However the majority of these papers are concerned with the combinatorial structure and determinant representation of the polynomials, often related to the Hamiltonian structure and affine Weyl symmetries of the Painlevé equations. Typically these polynomials arise as the "τ-functions" for special solutions of the Painlevé equations and are generated through nonlinear, three-term recurrence relations which are Toda equations that arise from the associated Bäcklund transformations of the Painlevé equations. The coefficients of these special polynomials have some interesting, indeed somewhat mysterious, combinatorial properties (see (Noumi M. and H., 1998; Umemura, 1998; Umemura, 2001; Umemura, 2003)). Additionally these polynomials have been expressed as special cases of *Schur polynomials*, which are irreducible polynomial representations of the general linear group $GL(n)$ and arise as τ-functions of the Kadomtsev-Petviashvili (KP) hierarchy (Jimbo and Miwa, 1983). The Yablonskii-Vorob'ev polynomials associated with P_{II} are expressible in terms of 2-reduced Schur functions (Kajiwara and Masuda, 1999a; Kajiwara and Ohta, 1996), and are related to the τ-function for the rational solution of the modified Korteweg de Vries (mKdV) equation since P_{II} arises as a similarity reduction of the mKdV equation. The Okamoto polynomials associated with P_{IV} are expressible in terms of 3-reduced Schur functions (Kajiwara and Ohta, 1998; Noumi and Yamada, 1999) since P_{IV} arises as a similarity reduction of the Boussinesq equation (cf. (Clarkson and Kruskal, 1989)), which belongs to the so-called 3-reduction of the KP hierarchy (Jimbo and Miwa, 1983).

It is also well-known that P_{II}–P_{VI} possess solutions which are expressible in terms of the classical special functions; these are often referred to as "one-parameter families of solutions". For P_{II} these special function solutions are expressed in terms of Airy functions $\mathrm{Ai}(z)$ (Airault, 1979; Flaschka and Newell, 1980; Gambier, 1910; Okamoto, 1986), for P_{III} they are expressed in terms of Bessel functions $J_\nu(z)$ (Lukashevich, 1967a; Milne et al., 1997; Murata, 1995; Okamoto, 1987c), for P_{IV} they are expressed in terms of Weber-Hermite (parabolic cylinder) functions $D_\nu(z)$ (Bassom et al., 1995; Gromak, 1987; Lukashevich, 1967b; Murata, 1985; Okamoto, 1986), for P_V they are expressed in terms of Whittaker functions $M_{\kappa,\mu}(z)$, or equivalently confluent hypergeometric functions ${}_1F_1(a;c;z)$ (Lukashevich, 1968; Gromak, 1976; Okamoto, 1987b; Watanabe, 1995), and for P_{VI} they are expressed in terms of hypergeometric functions ${}_2F_1(a,b;c;z)$ (Fokas and Yortsos, 1981; Lukashevich and Yablonskii, 1967; Okamoto, 1987a); see also (Ablowitz and Clarkson, 1991; Gromak, 1978b; Gromak, 1999; Gromak and Lukashevich, 1982; Tamizhmani et al., 2001). Some classical orthogonal polynomials arise as particular cases of these special function solutions and thus yield rational solutions of the associated Painlevé equations, especially in the representation of rational solutions through determinants.

For P_{III} and P_V these are in terms of associated Laguerre polynomials $L_n^{(k)}(z)$ (Charles, 2002; Kajiwara and Masuda, 1999b; Masuda et al., 2002; Noumi and Yamada, 1998b), for P_{IV} in terms of Hermite polynomials $H_n(z)$ (Bassom et al., 1995; Kajiwara and Ohta, 1998; Murata, 1985; Okamoto, 1986), and for for P_{VI} in terms of Jacobi polynomials $P_n^{(\alpha,\beta)}(z)$ (Masuda, 2002; Taneda, 2001b). In fact all rational solutions of P_{VI} arise as particular cases of the special solutions given in terms of hypergeometric functions (Mazzocco, 2001).

This paper is organised as follows. The Yablonskii-Vorob'ev polynomials and rational solutions for P_{II} are studied in §2. We compare the properties of these special polynomials with properties of classical orthogonal polynomials. The analogous special polynomials associated with rational solutions of P_{III}, which occur in the generic case when $\gamma\delta \neq 0$, are studied in §3. Further, in §3 we study the special polynomials associated with algebraic solutions of P_{III}, which occur in the cases when either $\gamma = 0$ and $\alpha\delta \neq 0$, or $\delta = 0$ and $\beta\gamma \neq 0$. In §4 the special polynomials associated with rational solutions for P_{IV}. Here there are four types of special polynomials, two classes of Okamoto polynomials, which were introduced by Okamoto (Okamoto, 1986), generalized Hermite polynomials and generalized Okamoto polynomials, both of which were introduced by Noumo and Yamada (Noumi and Yamada, 1999). Finally in §5 we discuss our results and pose some open questions.

2. Second Painlevé equation

2.1 Rational solutions of P_{II}

The rational solutions of P_{II} (1.1) are summarized in the following Theorem due to Vorob'ev (Vorob'ev, 1965) and Yablonskii (Yablonskii, 1959); see also (Fukutani et al., 2000; Umemura, 1998; Umemura and Watanabe, 1997; Taneda, 2000).

Theorem 2.1. *Rational solutions of P_{II} exist if and only if $\alpha = n \in \mathbb{Z}$, which are unique, and have the form*

$$w(z;n) = \frac{\mathrm{d}}{\mathrm{d}z}\left\{\ln\left[\frac{Q_{n-1}(z)}{Q_n(z)}\right]\right\}, \tag{2.1}$$

for $n \geq 1$, where the polynomials $Q_n(z)$ satisfy the differential-difference equation

$$Q_{n+1}Q_{n-1} = zQ_n^2 - 4\left[Q_nQ_n'' - (Q_n')^2\right], \tag{2.2}$$

with $Q_0(z) = 1$ and $Q_1(z) = z$. The other rational solutions are given by

$$w(z;0) = 0, \qquad w(z;-n) = -w(z;n). \tag{2.3}$$

The polynomials $Q_n(z)$ are monic polynomials of degree $\frac{1}{2}n(n+1)$ with integer coefficients, and are called the *Yablonskii-Vorob'ev polynomials.* The first few polynomials $Q_n(z)$ are

$$\begin{aligned}
Q_2(z) &= z^3 + 4,\\
Q_3(z) &= z^6 + 20z^3 - 80,\\
Q_4(z) &= (z^9 + 60z^6 + 11200)z,\\
Q_5(z) &= z^{15} + 140z^{12} + 2800z^9 + 78400z^6 - 313600z^3 - 6272000\\
Q_6(z) &= z^{21} + 280z^{18} + 18480z^{15} + 627200z^{12} - 17248000z^9\\
&\quad + 1448832000z^6 + 19317760000z^3 - 38635520000,\\
Q_7(z) &= (z^{27} + 504z^{24} + 75600z^{21} + 5174400z^{18} + 62092800z^{15}\\
&\quad + 13039488000z^{12} - 828731904000z^9\\
&\quad - 49723914240000z^6 - 3093932441600000)z\\
Q_8(z) &= z^{36} + 840z^{33} + 240240z^{30} + 32771200z^{27} + 2018016000z^{24}\\
&\quad + 124309785600z^{21} - 6629855232000z^{18}\\
&\quad + 407736096768000z^{15} + 126696533483520000z^{12}\\
&\quad + 17697293565952000000z^9 + 371643164884992000000z^6\\
&\quad - 7432863297699840000000z^3 - 99104843969331200000000.
\end{aligned} \tag{2.4}$$

Remarks 2.2.

1. The hierarchy of rational solutions for $\mathrm{P_{II}}$ given by (2.1) can also be derived using the Bäcklund transformation of $\mathrm{P_{II}}$
$$w(z;\alpha+1) = -\,w(z;\alpha) - \frac{2\alpha+1}{2w^2(z;\alpha)+2w'(z;\alpha)+z}, \tag{2.5}$$
(Lukashevich, 1971), with "seed solution" $w_0 = w(z;0) = 0$.

2. It is clear from the recurrence relation (2.2) that the $Q_n(z)$ are rational functions, though it is not obvious that in fact they are polynomials since one is dividing by $Q_{n-1}(z)$ at every iteration. Indeed it is somewhat remarkable that the $Q_n(z)$ defined by (2.2) are polynomials.

3. Letting $Q_n(z) = c_n\tau_n(z)\exp(z^3/24)$, with $c_n = (2\mathrm{i})^{n(n+1)}$, in (2.2) yields the Toda equation
$$\tau_n\tau_n'' - \left(\tau_n'\right)^2 = \tau_{n+1}\tau_{n-1}. \tag{2.6}$$

4. The Yablonskii-Vorob'ev polynomials $Q_n(z)$ possess the discrete symmetry
$$Q_n(\omega z) = \omega^{n(n+1)/2} Q_n(z), \tag{2.7}$$
where $\omega^3 = 1$ and $\frac{1}{2}n(n+1)$ is the degree of $Q_n(z)$.

Fukutani, Okamoto and Umemura (Fukutani et al., 2000) and Taneda (Taneda, 2000) have proved Theorems 2.3 and 2.4 below, respectively, concerning the roots of the Yablonskii-Vorob'ev polynomials. Further these authors also give a purely algebraic proof of Theorem 2.1.

Theorem 2.3. *For every positive integer n, the polynomial $Q_n(z)$ has simple roots. Further the polynomials $Q_n(z)$ and $Q_{n+1}(z)$ do not have a common root.*

Theorem 2.4. *The polynomial $Q_n(z)$ is divisible by z if and only if $n \equiv 1 \bmod 3$. Further $Q_n(z)$ is a polynomial in z^3 if $n \not\equiv 1 \bmod 3$ and $Q_n(z)/z$ is a polynomial in z^3 if $n \equiv 1 \bmod 3$.*

Remarks 2.5.

1. From these theorems, since each polynomial $Q_n(z)$ has only simple roots then it can be written as
$$Q_n(z) = \prod_{k=1}^{n(n+1)/2} (z - a_{n,k}), \tag{2.8}$$
where $a_{n,k}$, for $k = 1, \ldots, \frac{1}{2}n(n+1)$, are the roots. Thus the rational solution of P_{II} can be written as
$$w(z;n) = \frac{Q'_{n-1}(z)}{Q_{n-1}(z)} - \frac{Q'_n(z)}{Q_n(z)} = \sum_{k=1}^{n(n-1)/2} \frac{1}{z - a_{n-1,k}} - \sum_{k=1}^{n(n+1)/2} \frac{1}{z - a_{n,k}}, \tag{2.9}$$
and so $w(z;n)$ has n roots, $\frac{1}{2}n(n-1)$ with residue $+1$ and $\frac{1}{2}n(n+1)$ with residue -1; see also (Gromak, 2001).

2. The roots $a_{n,k}$ of the polynomial $Q_n(z)$ satisfy
$$\sum_{k=1,k\neq j}^{n(n+1)/2} \frac{1}{(a_{n,j} - a_{n,k})^3} = 0, \qquad j = 1, 2, \ldots, \frac{1}{2}n(n+1). \tag{2.10}$$

This follows from the study of rational solutions of the Korteweg-de Vries (KdV) equation
$$u_t + 6uu_x + u_{xxx} = 0 \tag{2.11}$$

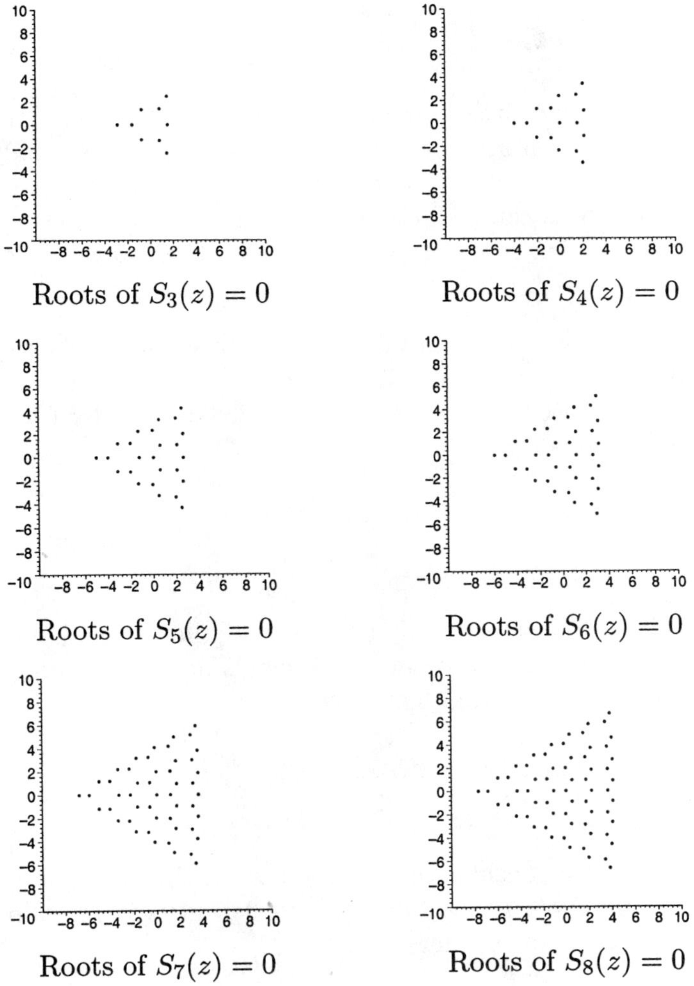

Figure 2. Poles of rational solutions of $\mathrm{P_{II}}$

1. From Figure 1 we see that the roots of the Yablonskii-Vorob'ev polynomials form approximately equilateral triangles, in fact approximate "Pascal triangles." The values of the roots in Table 2.1 show that they actually lie on curves rather than straight lines.

2. The roots of $Q_n(z) = 0$ lie on circles with centre the origin. If we define

$$q_n(\zeta) = \begin{cases} Q_n(\zeta^{1/3}) & \text{if} \quad n \not\equiv 1 \bmod 3, \\ Q_n(\zeta^{1/3})/\zeta^{1/3} & \text{if} \quad n \equiv 1 \bmod 3. \end{cases}$$

The radii of the circles are given by the third roots of the absolute values of the non-zero roots of $q_n(\zeta) = 0$, with three equally spaced roots of $Q_n(z) = 0$ on circles for the real roots of $q_n(\zeta) = 0$ and six roots, three complex conjugate pairs, of $Q_n(z) = 0$ on a circles for the complex roots of $q_n(\zeta) = 0$ (see (Clarkson and Mansfield, 2003)).

3. The plots in Figures 1 and 2 are invariant under rotations through $\frac{2}{3}\pi$ and reflections in the real z-axis and the lines $\arg(z) = \pm\frac{1}{3}\pi, \pm\frac{2}{3}\pi$. This is because $\mathrm{P_{II}}$ admits the finite group of order 6 of scalings and reflections

$$w \to \varepsilon\mu^2 w, \qquad z \to \mu z, \qquad \alpha \to \varepsilon\alpha, \tag{2.13}$$

where $\mu^3 = 1$ and $\varepsilon^2 = 1$.

4. From Figure 2 we see that the poles of the rational solutions of $\mathrm{P_{II}}$ that the location of the poles yields an approximate triangle structure, with internal hexagons.

3. Third Painlevé equation

3.1 Rational solutions of $\mathrm{P_{III}}$

In this section we consider the generic case of $\mathrm{P_{III}}$ when $\gamma\delta \neq 0$, then we set $\gamma = 1$ and $\delta = -1$, without loss of generality (by rescaling w and z if necessary), and so consider

$$w'' = \frac{(w')^2}{w} - \frac{w'}{z} + \frac{\alpha w^2 + \beta}{z} + w^3 - \frac{1}{w}. \tag{3.1}$$

The location of rational solutions for the generic case of $\mathrm{P_{III}}$ given by (3.1) is stated in the following theorem.

Theorem 3.1. *Equation* (3.1), *i.e.,* $\mathrm{P_{III}}$ *with* $\gamma = -\delta = 1$, *has rational solutions if and only if* $\alpha + \varepsilon\beta = 4n$, *with* $n \in \mathbb{Z}$ *and* $\varepsilon = \pm 1$. *These* *rational solutions have the form* $w = P_m(z)/Q_m(z)$, *where* $P_m(z)$ *and* $Q_m(z)$ *and polynomials of degree* m *with no common roots.*

Proof See Gromak, Laine and Shimomura (Gromak et al., 2002), p. 174 (see also (Milne et al., 1997; Murata, 1995; Umemura and Watanabe, 1998)).

We remark that the rational solutions of the generic case of P_{III} (3.1) lie on lines in the α-β plane, rather than isolated points as is the case for P_{IV} (see §4). Further, equation (3.1) is of type D_6 in the terminology of Sakai (Sakai, 2001), who studied the Painlevé equations through a geometric approach based on rational surfaces.

Umemura (Umemura, 2003), see also (Kajiwara and Masuda, 1999b; Noumi M. and H., 1998; Umemura, 1998; Umemura, 2001), derived special polynomials associated with rational solutions of P_{III}, which are defined in Theorem 3.2 below. Further Umemura states that these polynomials are the analogues of the Yablonskii-Vorob'ev polynomials associated with rational solutions of P_{II} and the Okamoto polynomials associated with rational solutions of P_{IV}.

Theorem 3.2. *Suppose that* $T_n(z;\mu)$ *satisfies the recursion relation*

$$zT_{n+1}T_{n-1} = -z\left[T_nT_n'' - (T_n')^2\right] - T_nT_n' + (z+\mu)T_n^2, \tag{3.2}$$

with $T_{-1}(z;\mu) = 1$ *and* $T_0(z;\mu) = 1$. *Then*

$$\begin{aligned} w_n(z;\mu) &\equiv w(z;\alpha_n,\beta_n,1,-1) \\ = 1 + \frac{\mathrm{d}}{\mathrm{d}z}\left\{\ln\left[\frac{T_{n-1}(z;\mu-1)}{z^nT_n(z;\mu)}\right]\right\} &= \frac{T_n(z;\mu-1)\,T_{n-1}(z;\mu)}{T_n(z;\mu)\,T_{n-1}(z;\mu-1)}, \end{aligned} \tag{3.3}$$

satisfies P_{III}, *with* $\alpha_n = 2n + 2\mu - 1$ *and* $\beta_n = 2n - 2\mu + 1$.

The first few polynomials for P_{III} defined by (3.2) are

$$\begin{aligned}
T_1(z;\mu) &= 1+\frac{\mu}{z},\\
T_2(z;\mu) &= 1+\frac{3\mu}{z}+\frac{3\mu^2}{z^2}+\frac{\mu(\mu^2-1)}{z^3},\\
T_3(z;\mu) &= 1+\frac{6\mu}{z}+\frac{15\mu^2}{z^2}+\frac{5\mu(4\mu^2-1)}{z^3}+\frac{15\mu^2(\mu^2-1)}{z^4}\\
&\quad+\frac{3\mu(\mu^2-1)(2\mu^2-3)}{z^5}+\frac{\mu^2(\mu^2-1)(\mu^2-4)}{z^6},\\
T_4(z;\mu) &= 1+\frac{10\mu}{z}+\frac{45\mu^2}{z^2}+\frac{15\mu(8\mu^2-1)}{z^3}+\frac{105\mu^2(2\mu^2-1)}{z^4}\\
&\quad+\frac{63\mu(\mu^2-1)(4\mu^2-1)}{z^5}+\frac{105\mu^2(\mu^2-1)(2\mu^2-3)}{z^6}\\
&\quad+\frac{15\mu(\mu^2-1)(8\mu^4-27\mu^2+15)}{z^7}\\
&\quad+\frac{45\mu^2(\mu^2-1)(\mu^2-2)(\mu^2-4)}{z^8}\\
&\quad+\frac{5\mu^3(\mu^2-1)(\mu^2-4)(2\mu^2-11)}{z^9}\\
&\quad+\frac{\mu^2(\mu^2-1)^2(\mu^2-4)(\mu^2-9)}{z^{10}},
\end{aligned} \tag{3.4}$$

and associated rational solutions of P_{III} are

$$\begin{aligned}
w_0(z;\mu) &= 1,\\
w_1(z;\mu) &= 1-\frac{1}{z+\mu},\\
w_2(z;\mu) &= 1+\frac{1}{z+\mu-1}-\frac{3(z+\mu)^2}{(z+\mu)^3-\mu},\\
w_3(z;\mu) &= 1+\frac{3(z+\mu-1)^2}{(z+\mu-1)^3-\mu+1}\\
&\quad-\frac{6(z+\mu)^5-15\mu(z+\mu)^2+9\mu}{(z+\mu)^6-5\mu(z+\mu)^3+9\mu(z+\mu)-5\mu^2}.
\end{aligned} \tag{3.5}$$

The hierarchy of rational solutions of P_{III} given in (3.5) can also be derived using the Bäcklund transformation of P_{III} given by

$$\widetilde{w}(z;\widetilde{\alpha},\widetilde{\beta},1,-1)=\frac{zw'+zw^2-\beta w-w+z}{w[zw'+zw^2+\alpha w+w+z]}, \tag{3.6}$$

where $w \equiv w(z;\alpha,\beta,1,-1)$, $\widetilde{\alpha}=\alpha+2$ and $\widetilde{\beta}=\beta+2$ (Gromak, 1973; Gromak, 1975) (see also (Milne et al., 1997; Murata, 1995; Umemura and Watanabe, 1998)), i.e.,

$$w_{n+1}(z;\alpha_{n+1},\beta_{n+1};1;-1) = \frac{zw_n' + zw_n^2 + (\beta_n+1)w_n + z}{w_n[zw_n' + zw_n^2 + (\alpha_n+1)w_n + z]}, \tag{3.7}$$

where $w_n \equiv w(z;\alpha_n,\beta_n,1,-1)$, $\alpha_n = 2n+2\mu-1$ and $\beta_n = 2n-2\mu+1$, with "seed solution" $w_0(z;\alpha_0,\beta_0;1;-1) = 1$ where $\alpha_0 = 2\mu-1$ and $\beta_0 = -2\mu+1$.

The polynomials $T_n(z)$ are somewhat unsatisfactory since they are polynomials in $\xi = 1/z$ rather than polynomials in z, which would be more natural and is the case for the Yablonskii-Vorob'ev polynomials and Okamoto polynomials associated with rational solutions of $\mathrm{P_{II}}$ and $\mathrm{P_{IV}}$, respectively. However it is straightforward to determine a sequence of functions $S_n(z)$ which are generated through a Toda equation that are polynomials in z. These are given in the following theorem.

Theorem 3.3. *Suppose that $S_n(z;\mu)$ satisfies the recursion relation (Toda equation)*

$$S_{n+1}S_{n-1} = -z\left[S_nS_n'' - \left(S_n'\right)^2\right] - S_nS_n' + (z+\mu)S_n^2, \tag{3.8}$$

with $S_{-1}(z;\mu) = S_0(z;\mu) = 1$. Then

$$\begin{aligned} w_n = w(z;\alpha_n,\beta_n,1,-1) &= 1 + \frac{\mathrm{d}}{\mathrm{d}z}\left\{\ln\left[\frac{S_{n-1}(z;\mu-1)}{S_n(z;\mu)}\right]\right\} \\ &= \frac{S_n(z;\mu-1)\,S_{n-1}(z;\mu)}{S_n(z;\mu)\,S_{n-1}(z;\mu-1)}, \end{aligned} \tag{3.9}$$

satisfies $\mathrm{P_{III}}$ with $\alpha_n = 2n+2\mu-1$ and $\beta_n = 2n-2\mu+1$.

Proof. This essentially follows from Theorem 1 due to Kajiwara and Masuda (Kajiwara and Masuda, 1999b) since the Toda equation (3.8), modulo a scaling factor, is equation (16) in Proposition 3 of (Kajiwara and Masuda, 1999b).

The first few polynomials $S_n(z;\mu)$, which are monic polynomials of degree $\frac{1}{2}n(n+1)$, are

$$\begin{aligned}
S_1(z;\mu) &= z+\mu,\\
S_2(z;\mu) &= (z+\mu)^3-\mu,\\
S_3(z;\mu) &= (z+\mu)^6-5\mu(z+\mu)^3+9\mu(z+\mu)-5\mu^2,\\
S_4(z;\mu) &= (z+\mu)^{10}-15\mu(z+\mu)^7+63\mu(z+\mu)^5-225\mu(z+\mu)^3\\
&\quad +315\mu^2(z+\mu)^2-175\mu^3(z+\mu)+36\mu^2,\\
S_5(z;\mu) &= (z+\mu)^{15}-35\mu(z+\mu)^{12}+252\mu(z+\mu)^{10}\\
&\quad +175\mu^2(z+\mu)^9-2025\mu(z+\mu)^8+945\mu^2(z+\mu)^7\\
&\quad -1225\mu(\mu^2-9)(z+\mu)^6-26082\mu^2(z+\mu)^5\\
&\quad +33075\mu^3(z+\mu)^4-350\mu^2(35\mu^2+36)(z+\mu)^3\\
&\quad +11340\mu^3(z+\mu)^2-225\mu^2(49\mu^2-36)(z+\mu)\\
&\quad +7\mu^3(875\mu^2-828).
\end{aligned} \tag{3.10}$$

The associated rational solutions of P_{III} are in (3.5). It is clear from the recurrence relation (3.8) that the $S_n(z;\mu)$, are rational functions, though it is not obvious that in fact they are polynomials since one is dividing by $S_{n-1}(z;\mu)$ at every iteration. Indeed it is somewhat remarkable that the $S_n(z;\mu)$ defined by (3.8) are polynomials. The polynomials $S_n(z;\mu)$ defined by (3.8) are related to the polynomials $R_n(z;\mu)$ defined by (3.2) through $S_n(z;\mu) = z^{n(n+1)/2}R_n(z;\mu)$. The polynomials $S_n(z;\mu)$ have the property that $S_n(z;\mu) = S_n(-z;-\mu)$.

In Figure 3 plots of the roots of the polynomial $S_4(\xi-\mu,\mu)$ defined by (3.10) for various μ are given. Initially for $\mu=-3$ there is an approximate triangle of roots with 4 roots on each side. As μ increases, sets of roots then in turn coalesce until there is a multiple root of order 10 for $m=0$. Then as μ another approximate triangle appears which is "turned round" from the configuration for $\mu=-3$ since the symmetry is $S_n(z;\mu)=S_n(-z;\mu)$ implies that the roots for $S_n(z;\mu)$ is a reflection of those for $S_n(z;-\mu)$ in the imaginary axis. □

3.2 Algebraic solutions of P_{III}

In this section we consider the special case of P_{III} when either (i), $\gamma=0$ and $\alpha\delta\neq0$, or (ii), $\delta=0$ and $\beta\gamma\neq0$. In case (i), we make the

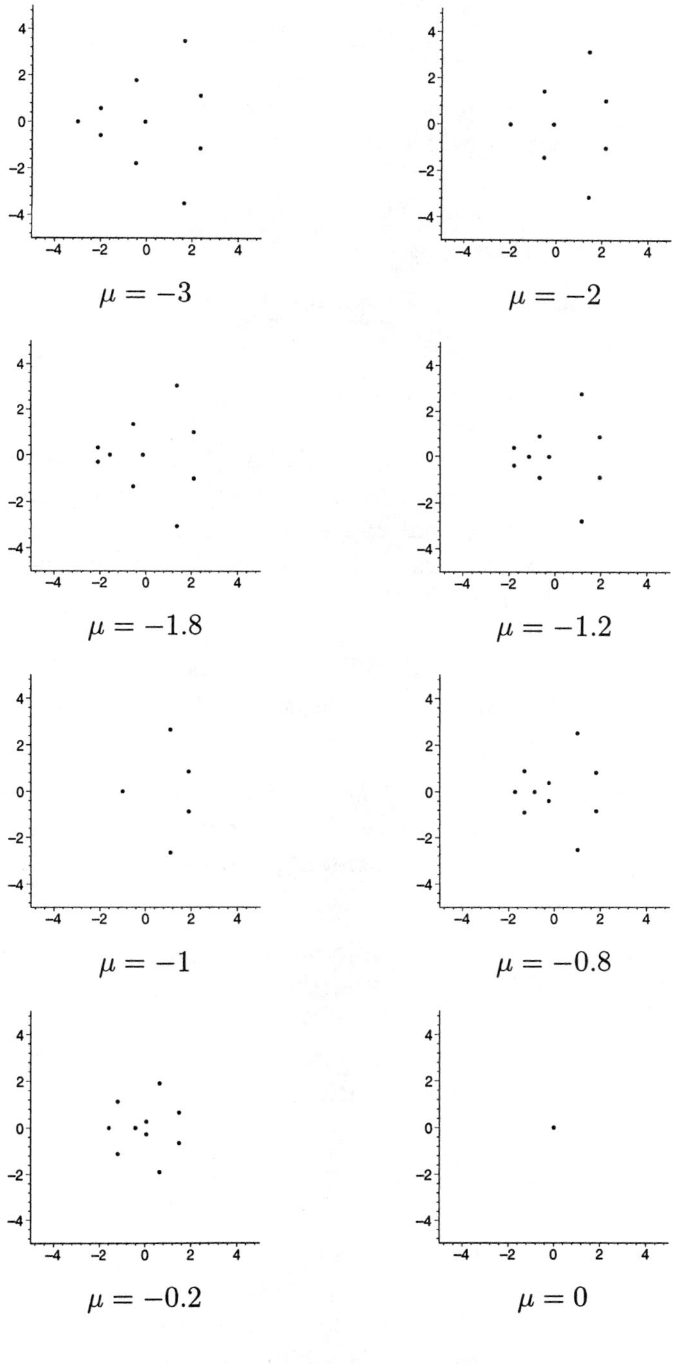

Figure 3. Roots of the polynomial $S_4(\xi - \mu, \mu)$ defined by (3.10) for various μ

transformation

$$w(z) = \left(\frac{2}{3}\right)^{1/2} u(\zeta), \qquad z = \left(\frac{2}{3}\right)^{3/2} \zeta^3, \tag{3.11}$$

and set $\alpha = 1$, $\beta = 2\mu$ and $\delta = -1$, with μ an arbitrary constant, without loss of generality, which yields

$$\frac{\mathrm{d}^2u}{\mathrm{d}\zeta^2} = \frac{1}{u}\left(\frac{\mathrm{d}u}{\mathrm{d}\zeta}\right)^2 - \frac{1}{\zeta}\frac{\mathrm{d}u}{\mathrm{d}\zeta} + 4\zeta u^2 + 12\mu\zeta - \frac{4\zeta^4}{u}. \tag{3.12}$$

In case (ii), we make the transformation

$$w(z) = \left(\frac{3}{2}\right)^{1/2} / u(\zeta), \qquad z = \left(\frac{2}{3}\right)^{3/2} \zeta^3, \tag{3.13}$$

and set $\alpha = 2\mu$, $\beta = -1$ and $\gamma = 1$, with μ an arbitrary constant, without loss of generality, which again yields (3.12). The scalings in (3.11) and (3.13) have been chosen so that the associated special polynomials are monic polynomials. We remark that equation (3.12) is of type D_7 in the terminology of Sakai (Sakai, 2001).

Rational solutions of (3.12) correspond to algebraic solutions of $\mathrm{P_{III}}$ with $\gamma = 0$ and $\alpha\delta \neq 0$, or $\delta = 0$ and $\beta\gamma \neq 0$. Lukashevich (Lukashevich, 1965; Lukashevich, 1967a) obtained algebraic solutions of $\mathrm{P_{III}}$, which are classified in the following theorem.

Theorem 3.4. *Equation* (3.12) *has rational solutions if and only if* $\mu = n$, *with* $n \in \mathbb{Z}$. *These rational solutions have the form* $u(\zeta) = P_{n^2+1}(\zeta)/Q_{n^2}(\zeta)$, *where* $P_{n^2+1}(\zeta)$ *and* $Q_{n^2}(\zeta)$ *and monic polynomials of degree* n^2+1 *and* n^2, *respectively.*

Proof. See Gromak, Laine and Shimomura (Gromak et al., 2002), p. 164 (see also (Gromak, 1973; Gromak, 1978a; Milne et al., 1997; Murata, 1995)).

A straightforward method for generating rational solutions of (3.12) is through the Bäcklund transformation

$$u_{\mu+\varepsilon} = \frac{\zeta^3}{u_\mu^2} + \frac{\varepsilon\zeta}{2u_\mu^2}\frac{\mathrm{d}u_\mu}{\mathrm{d}\zeta} - \frac{3(2\mu+\varepsilon)}{2u_\mu}, \tag{3.14}$$

where $\varepsilon^2 = 1$ and u_μ is the solution of (3.12) for parameter μ, using the "seed solution" $u_0(\zeta) = \zeta$ for $\mu = 0$ (see Gromak, Laine and Shimomura (Gromak et al., 2002), p. 164 — see also (Gromak, 1973; Gromak, 1978a; Milne et al., 1997; Murata, 1995)). Further we note that $u_{-\mu}(\zeta) = -\mathrm{i}u_\mu(\mathrm{i}\zeta)$. Therefore the transformation group for (3.12) is isomorphic to the affine Weyl group $\widetilde{A}_1$, which also is the transformation group for $\mathrm{P_{II}}$ (Okamoto, 1986; Umemura, 2000; Umemura and Watanabe, 1997). □

Ohyama (Ohyama, 2001) derived special polynomials associated with the rational solutions of (3.12). These are essentially described in Theorem 3.5 below, though here the variables have been scaled and the expression of the rational solutions of (3.12) in terms of these special polynomials is explicitly given.

Theorem 3.5. *Suppose that $R_n(\zeta)$ satisfies the recursion relation (Toda equation)*

$$2\zeta R_{n+1}R_{n-1} = -R_n\frac{\mathrm{d}^2R_n}{\mathrm{d}\zeta} + \left(\frac{\mathrm{d}R_n}{\mathrm{d}\zeta}\right)^2 - \frac{R_n}{\zeta}\frac{\mathrm{d}R_n}{\mathrm{d}\zeta} + 2\left(\zeta^2 - n\right)R_n^2, \quad (3.15)$$

with $R_0(\zeta) = 1$ and $R_1(\zeta) = \zeta^2$. Then

$$u_n(\zeta) = \frac{R_{n+1}(\zeta)\,R_{n-1}(\zeta)}{R_n^2(\zeta)}, \quad (3.16)$$

satisfies (3.12) with $\mu = n$. Additionally $u_{-n}(\zeta) = -\mathrm{i}u_n(\mathrm{i}\zeta)$.

The first few polynomials $R_n(\zeta)$ defined by (3.15) are

$$\begin{aligned}
R_2(\zeta) &= (\zeta^2 - 1)\,\zeta^3,\\
R_3(\zeta) &= (\zeta^4 - 4\zeta^2 + 5)\,\zeta^5,\\
R_4(\zeta) &= (\zeta^8 - 10\zeta^6 + 40\zeta^4 - 70\zeta^2 + 35)\,\zeta^6,\\
R_5(\zeta) &= (\zeta^{12} - 20\zeta^{10} + 175\zeta^8 - 840\zeta^6 + 2275\zeta^4\\
&\qquad -3220\zeta^2 + 1925)\,\zeta^8,\\
R_6(\zeta) &= (\zeta^{18} - 35\zeta^{16} + 560\zeta^{14} - 5320\zeta^{12} - 32690\zeta^{10}\\
&\qquad + 133070\zeta^8 - 354200\zeta^6 + 585200\zeta^4\\
&\qquad - 525525\zeta^2 + 175175)\zeta^9,\\
R_7(\zeta) &= (\zeta^{24} - 56\zeta^{22} + 1470\zeta^{20} - 23800\zeta^{18} + 263375\zeta^{16}\\
&\qquad - 2088240\zeta^{14} + 12105940\zeta^{12} - 51466800\zeta^{10}\\
&\qquad + 158533375\zeta^8 - 343343000\zeta^6 + 493643150\zeta^4\\
&\qquad - 421821400\zeta^2 + 163788625)\zeta^{11},\\
R_8(\zeta) &= (\zeta^{32} - 84\zeta^{30} + 3360\zeta^{28} - 84700\zeta^{26} + 1501500\zeta^{24}\\
&\qquad - 19787460\zeta^{22} + 199916640\zeta^{20} - 1574673100\zeta^{18}\\
&\qquad + 9741481750\zeta^{16} - 47328781500\zeta^{14}\\
&\qquad + 179306327200\zeta^{12} - 521782561300\zeta^{10}\\
&\qquad + 1136861225500\zeta^8 - 1778744467500\zeta^6\\
&\qquad + 1860638780000\zeta^4 - 1132762130500\zeta^2\\
&\qquad + 283190532625)\zeta^{12},
\end{aligned} \tag{3.17}$$

and associated rational solutions of (3.12) are

$$\begin{aligned}
u_1(\zeta) &= \frac{\zeta^2 - 1}{\zeta},\\
u_2(\zeta) &= \frac{\zeta(\zeta^4 - 4\zeta^2 + 5)}{(\zeta^2 - 1)^2},\\
u_3(\zeta) &= \frac{(\zeta^2 - 1)(\zeta^8 - 10\zeta^6 + 40\zeta^4 - 70\zeta^2 + 35)}{\zeta(\zeta^4 - 4\zeta^2 + 5)^2}.
\end{aligned} \tag{3.18}$$

The polynomial $R_n(\zeta)$ is a monic polynomial of degree $\frac{1}{2}n(n+3)$ with integer coefficients. Further it has the form $R_n(\zeta) = S_n(\zeta)\zeta^{\kappa_n}$, where $\kappa_n = \frac{1}{2}n^2 - \frac{1}{4}[1-(-1)^n]$ and $S_n(\zeta)$ a monic polynomial of degree $\frac{3}{2}n + \frac{1}{4}[1-(-1)^n]$ with simple zeros and $S_n(0) \neq 0$.

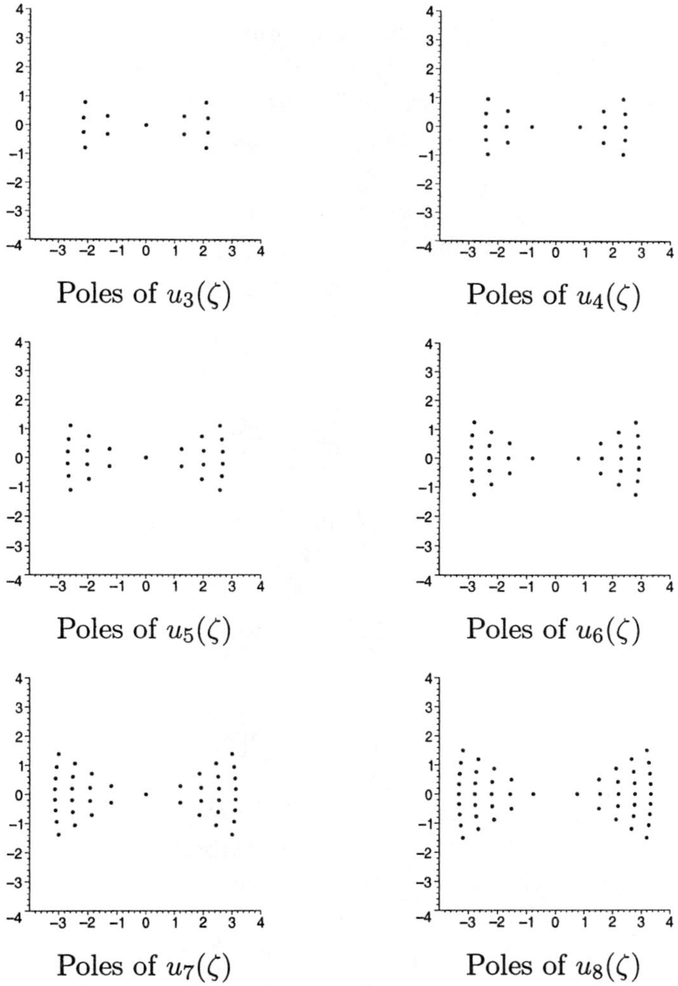

Figure 4. Poles of algebraic solutions of $\mathrm{P_{III}}$-$\mathrm{D_7}$ (3.12)

In Figure 4 plots of the locations of the poles of the algebraic solutions of $\mathrm{P_{III}}$-$\mathrm{D_7}$ (3.12) given by $u_n(z)$, for $n = 3, 4, \ldots, 8$, as defined in (3.18) are given. These take the form of two "triangles" in a "bow-tie" shape.

4. Fourth Painlevé equation

4.1 Rational solutions of $\mathrm{P_{IV}}$

Lukashevich (Lukashevich, 1967b), Gromak (Gromak, 1987) and Murata (Murata, 1985) (see also (Bassom et al., 1995; Gromak et al., 2002; Umemura and Watanabe, 1997)), have proved the following theorem

Theorem 4.1. *$\mathrm{P_{IV}}$ has rational solutions if and only if*

$$\alpha = m, \qquad \beta = -2(1 + 2n - m)^2, \tag{4.1}$$

or

$$\alpha = m, \qquad \beta = -\frac{2}{9}(1 + 6n - 3m)^2, \tag{4.2}$$

with $m, n \in \mathbb{Z}$. Further the rational solutions for these parameter values are unique.

Three simple rational solutions of $\mathrm{P_{IV}}$ are

$$w_1(z; \pm 2, -2) = \pm 1/z, \quad w_2(z; 0, -2) = -2z, \quad w_3\left(z; 0, -\frac{2}{9}\right) = -\frac{2}{3}z. \tag{4.3}$$

It is known that there are three families of unique rational solutions of $\mathrm{P_{IV}}$, which have the solutions (4.3) as the simplest members. These are summarized in the following theorem (see (Bassom et al., 1995; Murata, 1985; Umemura and Watanabe, 1997) for further details).

Theorem 4.2. *There are three families of rational solutions of $\mathrm{P_{IV}}$, which have the forms*

$$w_1\left(z; \alpha_1, \beta_1\right) = P_{1,n-1}(z)/Q_{1,n}(z), \tag{4.4a}$$

$$w_2\left(z; \alpha_2, \beta_2\right) = -2z + P_{2,n-1}(z)/Q_{2,n}(z), \tag{4.4b}$$

$$w_3\left(z; \alpha_3, \beta_3\right) = -\frac{2}{3}z + P_{3,n-1}(z)/Q_{3,n}(z), \tag{4.4c}$$

where $P_{j,n}(z)$ *and* $Q_{j,n}(z)$, $j = 1, 2, 3$, *are polynomials of degree* n, *and*

$$(\alpha_1, \beta_1) = \left(\pm m, -2(1+2n+m)^2\right), \quad m, n \in \mathbb{Z}, \quad n \le -1, \quad m \ge -2n, \tag{4.5a}$$

$$(\alpha_2, \beta_2) = \left(m, -2(1+2n+m)^2\right), \quad m, n \in \mathbb{Z}, \quad n \ge 0, \quad m \ge -n, \tag{4.5b}$$

$$(\alpha_3, \beta_3) = \left(m, \frac{2}{9}(1+6n-3m)^2\right), \quad m, n \in \mathbb{Z}. \tag{4.5c}$$

The three hierarchies given in this theorem are known as the "$-1/z$ hierarchy", the "$-2z$ hierarchy" and the "$-\frac{2}{3}z$ hierarchy", respectively (see (Bassom et al., 1995) where the terminology was introduced). The "$-1/z$ hierarchy" and the "$-2z$ hierarchy" form the set of rational solutions of $\mathrm{P_{IV}}$ with parameter values given by (4.1) and the "$-\frac{2}{3}z$ hierarchy" forms the set with parameter values given by (4.2). The rational solutions of $\mathrm{P_{IV}}$ with parameter values given by (4.1) lie at the vertexes of the "Weyl chambers" and those with parameter values given by (4.2) lie at the vertexes of the "Weyl chamber" (Umemura and Watanabe, 1997).

4.2 Okamoto polynomials

In a comprehensive study of the fourth Painlevé equation $\mathrm{P_{IV}}$, Okamoto (Okamoto, 1986), see also (Kajiwara and Ohta, 1998; Noumi and Yamada, 1999; Umemura, 1998) defined two sets of polynomials analogous to the Yablonskii-Vorob'ev polynomials associated with $\mathrm{P_{II}}$. These polynomials are defined in Theorems 4.3 and 4.5 below, where they have been scaled compared to Okamoto's original definition, where the polynomials were monic, so that they are for the standard version of $\mathrm{P_{IV}}$.

Theorem 4.3. *Suppose that* $Q_n(z)$ *satisfies the recursion relation*

$$Q_{n+1}Q_{n-1} = \frac{9}{2}\left[Q_n Q_n'' - (Q_n')^2\right] + \left[2z^2 + 3(2n-1)\right] Q_n^2, \tag{4.6}$$

with $Q_0(z) = Q_1(z) = 1$. *Then*

$$w_n = w(z; \alpha_n, \beta_n) = -\frac{2}{3}z + \frac{d}{dz}\left\{\ln\left[\frac{Q_{n+1}(z)}{Q_n(z)}\right]\right\}, \tag{4.7}$$

satisfies $\mathrm{P_{IV}}$ *with* $(\alpha_n, \beta_n) = \left(2n, -\frac{2}{9}\right)$.

The first few polynomials $Q_n(z)$, which are referred to as the *Okamoto polynomials*, are

$$\begin{aligned}
Q_0 &= Q_1 = 1,\\
Q_2 &= 2z^2 + 3,\\
Q_3 &= 8z^6 + 60z^4 + 90z^2 + 135,\\
Q_4 &= 64z^{12} + 1344z^{10} + 9360z^8 + 30240z^6 + 56700z^4\\
&\qquad + 170100z^2 + 127575,\\
Q_5 &= 1024z^{20} + 46080z^{18} + 817920z^{16} + 7603200z^{14}\\
&\qquad + 41731200z^{12} + 155675520z^{10} + 493970400z^8\\
&\qquad + 1886068800z^6 + 5304568500z^4 + 5304568500z^2\\
&\qquad + 3978426375,\\
Q_6 &= 32768z^{30} + 2703360z^{28} + 95477760z^{26} + 1916006400z^{24}\\
&\qquad + 24472627200z^{22} + 212580910080z^{20}\\
&\qquad + 1332821952000z^{18} + 6627106886400z^{16}\\
&\qquad + 30481566192000z^{14} + 148952283480000z^{12}\\
&\qquad + 702723772951200z^{10} + 2375788921506000z^8\\
&\qquad + 4874463476883000z^6 + 6451495778227500z^4\\
&\qquad + 9677243667341250z^2 + 4838621833670625.
\end{aligned}$$

Remarks 4.4.

1. The polynomials $Q_n(z)$ are polynomials of degree $n(n-1)$, in fact they are monic polynomials in $\zeta = \sqrt{2}\,z$ with integer coefficients, which is the form in which Okamoto (Okamoto, 1986) originally defined these polynomials.

2. The hierarchy of rational solutions of $\mathrm{P_{IV}}$ defined by (4.7) can be derived using the following Bäcklund transformation of $\mathrm{P_{IV}}$

$$\widetilde{w}\left(z;\widetilde{\alpha},\widetilde{\beta}\right) = \frac{\left(w' - w^2 - 2zw\right)^2 + 2\beta}{2w\left[w' - w^2 - 2zw + 2(\alpha+1)\right]}, \qquad (4.8)$$
$$\widetilde{\alpha} = \alpha + 2, \qquad \widetilde{\beta} = \beta,$$

where $w \equiv w(z;\alpha,\beta)$, which is the Bäcklund transformation $\mathcal{T}_+$ derived by Murata (Murata, 1985) and the Schlesinger transformation $\mathcal{R}^{[5]}$ derived by Fokas, Mugan and Ablowitz (Fokas et al.,

1988). Specifically

$$w_{n+1} = \frac{9\left[w_n' - w_n^2 - 2zw_n\right]^2 - 4}{18w_n\left[w_n' - w_n^2 - 2zw_n + 4n + 2\right]}, \tag{4.9}$$

where $w_n = w\left(z; 2n, -\frac{2}{9}\right)$, with "seed solution" $w_0 = w\left(z; 0, -\frac{2}{9}\right) = -\frac{2}{3}z$.

3. The solutions w_n are members of the so-called "$-\frac{2}{3}z$" hierarchy of rational solutions of $\mathrm{P_{IV}}$, recall Theorem 4.2, which is one of three hierarchies of rational solutions of $\mathrm{P_{IV}}$ (see, for example, (Bassom et al., 1995; Murata, 1985) for further details).

The second set of polynomials introduced by Okamoto (Okamoto, 1986) are defined in the following theorem.

Theorem 4.5. *Suppose that $S_n(z)$ satisfies the recursion relation*

$$S_{n+1}S_{n-1} = \frac{9}{2}\left[S_n S_n'' - \left(S_n'\right)^2\right] + 2\left(z^2 + 3n\right)S_n^2, \tag{4.10}$$

with $S_0(z) = 1$ and $S_1(z) = \sqrt{2}\,z$. Then

$$w\left(z; \alpha_n, \beta_n\right) = -\frac{2}{3}z + \frac{\mathrm{d}}{\mathrm{d}z}\left\{\ln\left[\frac{S_{n+1}(z)}{S_n(z)}\right]\right\}, \tag{4.11}$$

for $n \geq 0$, satisfies $\mathrm{P_{IV}}$ with $(\alpha_n, \beta_n) = \left(2n+1, -\frac{8}{9}\right)$.

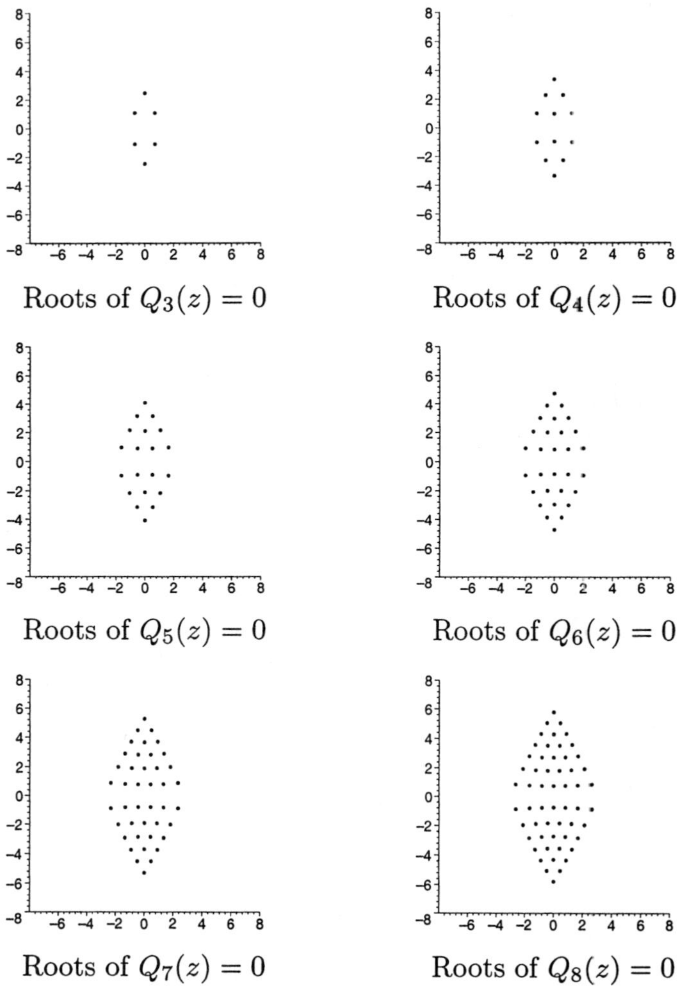

Figure 5. Roots of the Okamoto polynomials I defined by (4.6)

The first few polynomials $S_n(z)$, are

$$S_0 = 1,$$
$$S_1 = \sqrt{2}\,z,$$
$$S_2 = 4z^4 + 12z^2 - 9,$$
$$S_3 = \sqrt{2}\,z\left(16z^8 + 192z^6 + 504z^4 - 2835\right),$$
$$\begin{aligned}S_4 = {} & 256z^{16} + 7680z^{14} + 80640z^{12} + 362880z^{10} + 453600z^8 \\ & - 1905120z^6 - 14288400z^4 - 21432600z^2 + 8037225,\end{aligned}$$
$$\begin{aligned}S_5 = {} & \sqrt{2}\,z\,(4096z^{24} + 245760z^{22} + 5990400z^{20} + 77414400z^{18} \\ & + 569721600z^{16} + 2246952960z^{14} + 1600300800z^{12} \\ & - 35663846400z^{10} - 275837562000z^8 - 1103350248000z^6 \\ & -1737776640600z^4 + 3258331201125),\end{aligned}$$
$$\begin{aligned}S_6 = {} & 262144z^{36} + 27525120z^{34} + 1259274240z^{32} \\ & + 33195294720z^{30} + 560170598400z^{28} + 6324632616960z^{26} \\ & + 47742002380800z^{24} + 219281707008000z^{22} \\ & + 228319944652800z^{20} - 5825689309440000z^{18} \\ & - 63304058468851200z^{16} - 412776567979776000z^{14} \\ & - 1810902281636448000z^{12} - 4651958555820576000z^{10} \\ & - 4025733365613960000z^8 + 11272053423719088000z^6 \\ & + 47553975381314902500z^4 + 47553975381314902500z^2 \\ & - 11888493845328725625.\end{aligned}$$

Remarks 4.6.

1. The polynomials $S_n(z)$ are polynomials of degree n^2, in fact they are monic polynomials in $\zeta = 2z$ with integer coefficients, which is the form in which Okamoto (Okamoto, 1986) originally defined these polynomials.

2. The hierarchy of rational solutions of $\mathrm{P_{IV}}$ defined by (4.11) can be derived using the Bäcklund transformation (4.8) of $\mathrm{P_{IV}}$, derived by Murata (Murata, 1985) and Fokas, Mugan and Ablowitz (Fokas et al., 1988). Hence

$$\widehat{w}_{n+1} = \frac{9\left[\widehat{w}_n' - \widehat{w}_n^2 - 2z\widehat{w}_n\right]^2 - 16}{18\widehat{w}_n\left[\widehat{w}_n' - \widehat{w}_n^2 - 2z\widehat{w}_n + 4n + 4\right]}, \tag{4.12}$$

where $\widehat{w}_n = w\left(z; 2n+1, 2n - \frac{8}{9}\right)$, with "seed solution" $\widehat{w}_0 = w\left(z; 1, -\frac{8}{9}\right) = -\frac{2}{3}z + 1/z$.

3. The solutions $\widehat{w}_n$ are also members of the so-called "$-\frac{2}{3}z$" hierarchy of rational solutions of $\mathrm{P_{IV}}$, recall Theorem 4.2.

4. The two hierarchies of rational solutions of $\mathrm{P_{IV}}$ given by (4.7) and (4.11) are linked by the Schlesinger transformations $\mathcal{R}^{[1]}$ and $\mathcal{R}^{[3]}$ for $\mathrm{P_{IV}}$ given by Fokas, Mugan and Ablowitz (Fokas et al., 1988).

$$\mathcal{R}^{[1]}: \quad w^{[1]} = \frac{\left(w' + \sqrt{-2\beta}\right)^2 - w^2\left[2\sqrt{-2\beta} - 4\alpha - 4 + (w+2z)^2\right]}{2w\left(w^2 + 2zw - w' - \sqrt{-2\beta}\right)},$$
$$\alpha^{[1]} = \alpha + 1, \qquad \beta^{[1]} - \frac{1}{2}\left(-2 + \sqrt{-2\beta}\right)^2, \tag{4.13}$$

$$\mathcal{R}^{[3]}: \quad w^{[3]} = \frac{\left(w' - \sqrt{-2\beta}\right)^2 + w^2\left[2\sqrt{-2\beta} + 4\alpha + 4 - (w+2z)^2\right]}{2w\left(w^2 + 2zw - w' + \sqrt{-2\beta}\right)},$$
$$\alpha^{[3]} = \alpha + 1, \qquad \beta^{[3]} = -\frac{1}{2}\left(2 + \sqrt{-2\beta}\right)^2, \tag{4.14}$$

where $w \equiv w(z;\alpha,\beta)$, $w^{[j]} \equiv w\left(z;\alpha^{[j]},\beta^{[j]}\right)$. Specifically, for $n \geq 0$

$$\widehat{w}_n = \frac{\left(w_n' + \frac{2}{3}\right)^2 - w_n^2\left[8n + \frac{8}{3} - (w_n + 2z)^2\right]}{2w_n\left(w_n^2 + 2zw_n - w_n' - \frac{2}{3}\right)}, \tag{4.15}$$

$$w_{n+1} = \frac{\left(\widehat{w}_n' + \frac{4}{3}\right)^2 + \widehat{w}_n^2\left[8n + \frac{16}{3} - (\widehat{w}_n + 2z)^2\right]}{2\widehat{w}_n\left(\widehat{w}_n^2 + 2z\widehat{w}_n - \widehat{w}_n' - \frac{4}{3}\right)}. \tag{4.16}$$

5. The Schlesinger transformations $\mathcal{R}^{[1]}$, $\mathcal{R}^{[3]}$ and $\mathcal{R}^{[5]}$ are related by $\mathcal{R}^{[1]}\mathcal{R}^{[3]} = \mathcal{R}^{[3]}\mathcal{R}^{[1]} = \mathcal{R}^{[5]}$, from the definition given by Fokas, Mugan and Ablowitz (Fokas et al., 1988).

In Figures 5 and 6 plots of the locations of the roots for the Okamoto polynomials $Q_n(z) = 0$, defined by (4.6), and $S_n(z) = 0$, defined by (4.10), for $n = 3, 4, \ldots, 8$, respectively, are given. These both take the form of two "triangles" with the polynomials $R_n(z)$ having an additional row of roots on a straight line between the two "triangles."

4.3 Generalized Hermite polynomials

Noumi and Yamada (Noumi and Yamada, 1999) generalized the results of Okamoto (Okamoto, 1986) and introduced the *generalized Hermite polynomials* $H_{m,n}(z)$, which are defined in Theorem 4.7 and discussed below in this section, and *generalized Okamoto polynomials* $Q_{m,n}(z)$

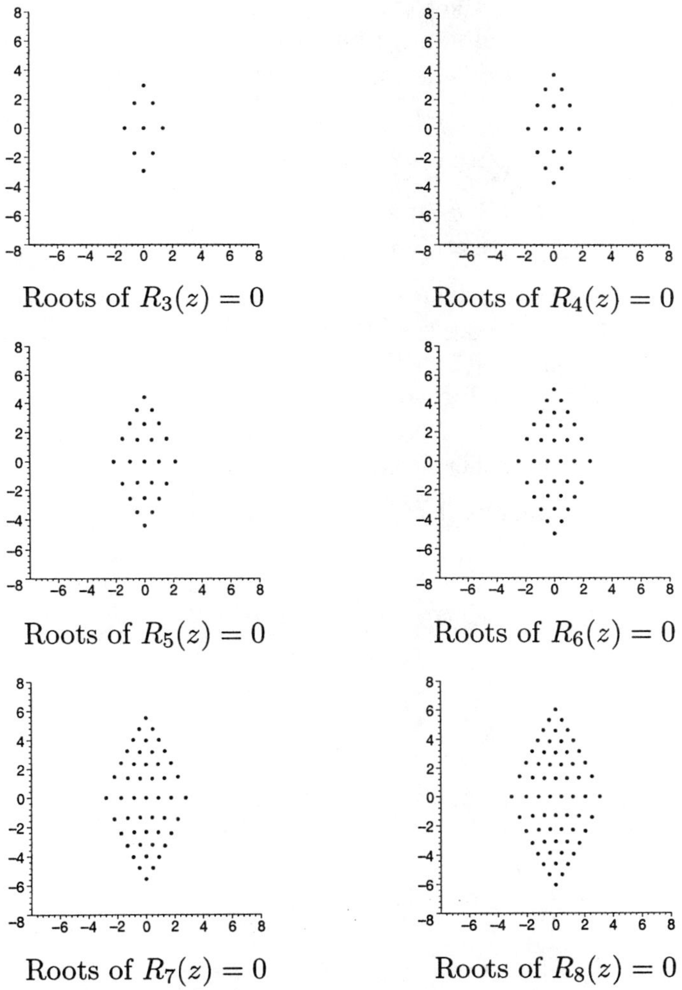

Roots of $R_3(z) = 0$ Roots of $R_4(z) = 0$

Roots of $R_5(z) = 0$ Roots of $R_6(z) = 0$

Roots of $R_7(z) = 0$ Roots of $R_8(z) = 0$

Figure 6. Roots of the Okamoto polynomials II defined by (4.10)

which are defined in Theorem 4.9 and discussed in §4.4. Noumi and Yamada (Noumi and Yamada, 1999) expressed both the generalized Hermite polynomials and the generalized Okamoto polynomials in terms of Schur functions related to the so-called modified Kadomtsev-Petviashvili (mKP) hierarchy. Kajiwara and Ohta (Kajiwara and Ohta, 1998) also expressed rational solutions of $\mathrm{P_{IV}}$ in terms of Schur functions by expressing the solutions in the form of determinants. Further Noumi and Yamada (Noumi and Yamada, 1999) obtained their results on rational solutions of $\mathrm{P_{IV}}$ by considering the symmetric representation of $\mathrm{P_{IV}}$ given by the system

$$\varphi_1' + \varphi_1 (\varphi_2 - \varphi_3) + 2\mu_1 = 0, \tag{4.17a}$$
$$\varphi_2' + \varphi_2 (\varphi_3 - \varphi_1) + 2\mu_2 = 0, \tag{4.17b}$$
$$\varphi_3' + \varphi_3 (\varphi_1 - \varphi_2) + 2\mu_3 = 0, \tag{4.17c}$$

where μ_1, μ_2 and μ_3 are arbitrary constants, with $\mu_1 + \mu_2 + \mu_3 = 1$, and the constraint $\varphi_1 + \varphi_2 + \varphi_3 = -2z$. Then eliminating $\varphi_2(z)$ and $\varphi_3(z)$, $w(z) = \varphi_1(z)$ satisfies $\mathrm{P_{IV}}$ with $(\alpha, \beta) = \left(\mu_3 - \mu_2, -2\mu_1^2\right)$, which was first derived by Bureau (Bureau, 1992) — see also (Adler, 1994; Noumi and Yamada, 1998a; Schiff, 1995; Veselov and Shabat, 1993).

First we discuss the generalized Hermite polynomials $H_{m,n}(z)$.

Theorem 4.7. *Suppose that $H_{m,n}(z)$ satisfies the recurrence relations*

$$2mH_{m+1,n}H_{m-1,n} = H_{m,n}H_{m,n}'' - \left(H_{m,n}'\right)^2 + 2mH_{m,n}^2, \tag{4.18a}$$
$$2nH_{m,n+1}H_{m,n-1} = -H_{m,n}H_{m,n}'' + \left(H_{m,n}'\right)^2 + 2nH_{m,n}^2, \tag{4.18b}$$

with

$$H_{0,0} = H_{1,0} = H_{0,1} = 1, \qquad H_{1,1} = 2z, \tag{4.18c}$$

and $m, n \geq 0$, then

$$\begin{aligned} w_{m,n}^{(\mathrm{I})} = w\left(z; \alpha_{m,n}^{(\mathrm{I})}, \beta_{m,n}^{(\mathrm{I})}\right) &= -\frac{\mathrm{d}}{\mathrm{d}z}\left\{\ln\left(\frac{H_{m,n+1}}{H_{m,n}}\right)\right\} \\ &\equiv -2m\,\frac{H_{m+1,n}\,H_{m-1,n+1}}{H_{m,n+1}\,H_{m,n}}, \end{aligned} \tag{4.19}$$

$$\begin{aligned} w_{m,n}^{(\mathrm{II})} = w\left(z; \alpha_{m,n}^{(\mathrm{II})}, \beta_{m,n}^{(\mathrm{II})}\right) &= \frac{\mathrm{d}}{\mathrm{d}z}\left\{\ln\left(\frac{H_{m+1,n}}{H_{m,n}}\right)\right\} \\ &\equiv 2n\,\frac{H_{m,n+1}\,H_{m+1,n-1}}{H_{m+1,n}\,H_{m,n}}, \end{aligned} \tag{4.20}$$

is a solution of $\mathrm{P_{IV}}$, respectively for the parameters

$$\alpha_{m,n}^{(\mathrm{I})} = -(m + 2n + 1), \qquad \beta_{m,n}^{(\mathrm{I})} = -2m^2, \tag{4.21}$$
$$\alpha_{m,n}^{(\mathrm{II})} = 2m + n + 1, \qquad \beta_{m,n}^{(\mathrm{II})} = -2n^2. \tag{4.22}$$

Remarks 4.8.

1. The rational solutions of P_{IV} defined by (4.19) and (4.20) include all the solutions in the "$-1/z$" and "$-2z$" hierarchies, as is easily verified by comparing the parameters in (4.21) and (4.22) with those in (4.5a) and (4.5b). Further they are the set of rational solutions of P_{IV} with parameter values given by (4.1).

2. Each generalized Hermite polynomial $H_{m,n}(z)$ is a polynomial of degree mn with integer coefficients (Noumi and Yamada, 1999). In fact $H_{m,n}\left(\frac{1}{2}x\right)$ is a monic polynomial in x of degree mn with integer coefficients.

3. The polynomials $H_{m,n}(z)$ possess the symmetry $H_{m,n}(\mathrm{i}z) = \mathrm{i}^{mn} H_{n,m}(z)$, where mn is the degree of $H_{m,n}(z)$.

4. $H_{n,1}(z) = H_n(z)$ and $H_{1,n}(z) = \mathrm{i}^{-n} H_n(\mathrm{i}z)$, where $H_n(z)$ is the usual Hermite polynomial defined by

$$H_n(z) = (-1)^n \exp(z^2) \frac{\mathrm{d}^n}{\mathrm{d}z^n} \left\{\exp(-z^2)\right\}. \tag{4.23}$$

Some generalized Hermite polynomials $H_{m,n}(z)$ are given in Appendix A. Plots of the locations of the roots of the polynomials and $H_{m,7}(z)$, $H_{7,n}(z)$ for $4 \leq m \leq 6$ and $4 \leq n \leq 6$, are given in Figure 7. These plots, which are invariant under reflections in the real and imaginary z-axes, take the form of $m \times n$ "rectangles", though these are only approximate rectangles as can be seen by looking at the actual values of the roots.

4.4 Generalized Okamoto polynomials

In this section we discuss the *generalized Okamoto polynomials* $P_{m,n}(z)$ which were introduced by Noumi and Yamada (Noumi and Yamada, 1999) and are defined in Theorem 4.9 below. We have reindexed these polynomials by setting $Q_{m,n}(z) = P_{m-n,n}(z)$, i.e., $Q_{m+n,n}(z) = P_{m,n}(z)$, where $Q_{m+n,n}(z)$ is the polynomial defined Noumi and Yamada (Noumi and Yamada, 1999), since we feel that $P_{m,n}(z)$ is more natural, especially when one studies the plots of the locations of the roots for various generalized Okamoto polynomials in Figure 8.

Theorem 4.9. *Suppose that $P_{m,n}(z)$ satisfies the recurrence relations*

$$P_{m+1,n}P_{m-1,n} = \frac{9}{2}\left\{P_{m,n}P''_{m,n} - \left(P'_{m,n}\right)^2\right\} + \left[2z^2 + 3(2m+n-1)\right] P^2_{m,n}, \tag{4.24a}$$

$$P_{m,n+1}P_{m,n-1} = \frac{9}{2}\left\{P_{m,n}P''_{m,n} - \left(P'_{m,n}\right)^2\right\} + \left[2z^2 + 3(1-m-2n)\right] P^2_{m,n}, \tag{4.24b}$$

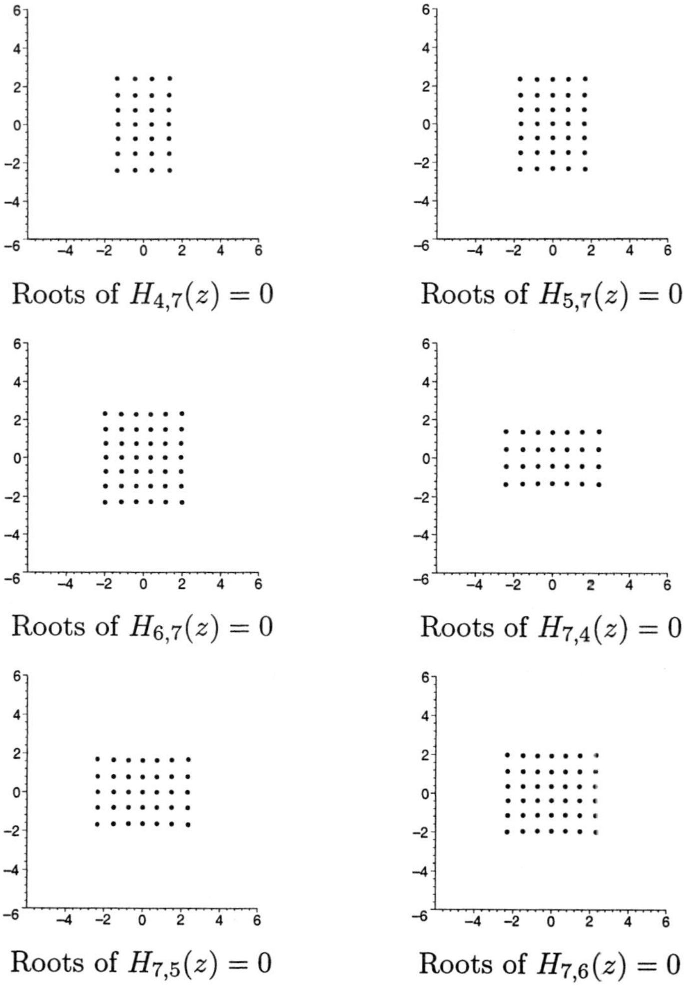

Figure 7. Roots of generalized Hermite polynomials defined by (4.18)

with

$$P_{0,0} = P_{1,0} = P_{0,1} = 1, \qquad P_{1,1} = \sqrt{2}\,z, \tag{4.24c}$$

then

$$w_{m,n}^{(\mathrm{I})} = w\left(z; \alpha_{m,n}^{(\mathrm{I})}, \beta_{m,n}^{(\mathrm{I})}\right) = -\frac{2}{3}z - \frac{\mathrm{d}}{\mathrm{d}z}\left\{\ln\left(\frac{P_{m,n+1}}{P_{m,n}}\right)\right\}, \tag{4.25}$$

$$w_{m,n}^{(\mathrm{II})} = w\left(z; \alpha_{m,n}^{(\mathrm{II})}, \beta_{m,n}^{(\mathrm{II})}\right) = -\frac{2}{3}z + \frac{\mathrm{d}}{\mathrm{d}z}\left\{\ln\left(\frac{P_{m+1,n}}{P_{m,n}}\right)\right\}, \tag{4.26}$$

are solutions of $\mathrm{P_{IV}}$*, respectively for the parameters*

$$\alpha_{m,n}^{(\mathrm{I})} = -2n - m, \qquad \beta_{m,n}^{(\mathrm{I})} = -\frac{2}{9}(3m-1)^2, \tag{4.27}$$

$$\alpha_{m,n}^{(\mathrm{II})} = 2m + n, \qquad \beta_{m,n}^{(\mathrm{II})} = -\frac{2}{9}(3n-1)^2. \tag{4.28}$$

Remarks 4.10.

1. The rational solutions of $\mathrm{P_{IV}}$ defined by (4.25) and (4.26) include all the solutions in the "$-\frac{2}{3}z$" hierarchy, as is easily verified by comparing the parameters in (4.27) and (4.28) with those in (4.5c). Further they are the set of rational solutions of $\mathrm{P_{IV}}$ with parameter values given by (4.2).

2. Each polynomial $P_{m,n}(z)$ is a polynomial of degree $d_{m,n} = m^2 + n^2 + mn - m - n$ with integer coefficients (Noumi and Yamada, 1999). Further $P_{m,n}(z)$ is a monic polynomial in $\zeta = \sqrt{2}\,z$ of degree $d_{m,n}$ with integer coefficients.

3. The original Okamoto polynomials defined in Theorems 4.3 and 4.5 are respectively given by $Q_m(z) = P_{m,0}(z)$ and $R_m(z) = P_{m,1}(z)$.

4. The polynomials $P_{m,n}(z)$ possess the symmetry $P_{m,n}(\mathrm{i}z) = \exp(\frac{1}{2}\pi\mathrm{i}d_{m,n})\,P_{n,m}(z)$, where $d_{m,n} = m^2 + n^2 + mn - m - n$ is the degree of $P_{m,n}(z)$.

5. The hierarchies of rational solutions of $\mathrm{P_{IV}}$ generated from the generalized Hermite polynomials $H_{m,n}(z)$ defined in Theorem 4.7 and the generalized Okamoto polynomials $P_{m,n}(z)$ defined in Theorem 4.9 are linked by the Schlesinger transformations $\mathcal{R}^{[2]}$ (or $\mathcal{R}^{[4]}$) and $\mathcal{R}^{[5]} \equiv \mathcal{R}^{[1]}\mathcal{R}^{[3]}$ given by Fokas, Mugan and Ablowitz (Fokas et al., 1988).

Some generalized Okamoto polynomials $P_{m,n}(z)$ are given in Appendix B. Plots of the locations of the roots of the polynomials and

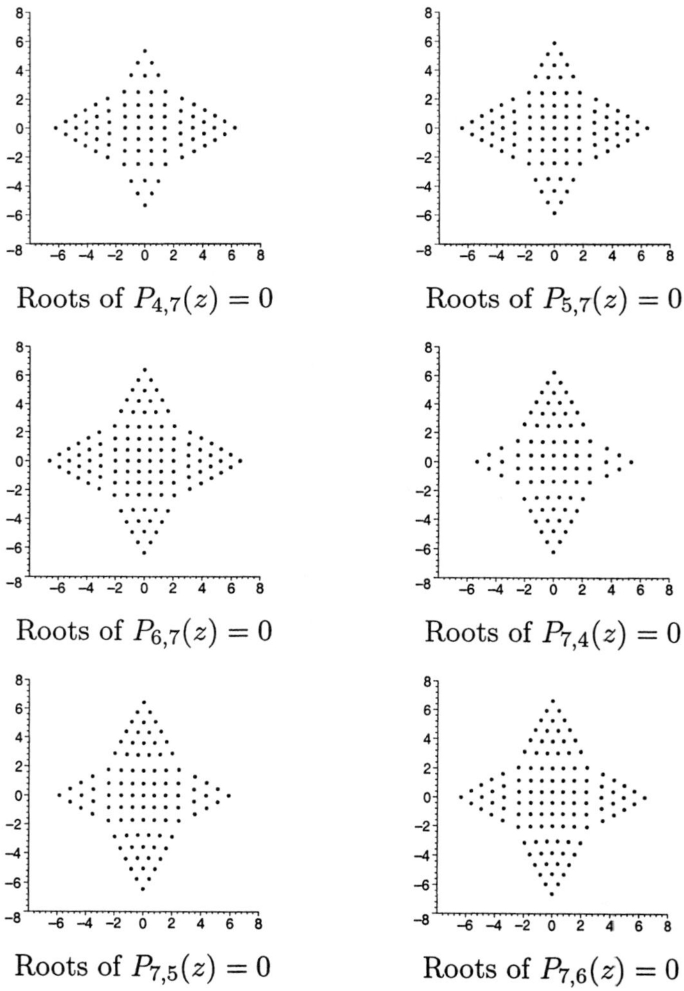

Roots of $P_{4,7}(z) = 0$ Roots of $P_{5,7}(z) = 0$

Roots of $P_{6,7}(z) = 0$ Roots of $P_{7,4}(z) = 0$

Roots of $P_{7,5}(z) = 0$ Roots of $P_{7,6}(z) = 0$

Figure 8. Roots of generalized Okamoto polynomials defined by (4.24)

$P_{m,7}(z)$, $P_{7,n}(z)$ for $4 \le m \le 6$ and $4 \le n \le 6$, are given in Figure 8. The roots of the polynomial $P_{m,n}(z)$ takes the form of $m \times n$ "rectangles" with an "equilateral triangle," which have either $m-1$ or $n-1$ roots on each of its sides. These are only approximate rectangles and equilateral triangles as can be seen by looking at the actual values of the roots. The triangles We remark that as for the Bi-Hermite polynomials above, the plots are invariant under reflections in the real and imaginary z-axes.

5. Conclusions

An important, well-known property of classical orthogonal polynomials, such as the Hermite, Laguerre or Legendre polynomials whose roots all lie on the real line (cf. (Abramowitz and Stegun, 1972; Andrews et al., 1999; Temme, 1996)), is that the roots of successive polynomials interlace. Thus for a set of orthogonal polynomials $\varphi_n(z)$, for $n = 0, 1, 2, \ldots$, if $z_{n,m}$ and $z_{n,m+1}$ are two successive roots of $\varphi_n(z)$, i.e., $\varphi_n(z_{n,m}) = 0$ and $\varphi_n(z_{n,m+1}) = 0$, then $\varphi_{n-1}(\zeta_{n-1}) = 0$ and $\varphi_{n+1}(\zeta_{n+1}) = 0$ for some ζ_{n-1} and ζ_{n+1} such that $z_{n,m} < \zeta_{n-1}, \zeta_{n+1} < z_{n,m+1}$. Further the derivatives $\varphi_n'(z)$ and $\varphi_{n+1}'(z)$ also have roots in the interval $(z_{n,m}, z_{n,m+1})$, that is $\varphi_n'(\xi_n) = 0$ and $\varphi_{n+1}'(\xi_{n+1}) = 0$ for some ξ_n and ξ_{n+1} such that $z_{n,m} < \xi_n, \xi_{n+1} < z_{n,m+1}$.

An interesting open question is whether there are analogous results for the polynomials associated with rational and algebraic solutions of the Painlevé equations discussed here. Clearly there are notable differences since these special polynomials have complex roots whereas classical orthogonal polynomials $\varphi_n(z)$, have real roots. The pattern of the roots of the special polynomials associated with the Painlevé equations are highly symmetric and structured, suggesting that they have interesting properties. A particularly interesting question is whether there is any "interlacing of roots" analogous to that for classical orthogonal polynomials. Clearly this warrants further analytical study as does an investigation of the relative locations of the roots for the special polynomials and their derivatives. We shall not pursue these questions further here.

Acknowledgments

I wish to thank Mark Ablowitz, Frédérick Cheyzak, Chris Cosgrove, Andy Hone, Nalini Joshi, Arieh Iserles, Alexander Its, Elizabeth Mansfield and Marta Mazzocco for their helpful comments and illuminating discussions. This research arose whilst involved with the "Digital Library of Mathematical Functions" project (see `http://dlmf.nist.gov/`), in particular I thank Frank Olver and Dan Lozier for the opportunity to participate in this project.

References

Ablowitz, M. J. and Clarkson, P. A. (1991). *Solitons, Nonlinear Evolution Equations and Inverse Scattering*, volume 149 of *Lecture Notes in Mathematics*. Cambridge University Press, Cambridge.

Ablowitz, M. J. and Satsuma, J. (1978). Solitons and rational solutions of nonlinear evolution equations. *J. Math. Phys.*, 19:2180–2186.

Abramowitz, M. and Stegun, I. A. (1972). *Handbook of Mathematical Functions*. Dover, New York, tenth edition.

Adler, M. and Moser, J. (1978). On a class of polynomials associated with the Korteweg-de Vries equation. *Commun. Math. Phys.*, 61:1–30.

Adler, V. E. (1994). Nonlinear chains and Painlevé equations. *Physica*, D73:335–351.

Airault, H. (1979). Rational solutions of Painlevé equations. *Stud. Appl. Math.*, 61:31–53.

Airault, H., McKean, H. P., and Moser, J. (1977). Rational and elliptic solutions of the KdV equation and related many-body problems. *Commun. Pure Appl. Math.*, 30:95–148.

Albrecht, D. W., Mansfield, E. L., and Milne, A. E. (1996). Algorithms for special integrals of ordinary differential equations. *J. Phys. A: Math. Gen.*, 29:973–991.

Andrews, G., Askey, R., and Roy, R. (1999). *Special Functions*, volume 71 of *Encyclopedia of Mathematics and its Applications*. Cambridge University Press, Cambridge.

Bassom, A. P., Clarkson, P. A., and Hicks, A. C. (1995). Bäcklund transformations and solution hierarchies for the fourth Painlevé equation. *Stud. Appl. Math.*, 95:1–71.

Bureau, F. (1992). Differential equations with fixed critical points. In *Painlevé transcendents (Sainte-Adèle, PQ, 1990)*, volume 278 of *NATO Adv. Sci. Inst. Ser. B Phys.*, pages 103–123. Plenum, New York.

Charles, P. J. (2002). *Painlevé analysis and the study of continuous and discrete Painlevé equations*. PhD thesis, Institute of Mathematics and Statistics, University of Kent, UK.

Clarkson, P. A. and Kruskal, M. D. (1989). New similarity solutions of the Boussinesq equation. *J. Math. Phys.*, 30:2201–2213.

Clarkson, P. A. and Mansfield, E. L. (2003). The second Painlevé equation, its hierarchy and associated special polynomials. *Nonlinearity*, 16:R1–R26.

Flaschka, H. and Newell, A. C. (1980). Monodromy-and spectrum preserving deformations. I. *Commun. Math. Phys.*, 76:65–116.

Fokas, A. S. and Ablowitz, M. J. (1982). On a unified approach to transformations and elementary solutions of Painlevé equations. *J. Math. Phys.*, 23:2033–2042.

Fokas, A. S., Muğan, U., and Ablowitz, M. J. (1988). A method of linearisation for Painlevé equations: Painlevé IV, V. *Physica*, D30:247–283.

Fokas, A. S. and Yortsos, Y. C. (1981). The transformation properties of the sixth painlevé equation and one-parameter families of solutions. *Lett. Nuovo Cim.*, 30:539–544.

Fukutani, S., Okamoto, K., and Umemura, H. (2000). Special polynomials and the Hirota bilinear relations of the second and fourth Painlevé equations. *Nagoya Math. J.*, 159:179–200.

Gambier, B. (1910). Sur les équations différentielles du second ordre et du premeir degre dont l'intégrale générale est à points critiques fixés. *Acta Math.*, 33:1–55.

Gromak, V. I. (1973). The solutions of Painlevé's third equation. *Diff. Eqns.*, 9:1599–1600.

Gromak, V. I. (1975). Theory of Painlevé's equation. *Diff. Eqns.*, 11:285–287.

Gromak, V. I. (1976). Solutions of Painlevé's fifth equation. *Diff. Eqns.*, 12:519–521.

Gromak, V. I. (1978a). Algebraic solutions of the third Painlevé equation. *Dokl. Akad. Nauk BSSR*, 23:499–502.

Gromak, V. I. (1978b). One-parameter systems of solutions of Painlevé's equations. *Diff. Eqns.*, 14:1510–1513.

Gromak, V. I. (1987). Theory of the fourth painlevé equation. *Diff. Eqns.*, 23:506–513.

Gromak, V. I. (1999). Bäcklund transformations of Painlevé equations and their applications. In Conte, R., editor, *The Painlevé Property, One Century Later*, CRM Series in Mathematical Physics, pages 687–734. Springer, New York.

Gromak, V. I. (2001). Bäcklund transformations of the higher order Painlevé equations. In Coley, A., Levi, D., Milson, R., Rogers, C., and Winternitz, P., editors, *Bäcklund and Darboux transformations. The geometry of solitons (Halifax, NS, 1999)*, volume 29 of *CRM Proc. Lecture Notes*, pages 3–28. Amer. Math. Soc., Providence, RI.

Gromak, V. I., Laine, I., and Shimomura, S. (2002). *Painlevé Differential Equations in the Complex Plane*, volume 28 of *Studies in Mathematics*. de Gruyter, Berlin, New York.

Gromak, V. I. and Lukashevich, N. A. (1982). Special classes of solutions of Painlevé's equations. *Diff. Eqns.*, 18:317–326.

Iwasaki, K., Kajiwara, K., and Nakamura, T. (2002). Generating function associated with the rational solutions of the Painlevé II equation. *J. Phys. A: Math. Gen.*, 35:L207–L211.

Iwasaki, K., Kimura, H., Shimomura, S., and Yoshida, M. (1991). *From Gauss to Painlevé: a Modern Theory of Special Functions*, volume 16 of *Aspects of Mathematics E*. Viewag, Braunschweig, Germany.

Jimbo, M. and Miwa, T. (1983). Solitons and infinite dimensional Lie algebras. *Publ. RIMS, Kyoto Univ.*, 19:943–1001.

Kajiwara, K. and Masuda, T. (1999a). A generalization of determinant formulae for the solutions of Painlevé II and XXXIV equations. *J. Phys. A: Math. Gen.*, 32:3763–3778.

Kajiwara, K. and Masuda, T. (1999b). On the Umemura polynomials for the Painlevé III equation. *Phys. Lett.*, A260:462–467.

Kajiwara, K. and Ohta, Y. (1996). Determinantal structure of the rational solutions for the painlevé II equation. *J. Math. Phys.*, 37:4393–4704.

Kajiwara, K. and Ohta, Y. (1998). Determinant structure of the rational solutions for the Painlevé IV equation. *J. Phys. A: Math. Gen.*, 31:2431–2446.

Kametaka, Y. (1983). On poles of the rational solution of the Toda equation of Painlevé-II type. *Proc. Japan Acad. Ser. A Math. Sci.*, 59:358–360.

Kametaka, Y. (1985). On the irreducibility conjecture based on computer calculation for Yablonskiĭ-Vorobev polynomials which give a rational solution of the Toda equation of Painlevé-II type. *Japan J. Appl. Math.*, 2:241–246.

Kaneko, M. and Ochiai, H. (2002). On coefficients of Yablonskii-Vorob'ev polynomials. Preprint, arXiv:math.QA/0205178.

Kirillov, A. N. and Taneda, M. (2002a). Generalized Umemura polynomials. *Rocky Mount. J. Math.*, 32:691–702.

Kirillov, A. N. and Taneda, M. (2002b). Generalized Umemura polynomials and Hirota-Miwa equations. In Kashiwara, M. and Miwa, T., editors, *MathPhys Odyssey, 2001*, volume 23 of *Prog. Math. Phys.*, pages 313–331. Birkhauser-Boston, Boston, MA.

Lukashevich, N. A. (1965). Elementary solutions of certain Painlevé equations. *Diff. Eqns.*, 1:561–564.

Lukashevich, N. A. (1967a). On the theory of the third Painlevé equation. *Diff. Eqns.*, 3:994–999.

Lukashevich, N. A. (1967b). Theory of the fourth Painlevé equation. *Diff. Eqns.*, 3:395–399.

Lukashevich, N. A. (1968). Solutions of the fifth equation of painlevé equation. *Diff. Eqns.*, 4:732–735.

Lukashevich, N. A. (1971). The second painlevé equation. *Diff. Eqns.*, 6:853–854.

Lukashevich, N. A. and Yablonskii, A. I. (1967). On a class of solutions of the sixth painlevé equation. *Diff. Eqns.*, 3:264–266.

Masuda, T. (2002). On a class of algebraic solutions to Painlevé VI equation, its determinant formula and coalescence cascade. preprint, arXiv:nlin.SI/0202044.

Masuda, T., Ohta, Y., and Kajiwara, K. (2002). A determinant formula for a class of rational solutions of Painlevé V equation. *Nagoya Math. J.*, 168:1–25.

Mazzocco, M. (2001). Rational solutions of the Painlevé VI equation. *J. Phys. A: Math. Gen.*, 34:2281–2294.

Milne, A. E., Clarkson, P. A., and Bassom, A. P. (1997). Bäcklund transformations and solution hierarchies for the third Painlevé equation. *Stud. Appl. Math.*, 98:139–194.

Murata, Y. (1985). Rational solutions of the second and the fourth Painlevé equations. *Funkcial. Ekvac.*, 28:1–32.

Murata, Y. (1995). Classical solutions of the third Painlevé equations. *Nagoya Math. J.*, 139:37–65.

Noumi, M. and Yamada, Y. (1998a). Affine Weyl groups, discrete dynamical systems and Painlevé equations. *Commun. Math. Phys.*, 199:281–295.

Noumi, M. and Yamada, Y. (1998b). Umemura polynomials for the Painlevé V equation. *Phys. Lett.*, A247:65–69.

Noumi, M. and Yamada, Y. (1999). Symmetries in the fourth Painlevé equation and Okamoto polynomials. *Nagoya Math. J.*, 153:53–86.

Noumi M., Okada S., O. K. and H., U. (1998). Special polynomials associated with the Painlevé equations. II. In Saito, M.-H., Shimizu, Y., and Ueno, R., editors, *Integrable Systems and Algebraic Geometry*, pages 349–372. World Scientific, Singapore.

Ohyama, Y. (2001). On the third Painlevé equations of type d_7. Preprint, Department of Pure and Applied Mathematics, Graduate School of Information Science and Technology, Osaka University.

Okamoto, K. (1986). Studies on the Painlevé equations III. Second and fourth Painlevé equations, $P_{\rm II}$ and $P_{\rm IV}$. *Math. Ann.*, 275:221–255.

Okamoto, K. (1987a). Studies on the Painlevé equations I. Sixth Painlevé equation $P_{\rm VI}$. *Ann. Mat. Pura Appl. (4)*, 146:337–381.

Okamoto, K. (1987b). Studies on the Painlevé equations II. Fifth Painlevé equation $P_{\rm V}$. *Japan. J. Math.*, 13:47–76.

Okamoto, K. (1987c). Studies on the Painlevé equations IV. Third Painlevé equation $P_{\rm III}$. *Funkcial. Ekvac.*, 30:305–332.

Sakai, H. (2001). Rational surfaces associated with affine root systems and geometry of the Painlevé equations. *Commun. Math. Phys.*, 220:165–229.

Schiff, J. (1995). Bäcklund transformations of MKdV and Painlevé equations. *Nonlinearity*, 7:305–312.

Tamizhmani, T., Grammaticos, B., Ramani, A., and Tamizhmani, K. M. (2001). On a class of special solutions of the Painlevé equations. *Physica*, A295:359–370.

Taneda, M. (2000). Remarks on the Yablonskii-Vorob'ev polynomials. *Nagoya Math. J.*, 159:87–111.

Taneda, M. (2001a). Polynomials associated with an algebraic solution of the sixth Painlevé equation. *Japan. J. Math.*, 27:257–274.

Taneda, M. (2001b). Representation of Umemura polynomials for the sixth Painlevé equation by the generalized Jacobi polynomials. In Kirillov, A. N., Tsuchiya, A., and Umemura, H., editors, *Physics and Combinatorics*, pages 366–376. World Scientific, Singapore.

Temme, N. M. (1996). *Special Functions. An Introduction to the Classical Functions of Mathematical Physics.* Wiley, New York.

Umemura, H. (1998). Painlevé equations and classical functions. *Sugaku Expositions*, 11:77–100.

Umemura, H. (2000). On the transformation group of the second Painlevé equation. *Nagoya Math. J.*, 157:15–46.

Umemura, H. (2001). Painlevé equations in the past 100 years. *AMS Translations*, 204:81–110.

Umemura, H. (2003). Special polynomials associated with the Painlevé equations. I. In Vinet, L. and Winternitz, P., editors, *Theory of Nonlinear Special Functions: the Painlevé Transcendents (Montreal, Canada, 1996)*, CRM Series in Mathematical Physics. Springer-Verlag, New York. To appear.

Umemura, H. and Watanabe, H. (1997). Solutions of the second and fourth Painlevé equations I. *Nagoya Math. J.*, 148:151–198.

Umemura, H. and Watanabe, H. (1998). Solutions of the third Painlevé equation I. *Nagoya Math. J.*, 151:1–24.

Veselov, A. P. and Shabat, A. B. (1993). A dressing chain and the spectral theory of the Schrödinger operator. *Funct. Anal. Appl.*, 27:1–21.

Vorob'ev, A. P. (1965). On rational solutions of the second Painlevé equation. *Diff. Eqns.*, 1:58–59.

Watanabe, H. (1995). Solutions of the fifth Painlevé equation I. *Hokkaido Math. J.*, 24:231–267.

Yablonskii, A. I. (1959). On rational solutions of the second Painlevé equation. *Vesti Akad. Navuk. BSSR Ser. Fiz. Tkh. Nauk.*, 3:30–35. In Russian.

Yamada, Y. (2000). Special polynomials and generalized Painlevé equations. In Koike, K., Kashiwara, M., Okada, S., Terada, I., and Yamada, H. F., editors, *Combina-*

torial Methods in Representation Theory (Kyoto, 1998), volume 28 of *Adv. Stud. Pure Math.*, pages 391–400. Kinokuniya, Tokyo.

Yuan, W.-J. and Li, Y.-Z. (2002). Rational solutions of painlevé equations. *Canad. J. Math.*, 54:648–670.

Appendix: The Generalized Hermite polynomials

$$H_{n,0} = H_{0,n} = 1 \qquad n \geq 0$$
$$H_{1,1} = 2z$$
$$H_{1,2} = 4z^2 + 2$$
$$H_{1,3} = 8z^3 + 12z$$
$$H_{1,4} = 16z^4 + 48z^2 + 12$$
$$H_{1,5} = 32z^5 + 160z^3 + 120z$$
$$H_{2,1} = 4z^2 - 2$$
$$H_{2,2} = 16z^4 + 12$$
$$H_{2,3} = 64z^6 + 96z^4 + 144z^2 - 72$$
$$H_{2,4} = 256z^8 + 1024z^6 + 1920z^4 + 720$$
$$H_{2,5} = 1024z^{10} + 7680z^8 + 23040z^6 + 19200z^4 + 14400z^2 - 7200$$
$$H_{3,1} = 8z^3 - 12z$$
$$H_{3,2} = 64z^6 - 96z^4 + 144z^2 + 72$$
$$H_{3,3} = 512z^9 + 2304z^5 - 4320z$$
$$H_{3,4} = 4096z^{12} + 12288z^{10} + 46080z^8 + 30720z^6 - 57600z^4 + 172800z^2 + 43200$$
$$H_{3,5} = 32768z^{15} + 245760z^{13} + 1105920z^{11} + 2150400z^9 + 1382400z^7 + 4838400z^5 - 4032000z^3 - 6048000z$$
$$H_{4,1} = 16z^4 - 48z^2 + 12$$
$$H_{4,2} = 256z^8 - 1024z^6 + 1920z^4 + 720$$
$$H_{4,3} = 4096z^{12} - 12288z^{10} + 46080z^8 - 30720z^6 - 57600z^4 - 172800z^2 + 43200$$
$$H_{4,4} = 65536z^{16} + 983040z^{12} - 1843200z^8 + 32256000z^4 + 6048000$$
$$H_{4,5} = 1048576z^{20} + 5242880z^{18} + 35389440z^{16} + 78643200z^{14} + 68812800z^{12} + 206438400z^{10} + 1290240000z^8 - 3612672000z^6 - 3386880000z^4 - 3386880000z^2 + 846720000$$
$$H_{5,1} = 32z^5 - 160z^3 + 120z$$
$$H_{5,2} = 1024z^{10} - 7680z^8 + 23040z^6 - 19200z^4 + 14400z^2 + 7200 + 1024z$$
$$H_{5,3} = 32768z^{15} - 245760z^{13} + 1105920z^{11} - 2150400z^9 + 1382400z^7 - 4838400z^5 - 4032000z^3 + 6048000z$$
$$H_{5,4} = 1048576z^{20} - 5242880z^{18} + 35389440z^{16} - 78643200z^{14} + 68812800z^{12} - 206438400z^{10} + 1290240000z^8 + 3612672000z^6 - 3386880000z^4 + 3386880000z^2 + 846720000$$
$$H_{5,5} = 33554432z^{25} + 1258291200z^{21} + 3303014400z^{17} + 115605504000z^{13} - 2059223040000z^9 - 3413975040000z^5 + 2133734400000z$$

Appendix: The Generalized Okamoto polynomials

$$Q_{0,0} = Q_{0,1} = 1$$

$$Q_{0,2} = 2z^2 - 3$$

$$Q_{0,3} = 8z^6 - 60z^4 + 90z^2 - 135$$

$$Q_{0,4} = 64z^{12} - 1344z^{10} + 9360z^8 - 30240z^6 + 56700z^4 - 170100z^2 + 127575$$

$$\begin{aligned} Q_{0,5} = {} & 1024z^{20} - 46080z^{18} + 817920z^{16} - 7603200z^{14} + 41731200z^{12} \\ & - 155675520z^{10} + 493970400z^8 - 1886068800z^6 + 5304568500z^4 \\ & - 5304568500z^2 + 3978426375 \end{aligned}$$

$$Q_{1,0} = 1$$

$$Q_{1,1} = \sqrt{2}\,z$$

$$Q_{1,2} = 4z^4 - 12z^2 - 9$$

$$Q_{1,3} = \sqrt{2}\,z(16z^8 - 192z^6 + 504z^4 - 2835)$$

$$\begin{aligned} Q_{1,4} = {} & 256z^{16} - 7680z^{14} + 80640z^{12} - 362880z^{10} + 453600z^8 \\ & + 1905120z^6 - 14288400z^4 + 21432600z^2 + 8037225 \end{aligned}$$

$$\begin{aligned} Q_{1,5} = {} & \sqrt{2}\,z(4096z^{24} - 245760z^{22} + 5990400z^{20} - 77414400z^{18} \\ & + 569721600z^{16} - 2246952960z^{14} + 1600300800z^{12} \\ & + 35663846400z^{10} - 275837562000z^8 + 1103350248000z^6 \\ & - 1737776640600z^4 + 3258331201125) \end{aligned}$$

$$Q_{2,0} = 2z^2 + 3$$

$$Q_{2,1} = 4z^4 + 12z^2 - 9$$

$$Q_{2,2} = 16z^8 - 504z^4 - 567$$

$$\begin{aligned} Q_{2,3} = {} & 128z^{14} - 1344z^{12} - 6048z^{10} + 75600z^8 - 158760z^6 - 238140z^4 \\ & - 1071630z^2 + 535815 \end{aligned}$$

$$\begin{aligned} Q_{2,4} = {} & 2048z^{22} - 64512z^{20} + 483840z^{18} + 3144960z^{16} - 61689600z^{14} \\ & + 297198720z^{12} - 445798080z^{10} - 1114495200z^8 - 5851099800z^6 \\ & + 43883248500z^4 - 13164974550z^2 + 19747461825 \end{aligned}$$

$$\begin{aligned} Q_{2,5} = {} & 65536z^{32} - 4325376z^{30} + 106168320z^{28} - 1021870080z^{26} \\ & - 3019161600z^{24} + 169374965760z^{22} - 1749906063360z^{20} \\ & + 8630650828800z^{18} - 16958158963200z^{16} - 26480405952000z^{14} \\ & - 87626070604800z^{12} + 397603295369280z^{10} \\ & - 1897652091535200z^8 + 2846478137302800z^6 \\ & + 640457580893130z^2 - 24017159283492375 \end{aligned}$$

$$
\begin{aligned}
Q_{3,0} &= 8z^6 + 60z^4 + 90z^2 + 135 \\
Q_{3,1} &= \sqrt{2}\,z(16z^8 + 192z^6 + 504z^4 - 2835) \\
Q_{3,2} &= 128z^{14} + 1344z^{12} - 6048z^{10} - 75600z^8 - 158760z^6 + 238140z^4 \\
&\qquad - 1071630z^2 - 535815 \\
Q_{3,3} &= \sqrt{2}\,z(1024z^{20} - 241920z^{16} + 12700800z^{12} - 371498400z^8 \\
&\qquad - 2925549900z^4 + 6582487275) \\
Q_{3,4} &= 32768z^{30} - 737280z^{28} - 7741440z^{26} + 251596800z^{24} \\
&\qquad - 377395200z^{22} - 21398307840z^{20} + 89159616000z^{18} \\
&\qquad + 740024812800z^{16} - 6753840912000z^{14} + 9127715688000z^{12} \\
&\qquad - 35598091183200z^{10} + 390209845662000z^8 \\
&\qquad + 646926849387000z^6 - 1617317123467500z^4 \\
&\qquad + 2425975685201250z^2 + 727792705560375 \\
Q_{3,5} &= \sqrt{2}\,z(1048576z^{40} - 62914560z^{38} + 920125440z^{36} + 14722007040z^{34} \\
&\qquad - 543449088000z^{32} + 3594009968640z^{30} + 44997584486400z^{28} \\
&\qquad - 748655463628800z^{26} + 1257364568678400z^{24} \\
&\qquad + 46066277117952000z^{22} - 433373509191168000z^{20} \\
&\qquad + 1556239013941248000z^{18} - 2809510888766400000z^{16} \\
&\qquad + 3686078286061516800 0z^{14} - 20059907745792096000 0z^{12} \\
&\qquad - 368158306863949056000\,z^{10} + 427121160697628397000 0z^8 \\
&\qquad - 569494880930171196000 0z^6 + 1494924062441699389500 0z^4 \\
&\qquad - 16817895702469118131875)
\end{aligned}
$$

ZETA FUNCTIONS OF HEISENBERG GRAPHS OVER FINITE RINGS

Michelle DeDeo, María Martínez, Archie Medrano, Marvin Minei, Harold Stark, and Audrey Terras
Department of Mathematics
University of California, San Diego
La Jolla, CA 92092-0112
aterras@ucsd.edu

Abstract We investigate Ihara-Selberg zeta functions of Cayley graphs for the Heisenberg group over finite rings $\mathbb{Z}/p^n\mathbb{Z}$, where p is a prime. In order to do this, we must compute the Galois group of the covering obtained by reducing coordinates in $\mathbb{Z}/p^{n+1}\mathbb{Z}$ modulo p^n. The Ihara-Selberg zeta functions of the Heisenberg graph mod p^{n+1} factor as a product of Artin L-functions corresponding to the irreducible representations of the Galois group of the covering. Emphasis is on graphs of degree four. These zeta functions are compared with zeta functions of finite torus graphs which are Cayley graphs for the abelian groups $(\mathbb{Z}/p^n\mathbb{Z})^r$.

1. Introduction

The aim of this paper is to study the special functions known as Ihara-Selberg zeta functions for Cayley graphs of finite Heisenberg groups as well as their factorizations into products of Artin-Ihara L-functions. The **Heisenberg group** $H(R)$ over a ring R consists of upper triangular 3×3 matrices with entries in R and ones on the diagonal. The Ihara-Selberg zeta function is analogous to the Riemann zeta function with primes replaced by certain closed paths in a graph. This paper is a continuation of (DeDeo et al., 2004) where we presented a study of the statistics of the spectra of adjacency matrices of finite Heisenberg graphs.

When R is the field of real numbers $\mathbb{R}$, the group is well known for its connection with the uncertainty principle in quantum physics. When the ring R is $\mathbb{Z}$, the ring of integers, there are degree 4 and 6 Cayley graphs (see the next paragraph) associated to $H(\mathbb{Z})$ whose spectra (i.e., eigenvalues of the adjacency matrix) have been much studied starting with D. R. Hofstadter's work on energy levels of Bloch electrons (Hofstadter, 1976) which includes a picture of the Hofstadter butterfly. This subject also goes under the name of the spectrum of the almost Mathieu operator or the Harper operator. See (DeDeo et al., 2004) and (Terras, 1999) for more information on the Heisenberg group. See also (Kotani and Sunada, 2000).

If S is a subset of a finite group G, the **Cayley graph** $X(G, S)$ has as its vertex set the set G. Edges connect vertices $g \in G$ and gs, for all $s \in S$. Usually we will assume that $s \in S$ implies $s^{-1} \in S$ so that the graph is undirected. And we will normally assume that S is a set of generators of G so that the graph will be connected. It is not hard to see that the spectrum of the adjacency matrix of $X(G, S)$ is contained in the interval $[-k, k]$, if $k = |S|$.

Heisenberg groups over finite fields have provided examples of random number generators (see (Zack, 1990)) as well Ramanujan graphs (see (Myers, 1995)). **Ramanujan graphs** were defined by (Lubotzky et al., 1988) to be finite connected k-regular graphs such that the eigenvalues λ of the adjacency matrix (not equal to k or $-k$) satisfy $|\lambda| \leq 2\sqrt{k-1}$. Other references are (Diaconis and Saloff-Coste, 1994) and (Terras, 1999). As shown in the last reference, the size of the second largest (in absolute value) eigenvalue of the adjacency matrix governs the speed of convergence to uniform for the standard random walk on a connected regular graph. Ramanujan graphs have the best possible eigenvalue bound for connected regular graphs of fixed degree in an infinite sequence of graphs with number of vertices going to infinity. For such graphs, the random walker gets lost as quickly as possible. Equivalently, this says that such graphs can be used to build efficient communication networks.

There are more reasons to study the Heisenberg group. First, as a nilpotent group (see (Terras, 1999) for the definition), it may be viewed as the closest to abelian. Second, it is an important subgroup of $GL(3, R)$ (the general linear group of matrices x such that x and x^{-1} have entries in the ring R) for those interested in creating a finite model of the symmetric space of the real $GL(3, \mathbb{R})$ analogous to the finite upper half plane model of the Poincaré upper half plane.

Some of our motivation comes from quantum chaoticists who investigate the statistics of various spectra as well as zeros of zeta functions. This MSRI website (`http://www.msri.org/`) has movies and transparencies of many talks from 1999 on the subject. See, for example, the talks of Sarnak from Spring, 1999. Other references are (Sarnak, 1995) and (Terras, 2000; Terras, 2002).

The Ihara-Selberg zeta function is an analogue of the **Riemann zeta function** $\zeta(s)$. The latter is defined for $\text{Re}(s) > 1$ by

$$\zeta(s) = \sum_{n=1}^{\infty} \frac{1}{n^s} = \prod_{p \text{ prime}} \left(1 - \frac{1}{p^s}\right)^{-1}.$$

Thanks to this Euler product, the zeros of zeta are important for any work on the statistics of the primes. The earliest results on the location of zeta zeros led to a proof of the prime number theorem which says that the number of primes less than or equal to x is asymptotic to $x/\log x$, as x goes to infinity. Now there is a million dollar prize problem to prove the Riemann hypothesis which says that the non-real zeros of the analytic continuation of $\zeta(s)$ lie on the line $\text{Re}(s) = 1/2$. This would give the best possible error estimate in the prime number theorem. For a report of experimental verification for the first 100 billion zeros, see the web site `http://www.hipilib.de/zeta/index.html`. Quantum chaoticists have experimental evidence that the zeros of zeta behave analogously to the eigenvalues of a random Hermitian matrix. See (Katz and Sarnak, 1999) for a discussion of various zeta functions whose zeros and poles have been studied in the same manner that the physicists study energy levels of physical systems.

To define a graph-theoretical analogue of $\zeta(s)$, we must define "prime" in a graph X. Modelling the idea of the Selberg zeta function of a Riemannian manifold, we use the prime cycles $[C]$ in X. Orient the edges of X, which we assume is a finite connected graph. A **prime** $[C]$ in X is an equivalence class of tailless backtrackless primitive cycles C in X. Here $C = a_1 a_2 \cdots a_s$, where the a_j are oriented edges of X. The **length** of C is $\nu(C) = s$. "Backtrackless" means that $a_{i+1} \neq a_i^{-1}$, for all i. "Tailless" means that $a_s^{-1} \neq a_1$. The "equivalence class" of C is $[C]$ which consists of all cycles $a_i a_{i+1} \cdots a_s a_1 a_2 \cdots a_{i-1}$; i.e., the same path with all possible starting points. We call the class $[C]$ "primitive" if you only go around once; i.e., $C \neq D^m$, for all integers $m \geq 2$ and all paths D in X.

The **Ihara zeta function** of a connected graph X is defined for $u \in \mathbb{C}$, with $|u|$ sufficiently small, by

$$\zeta_X(u) = \prod_{\substack{[C] \text{ prime} \\ \text{cycle in } X}} \left(1 - u^{\nu(C)}\right)^{-1}. \tag{1.1}$$

The connection with the adjacency matrix A of X is given by **Ihara's theorem** which says

$$\zeta_X(u)^{-1} = \left(1 - u^2\right)^{r-1} \det\left(I - Au + Qu^2\right), \tag{1.2}$$

where $r = |E|-|V|-1$ = rank of fundamental group of X and Q is the diagonal matrix whose jth diagonal entry is Q_{jj} =(-1+ degree of jth vertex). Proofs of the Ihara theorem can be found in (Stark and Terras, 1996; Stark and Terras, 2000), (Terras, 1999). In the first two papers, edge and path zeta functions of more than one variable are also discussed. The most elementary proof of formula (1.2) was found by Bass and involves the edge zeta function associated to more than one variable for which the analogous determinant formula is easy to prove. See (Stark and Terras, 2000) pages 168 and 172.

Remark 1.1. *The Ihara zeta function is related to walk generating functions of graphs, in particular, that for reduced walks considered by (Godsil, 1993, p. 72), but it is not the same. Differences come from not counting tails and the fact that a prime can pass through a given vertex more than once. Related generating functions have also been considered by probabilists studying first passage times for random walks but again they are different. See (Kemperman, 1961).*

We believe that it is worth singling out this special function associated to graphs for several reasons. First, for number theorists, it provides a new analogue of the Riemann zeta function which is easier to experiment on than the zeta functions of number or function fields. Secondly, it connects the zeta functions from many disparate areas such as number theory, differential geometry, and dynamical systems. Thirdly, this zeta function has a generalization to analogues of Artin L-functions. See the definition in formula (2.7)*. Thus we can make use of the Galois theory of normal covering graphs to obtain factorizations of the zeta function.*

It follows from formula (1.2) that there is an analogue of the prime number theorem for primes in a graph. This says that if X is a connected $(q+1)$-regular graph and $\pi(n)$ is the number of prime paths $[C]$ of length n in X, then

$$\pi(n) \sim \frac{q^n}{n}, \text{ as } n \to \infty. \tag{1.3}$$

The proof is easy. One can simply imitate the proof of the analogous result for function fields over finite fields in (Rosen, 2002), pp. 56–57.

From (1.2), we know that these zeta functions are reciprocals of polynomials. When the graph is connected and $(q+1)$-regular, one sees that it is a Ramanujan graph if (and only if) the Ihara-Selberg zeta function satisfies **the Riemann hypothesis** in the sense that the zeros of the polynomial satisfy $|u| = q^{-1/2}$. See (Terras, 1999, p. 418). When the zeta function satisfies the Riemann hypothesis, the error estimate in the prime number theorem (1.3) is best possible. While the Ihara zeta function of a random regular graph may satisfy the Riemann hypothesis, the zeta functions that we encounter here in the study of finite Heisenberg graphs are not Ramanujan in general. See (DeDeo et al., 2004), where it is shown that the spectrum of the adjacency matrix of the degree 4 Heisenberg graph over a finite ring with q elements approaches the interval $[-4, 4]$ as q approaches infinity.

Special values or residues of the Ihara-Selberg zeta function give graph theoretic constants such as the number of spanning trees. There are connections with famous polynomials such as the Alexander polynomials of knots. See (Lin and Wang, 2001).

Here we consider Cayley graphs $\mathcal{H}_S(q) = X(G, S)$ with vertex set the **Heisenberg group** $G = \text{Heis}(\mathbb{Z}/q\mathbb{Z})$ consisting of matrices $(x, y, z) = \begin{pmatrix} 1 & x & z \\ 0 & 1 & y \\ 0 & 0 & 1 \end{pmatrix}$, where $x, y, z \in \mathbb{Z}/q\mathbb{Z}$, $q = p^n$ and p is prime. The edge set S is chosen to have 4 elements $S = \{\ X^{\pm 1}, A^{\pm 1}\ \}$, where $X = (x, y, z)$ and $A = (a, b, c)$. We assume that $ay \not\equiv bx \pmod{p}$ to insure that the graph is connected (see (DeDeo et al., 2004)). For p odd, all these graphs are isomorphic. When $p = 2$, there are only two isomorphism classes. These facts are proved in (DeDeo et al., 2004). Define the **degree 4 Heisenberg graph**

$$\mathcal{H}(q) = X(\text{Heis}(\mathbb{Z}/q\mathbb{Z}),\ \{(\pm 1, 0, 0), (0, \pm 1, 0)\}). \tag{1.4}$$

When $p = 2$, define a second Cayley graph

$$\mathcal{H}(2^n)' = X\left(\text{Heis}(\mathbb{Z}/2^n\mathbb{Z}), \{(1, 1, 0)^{\pm 1}, (\pm 1, 0, 0)\}\right).$$

Histograms of the spectra of the degree 4 Heisenberg graphs were studied in (DeDeo et al., 2004). These figures were made using the representations of the Heisenberg group to block diagonalize the adjacency matrix of $\mathcal{H}_S(q)$. This changes the size of the eigenvalue problem from a $p^{3n} \times p^{3n}$ matrix problem to a collection of $p^n \times p^n$ matrix problems. The histograms were compared with those for the **finite torus graphs**

$$\mathcal{T}^{(n)}(q) = X\left((\mathbb{Z}/q\mathbb{Z})^n, \{\pm e_1, \pm e_2, \ldots, \pm e_n\}\right), \tag{1.5}$$

where e_i denotes a unit vector with ith component 1 and the rest 0.

Here we investigate the Ihara-Selberg zeta functions of these Heisenberg graphs. Taking $S = \{\pm(1, 0, 0), \pm(0, 1, 0)\}$, the graph $\mathcal{H}_S\left(p^{n+1}\right)$ covers the graph $\mathcal{H}_S\left(p^n\right)$ in the usual sense of covering spaces in topology. See Theorem 2.2. The covering is unramified and normal or Galois with Abelian Galois group isomorphic to the subgroup of (x, y, z) in $\text{Heis}\left(\mathbb{Z}/p^{n+1}\mathbb{Z}\right)$ such that x, y, and z are all congruent to 0 modulo p^n. This implies that the spectrum of the adjacency operator on

$\mathcal{H}_S(p^n)$ is contained in that of $\mathcal{H}_S(p^{n+1})$ and that $\zeta_{\mathcal{H}(p^n)}(u)^{-1}$ divides $\zeta_{\mathcal{H}(p^{n+1})}(u)^{-1}$. Moreover it says that the adjacency matrix of $\mathcal{H}_S(p^{n+1})$ can be block diagonalized with blocks the size of the adjacency matrix of $\mathcal{H}_S(p^n)$ associated to the characters of the Galois group. See Proposition 2.1.

The same result that implies Proposition 2.1 implies that the Ihara zeta function of $\mathcal{H}_S(p^{n+1})$ factors as a product of Artin-Ihara L-functions $L(u,\chi)$ corresponding to the characters χ of irreducible representations ρ of the Galois group of the covering. See (Hashimoto, 1990) or (Stark and Terras, 2000). We use this factorization to compute the Ihara-Selberg zeta function explicitly for the smallest Heisenberg graphs. See formulas (2.11) and (2.12). Contour maps of (powers of) the absolute value of $\zeta_{\mathcal{H}(2)}(u)^{-1}$ and $\zeta_{\mathcal{H}(4)}(u)^{-1}$ can be found in Figures 3 and 4.

The last part of this paper concerns comparisons of zeta functions for Cayley graphs of the Heisenberg group with analogous Cayley graphs for finite torus groups. We find, for example, that the zeta functions of the smallest degree four Heisenberg and torus graphs can be compared using the following formula

$$\begin{aligned} \zeta_{\mathcal{H}(4)}(u)^{-1}/\zeta_{T^{(2)}(4)}(u)^{-1} &= \left(1-u^2\right)^{48}\left(3u^2+1\right)^{20} \\ &\times \left(3u^2-2u+1\right)^4\left(3u^2+2u+1\right)^4\left(9u^4-2u^2+1\right)^{10}. \end{aligned} \tag{1.6}$$

2. Ihara-Selberg Zeta Functions

We say that Y is an **unramified finite covering** of a finite graph X if there is a covering map $\pi : Y \to X$ which is an onto graph map (i.e., taking adjacent vertices to adjacent vertices) such that for every $x \in X$ and for every $y \in \pi^{-1}(x)$, the set of points adjacent to y in Y is mapped by π one-to-one, onto the points in X which are adjacent to x. Note that when graphs have loops and multiple edges, one must be a bit more careful with this definition if one wants Galois theory to work properly. See (Stark and Terras, 2000, p. 137). A d-sheeted covering is a **normal** covering iff there are d graph automorphisms $\sigma : Y \to Y$ such that $\pi(\sigma(y)) = \pi(y)$ for all $y \in Y$. These automorphisms form the Galois group $G(Y/X)$. See (Stark and Terras, 1996; Stark and Terras, 2000) for examples of normal and non-normal coverings and the factorization of their zeta functions.

Take a spanning tree T in X. View Y as $|G|$ sheets, where each sheet is a copy of T labeled by the elements of the Galois group G. So the points of Y are (x,g), with $x \in X$ and $g \in G$. Then an element $a \in G$ acts on the cover by $a(x,g) = (x,ag)$.

Suppose the graph X has m vertices. Define the $m \times m$ matrix $A(g)$ for $g \in G$ by defining the i, j entry to be

$$A(g)_{i,j} = \text{the number of edges in } Y \text{ between } (i, e) \text{ and } (j, g), \qquad (2.1)$$

where e denotes the identity in G. Using these $m \times m$ matrices, we can find a block diagonalization of the adjacency matrix of Y as follows.

Proposition 2.1. *If Y is a normal d-sheeted covering of X with Galois group G, then the adjacency matrix of Y can be block diagonalized where the blocks are of the form*

$$M_\rho = \sum_{g \in G}^{\oplus} A(g) \otimes \rho(g),$$

each taken d_ρ = degree of ρ times, as the representations ρ run through $\widehat{G}$. Here $A(g)$ is defined in formula (2.1).

Proof. The adjacency matrix A_Y of Y has the $(i, g), (j, h)$ entry for $i, j \in X$ and $g, h \in G$ given by

$$(A_Y)_{(i,a),(j,b)} = \text{the number of edges between } (i, a) \text{ and } (j, b). \qquad (2.2)$$

and this is the same as the number of edges between (i, e) and $\left(j, a^{-1}b\right)$, if e is the identity of G.

Also define the $|G| \times |G|$ matrix $\sigma(g)$ indexed by elements $a, b \in G$:

$$(\sigma(g))_{a,b} = \begin{cases} 1, \text{ if } a^{-1}b = g, \\ 0, \text{ otherwise.} \end{cases} \qquad (2.3)$$

Note that σ is essentially the matrix of the right regular representation of G, since if δ_a is the vector with 1 in the a position and 0 everywhere else, we have $\sigma(g)\delta_a = \delta_{ag^{-1}}$.

It follows from (2.1), (2.2), and (2.3) that

$$A_Y = \sum_{g \in G} A(g) \otimes \sigma(g). \qquad (2.4)$$

One of the fundamental theorems of representation theory (see (Terras, 1999, p. 256)) says that

$$\sigma(g) \cong \sum_{\rho \in \widehat{G}}^{\oplus} d_\rho \rho(g). \qquad (2.5)$$

It follows that $A_Y \cong \sum_{\rho\in\widehat{G}}^{\oplus} d_\rho M_\rho$. This completes the proof of Proposition 2.1. □

Theorem 2.2. *Assume p is odd. $\mathcal{H}(p^{n+1})$ is an unramified graph covering of $\mathcal{H}(p^n)$. Moreover it is a normal covering with abelian Galois group*

$$\mathrm{Gal}\left(\mathcal{H}\left(p^{n+1}\right)/\mathcal{H}\left(p^{n}\right)\right) \cong \Gamma$$
$$\cong \left\{(a,b,c) \in \mathrm{Heis}\left(\mathbb{Z}/p^{n+1}\mathbb{Z}\right) | (a,b,c) \equiv 0 \pmod{p}^n\right\}.$$

Proof. The projection $\pi : \mathcal{H}\left(p^{n+1}\right) \to \mathcal{H}(p^n)$ is just the reduction of the coordinates mod p^{n+1} to coordinates mod p^n. Clearly this preserves adjacency. Moreover, given $g \in \mathcal{H}\left(p^n\right)$, if we take a point $g' \in \mathcal{H}\left(p^{n+1}\right)$ in $\pi^{-1}g$, we see that the points in $\mathcal{H}\left(p^{n+1}\right)$ adjacent to g' have the form $g's$, for $s \in S_0 = \{\ (\pm 1,0,0),(0,\pm 1,0)\ \}$. The points adjacent to g in $\mathcal{H}\left(p^n\right)$ are of the same form except computed mod p^n. And π maps these adjacent points in $\mathcal{H}\left(p^{n+1}\right)$ one-to-one, onto those in $\mathcal{H}\left(p^n\right)$.

If $(a,b,c) \in \Gamma$ defined in the statement of Theorem 2.2, we define the Galois group element

$$\gamma_{(a,b,c)}\left((x,y,z) \bmod p^{n+1}\right) = (a,b,c)(x,y,z) \bmod p^{n+1}.$$

It follows that $\pi\circ\gamma = \pi$, since $(a,b,c) \equiv 0 \,(\bmod\, p^n)$ and π reduces things mod p^n. Moreover, it is easy to see that Γ is abelian since if (a,b,c) and (u,v,w) are both $\equiv 0\,(\bmod\, p^n)$, then $(a,b,c)(u,v,w) = (a+u, b+v, c+w+av)$ and p^n divides both a and v so that $av \equiv 0 \left(\bmod\, p^{n+1}\right)$. □

Corollary 2.3. *The spectrum of $\mathcal{H}\left(p^n\right)$ is contained in the spectrum of $\mathcal{H}\left(p^{n+1}\right)$. Moreover $\zeta_{\mathcal{H}(p^n)}(u)^{-1}$ divides $\zeta_{\mathcal{H}(p^{n+1})}(u)^{-1}$.*

Proof. Use Proposition 2.1 or see (Stark and Terras, 1996, p. 131). □

Example. The last Theorem and Corollary also work if $p = 2$, except that then the graph at the bottom of the cover can be a multi-graph when $n = 1$, as in Figure 1. Consider the covering $\mathcal{H}(4)$ over $\mathcal{H}(2)$. Note that $\mathcal{H}(2)$ is a multigraph with 2 edges between any vertices that are adjacent, because $1 \equiv -1 \pmod 2$ and we want the graph to have degree 4. So the graph of $\mathcal{H}(2)$ is a cycle graph as in Figure 1. We label the vertices using the following table.

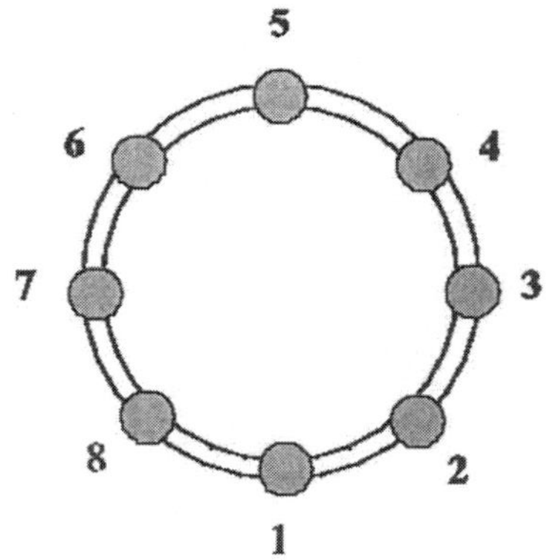

Figure 1. The Cayley Graph $\mathcal{H}(2) = X(\text{Heis}(\mathbb{Z}/2\mathbb{Z}), \{(\pm1,0,0),(0,\pm1,0)\})$.

Table 1. Vertex Labeling for $\mathcal{H}(2)$.

label	1	2	3	4	5	6	7	8
vertex	$(0,0,0)$	$(1,0,0)$	$(1,1,1)$	$(0,1,1)$	$(0,0,1)$	$(1,0,1)$	$(1,1,0)$	$(0,1,0)$

We obtain a spanning tree for $\mathcal{H}(2)$ by cutting one of each pair of double edges and then cutting both edges between vertices 6 and 7. This really gives a line graph but we will draw it as a circle cut between vertices 6 and 7. So we draw the covering graph $\mathcal{H}(4)$ by placing 8 copies of the cut circle which is the spanning tree of $\mathcal{H}(2)$ and labeling each with a group element from $\text{Gal}(\mathcal{H}(4)/\mathcal{H}(2))$. We know that this can be identified with the subgroup of $\text{Heis}(\mathbb{Z}/4\mathbb{Z})$ consisting of (u,v,w) where u,v,w are all even. We label the Galois group elements using the following table.

The covering graph $\mathcal{H}(4)/\mathcal{H}(2)$ has 8 sheets and each sheet is a copy of the spanning tree of $\mathcal{H}(2)$. So every point on $\mathcal{H}(4)$ has a label (n,v), where $1 \le n \le 8$ and $v \in \{a,b,c,d,e,f,g,h\}$. We will just write nv. See Figure 2 for a picture of the tree with connections between level a and the rest. You can use the action of the Galois group to find all the edges of $\mathcal{H}(4)$. It makes a pretty complicated figure. The following table

Table 2. Galois Group Labeling for $\mathrm{Gal}(\mathcal{H}(4)/\mathcal{H}(2))$. In this labeling, a not e is the identity of the group.

label	a	b	c	d	e	f	g	h
Galois group element	(0,0,0)	(2,0,0)	(2,2,2)	(0,2,2)	(0,0,2)	(2,0,2)	(2,2,0)	(0,2,0)

shows which connections are made in Figure 2. This table allows one to compute the matrices $A(g), g \in G = \mathrm{Gal}(\mathcal{H}(4)/\mathcal{H}(2))$.

Table 3. Table of Connections Between Sheet a in $\mathcal{H}(4)$ and the other sheets.

vertex	adjacent vertices in $\mathcal{H}(4)$
$1a$	$2b, 8h, 2a, 8a$
$2a$	$1b, 3d, 1a, 3a$
$3a$	$2d, 4f, 2a, 4a$
$4a$	$3f, 5h, 3a, 5a$
$5a$	$4h, 6b, 4a, 6a$
$6a$	$5b, 7e, 7h, 5a$
$7a$	$6e, 6h, 8f, 8a$
$8a$	$1h, 7f, 7a, 1a$

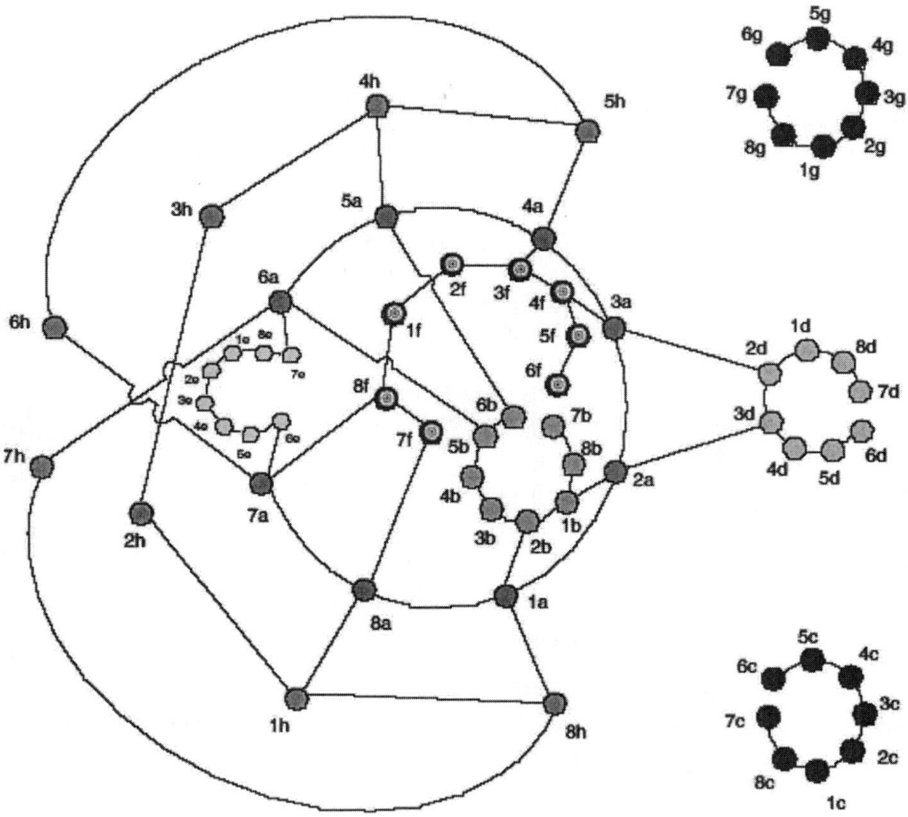

Figure 2. Connections Between Level a and the Rest of the Cayley Graph $\mathcal{H}(4) = X(\text{Heis}(\mathbb{Z}/4\mathbb{Z}), \{(\pm 1, 0, 0), (0, \pm 1, 0\})$

The representations of the abelian Galois group have the form $\chi_{r,s,t}(a,b,c) = \exp\left(\frac{2\pi i(ra+sb+tc)}{4}\right)$, for $r, s, t \pmod 2$. Then one must compute the matrices $M_{\chi_{r,s,t}}$ appearing in Proposition 2.1. For example

$M_{\chi_{0,0,0}}$ is the adjacency matrix of $\mathcal{H}(2)$ and

$$M_{\chi_{0,1,1}} = \begin{pmatrix} 0 & 2 & 0 & 0 & 0 & 0 & 0 & 0 \\ 2 & 0 & 2 & 0 & 0 & 0 & 0 & 0 \\ 0 & 2 & 0 & 0 & 0 & 0 & 0 & 0 \\ 0 & 0 & 0 & 0 & 0 & 0 & 0 & 0 \\ 0 & 0 & 0 & 0 & 0 & 2 & 0 & 0 \\ 0 & 0 & 0 & 0 & 2 & 0 & -2 & 0 \\ 0 & 0 & 0 & 0 & 0 & -2 & 0 & 0 \\ 0 & 0 & 0 & 0 & 0 & 0 & 0 & 0 \end{pmatrix},$$
$$M_{\chi_{1,0,0}} = \begin{pmatrix} 0 & 0 & 0 & 0 & 0 & 0 & 0 & 2 \\ 0 & 0 & 2 & 0 & 0 & 0 & 0 & 0 \\ 0 & 2 & 0 & 0 & 0 & 0 & 0 & 0 \\ 0 & 0 & 0 & 0 & 2 & 0 & 0 & 0 \\ 0 & 0 & 0 & 2 & 0 & 0 & 0 & 0 \\ 0 & 0 & 0 & 0 & 0 & 0 & 2 & 0 \\ 0 & 0 & 0 & 0 & 0 & 2 & 0 & 0 \\ 2 & 0 & 0 & 0 & 0 & 0 & 0 & 0 \end{pmatrix}. \tag{2.6}$$

The eigenvalues of the M_χ are to be found in the following table.

Table 4. Eigenvalues of $M_{r,s,t} = M_{\chi_{r,s,t}}$.

(r, s, t)	Eigenvalues of $M_{r,s,t}$
$(0,0,0)$	$-4,0,0,4,-2\sqrt{2},-2\sqrt{2},2\sqrt{2},2\sqrt{2}$
$(1,0,0)$ and $(0,1,0)$	$-2,-2,-2,-2,2,2,2,2$
$(1,1,0)$	$0,0,0,0,0,0,0,0$
$(1,1,1),(0,1,1),(0,0,1)$, and $(1,0,1)$	$0,0,0,0,-2\sqrt{2},-2\sqrt{2},2\sqrt{2},2\sqrt{2}$

So we see that the spectrum of $\mathcal{H}(4)$ for $p=2$ is given in Table 5.

Table 5. Spectrum of $X(\mathrm{Heis}(\mathbb{Z}/4\mathbb{Z})), \{(\pm1,0,0),(0,\pm1,0)\}$.

eigenvalue	multiplicity
±4	1
0	26
±2	8
$\pm2\sqrt{2}$	10

The **Artin L-function** associated to the representation ρ of $G = \mathrm{Gal}(Y/X)$ can be defined by a product over prime cycles in X as

$$L(u, \rho, Y/X) = \prod_{[C] \text{ prime in } X} \det\left(I - \rho\left(\mathrm{Frob}\left(\tilde{C}\right)\right) u^{\nu(C)}\right)^{-1}, \quad (2.7)$$

where $\tilde{C}$ denotes any lift of C to Y and $\mathrm{Frob}\left(\tilde{C}\right)$ denotes the Frobenius automorphism defined by

$$\mathrm{Frob}\left(\tilde{C}\right) = ji^{-1},$$

if $\tilde{C}$ starts on Y-sheet labeled by $i \in G$ and ends on Y-sheet labeled by $j \in G$. As in Proposition 2.1, define

$$M_\rho = \sum_{g \in G} A(g) \otimes \rho(g). \quad (2.8)$$

Then, setting $Q_\rho = Q \otimes I_{d_\rho}$, with $d_\rho = d = \deg \rho$, we have the following analogue of formula (1.2):

$$L(u, \rho, Y/X)^{-1} = \left(1 - u^2\right)^{(r-1)d_\rho} \det\left(I - M_\rho u + Q_\rho u^2\right). \quad (2.9)$$

See (Stark and Terras, 1996) for an elementary proof and more information.

Formula (2.5) implies that the zeta function of Y factors as follows

$$\zeta_X(u) = \prod_{\rho \in \widehat{G}} L(u, \rho, Y/X)^{d_\rho}. \quad (2.10)$$

See (Stark and Terras, 2000).

For our example, the Galois group is abelian and all degrees are 1. We obtain a factorization of the Ihara-Selberg zeta function of $\mathcal{H}(4)$ as a product of Artin L-functions of the Galois group of $\mathcal{H}(4)/\mathcal{H}(2)$. We use definition (2.8)) and Table 3 to compute the matrices $M_{\chi_{r,s,t}}$ as in formula (2.6). Then formula (2.9) gives the following list of L-functions. Here $Q = 3I_8$, $r = 9$.

Reciprocals of L-functions for $H(4)/H(2)$.

1) For $\chi = \chi_{0,0,0}$, $A =$ adjacency matrix of $\mathcal{H}(2)$, and

$$\begin{aligned}\zeta_{\mathcal{H}(2)}(u)^{-1} = L(u, 1)^{-1} &= \left(1 - u^2\right)^8 (u - 1)(u + 1) \\ &\times (3u - 1)(3u + 1)\left(3u^2 + 1\right)^2 \left(9u^4 - 2u^2 + 1\right)^2. \end{aligned} \quad (2.11)$$

2) $L\left(u,\chi_{1,0,0}\right)^{-1} = L\left(u,\chi_{0,1,0}\right)^{-1}$
$= \left(1-u^2\right)^8\left(3u^2+2u+1\right)^4\left(3u^2-2u+1\right)^4.$

3) $L\left(u,\chi_{1,1,1}\right)^{-1} = L\left(u,\chi_{0,1,1}\right)^{-1} = L\left(u,\chi_{0,0,1}\right)^{-1} = L\left(u,\chi_{1,0,1}\right)^{-1}$
$= \left(1-u^2\right)^8\left(9u^4-2u^2+1\right)^2\left(3u^2+1\right)^4.$

4) When $\rho = \chi_{1,1,0}$ we find that $M_{\chi_{1,1,0}} = 0$, so that

$$L\left(u,\chi_{1,1,0}\right)^{-1} = \left(1-u^2\right)^{(r-1)d}\det\left(I+Q_\rho u^2\right) = \left(1-u^2\right)^8\left(1+3u^2\right)^8.$$

It follows from these computations and (2.10) that the Ihara zeta function of $\mathcal{H}(4)$ is

$$\zeta_{\mathcal{H}(4)}(u)^{-1} = -\left(1-u^2\right)^{65}\left(9u^2-1\right)\left(3u^2+1\right)^{26} \times\left(9u^4-2u^2+1\right)^{10}\left(3u^2+2u+1\right)^8\left(3u^2-2u+1\right)^8. \tag{2.12}$$

Consider the torus graphs

$$\mathcal{T}^{(n)}(q) = X\left((\mathbb{Z}/q\mathbb{Z})^n, \{\pm e_1, \pm e_2, \cdots, \pm e_n\}\right),$$

where e_i denotes the vector with 1 in the ith coordinate and 0 elsewhere. Because the torus groups $(\mathbb{Z}/q\mathbb{Z})^n$ are abelian, it is relatively easy to generate spectra. In fact, the eigenvalues of the adjacency matrix of $\mathcal{T}^{(n)}(q)$ are

$$\lambda_a = 2\left(\cos\left(\frac{2\pi i a_1 b_1}{q}\right)+\cos\left(\frac{2\pi i a_2 b_2}{q}\right)+\cdots+\cos\left(\frac{2\pi i a_n b_n}{q}\right)\right),$$

for $a, b \in (\mathbb{Z}/q\mathbb{Z})^n$. Note that, by a result of our earlier paper (DeDeo et al., 2004), the part of the spectrum of the degree 4 Heisenberg graph $\mathcal{H}(4)$ corresponding to 1-dimensional representations of $\mathcal{H}(4)$ contains the spectrum of $\mathcal{T}^{(2)}(4)$. One obtains a second proof of this fact by noting that $\mathcal{H}(4)$ is actually a covering graph of $\mathcal{T}^{(2)}(4)$, via the covering map sending (x, y, z) to (x, y).

We can easily compute the Selberg-Ihara zeta functions of the small torus graphs using covering graph theory. As in Theorem 2.2, the Galois group of $\mathcal{T}^{(n)}\left(p^{r+1}\right)/\mathcal{T}^{(n)}\left(p^r\right)$ is

$$\Gamma \cong \left\{x \in \left(\mathbb{Z}/p^{r+1}\mathbb{Z}\right)^n \mid x \equiv 0 \pmod{p}^r\right\}.$$

Since the 1-dimensional graphs are cycles, we know that

$$\zeta_{\mathcal{T}^{(1)}(q)}(u)^{-1} = \left(1-u^q\right)^2, \text{ for all } q.$$

In 2-dimensions, we consider only the smallest values of q (namely $q = 2$ and $q = 4$) and find that if $\Gamma = \mathrm{Gal}\left(\mathcal{T}^{(2)}(4)/\mathcal{T}^{(2)}(2)\right)$, the representations of Γ have the form $\chi_{r,s}(x,y) = \exp\left(\frac{2\pi i(rx+sy)}{4}\right)$, for $(x,y) \in \Gamma$, $(r,s) \in (\mathbb{Z}/2\mathbb{Z})^2$. Therefore $(x,y) \equiv 0 \pmod 2$. It follows that

$$\zeta_{\mathcal{T}^{(2)}(4)}(u)^{-1} = \zeta_{\mathcal{T}^{(2)}(2)}(u)^{-1} L\left(u,\chi_{0,1}\right)^{-1} L\left(u,\chi_{1,1}\right)^{-1} L\left(u,\chi_{1,0}\right)^{-1}$$
$$= -\left(1-u^2\right)^{17}\left(9u^2-1\right)\left(3u^2+1\right)^6\left(3u^2-2u+1\right)^4\left(3u^2+2u+1\right)^4.$$

Here $\zeta_{\mathcal{T}^{(2)}(2)}(u)^{-1} = -\left(1-u^2\right)^5\left(9u^2-1\right)\left(3u^2+1\right)^2$.

From these results plus (2.11) and (2.12) we see that

$$\begin{aligned}\zeta_{\mathcal{H}(4)}(u)^{-1}/\zeta_{\mathcal{T}^{(2)}(4)}(u)^{-1} &= \left(1-u^2\right)^{48}\left(3u^2+1\right)^{20} \\ &\times \left(3u^2-2u+1\right)^4\left(3u^2+2u+1\right)^4\left(9u^4-2u^2+1\right)^{10}\end{aligned} \tag{2.13}$$

and

$$\zeta_{\mathcal{H}(2)}(u)^{-1}/\zeta_{\mathcal{T}^{(2)}(2)}(u)^{-1} = \left(1-u^2\right)^4\left(9u^4-2u^2+1\right)^2. \tag{2.14}$$

Figure 3 shows a contour plot of the absolute value of $\zeta_{\mathcal{H}(2)}(x+iy)^{-1}$ made using the Mathematica command Plot3D. It should be compared with Figure 4 showing a contour plot of the 1/10 power of the absolute value of $\zeta_{\mathcal{H}(4)}(x+iy)^{-1}$. The roots of $\zeta_{\mathcal{H}(4)}(x+iy)^{-1}$ (not counting multiplicity) are approximately the following 14 numbers:

$$\begin{gathered}-1, -0.333333, 0.57735i, -0.471405-0.333333i, \\ -0.471405+0.333333i, -0.333333-0.471405i, \\ -0.333333+0.471405i, 0.333333-0.471405i, \\ 0.333333+0.471405i, 0.471405-0.333333i, \\ 0.471405+0.333333i, -0.57735i, 0.333333, 1.\end{gathered}$$

Future Work. There are many other questions one can ask in this context. One should study the zeros of Ihara-Selberg zeta functions $\mathcal{H}(q)$ for large q. One should consider these questions for Cayley graphs of other finite groups and even for irregular graphs for which there is no obvious relation between the spectrum of the adjacency matrix and the zeros of the Ihara-Selberg zeta function.

Can such zeta functions be used to recognize groups involved in Cayley graphs? In particular, one wonders whether you can see the shape of a group by staring at zeros of the zeta function of Cayley graphs associated to the group? This is an analogous question to that of Mark Kac about hearing the shape of a drum (as the Dirichlet spectrum of

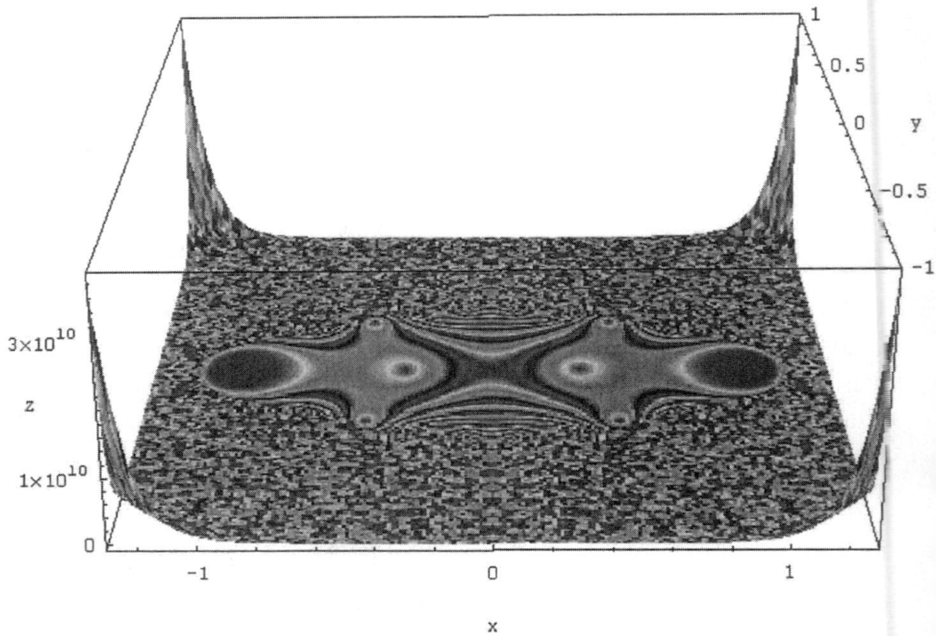

Figure 3. A contour plot of the absolute value of $\zeta_{\mathcal{H}(2)}(x+iy)^{-1}$.

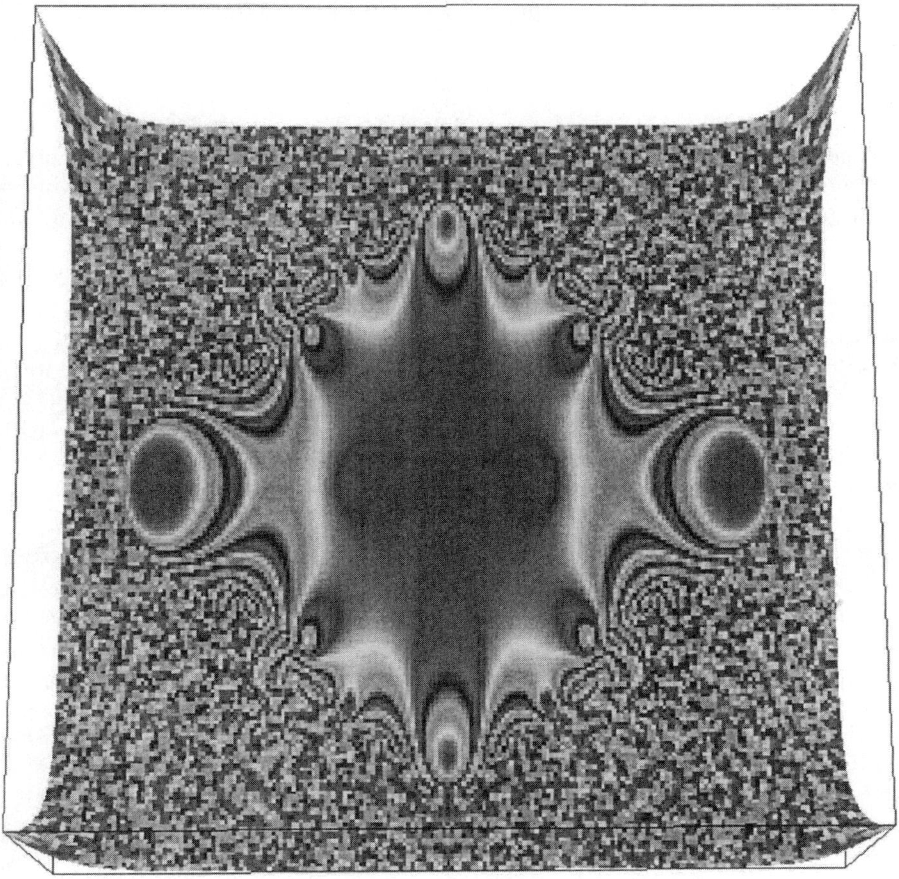

Figure 4. A contour plot of 1/10 power of the absolute value of $\zeta_{\mathcal{H}(4)}(x+iy)^{-1}$.

the Laplace operator on a plane drum determines the fundamental frequencies of vibration). Here we wonder if one can somehow recognize groups from properties of the zero set of zeta functions of associated Cayley graphs with some sort of condition on the generating sets S. Instead of hearing the drum in its spectrum, we are trying to see it. Of course, it is known that there are graphs with the same zeta function that are not isomorphic. See (Stark and Terras, 2000) for examples that are connected, regular, without loops or multiple edges.

References

DeDeo, M., Martínez, M., Medrano, A., Minei, M., Stark, H., and Terras, A. (2004). Spectra of Heisenberg graphs over finite rings. In Feng, W., Hu, S., and Lin, X., editors, *Discrete and Continuous Dynamical Systems, 2003 Supplement Volume*, pages 213–222. Proc. of 4th Internatl. Conf. on Dynamical Systems and Differential Equations.

Diaconis, P. and Saloff-Coste, L. (1994). Moderate growth and random walk on finite groups. *Geom. Funct. Anal.*, 4:1–36.

Godsil, C. D. (1993). *Algebraic Combinatorics.* Chapman and Hall, New York.

Hashimoto, K. (1990). On Zeta and L-functions of finite graphs. *Intl. J. Math.*, 1(4):381–396.

Hofstadter, D. R. (1976). Energy levels and wave functions of Bloch electrons in rational and irrational magnetic fields. *Physical Review B*, 14:2239–2249.

Katz, N. and Sarnak, P. (1999). Zeros of zeta functions and symmetry. *Bull. Amer. Math. Soc.*, 36(1):1–26.

Kemperman, J. H. B. (1961). *The Passage Problem for a Stationary Markov Chain.* Univ. of Chicago Press, Chicago, IL.

Kotani, M. and Sunada, T. (2000). Spectral geometry of crystal lattices. Preprint.

Lin, X.-S. and Wang, Z. (2001). Random walk on knot diagrams, colored Jones polynomial and Ihara-Selberg zeta function. In Gillman, J., Menasco, W., and Lin, X.-S., editors, *Knots, Braids, and Mapping Class Groups–Papers dedicated to Joan S. Birman (New York, 1998)*, volume 24 of *AMS/IP Studies in Adv. Math.*, pages 107–121. Amer. Math. Soc., Providence, RI.

Lubotzky, A., Phillips, R., and Sarnak, P. (1988). Ramanujan graphs. *Combinatorica*, 8:261–277.

Myers, P. (1995). *Euclidean and Heisenberg Graphs: Spectral Properties and Applications.* PhD thesis, Univ. of California, San Diego.

Rosen, M. (2002). *Number Theory in Function Fields*, volume 210 of *Graduate Texts in Mathematics.* Springer-Verlag, New York.

Sarnak, P. (1995). Arithmetic quantum chaos. In *The Schur lectures (1992) (Tel Aviv)*, volume 8 of *Israel Math. Soc. Conf. Proc.*, pages 183–236. Bar-Ilan Univ., Ramat-Gan, Israel.

Stark, H. M. and Terras, A. (1996). Zeta functions of finite graphs and coverings. *Adv. in Math.*, 121:124–165.

Stark, H. M. and Terras, A. (2000). Zeta functions of finite graphs and coverings. II. *Adv. in Math.*, 154(1):132–195.

Terras, A. (1999). *Fourier Analysis on Finite Groups and Applications.* Cambridge Univ. Press, Cambridge, UK.

Terras, A. (2000). Statistics of graph spectra for some finite matrix groups: finite quantum chaos. In Dunkl, C., Ismail, M., and Wong, R., editors, *Special Functions (Hong Kong, 1999)*, pages 351–374. World Scientific, Singapore.

Terras, A. (2002). Finite quantum chaos. *Amer. Math. Monthly*, 109(2):121–139.

Zack, M. (1990). Measuring randomness and evaluating random number generators using the finite Heisenberg group. In *Limit Theorems in Probability and Statistics (Pécs, 1989)*, volume 57 of *Colloq. Math. Soc. János Bolyai*, pages 537–544. North-Holland, Amsterdam.

q-ANALOGUES OF SOME MULTIVARIABLE BIORTHOGONAL POLYNOMIALS

George Gasper
Department of Mathematics
Northwestern University
Evanston, IL 60208

Mizan Rahman*
School of Mathematics and Statistics
Carleton University
Ottawa, Ontario
CANADA K1S 5B6

Abstract In 1989, M. V. Tratnik found a pair of multivariable biorthogonal polynomials $P_{\mathbf{n}}(\mathbf{x})$ and $\bar{P}_{\mathbf{m}}(\mathbf{x})$, which is not necessarily the complex conjugate of $P_{\mathbf{m}}(\mathbf{x})$, such that

$$\int_{-\infty}^{\infty} \cdots \int_{-\infty}^{\infty} w(\mathbf{x}) P_{\mathbf{n}}(\mathbf{x}) \bar{P}_{\mathbf{m}}(\mathbf{x}) \prod_{j=1}^{p} dx_j = \mu_{\mathbf{n},\mathbf{m}} \delta_{N,M},$$

where $\mathbf{x} = (x_1, \ldots, x_p)$, $\mathbf{n} = (n_1, \ldots, n_p)$, $\mathbf{m} = (m_1, \ldots, m_p)$, $N = \sum_{j=1}^{p} n_j$, $M = \sum_{j=1}^{p} m_j$, $\mu_{\mathbf{n},\mathbf{m}}$ is the constant of biorthogonality (which Tratnik did not evaluate),

$$w(\mathbf{x}) = \Gamma(A - iX)\Gamma(B + iX) \left| \frac{\Gamma(c + iX)\Gamma(d + iX)}{\Gamma(2iX)} \right|^2 \prod_{k=1}^{p} \Gamma(a_k + ix_k)\, \Gamma(b_k - ix_k),$$

$$X = \sum_{k=1}^{p} x_k, \quad A = \sum_{k=1}^{p} a_k, \quad B = \sum_{k=1}^{p} b_k,$$

and the a's, b's, x's, c and d are real. In the q-case we find that the appropriate weight function is a product of a multivariable version of the integrand in the Askey-Roy integral and of the Askey-Wilson weight function in a single variable that depends on $x_1, \ldots, x_p$.

*Supported in part by the NSERC grant #A6197.

In a related problem we find a discrete 2-variable Racah type biorthogonality:

$$\sum_{x=0}^{N}\sum_{y=0}^{N} w_N(x,y)F_{m,n}(x,y)G_{m',n'}(x,y) = \nu_{m,n}\delta_{m,m'}\delta_{n,n'},$$

where

$$w_N(x,y) = \frac{(\alpha q/\gamma\gamma', \gamma'/c, \alpha cq/\gamma'; q)_N}{(\alpha q, 1/c, \alpha cq/\gamma\gamma'; q)_N}$$
$$\times \frac{\left(1-\frac{\gamma\gamma' q^{2x-N-1}}{\alpha c}\right)\left(1-cq^{2y-N}\right)\left(\frac{\gamma\gamma' q^{-N-1}}{\alpha c}, \gamma; q\right)_x \left(cq^{-N}, \gamma'; q\right)_y}{\left(1-\frac{\gamma\gamma' q^{-N-1}}{\alpha c}\right)\left(1-cq^{-N}\right)\left(q, \frac{\gamma' q^{-N}}{\alpha c}; q\right)_x \left(q, \frac{cq^{1-N}}{\gamma'}; q\right)_y}$$
$$\times \frac{(1/c;q)_{x-y}(q^{-N};q)_{x+y}}{\left(\frac{\gamma\gamma'}{\alpha c};q\right)_{x-y}\left(\frac{\gamma\gamma' q^{-N}}{\alpha};q\right)_{x+y}}\alpha^{-x}(\gamma')^{x-y},$$

and $F_{m,n}(x,y)$, $G_{m',n'}(x,y)$ are certain bivariate extensions of the q-Racah polynomials.

1. Introduction

Wilson polynomials (Wilson, 1980), defined by

$$P_n(x) = (a+b)_n(a+c)_n(a+d)_n \times {}_4F_3\left[\begin{matrix} -n, n+a+b+c+d-1, a-ix, a+ix \\ a+b, a+c, a+d \end{matrix};1\right] \quad (1.1)$$

satisfy an orthogonality relation on the real line

$$\int_{-\infty}^{\infty} P_n(x)P_m(x)w(x)dx = h_n\delta_{n,m}, \quad (1.2)$$

where

$$w(x) = \left|\frac{\Gamma(a+ix)\Gamma(b+ix)\Gamma(c+ix)\Gamma(d+ix)}{\Gamma(2ix)}\right|^2 \quad (1.3)$$

is the positive weight function (under the assumption that a, b, c, d are real or occur in complex conjugate pairs), and

$$h_n = 4\pi n!\,(n+a+b+c+d-1)_n\Gamma(n+a+b) \times \frac{\Gamma(n+a+c)\Gamma(n+a+d)\Gamma(n+b+c)\Gamma(n+b+d)\Gamma(n+c+d)}{\Gamma(2n+a+b+c+d)} \quad (1.4)$$

is the normalization constant. By Whipple's transformation it is easy to see that $P_n(x)$ is symmetric in a, b, c, d, and that

$$\begin{aligned} P_n(x) &= (a+b)_n(c-ix)_n(d-ix)_n \\ &\quad \times {}_4F_3\left[\begin{matrix} -n, 1-c-d-n, a+ix, b+ix \\ a+b, 1-c-n+ix, 1-d-n+ix \end{matrix}; 1\right] \\ &= (b+a)_n(c+ix)_n(d+ix)_n \\ &\quad \times {}_4F_3\left[\begin{matrix} -n, 1-c-d-n, a-ix, b-ix \\ a+b, 1-c-n-ix, 1-d-n-ix \end{matrix}; 1\right]. \end{aligned} \tag{1.5}$$

Corresponding to each of these forms M. V. Tratnik (Tratnik, 1989b) introduced a multivariable polynomial:

$$\begin{aligned} P_n(\mathbf{x}) &= (A+c)_N(A+d)_N \prod_{k=1}^{p} (a_k+b_k)_{n_k} \\ &\quad \times \sum_{\mathbf{j}} \frac{(N+A+B+c+d-1)_J(A-iX)_J}{(A+c)_J(A+d)_J} \\ &\quad \times \prod_{k=1}^{p} \frac{(-n_k)_{j_k}(a_k+ix_k)_{j_k}}{(a_k+b_k)_{j_k}\, j_k!}, \end{aligned} \tag{1.6}$$

$$\begin{aligned} \bar{P}_n(\mathbf{x}) &= (B+c)_N(B+d)_N \prod_{k=1}^{p} (b_k+a_k)_{n_k} \\ &\quad \times \sum_{\mathbf{j}} \frac{(N+A+B+c+d-1)_J(B+iX)_J}{(B+c)_J(B+d)_J} \\ &\quad \times \prod_{k=1}^{p} \frac{(-n_k)_{j_k}(b_k-ix_k)_{j_k}}{(b_k+a_k)_{j_k}\, j_k!}, \end{aligned} \tag{1.7}$$

$$\begin{aligned} Q_{\mathbf{n}}(\mathbf{x}) &= (c-iX)_N(d-iX)_N \prod_{k=1}^{p} (a_k+b_k)_{n_k} \\ &\quad \times \sum_{\mathbf{j}} \frac{(1-c-d-N)_J(B+iX)_J}{(1-c-N+iX)_J(1-d-N+iX)_J} \\ &\quad \times \prod_{k=1}^{p} \frac{(-n_k)_{j_k}(a_k+ix_k)_{j_k}}{(a_k+b_k)_{j_k}\, j_k!}, \end{aligned} \tag{1.8}$$

$$\begin{aligned}\bar{Q}_{\mathbf{n}}(\mathbf{x}) &= (c+iX)_N(d+iX)_N \prod_{k=1}^{p}(b_k+a_k)_{n_k} \\ &\times \sum_{\mathbf{j}} \frac{(1-c-d-N)_J(A-iX)_J}{(1-c-N-iX)_J(1-d-N-iX)_J} \\ &\times \prod_{k=1}^{p} \frac{(-n_k)_{j_k}(b_k-ix_k)_{j_k}}{(b_k+a_k)_{j_k}\, j_k!},\end{aligned} \tag{1.9}$$

where $\mathbf{x} = (x_1, x_2, \ldots, x_p)$, $\mathbf{n} = (n_1, n_2, \ldots, n_p)$, $\mathbf{j} = (j_1, j_2, \ldots, j_p)$, and

$$X = \sum_{k=1}^{p} x_k, \quad N = \sum_{k=1}^{p} n_k, \quad M = \sum_{k=1}^{p} m_k,$$
$$A = \sum_{k=1}^{p} a_k, \quad B = \sum_{k=1}^{p} b_k, \quad J = \sum_{k=1}^{p} j_k,$$

and the sums in (1.6)–(1.9) are from $j_k = 0$ to n_k, $k = 1, \ldots, p$. Each of the polynomials in (1.6)–(1.9) is of (total) degree $2N$ in the variables $x_1, x_2, \ldots, x_p$. The overbars in (1.7), (1.9), and in (1.21) below are used to denote distinct systems of polynomials and should not be confused with complex conjugation. Tratnik proved that

$$\int_{-\infty}^{\infty} \cdots \int_{-\infty}^{\infty} P_{\mathbf{n}}(\mathbf{x})\bar{P}_{\mathbf{m}}(\mathbf{x})w(\mathbf{x}) \prod_{k=1}^{p} dx_k = 0, \quad \text{if } N \neq M, \tag{1.10}$$

$$\int_{-\infty}^{\infty} \cdots \int_{-\infty}^{\infty} Q_{\mathbf{n}}(\mathbf{x})\bar{Q}_{\mathbf{m}}(\mathbf{x})w(\mathbf{x}) \prod_{k=1}^{p} dx_k = 0, \quad \text{if } N \neq M, \tag{1.11}$$

$$\int_{-\infty}^{\infty} \cdots \int_{-\infty}^{\infty} P_{\mathbf{n}}(\mathbf{x})Q_{\mathbf{m}}(\mathbf{x})w(\mathbf{x}) \prod_{k=1}^{p} dx_k = 0, \quad \text{if } \mathbf{n} \neq \mathbf{m}, \tag{1.12}$$

and

$$\int_{-\infty}^{\infty} \cdots \int_{-\infty}^{\infty} \bar{P}_{\mathbf{n}}(\mathbf{x})\bar{Q}_{\mathbf{m}}(\mathbf{x})w(\mathbf{x}) \prod_{k=1}^{p} dx_k = 0, \quad \text{if } \mathbf{n} \neq \mathbf{m}, \tag{1.13}$$

where

$$w(\mathbf{x}) = \left|\frac{\Gamma(c+iX)\Gamma(d+iX)}{\Gamma(2iX)}\right|^2 \Gamma(A-iX)\Gamma(B+iX) \times \prod_{k=1}^{p} \Gamma(a+ix_k)\Gamma(b-ix_k). \tag{1.14}$$

Note that in (1.12) and (1.13) the biorthogonality holds in all of the indices $n_1, n_2, \ldots, n_p$, while in (1.10) and (1.11) the biorthogonality is for polynomials of different degrees ($N \neq M$).

Since Whipple's ${}_4F_3$ transformation does not apply for $p \geq 2$ the P's and Q's are no longer equivalent and hence the orthogonality in a single variable becomes biorthogonality in many variables.

We were curious to see what their q-analogues would be. At first sight it might appear that they could be found in a pretty straightforward manner. We were in for a surprise. The first hurdle is an appropriate analogue of the weight function in (1.14). There are many possible candidates but the one that works for a q-analogue of (1.10) is:

$$\begin{aligned} w^{(p)}(\mathbf{x};q) &:= \frac{1}{(2\pi)^p} \frac{\left(e^{2i\Theta}, e^{-2i\Theta}; q\right)_\infty}{(Ae^{-i\Theta}, Be^{i\Theta}; q)_\infty h(\cos\Theta; c, d; q)\left(\frac{\beta b_1}{B}e^{i\Theta}, \frac{qB}{\beta b_1}e^{-i\Theta}; q\right)_\infty} \\ &\quad \times \prod_{k=1}^{p} \frac{\left(\beta_k e^{i\theta_k}, q\beta_k^{-1}e^{-i\theta_k}; q\right)_\infty}{(a_k e^{i\theta_k}, b_k e^{-i\theta_k}; q)_\infty}, \quad p \geq 2, \end{aligned} \tag{1.15}$$

where $-\pi \leq \theta_k \leq \pi$, $\theta_k = x_k \log q$ so that $e^{i\theta_k} = q^{ix_k}$ for $k = 1, \ldots, p$, $\Theta = \sum_{j=1}^{p} \theta_j$, $A = \prod_{j=1}^{p} a_j$, $B = \prod_{j=1}^{p} b_j$, $h(\cos\Theta; c, d; q)$ is defined as in (Gasper and Rahman, 1990a, (6.1.2)), β is an arbitrary complex parameter such that $\beta \neq q^{\pm n}$ for $n = 0, 1, \ldots$, and

$$\beta_{k+1} = \frac{\beta_k}{a_k b_{k+1}}, \quad k = 1, 2, \ldots, p-1, \tag{1.16}$$

with $\beta_1 = \beta$. By making repeated use of the Askey-Roy integral (Gasper and Rahman, 1990a, (4.11.1)) followed by the use of the Askey-Wilson integral, we shall prove in Section 2 that

$$W^{(p)}(q) := \int_{-\pi}^{\pi} \cdots \int_{-\pi}^{\pi} w^{(p)}(\mathbf{x};q) \prod_{k=1}^{p} d\theta_k$$
$$= \frac{2(ABcd;q)_\infty \prod_{k=2}^{p} (b_k\beta_k, q/b_k\beta_k;q)_\infty}{(q;q)_\infty^p (Ac, Ad, Bc, Bd, cd;q)_\infty \prod_{k=1}^{p} (a_k b_k;q)_\infty}, \tag{1.17}$$

which is also valid for $p = 1$. It is understood that the $(p-2)$-fold product in the numerator is taken to be 1 when $p = 1$.

Let

$$A_j = \prod_{k=j}^{p} a_k, B_j = \prod_{k=j}^{p} b_k, J_j = \sum_{k=j}^{p} j_k, K_j = \sum_{r=j}^{p} k_r,$$
$$N_j = \sum_{k=j}^{p} n_k, M_j = \sum_{k=j}^{p} m_k, \Theta_j = \sum_{k=j}^{p} \theta_k, \tag{1.18}$$

so that

$$A_1 = A, B_1 = B, J_1 = J, K_1 = K, N_1 = N, M_1 = M, \Theta_1 = \Theta. \tag{1.19}$$

Analogous to Tratnik's polynomials in (1.6) and (1.7) we introduce the functions

$$P_{\mathbf{n}}(\mathbf{x};q) = (Ac, Ad;q)_N \prod_{k=1}^{p} (a_k b_k;q)_{n_k}$$
$$\times \sum_{\mathbf{j}} \frac{(ABcdq^{N-1}, Ae^{-i\Theta};q)_J}{(Ac, Ad;q)_J} q^J \prod_{k=1}^{p} \frac{(q^{-n_k}, a_k e^{i\theta_k};q)_{j_k}}{(q, a_k b_k;q)_{j_k}} \tag{1.20}$$
$$\times \frac{e^{i(j_1\Theta_2+\cdots+j_{p-1}\Theta_p)}}{B_2^{j_1}\cdots B_p^{j_{p-1}}} q^{-(N_2 j_1 + N_3 j_2 + \cdots + N_p j_{p-1})},$$

and

$$\bar{P}_{\mathbf{m}}(\mathbf{x};q) = (Bc, Bd; q)_M \prod_{k=1}^{p} (a_k b_k; q)_{m_k}$$
$$\times \sum_{\mathbf{k}} \frac{(ABcdq^{M-1}, Be^{i\Theta}; q)_K}{(Bc, Bd; q)_K} q^K \prod_{r=1}^{p} \frac{(q^{-m_r}, b_r e^{-i\theta_r}; q)_{k_r}}{(q, a_r b_r; q)_{k_r}}$$
$$\times \frac{e^{i(k_2(\Theta_2-\Theta)+\cdots+k_p(\Theta_p-\Theta))}}{a_1^{K_2} a_2^{K_3} \cdots a_{p-1}^{K_p}} q^{-\sum_{r=2}^{p} k_r(M-M_r)}. \tag{1.21}$$

Both $P_{\mathbf{n}}(\mathbf{x};q)$ and $\bar{P}_{\mathbf{m}}(\mathbf{x};q)$ are Laurent polynomials in the variables $q^{ix_1}, \ldots, q^{ix_p}$. Note that if we divide $P_{\mathbf{n}}(\mathbf{x};q)$ by $(1-q)^{3N}$ and replace its parameters $a_1, \ldots, a_p, b_1, \ldots, b_p, c, d$, respectively, by $q^{a_1}, \ldots, q^{a_p}, q^{b_1}, \ldots, q^{b_p}, q^c, q^d$, and then let $q \to 1$, we obtain $P_{\mathbf{n}}(\mathbf{x})$ as a limit case. Similarly, we see that $\bar{P}_{\mathbf{m}}(\mathbf{x})$ is limit case of $\bar{P}_{\mathbf{m}}(\mathbf{x};q)$. In Section 3 we shall do the integration and in Section 4 prove the following q-analogue of (1.10):

$$P_{\mathbf{n}} \cdot \bar{P}_{\mathbf{m}} := \int_{-\pi}^{\pi} \cdots \int_{-\pi}^{\pi} P_{\mathbf{n}}(\mathbf{x};q) \bar{P}_{\mathbf{m}}(\mathbf{x};q)\, w^{(p)}(\mathbf{x};q) \prod_{k=1}^{p} d\theta_k = 0, \quad \text{if } N \neq M, \tag{1.22}$$

where $w^{(p)}(\mathbf{x};q)$ is given by (1.15), and

$$P_{\mathbf{n}} \cdot \bar{P}_{\mathbf{m}} = L_p \sum_{k_1=0}^{m_1} \cdots \sum_{k_{p-1}=0}^{m_{p-1}} q^{\sum_{j=1}^{p-1} k_j \sum_{r=0}^{j-1} (n_r - m_r)}$$
$$\times \frac{\left(ABcdq^{N-1}, \frac{ABcdq^N}{a_p b_p}; q\right)_{k_1+\cdots+k_{p-1}}}{\left(ABcdq^{N+m_p}, \frac{ABcdq^{N-n_p}}{a_p b_p}; q\right)_{k_1+\cdots+k_{p-1}}} \prod_{r=1}^{p-1} \frac{(q^{-m_r}, a_r b_r q^{n_r}; q)_{k_r}}{(q, a_r b_r; q)_{k_r}}, \tag{1.23}$$

when $N = M$, with $n_0 = 1$ and $m_0 = 0$, and L_p is as defined in (3.7).

Discrete multivariable extensions of the Racah polynomials were considered in (Tratnik, 1991b) as well as in (van Diejen and Stokman, 1998) and in (Gustafson, 1990). For other related works see, for instance, (Granovskiĭ and Zhedanov, 1992), (Koelink and Van der Jeugt, 1998), (Tratnik, 1989a), (Tratnik, 1991b). We have found q-extensions of Tratnik's systems of multivariable Racah and Wilson polynomials, complete with their orthogonality relations, see this Proceedings (Gasper and Rahman, 1990b) for our multivariable extension of the Askey-Wilson polynomials.

However, there seems to be at least one more extension that, to our knowledge, has not yet been investigated. The seed of this extension lies in Rosengren's (Rosengren, 2001) multivariable extension of the q-Hahn polynomials as well as in Rahman's (Rahman, 1981) 2-variable discrete biorthogonal system. In Sections 5 and 6 we shall prove the following 2-variable extension of the q-Racah polynomial orthogonality (Gasper and Rahman, 1990a, (7.2.18)):

$$\sum_{x=0}^{N}\sum_{y=0}^{N} w_N(x,y)F_{m,n}(x,y)G_{m',n'}(x,y) = \nu_{m,n}\delta_{m,m'}\delta_{n,n'}, \tag{1.24}$$

where $0 \le m, n, m', n' \le N$,

$$\begin{aligned} &F_{m,n}(x,y)\\ &=\frac{\left(\frac{\alpha q^{N+1-x-y}}{\gamma\gamma'};q\right)_{m+n}(q^{x-y}/c;q)_n\left(\alpha c q^{1+y-x}/\gamma\gamma';q\right)_m}{(q^{-N};q)_{m+n}(\alpha cq/\gamma\gamma';q)_n(1/c;q)_m}c^{n-m}q^{mx+ny}\\ &\times\sum_{i=0}^{m}\sum_{j=0}^{n}\frac{\left(q^{-m},\gamma q^x,\gamma\gamma' q^{x-N-1}/\alpha c;q\right)_i\left(q^{-n},\gamma' q^y,cq^{y-N};q\right)_j}{(q,\gamma,\gamma\gamma' q^{x-y-m}/\alpha c;q)_i\,(q,\gamma',cq^{1+y-x-n};q)_j}\\ &\times\frac{(\gamma\gamma' q^{-m-n}/\alpha;q)_{i+j}}{(\gamma\gamma' q^{x+y-N-m-n}/\alpha;q)_{i+j}}q^{i+j}, \end{aligned} \tag{1.25}$$

$$\begin{aligned} &G_{m,n}(x,y)\\ &=\sum_{i=0}^{m}\sum_{j=0}^{n}\frac{\left(q^{-m},q^{-x},\gamma\gamma' q^{x-N-1}/\alpha c;q\right)_i\left(q^{-n},q^{-y},cq^{y-N};q\right)_j}{(q,\gamma,\gamma' q^n/c;q)_i\,(q,\gamma',\alpha c q^{m+1}/\gamma';q)_j}\\ &\times\frac{(\alpha q^{m+n};q)_{i+j}}{(q^{-N};q)_{i+j}}q^{i+j}, \end{aligned} \tag{1.26}$$

and the weight function is

$$w_N(x,y) = \frac{(\alpha q/\gamma\gamma', \gamma'/c, \alpha c q/\gamma'; q)_N}{(\alpha q, 1/c, \alpha c q/\gamma\gamma'; q)_N} \times \frac{\left(1-\gamma\gamma' q^{2x-N-1}/\alpha c\right)\left(1-cq^{2y-N}\right)}{\left(1-\gamma\gamma' q^{-N-1}/\alpha c\right)\left(1-cq^{-N}\right)} \times \frac{\left(\gamma\gamma' q^{-N-1}/\alpha c, \gamma; q\right)_x \left(cq^{-N}, \gamma'; q\right)_y}{\left(q, \gamma' q^{-N}/\alpha c; q\right)_x \left(q, cq^{1-N}/\gamma'; q\right)_y} \times \frac{(1/c; q)_{x-y}\left(q^{-N}; q\right)_{x+y}}{\left(\gamma\gamma'/\alpha c; q\right)_{x-y}\left(\gamma\gamma' q^{-N}/\alpha; q\right)_{x+y}} \alpha^{-x}\left(\gamma'\right)^{x-y}. \tag{1.27}$$

The normalization constant in (1.24) is given by

$$\nu_{m,n} = \frac{1-\alpha}{1-\alpha q^{2m+2n}} \frac{(q, \alpha c q/\gamma'; q)_m (q, \gamma'/c; q)_n}{(\gamma, 1/c; q)_m (\gamma', \alpha c q/\gamma\gamma'; q)_n} \times \frac{\left(\alpha q/\gamma\gamma', \alpha q^{N+1}; q\right)_{m+n}}{\left(\alpha, q^{-N}; q\right)_{m+n}} c^{n-m} q^{mn}. \tag{1.28}$$

Notice that both $F_{m,n}(x,y)$ and $G_{m,n}(x,y)$ are Laurent polynomials in the variables q^x and q^y, and $G_{m,n}(x,y)$ is a polynomial of (total) degree $n+m$ in the variables $q^{-x} + \gamma\gamma' q^{x-N-1}/\alpha c$ and $q^{-y} + cq^{y-N}$.

We wish to make the observation that the summation in (1.24) is over the square of length N, although the vanishing of the weight function above the main diagonal, because of the factor $\left(q^{-N}; q\right)_{x+y}$ in the numerator, makes it effectively over the triangle $0 \le x+y \le N$. A very innocuous observation but it will help simplify the calculations somewhat as we shall see in Section 6.

It seems reasonable to expect that there is a multivariable extension of (1.24), but we were unable to find it, mainly because an extension of the q-shifted factorials of the type $(a;q)_{x-y}$ doesn't appear too obvious to us.

2. Calculation of $W^{(p)}(q)$

The key to the proof of (1.17) is to observe that by periodicity we can change $\theta_1, \theta_2, \ldots, \theta_p$ to, say, $\Theta, \theta_2, \ldots, \theta_p$ (so that $\theta_1 = \Theta - \Theta_2$), with the limits of integration unchanged. So the total weight transforms to

$$W^{(p)}(q) = \frac{1}{(2\pi)^{p-1}} \int_{-\pi}^{\pi} \frac{\left(e^{2i\Theta}, e^{-2i\Theta}; q\right)_\infty d\Theta}{(Ae^{-i\Theta}, Be^{i\Theta}; q)_\infty \, h(\cos\Theta; c, d; q) \left(\frac{\beta}{B_2} e^{i\Theta}, \frac{qB_2}{\beta} e^{-i\Theta}; q\right)_\infty}$$
$$\times \int_{-\pi}^{\pi} \cdots \int_{-\pi}^{\pi} I_2(\theta_3, \ldots, \theta_p) \prod_{k=3}^{p} \frac{\left(\beta_k e^{i\theta_k}, q e^{-i\theta_k}/\beta_k; q\right)_\infty}{(a_k e^{i\theta_k}, b_k e^{-i\theta_k}; q)_\infty} d\theta_3 \cdots d\theta_p, \quad (2.1)$$

where

$$I_2(\theta_3, \ldots, \theta_p) = \frac{1}{2\pi} \int_{-\pi}^{\pi} \frac{\left(\beta_2 e^{i\theta_2}, q e^{i(\Theta_3 - \Theta) + i\theta_2}/\beta, q e^{-i\theta_2}/\beta_2, \beta e^{i(\Theta - \Theta_3) - i\theta_2}; q\right)_\infty}{\left(a_2 e^{i\theta_2}, b_1 e^{i(\Theta_3 - \Theta) + i\theta_2}, b_2 e^{-i\theta_2}, a_1 e^{i(\Theta - \Theta_3) - i\theta_2}; q\right)_\infty} d\theta_2. \quad (2.2)$$

However, this integral matches exactly with the Askey-Roy integral (Gasper and Rahman, 1990a, (4.11.1)), provided we assume that $\max(|a_1|, |b_1|, |a_2|, |b_2|) < 1$ (with, of course, $|q| < 1$). By (Gasper and Rahman, 1990a, (4.11.1)), it then follows that

$$I_2(\theta_3, \ldots, \theta_p) = \frac{\left(b_2\beta_2, q/b_2\beta_2, a_1a_2b_1b_2, a_1\beta_2 e^{i(\Theta - \Theta_3)}, q e^{i(\Theta_3 - \Theta)}/a_1\beta_2; q\right)_\infty}{\left(q, a_1b_1, a_2b_2, a_1a_2 e^{i(\Theta - \Theta_3)}, b_1b_2 e^{i(\Theta_3 - \Theta)}; q\right)_\infty}. \quad (2.3)$$

Substitution of (2.3) into (2.1) makes it clear that the integration over θ_4 presents exactly the same situation, and so does the remaining integrations up to and including θ_p. Finally, one is left with an Askey-Wilson integral over Θ:

$$W^{(p)}(q) = \frac{(AB; q)_\infty \prod_{k=2}^{p} (b_k\beta_k, q/b_k\beta_k; q)_\infty}{(q; q)_\infty^{p-1} \prod_{k=1}^{p} (a_k b_k; q)_\infty} \times \frac{1}{2\pi} \int_{-\pi}^{\pi} \frac{\left(e^{2i\Theta}, e^{-2i\Theta}; q\right)_\infty}{h(\cos\Theta;\ A, B, c, d; q)} d\Theta \quad (2.4)$$
$$= \frac{2(ABcd; q)_\infty \prod_{k=2}^{p} (b_k\beta_k, q/b_k\,\beta_k; q)_\infty}{(q; q)_\infty^{p} (Ac, Ad, Bc, Bd, cd; q)_\infty \prod_{k=1}^{p} (a_k b_k; q)_\infty},$$

by (Gasper and Rahman, 1990a, (6.1.1)), which completes the proof of (1.17).

3. Computation of the integral in (1.22)

We shall carry out the integrations in (1.22) in much the same way as we did in the previous section. We transform the integration variables $\theta_1, \ldots, \theta_p$ to $\theta_2, \ldots, \theta_p$ and Θ as before; then we isolate the θ_2-integral by observing that the factors $\left(a_1 e^{i(\Theta-\Theta_3)-i\theta_2}; q\right)_{j_1} \left(a_2 e^{i\theta_2}; q\right)_{j_2}$ $\times \left(b_1 e^{i(\Theta_3-\Theta)+i\theta_2}; q\right)_{k_1} \left(b_2 e^{-i\theta_2}; q\right)_{k_2} e^{i\theta_2(j_1+k_2)+ik_2(\Theta_3-\Theta)+ij_1\Theta_3}$ can be glued on to the integrand of $W^{(p)}(q)$, to get

$$(-\beta)^{j_1+k_2} q^{\binom{j_1+k_2}{2}} e^{ij_1\Theta} \frac{1}{2\pi}$$
$$\times \int_{-\pi}^{\pi} \frac{\left(\beta_2 e^{i\theta_2},\, q^{1-j_1-k_2} e^{i(\Theta_3-\Theta)+i\theta_2}/\beta,\, \beta q^{j_1+k_2} e^{i(\Theta-\Theta_3)-i\theta_2},\, q e^{-i\theta_2}/\beta_2; q\right)_\infty}{\left(a_2 q^{j_2} e^{i\theta_2},\, b_1 q^{k_1} e^{i(\Theta_3-\Theta)+i\theta_2},\, b_2 q^{k_2} e^{-i\theta_2},\, a_1 q^{j_1} e^{i(\Theta-\Theta_3)-i\theta_2}; q\right)_\infty} d\theta_2$$

which via (Gasper and Rahman, 1990a, (4.11.1)) equals, on a bit of simplification,

$$\begin{aligned} &a_1^{k_2} b_2^{j_1} q^{j_1 k_2} e^{ij_1\Theta_3} \frac{\left(b_2\beta_2,\, q/b_2\beta_2,\, a_1 a_2 b_1 b_2 q^{j_1+j_2+k_1+k_2}; q\right)_\infty}{(q,\, a_1 b_1 q^{j_1+k_1},\, a_2 b_2 q^{j_2+k_2}; q)_\infty} \\ &\times \frac{\left(a_1\beta_2 e^{i(\Theta-\Theta_3)},\, q e^{i(\Theta_3-\Theta)}/a_1\beta_2; q\right)_\infty}{\left(a_1 a_2 q^{j_1+j_2} e^{i(\Theta-\Theta_3)},\, b_1 b_2 q^{k_1+k_2} e^{i(\Theta_3-\Theta)}; q\right)_\infty}. \end{aligned} \tag{3.1}$$

Since $\Theta_3 = \theta_3 + \Theta_4$, we may now isolate the θ_3-integral in exactly the same way, carry out a similar integration, simplify, and obtain

$$\begin{aligned} &a_1^{k_2} (a_1 a_2)^{k_3} (b_2 b_3)^{j_1} b_3^{j_2} e^{i(j_1+j_2)\Theta_4} q^{j_1 k_2 + (j_1+j_2)k_3} \\ &\times \frac{\left(b_2\beta_2,\, q/b_2\beta_2,\, b_3\beta_3,\, q/b_3\beta_3,\, a_1 a_2 a_3 b_1 b_2 b_3 q^{j_1+j_2+j_3+k_1+k_2+k_3}; q\right)_\infty}{(q,\, q,\, a_1 b_1 q^{j_1+k_1},\, a_2 b_2 q^{j_2+k_2},\, a_3 b_3 q^{j_3+k_3}; q)_\infty} \\ &\times \frac{\left(a_1 a_2 \beta_3 e^{i(\Theta-\Theta_4)},\, q e^{i(\Theta_4-\Theta)}/a_1 a_2 \beta_3; q\right)_\infty}{\left(a_1 a_2 a_3 q^{j_1+j_2+j_3} e^{i(\Theta-\Theta_4)},\, b_1 b_2 b_3 q^{k_1+k_2+k_3} e^{i(\Theta_4-\Theta)}; q\right)_\infty}. \end{aligned} \tag{3.2}$$

A clear pattern is now emerging. The θ_p integral is

$$
\begin{aligned}
& q^{j_1k_2+(j_1+j_2)k_3+\cdots+(j_1+\cdots+j_{p-2})k_{p-1}} \left(a_1^{k_2+\cdots+k_{p-1}} a_2^{k_3+\cdots+k_{p-1}} \cdots a_{p-2}^{k_{p-1}}\right) \\
& \times \left(b_2^{j_1} b_3^{j_1+j_2} \cdots b_{p-1}^{j_1+\cdots+j_{p-2}}\right) \\
& \times \frac{\left(\frac{AB}{a_pb_p} q^{J+K-j_p-k_p};q\right)_\infty \prod_{r=2}^{p-1} (b_r\beta_r, q/b_r\beta_r;q)_\infty}{(q;q)_\infty^{p-2} \prod_{r=1}^{p-1} (a_rb_rq^{j_r+k_r};q)_\infty} \left[\frac{e^{-i\Theta k_p}}{2\pi}\right. \\
& \times \int_{-\pi}^{\pi} \frac{\left(\beta_p e^{i\theta_p}, \frac{qe^{i(\theta_p-\Theta)}}{(a_1\cdots a_{p-2})\beta_{p-1}}, qe^{-i\theta_p}/\beta_p, (a_1\cdots a_{p-2})\,\beta_{p-1}e^{i(\Theta-\theta_p)};q\right)_\infty}{\left(a_pq^{j_p}e^{i\theta_p}, \frac{B}{b_p}q^{K-k_p}e^{i(\theta_p-\Theta)}, b_pq^{k_p}e^{-i\theta_p}, \frac{A}{a_p}q^{J-j_p}e^{i(\Theta-\theta_p)};q\right)_\infty} \\
& \left.\qquad\qquad \times e^{i\theta_p(J-j_p+k_p)}\,d\theta_p\right].
\end{aligned}
\tag{3.3}
$$

The expression in [ ] above can, once again, be computed by use of (Gasper and Rahman, 1990a, (4.11.1)), and simplified to

$$
\left(\frac{Aq^{J-j_p}}{a_p}\right)^{k_p} \frac{b_p^{J-j_p}\left(b_p\beta_p, q/b_p\beta_p, \frac{A\beta_pe^{i\Theta}}{a_p}, \frac{qa_p}{A\beta_p}e^{-i\Theta}, ABq^{J+K};q\right)_\infty}{\left(q, a_pb_pq^{j_p+k_p}, Aq^Je^{i\Theta}, Bq^Ke^{-i\Theta}, \frac{AB}{a_pb_p}q^{J+K-j_p-k_p};q\right)_\infty}. \tag{3.4}
$$

Since, by repeated application of (1.16) we get $A\beta_p/a_p = \beta b_1/B$, the Θ-integral simply becomes the Askey-Wilson integral

$$
\begin{aligned}
& \frac{1}{2\pi}\int_{-\pi}^{\pi} \frac{\left(e^{2i\Theta}, e^{-2i\Theta};q\right)_\infty}{h(\cos\Theta; Aq^J, Bq^K, c, d;q)}\,d\Theta \\
& = \frac{2(ABcd\,q^{J+K};q)_\infty}{(q, cd, ABq^{J+K}, Acq^J, Adq^J, Bcq^K, Bdq^K;q)_\infty}.
\end{aligned}
\tag{3.5}
$$

Collecting these results and substituting into the integral in (1.22), we find that

$$P_{\mathbf{n}} \cdot \bar{P}_{\mathbf{m}} = L_p \sum_{\mathbf{j}} \sum_{\kappa} \frac{(ABcdq^{N-1};q)_J \, (ABcdq^{M-1};q)_K}{(ABcd;q)_{J+K}} q^{J+K}$$
$$\times \prod_{r=1}^{p} \frac{(q^{-n_r};q)_{j_r} \, (q^{-m_r};q)_{k_r} \, (a_r b_r;q)_{j_r+k_r}}{(q, a_r b_r;q)_{j_r} \, (q, a_r b_r;q)_{k_r}} \tag{3.6}$$
$$\times \; q^{\sum_{s=2}^{p}[j_{s-1}(K_s - N_s) + k_s(M_s - M)]},$$

where

$$L_p = (Ac, Ad;q)_N (Bc, Bd;q)_M W^{(p)}(q) \prod_{r=1}^{p} (a_r b_r;q)_{m_r} \, (a_r b_r;q)_{n_r}. \tag{3.7}$$

4. Biorthogonality

The sum over j_1 and k_1 in (3.6) gives

$$\frac{(ABcdq^{N-1};q)_{J_2} \, (ABcdq^{M-1};q)_{K_2}}{(ABcd;q)_{J_2+K_2}} q^{J_2+K_2}$$
$$\times \sum_{k_1=0}^{m_1} \frac{(q^{-m_1}, ABcdq^{M+K_2-1};q)_{k_1}}{(q, ABcdq^{J_2+K_2};q)_{k_1}} q^{k_1} \tag{4.1}$$
$$\times \, {}_3\varphi_2 \left[\begin{matrix} q^{-n_1}, & ABcdq^{N+J_2-1}, & a_1 b_1 q^{k_1} \\ & ABcdq^{J_2+K_2+k_1}, & a_1 b_1 \end{matrix} ; q, q^{1+K_2-N_2} \right].$$

Since, by (Gasper and Rahman, 1990a, (3.2.7)), the above ${}_3\varphi_2$ equals

$$\frac{(ABcd;q)_{J_2+K_2+k_1} \, (q^{1+K_2-N};q)_{n_1}}{(ABcd;q)_{J_2+K_2+n_1} \, (q^{1+K_2-N})_{k_1}}$$
$$\times \, {}_3\varphi_2 \left[\begin{matrix} q^{-k_1}, & ABcdq^{N+J_2-1}, & a_1 b_1 q^{n_1} \\ & ABcdq^{n_1+J_2+K_2}, & a_1 b_1 \end{matrix} ; q, q^{1+K-N} \right],$$

we can now do the summation over k_1 via (Gasper and Rahman, 1990a, (1.5.3)) to obtain that the expression in (4.1) reduces to

$$\frac{(ABcdq^{N-1};q)_{J_2} \, (ABcdq^{M-1};q)_{K_2} \, (ABcdq^{N+M_2-1};q)_{m_1} \, (q^{1+K_2-N};q)_{n_1}}{(ABcd;q)_{n_1+J_2+K_2} \, (q^{1+K_2-N};q)_{m_1}}$$
$$\times \, (-1)^{m_1} q^{\binom{m_1}{2} + (1+K_2-N)m_1 + J_2 + K_2}$$
$$\times \, {}_4\varphi_3 \left[\begin{matrix} q^{-m_1}, & a_1 b_1 q^{n_1}, & ABcdq^{N+J_2-1}, & ABcdq^{M+K_2-1} \\ & a_1 b_1, & ABcdq^{N+M_2-1}, & ABcdq^{n_1+J_2+K_2} \end{matrix} ; q, q \right]. \tag{4.2}$$

Note that the ${}_4\varphi_3$ series is balanced. Now, the sum over j_2 and k_2 gives

$$\frac{\left(ABcdq^{N-1};q\right)_{J_3}\left(ABcdq^{M-1};q\right)_{K_3}\left(ABcdq^{N+M_2-1};q\right)_{m_1}\left(q^{1+K_3-N};q\right)_{n_1}}{(ABcd;q)_{n_1+J_3+K_3}\left(q^{1+K_3-N};q\right)_{m_1}}$$
$$\times (-1)^{m_1} q^{\binom{m_1}{2}+(1+K_3-N)m_1+J_3+K_3}$$
$$\times \sum_{k_1=0}^{m_1} \frac{\left(q^{-m_1}, a_1b_1q^{n_1}, ABcdq^{N+J_3-1}, ABcdq^{M+K_3-1};q\right)_{k_1}}{\left(q, a_1b_1, ABcdq^{n_1+J_3+K_3}, ABcdq^{M_2+N-1};q\right)_{k_1}} q^{k_1}$$
$$\times \sum_{k_2=0}^{m_2} \frac{\left(q^{-m_2}, ABcdq^{M+K_3+k_1-1}, q^{1+K_3-N_2};q\right)_{k_2}}{\left(q, ABcdq^{n_1+J_3+K_3+k_1}, q^{1+K_3-N+m_1};q\right)_{k_2}} q^{k_2}$$
$$\times {}_3\varphi_2\left[\begin{matrix} q^{-n_2}, & ABcdq^{N+J_3+k_1-1}, & a_2b_2q^{k_2} \\ & ABcdq^{n_1+J_3+K}, & a_2b_2 \end{matrix} ;q, q^{1+K_3-N_3}\right]. \tag{4.3}$$

As in the previous step we apply (Gasper and Rahman, 1990a, (3.2.7)) to the ${}_3\varphi_2$ series above, use (Gasper and Rahman, 1990a, (1.5.3)) to do the k_2 sum and simplify the coefficients to reduce (4.3) to the following expression

$$\frac{\left(ABcdq^{N-1};q\right)_{J_3}\left(ABcdq^{M-1};q\right)_{K_3}}{(ABcd;q)_{n_1+n_2+J_3+K_3}}$$
$$\times \frac{\left(ABcdq^{N+M_3-1};q\right)_{m_1+m_2}\left(q^{1+K_3-N};q\right)_{n_1+n_2}}{\left(q^{1+K_3-N};q\right)_{m_1+m_2}}$$
$$\times (-1)^{m_1+m_2} q^{\binom{m_1+m_2}{2}+(1+K_3-N)(m_1+m_2)+J_3+K_3} \tag{4.4}$$
$$\times \sum_{k_1=0}^{m_1}\sum_{k_2=0}^{m_2} \frac{\left(q^{-m_1}, a_1b_1q^{n_1};q\right)_{k_1}\left(q^{-m_2}, a_2b_2q^{n_2};q\right)_{k_2}}{(q,a_1b_1;q)_{k_1}(q,a_2b_2;q)_{k_2}} q^{(n_1-m_1)k_2}$$
$$\times \frac{\left(ABcdq^{N+J_3-1}, ABcdq^{M+K_3-1};q\right)_{k_1+k_2}}{\left(ABcdq^{n_1+n_2+J_3+K_3}, ABcdq^{M_3+N-1};q\right)_{k_1+k_2}} q^{k_1+k_2}.$$

A clear pattern of terms is now emerging, and by induction we find that at the $(p-1)$-th step the sum over $j_1, k_1, \ldots, j_{p-1}, k_{p-1}$ in (3.6) equals

$$\begin{aligned}
&\frac{\left(ABcdq^{N-1};q\right)_{J_p}\left(ABcdq^{M-1};q\right)_{K_p}}{(ABcd;q)_{N-N_p+J_p+K_p}} \\
&\times \frac{\left(ABcdq^{N+M_p-1};q\right)_{M-m_p}\left(q^{1+K_p-N};q\right)_{N-n_p}}{\left(q^{1+K_p-N};q\right)_{M-m_p}} \\
&\times (-1)^{M-m_p} q^{\binom{M-m_p}{2}+(1+K_p-N)(M-m_p)+J_p+K_p} \\
&\times \sum_{k_1,\ldots,k_{p-1}} \left[\prod_{r=1}^{p-1} \frac{(q^{-m_r}, a_r b_r q^{n_r};q)_{k_r}}{(q, a_r b_r;q)_{k_r}}\right] \\
&\times \frac{\left(ABcdq^{N+J_p-1}, ABcdq^{M+K_p-1};q\right)_{K-k_p}}{\left(ABcdq^{N-n_p+J_p+K_p}, ABcdq^{M_p+N-1};q\right)_{K-k_p}} \\
&\times q^{k_1+k_2(1+n_1-m_1)+\cdots+k_{p-1}(1+n_1+\cdots+n_{p-2}-m_1-\cdots-m_{p-2})}.
\end{aligned} \tag{4.5}$$

Using (4.5) we obtain that the sum over $\mathbf{j}$ and $\mathbf{k}$ in (3.6) equals

$$\begin{aligned}
&\frac{\left(ABcdq^{N+m_p-1};q\right)_{M-m_p}}{(ABcd;q)_{N-n_p}}(-1)^{M-m_p} q^{\binom{M-m_p}{2}+(1-N)(M-m_p)} \\
&\times \sum_{k_1=0}^{m_1} \cdots \sum_{k_{p-1}=0}^{m_{p-1}} q^{k_1+k_2(1+n_1-m_1)+\cdots+k_{p-1}(1+n_1+\cdots+n_{p-2}-m_1-\cdots-m_{p-2})} \\
&\qquad \times \left[\prod_{r=1}^{p-1} \frac{(q^{-m_r}, a_r b_r q^{n_r};q)_{k_r}}{(q, a_r b_r;q)_{k_r}}\right] \\
&\qquad \times \frac{\left(ABcdq^{N-1}, ABcdq^{M-1};q\right)_{k_1+\cdots+k_{p-1}}}{\left(ABcdq^{N-n_p}, ABcdq^{N+m_p-1};q\right)_{k_1+\cdots+k_{p-1}}} S_p,
\end{aligned} \tag{4.6}$$

where

$$\begin{aligned}
S_p = &\sum_{k_p=0}^{m_p} \frac{\left(q^{-m_p}, ABcdq^{M+k_1+\cdots+k_{p-1}-1};q\right)_{k_p}\left(q^{1+M-N-m_p+k_p};q\right)_\infty}{\left(q, ABcdq^{N-n_p+k_1+\cdots+k_{p-1}};q\right)_{k_p}\left(q^{1+k_p-n_p};q\right)_\infty} q^{k_p} \\
&\times {}_3\varphi_2\left[\begin{matrix} q^{-n_p}, & ABcdq^{N+k_1+\cdots+k_{p-1}-1}, & a_p b_p q^{k_p} \\ & ABcdq^{N-n_p+k_1+\cdots+k_{p-1}+k_p}, & a_p b_p \end{matrix}; q, q\right].
\end{aligned} \tag{4.7}$$

Note that the ${}_3\varphi_2$ series is balanced, so by (Gasper and Rahman, 1990a, (II.12)) it has the sum

$$\frac{\left(q^{1+k_p-n_p}, \frac{ABcd}{a_pb_p}q^{N-n_p+k_1+\cdots+k_{p-1}}; q\right)_{n_p}}{\left(\frac{q^{1-n_p}}{a_pb_p}, ABcdq^{N-n_p+K}; q\right)_{n_p}}.$$

Hence,

$$S_p = \frac{\left(\frac{ABcd}{a_pb_p}q^{N-n_p+k_1+\cdots+k_{p-1}}; q\right)_{n_p}}{\left(\frac{q^{1-n_p}}{a_pb_p}, ABcdq^{N+k_1+\cdots+k_{p-1}-n_p}; q\right)_{n_p}} \times \sum_{k_p=0}^{m_p} \frac{\left(q^{-m_p}, ABcdq^{M+k_1+\cdots+k_{p-1}-1}; q\right)_{k_p}\left(q^{1+M-N-m_p+k_p}; q\right)_\infty}{\left(ABcdq^{N+k_1+\cdots+k_{p-1}}; q\right)_{k_p}(q;q)_\infty} q^{k_p}. \tag{4.8}$$

First, let us suppose that $N \geq M \geq 0$. Then it is clear from the right side of (4.8) that S_p is zero unless $k_p \geq N - M + m_p$, as well as $m_p \geq k_p$ So, we must have

$$m_p + (N - M) \leq k_p \leq m_p. \tag{4.9}$$

This is a contradiction unless $N = M$, and then $k_p = m_p$. In that case

$$S_p = q^{m_p}\frac{\left(\frac{ABcd}{a_pb_p}q^{N-n_p+k_1+\cdots+k_{p-1}}; q\right)_{n_p}}{\left(\frac{q^{1-n_p}}{a_pb_p}; q\right)_{n_p}} \times \frac{\left(q^{-m_p}, ABcdq^{N+k_1+\cdots+k_{p-1}-1}; q\right)_{m_p}}{\left(ABcdq^{N+k_1+\cdots+k_{p-1}-n_p}; q\right)_{m_p+n_p}}. \tag{4.10}$$

On the other hand, if $M \geq N \geq 0$ then

$$m_p - (M - N) \leq k_p \leq m_p. \tag{4.11}$$

So we get

$$S_p = q^{m_p+N-M}\frac{\left(\frac{ABcd}{a_pb_p}q^{N-n_p+k_1+\cdots+k_{p-1}}; q\right)_{n_p}}{\left(\frac{q^{1-n_p}}{a_pb_p}, ABcdq^{N-n_p+k_1+\cdots+k_{p-1}}; q\right)_{n_p}} \times \frac{\left(q^{-m_p}, ABcdq^{M+k_1+\cdots+k_{p-1}-1}; q\right)_{m_p+N-M}}{\left(ABcdq^{N+k_1+\cdots+k_{p-1}}; q\right)_{m_p+N-M}} \times {}_2\varphi_1\left[\begin{matrix} q^{N-M}, & ABcdq^{N+m_p+k_1+\cdots k_{p-1}-1} \\ & ABcdq^{2N-M+m_p+k_1+\cdots+k_{p-1}} \end{matrix}; q, q\right]. \tag{4.12}$$

However, the above ${}_2\varphi_1$ equals

$$\frac{\left(q^{1+N-M};q\right)_{M-N}}{\left(ABcdq^{2N-M+m_p+k_1+\cdots+k_{p-1}};q\right)_{M-N}}\left(ABcdq^{N+m_p+k_1+\cdots+k_{p-1}-1}\right)^{M-N}, \tag{4.13}$$

which vanishes unless $N = M$. This completes the proof of (1.22).

Also, with $N = M$, (3.6), (4.6) and (4.10) give

$$\begin{aligned} &P_{\mathbf{n}} \cdot \bar{P}_{\mathbf{m}} \\ &= L_p \frac{\left(ABcdq^{N-1};q\right)_N \left(\frac{a_p b_p q^{1-N}}{ABcd};q\right)_{n_p}}{(ABcd;q)_{N+m_p}\,(a_p b_p;q)_{n_p}} (-1)^N q^{-\binom{N}{2}-m_p-n_p}\left(ABcdq^N\right)^{n_p} \\ &\times \sum_{k_1=0}^{m_1}\cdots\sum_{k_{p-1}=0}^{m_{p-1}} q^{k_1+k_2(1+n_1-m_1)+\cdots+k_{p-1}(1+n_1+\cdots+n_{p-2}-m_1-m_2-\cdots-m_{p-2})} \\ &\times \frac{\left(ABcdq^{N-1}, \frac{ABcdq^N}{a_p b_p};q\right)_{k_1+\cdots+k_{p-1}}}{\left(ABcdq^{N+m_p}, \frac{ABcdq^{N-n_p}}{a_p b_p};q\right)_{k_1+\cdots+k_{p-1}}} \prod_{r=1}^{p-1} \frac{(q^{-m_r}, a_r b_r q^{n_r};q)_{k_r}}{(q, a_r b_r;q)_{k_r}}, \end{aligned} \tag{4.14}$$

which is, of course, the same as (1.23). By taking $p = 2$, e.g., in which case the series on the right hand side of (4.14) becomes a terminating balanced ${}_4\varphi_3$ series, it is easily seen that in general the above inner product does not vanish when $N = M$ and $\mathbf{n} \neq \mathbf{m}$.

In closing this section we would like to point out that unlike the $q \to 1$ case that corresponds to the Tratnik biorthogonalities, the q-analogues of $P_{\mathbf{n}} \cdot Q_{\mathbf{m}}$, $P_{\mathbf{n}} \cdot \bar{Q}_{\mathbf{m}}$ or $Q_{\mathbf{n}} \cdot \bar{Q}_{\mathbf{m}}$ do not seem to work out the same way as $P_{\mathbf{n}} \cdot \bar{P}_{\mathbf{m}}$.

5. Transformations of $F_{m,n}(x, y)$ and $G_{m,n}(x, y)$

We shall now address the problem of proving the biorthogonality relation (1.24). First of all, it is very simple to use (Gasper and Rahman, 1990a, (II.20)) to prove that

$$\sum_{x=0}^{N}\sum_{y=0}^{N} W_N(x, y) = 1. \tag{5.1}$$

The forms of $F_{m,n}(x,y)$ and $G_{m,n}(x,y)$ that turn out to be most convenient for the summations in (1.24) are as follows:

$$F_{m,n}(x,y) = \frac{\left(\frac{\gamma\gamma' q^{-N}}{\alpha};q\right)_{x+y}\left(\frac{\gamma\gamma'}{\alpha c};q\right)_{x-y}}{(q^{-N};q)_{x+y}\,(c^{-1};q)_{x-y}}\left(\frac{\alpha}{\gamma\gamma'}\right)^{x}\left(\frac{\alpha q^{N+n+1}}{\gamma\gamma'}\right)^{m} q^{Nn}$$
$$\times \sum_{j=0}^{x}\sum_{k=0}^{y}\frac{\left(\frac{\gamma\gamma'}{\alpha}q^{-m-n};q\right)_{j+k}\left(q^{-x},\frac{\gamma\gamma'}{\alpha c}q^{x-N-1},\gamma q^{m};q\right)_{j}}{\left(\frac{\gamma\gamma'}{\alpha}q^{-N};q\right)_{j+k}\left(q,\gamma,\frac{\gamma\gamma' q^{-n}}{\alpha c};q\right)_{j}}$$
$$\times \frac{(q^{-y},cq^{y-N},\gamma' q^{n};q)_{k}}{(q,cq^{1-m},\gamma';q)_{k}}q^{j+k}, \tag{5.2}$$

and

$$G_{m,n}(x,y) = \frac{(\alpha q^{N+1};q)_{m+n}\left(\frac{\alpha c q}{\gamma'};q\right)_{m}\left(\frac{\gamma'}{c};q\right)_{n}}{(q^{-N};q)_{m+n}\left(\frac{\gamma'}{c};q\right)_{m}\left(\frac{\alpha c q}{\gamma'};q\right)_{m}}\left(\frac{\gamma' q^{-N-1}}{\alpha c}\right)^{m}\left(\frac{cq^{-N}}{\gamma'}\right)^{n}$$
$$\times \sum_{j=0}^{m}\sum_{k=0}^{n}\frac{(\alpha q^{m+n};q)_{j+k}\left(q^{-m},\gamma q^{x},\frac{\alpha c}{\gamma'}q^{N-x+1};q\right)_{j}}{(\alpha q^{N+1};q)_{j+k}\left(q,\gamma,\frac{\alpha c}{\gamma'}q^{n+1};q\right)_{j}}$$
$$\times \frac{\left(q^{-n},\gamma' q^{y},\frac{\gamma' q^{N-y}}{c};q\right)_{k}}{\left(q,\gamma',\frac{\gamma' q^{m}}{c};q\right)_{k}}q^{j+k}, \quad \text{assuming } 0 \le m+n \le N. \tag{5.3}$$

Since

$${}_4\varphi_3\left[\begin{matrix} q^{-m}, & \alpha q^{j+m+n}, & q^{-x}, & \frac{\gamma\gamma'}{\alpha c}q^{x-N-1} \\ & \gamma, & \frac{\gamma' q^{n}}{c}, & q^{j-N} \end{matrix};q,q\right]$$
$$= \frac{\left(\frac{\alpha c q^{j+1}}{\gamma'},\alpha q^{N+n+1};q\right)_{m}}{\left(\frac{\gamma' q^{n}}{c},q^{j-N};q\right)_{m}}\left(\frac{\gamma' q^{-N-1}}{\alpha c}\right)^{m}$$
$$\times\, {}_4\varphi_3\left[\begin{matrix} q^{-m}, & \alpha q^{j+m+n}, & \gamma q^{x}, & \frac{\alpha c}{\gamma'}q^{N-x+1} \\ & \gamma, & \frac{\alpha c q^{j+1}}{\gamma'}, & \alpha q^{N+n+1} \end{matrix};q,q\right]$$

and

$$
{}_4\varphi_3\left[\begin{matrix} q^{-n}, & \alpha q^{i+m+n}, & q^{-y}, & cq^{y-N} \\ & \gamma', & \frac{\alpha c q^{i+1}}{\gamma'}, & q^{m-N} \end{matrix} ;q,q\right]
= \frac{\left(\frac{\gamma' q^m}{c}, \alpha q^{N+1+i}; q\right)_n}{\left(\frac{\alpha c q^{i+1}}{\gamma'}, q^{m-N}; q\right)_n} \left(\frac{cq^{-N}}{\gamma'}\right)^n
\times {}_4\varphi_3\left[\begin{matrix} q^{-n}, & \alpha q^{m+n+i}, & \gamma' q^y, & \frac{\gamma' q^{N-y}}{c} \\ & \gamma', & \frac{\gamma' q^m}{c}, & \alpha q^{N+1+i} \end{matrix} ;q,q\right]
$$

by (Gasper and Rahman, 1990a, (III.15)), (5.3) follows from (1.26) with a bit of simplification.

To derive (5.2) from (1.25) we need two applications of (Gasper and Rahman, 1990a, (III.15)) on each of the two ${}_4\varphi_3$ series involved in (1.25). First

$$
\begin{aligned}
&{}_4\varphi_3\left[\begin{matrix} q^{-m}, & \frac{\gamma\gamma'}{\alpha c} q^{x-N-1}, & \gamma q^x, & \frac{\gamma\gamma' q^{j-m-n}}{\alpha} \\ & \gamma, & \frac{\gamma\gamma' q^{x-y-m}}{\alpha c}, & \frac{\gamma\gamma'}{\alpha} q^{x+y-N-m-n+j} \end{matrix} ;q,q\right] \\
&= \frac{\left(q^{y-N}, c^{-1} q^{n-y-j}; q\right)_m}{\left(\frac{\alpha c}{\gamma\gamma'} q^{1+y-x}, \frac{\alpha q^{N-x-y+n+1-j}}{\gamma\gamma'}; q\right)_m} \left(\frac{\alpha c q^{N-x+1}}{\gamma\gamma'}\right)^m \\
&\times {}_4\varphi_3\left[\begin{matrix} q^{-x}, & \frac{\gamma\gamma' q^{x-N-1}}{\alpha c}, & q^{-m}, & \frac{\alpha}{\gamma'} q^{m+n-j} \\ & \gamma, & q^{y-N}, & c^{-1} q^{n-y-j} \end{matrix} ;q,q\right] \\
&= \frac{\left(\frac{\alpha c}{\gamma\gamma'} q^{1+y-x+m}, \frac{\alpha q^{N-x-y+1+m+n}}{\gamma\gamma'}; q\right)_{x-m}}{\left(q^{y+m-N}, c^{-1} q^{m+n-y}; q\right)_{x-m}} \left(\frac{\gamma\gamma'}{\alpha c} q^{x-N-1}\right)^{x-m} \\
&\times \frac{\left(\frac{\gamma\gamma'}{\alpha} q^{x+y-N-m-n}, c q^{1+y-x-n}; q\right)_j}{\left(\frac{\gamma\gamma'}{\alpha} q^{y-N-n}, c q^{1+y-m-n}; q\right)_j} \\
&\times {}_4\varphi_3\left[\begin{matrix} q^{-x}, & \frac{\gamma\gamma'}{\alpha c} q^{x-N-1}, & \gamma q^m, & \frac{\gamma\gamma' q^{j-m-n}}{\alpha} \\ & \gamma, & \frac{\gamma\gamma' q^{-y}}{\alpha c}, & \frac{\gamma\gamma'}{\alpha} q^{j+y-N-n} \end{matrix} ;q,q\right].
\end{aligned}
\tag{5.4}
$$

Substituted into (1.25) this leads to another balanced ${}_4\varphi_3$ series:

$$
{}_4\varphi_3\left[\begin{matrix} q^{-n}, & cq^{y-N}, & \gamma' q^y, & \frac{\gamma\gamma'}{\alpha} q^{i-m-n} \\ & \gamma', & cq^{y-m-n+1}, & \frac{\gamma\gamma'}{\alpha} q^{i+y-N-n} \end{matrix} ;q,q\right],
$$

which, when transformed twice in the same manner as in (5.4), leads to

$$\frac{\left(c^{-1}q^{m+n-y}, \frac{\alpha q^{N+1-y+n}}{\gamma\gamma'}; q\right)_{y-n}}{\left(q^{m+n-N}, \frac{\alpha c q^{n+1}}{\gamma\gamma'}; q\right)_{y-n}} \left(cq^{y-N}\right)^{y-n} \frac{\left(\frac{\gamma\gamma'}{\alpha} q^{y-N-n}, \frac{\gamma\gamma'}{\alpha c} q^{-y}; q\right)_i}{\left(\frac{\gamma\gamma'}{\alpha} q^{-N}, \frac{\gamma\gamma'}{\alpha c} q^{-n}; q\right)_i}$$
$$\times {}_4\varphi_3\left[\begin{matrix} q^{-y}, & cq^{y-N}, & \gamma' q^n, & \frac{\gamma\gamma'}{\alpha} q^{i-m-n} \\ & \gamma', \; cq^{1-m}, & \frac{\gamma\gamma'}{\alpha} q^{i-N} & \end{matrix}; q, q\right]. \tag{5.5}$$

After some simplifications (5.4) and (5.5) give (5.2). Denoting the left hand side of (1.24) by $F_{m,n} \cdot G_{m',n'}$, it follows that

$$F_{m,n} \cdot G_{m',n'} = A_{m,n,m',n'} \sum_{x=0}^{N} \sum_{y=0}^{N} \frac{\left(1 - \frac{\gamma\gamma'}{\alpha c} q^{2x-N-1}\right)\left(1 - cq^{2y-N}\right)}{\left(1 - \frac{\gamma\gamma'}{\alpha c} q^{-N-1}\right)\left(1 - cq^{-N}\right)}$$
$$\times \frac{\left(\frac{\gamma\gamma' q^{-N-1}}{\alpha c}, \gamma; q\right)_x}{\left(q, \frac{\gamma' q^{-N}}{\alpha c}; q\right)_x} \frac{\left(cq^{-N}, \gamma'; q\right)_y}{\left(q, \frac{cq^{1-N}}{\gamma'}; q\right)_y} \gamma^{-x} (\gamma')^{-y}$$
$$\times \sum_j \sum_k \frac{\left(\frac{\gamma\gamma' q^{-m-n}}{\alpha}; q\right)_{j+k} \left(q^{-x}, \frac{\gamma\gamma'}{\alpha c} q^{x-N-1}, \gamma q^m; q\right)_j}{\left(\frac{\gamma\gamma' q^{-N}}{\alpha}; q\right)_{j+k} \left(q, \gamma, \frac{\gamma\gamma' q^{-n}}{\alpha c}; q\right)_j}$$
$$\times \frac{\left(q^{-y}, cq^{y-N}, \gamma' q^n; q\right)_k}{\left(q, cq^{1-m}, \gamma'; q\right)_k} q^{j+k}$$
$$\times \sum_r \sum_s \frac{\left(\alpha q^{m'+n'}; q\right)_{r+s} \left(q^{-m'}, \gamma q^x, \frac{\alpha c q^{N-x+1}}{\gamma'}; q\right)_r}{\left(\alpha q^{N+1}; q\right)_{r+s} \left(q, \gamma, \frac{\alpha c}{\gamma'} q^{n'+1}; q\right)_r}$$
$$\times \frac{\left(q^{-n'}, \gamma' q^y, \frac{\gamma'}{c} q^{N-y}; q\right)_s}{\left(q, \gamma', \frac{\gamma' q^{m'}}{c}; q\right)_s} q^{r+s}, \tag{5.6}$$

where

$$A_{m,n,m',n'} = \frac{(\alpha q/\gamma\gamma', \gamma'/c, \alpha c q/\gamma'; q)_N \, (\alpha q^{N+1}; q)_{m'+n'} \left(\frac{\gamma'}{c}; q\right)_{n'}}{(\alpha q, 1/c, \alpha c q/\gamma\gamma'; q)_N \, (q^{-N}; q)_{m'+n'} \left(\frac{\alpha c q}{\gamma'}; q\right)_{n'}}$$
$$\times \frac{\left(\frac{\alpha c q}{\gamma'}; q\right)_{m'}}{\left(\frac{\gamma'}{c}; q\right)_{m'}} \left(\frac{\gamma' q^{-N-1}}{\alpha c}\right)^{m'} \left(c q^{-N}/\gamma'\right)^{n'} \left(\frac{\alpha q^{N+n+1}}{\gamma\gamma'}\right)^{m} q^{Nn}. \tag{5.7}$$

6. Proof of (1.24)

Since each term in the weight function can be glued on nicely with the x and y dependent terms of the two double series in (5.6), the x, y-sum can be isolated as

$$\begin{aligned} &{}_6W_5\left(\frac{\gamma\gamma'}{\alpha c} q^{2j-N-1}; \gamma q^{r+j}, \frac{\gamma\gamma'}{\alpha c} q^j, q^{j-N}; q, \gamma^{-1} q^{-j-r}\right) \\ &\times {}_6W_5\left(c q^{2k-N}; \gamma' q^{s+k}, c q^{k+1}, q^{k-N}; q, \left(\gamma' q^{k+s}\right)^{-1}\right) \\ &= \frac{\left(\frac{\gamma\gamma'}{\alpha c} q^{2j-N}, \frac{q^{-N-r}}{\gamma}; q\right)_{N-j} \left(c q^{2k-N+1}, \frac{q^{-N-s}}{\gamma'}; q\right)_{N-k}}{\left(\frac{\gamma' q^{j-r-N}}{\alpha c}, q^{j-N}; q\right)_{N-j} \left(\frac{c q^{1-N+k-s}}{\gamma'}, q^{k-N}; q\right)_{N-k}}, \end{aligned}$$

by (Gasper and Rahman, 1990a, (II.21)). The sum over j, k, r, s in (5.6) now reduces to

$$\begin{aligned} &F_{m,n} \cdot G_{m',n'} \\ &= A_{m,n,m',n'} \frac{(\gamma q, \gamma' q, \alpha c q/\gamma\gamma', 1/c; q)_N}{(q, q, \alpha c q/\gamma', \gamma'/c; q)_N} \\ &\times \sum_j \sum_k \sum_r \sum_s \frac{\left(\frac{\gamma\gamma' q^{-m-n}}{\alpha}; q\right)_{j+k} \left(\alpha q^{m'+n'}; q\right)_{r+s}}{\left(\frac{\gamma\gamma' q^{-N}}{\alpha}; q\right)_{j+k} (\alpha q^{N+1}; q)_{r+s}} \\ &\times \frac{\left(q^{-N}, \frac{\gamma\gamma'}{\alpha c}, \gamma q^m; q\right)_j (q^{-N}, \gamma' q^n, cq; q)_k \left(q^{-m'}, \frac{\alpha c q}{\gamma'}, \gamma q^{N+1}; q\right)_r}{\left(q, \gamma, \frac{\gamma\gamma' q^{-n}}{\alpha c}; q\right)_j (q, \gamma', c q^{1-m}; q)_k \left(q, \gamma, \frac{\alpha c q^{n'+1}}{\gamma'}; q\right)_r} \\ &\times \frac{\left(q^{-n'}, \gamma'/c, \gamma' q^{N+1}; q\right)_s}{\left(q, \gamma', \frac{\gamma' q^{m'}}{c}; q\right)_s} \frac{(\gamma; q)_{r+j} (\gamma'; q)_{s+k}}{(\gamma q; q)_{r+j} (\gamma' q; q)_{s+k}} q^{j+k+r+s}. \end{aligned} \tag{6.1}$$

The sum over j is a multiple of

$$
\begin{aligned}
&{}_5\varphi_4\left[\begin{matrix} q^{-N}, & \gamma q^r, & \frac{\gamma\gamma'}{\alpha c}, & \gamma q^m, & \frac{\gamma\gamma'}{\alpha}q^{k-m-n} \\ & \gamma q^{r+1}, & \frac{\gamma\gamma' q^{-n}}{\alpha c}, & \gamma, & \frac{\gamma\gamma'}{\alpha}q^{k-N}\end{matrix} ;q,q\right] \\
&= \frac{(q;q)_N \left(\frac{\gamma' q^{-n-r}}{\alpha c};q\right)_n (q^{-r};q)_m \left(\frac{\gamma' q^{k-N-r}}{\alpha};q\right)_{N-m-n}}{(\gamma q^{r+1};q)_N \left(\frac{\gamma\gamma' q^{-n}}{\alpha c};q\right)_n (\gamma;q)_m \left(\frac{\gamma\gamma'}{\alpha}q^{k-N};q\right)_{N-m-n}} (\gamma q^r)^N,
\end{aligned}
\tag{6.2}
$$

by (Gasper and Rahman, 1990a, (1.9.10)). Together with a similar expression for the sum over k we now have

$$
\begin{aligned}
&F_{m,n}\cdot G_{m',n'} \\
&= A_{m,n,m',n'} \frac{\left(\frac{\alpha c q}{\gamma\gamma'}, 1/c;q\right)_N}{(\alpha c q/\gamma', \gamma'/c;q)_N} \sum_r \sum_s \frac{\left(\alpha q^{m'+n'};q\right)_{r+s}}{(\alpha q^{N+1};q)_{r+s}} \\
&\times \frac{\left(q^{-m'}, \frac{\alpha c q}{\gamma'}, \gamma q^{N+1};q\right)_r \left(q^{-n'}, \frac{\gamma'}{c}, \gamma' q^{N+1};q\right)_s}{\left(q, \frac{\alpha c q^{n'+1}}{\gamma'};q\right)_r \left(q, \frac{\gamma' q^{m'}}{c};q\right)_s} q^{r+s} \\
&\times \frac{\left(\frac{\gamma' q^{-n-r}}{\alpha c};q\right)_n \left(\frac{\gamma' q^{-N-r}}{\alpha};q\right)_{N-m-n} (q^{-r};q)_m}{(\gamma q^{N+1};q)_r \left(\frac{\gamma\gamma' q^{-n}}{\alpha c};q\right)_n (\gamma;q)_m \left(\frac{\gamma\gamma'}{\alpha}q^{-N};q\right)_{N-m-n}} (\gamma q^r)^N \\
&\times \frac{\left(\frac{c q^{1-m-s}}{\gamma'};q\right)_m \left(\frac{q^{-N-r-s}}{\alpha};q\right)_{N-m-n} (q^{-s};q)_n}{(\gamma' q^{N+1};q)_s (cq^{1-m};q)_m (\gamma';q)_n \left(\frac{\gamma'}{\alpha}q^{-N-r};q\right)_{N-m-n}} (\gamma' q^s)^N \\
&= A_{m,n,m',n'} \frac{\left(\frac{\alpha c q}{\gamma\gamma'}, 1/c;q\right)_N}{\left(\frac{\alpha c q}{\gamma'}, \frac{\gamma'}{c};q\right)_N} (\gamma\gamma')^N \frac{\left(\frac{\gamma' q^{-n}}{\alpha c};q\right)_n}{\left(\gamma', \frac{\gamma\gamma' q^{-n}}{\alpha c};q\right)_n} \\
&\times \frac{\left(\frac{c q^{1-m}}{\gamma'};q\right)_m (q^{-N}/\alpha;q)_{N-m-n}}{(\gamma, cq^{1-m};q)_m \left(\frac{\gamma\gamma'}{\alpha}q^{-N};q\right)_{N-m-n}} \\
&\times \sum_r \sum_s \frac{\left(\alpha q^{m'+n'};q\right)_{r+s}}{(\alpha q^{m+n+1};q)_{r+s}} \frac{\left(q^{-m'}, \frac{\alpha c q^{n+1}}{\gamma};q\right)_r \left(q^{-n'}, \frac{\gamma' q^m}{c};q\right)_s}{\left(q, \frac{\alpha c q^{n'+1}}{\gamma'};q\right)_r \left(q, \frac{\gamma'}{c}q^{m'};q\right)_s} \\
&\qquad \times (q^{-r};q)_m (q^{-s};q)_n q^{(m+1)r+(n+1)s}.
\end{aligned}
\tag{6.3}
$$

The r, s sum is

$$(-1)^{m+n} q^{\binom{m+1}{2}+\binom{n+1}{2}} \frac{\left(\alpha q^{m'+n'};q\right)_{m+n}}{\left(\alpha q^{m+n+1};q\right)_{m+n}}$$
$$\times \frac{\left(q^{-m'}, \frac{\alpha c q^{n+1}}{\gamma'};q\right)_m \left(q^{-n'}, \frac{\gamma' q^m}{c};q\right)_n}{\left(\frac{\alpha c q^{n'+1}}{\gamma'};q\right)_m \left(\frac{\gamma' q^{m'}}{c};q\right)_n}$$
$$\times \sum_{r=0}^{m'-m} \sum_{s=0}^{n'-n} \frac{\left(\alpha q^{m+n+m'+n'};q\right)_{r+s} \left(q^{m-m'}, \frac{\alpha c q^{n+m+1}}{\gamma'};q\right)_r}{\left(\alpha q^{2m+2n+1};q\right)_{r+s} \left(q, \frac{\alpha c q^{m+n'+1}}{\gamma'};q\right)_r} \tag{6.4}$$
$$\times \frac{\left(q^{n-n'}, \frac{\gamma' q^{n+m}}{c};q\right)_s}{\left(q, \frac{\gamma' q^{n+m'}}{c};q\right)_s} q^{r+s}$$

which vanishes unless $m' \geq m$ and $n' \geq n$.

The sum in (6.4), via (Gasper and Rahman, 1990a, (II.12) and (II.6)), equals

$$\frac{\left(q^{1+m-m'+n-n'}, \frac{\alpha c}{\gamma'} q^{m+n+1};q\right)_{n'-n} \left(q^{1+m-m'};q\right)_{m'-m}}{\left(\alpha q^{2m+2n+1}, \frac{c q^{1-m'-n'}}{\gamma'};q\right)_{n'-n} \left(\alpha q^{2m+n+n'+1};q\right)_{m'-m}}$$
$$\times \left(\alpha q^{m+n+m'+n'}\right)^{m'-m},$$

which vanishes unless $m' \leq m$ and $n' \leq n$. Thus we must have

$$F_{m,n} \cdot G_{m',n'} = 0 \quad \text{unless } (m,n) = \left(m',n'\right), \text{ and then} \tag{6.5}$$

$$F_{m,n} \cdot G_{m,n}$$
$$= \frac{1-\alpha}{1-\alpha q^{2m+2n}} \frac{\left(q, \frac{\alpha c q}{\gamma'};q\right)_m \left(q, \frac{\gamma'}{c};q\right)_n \left(\frac{\alpha q}{\gamma\gamma'}, \alpha q^{N+1};q\right)_{m+n}}{\left(\gamma, 1/c;q\right)_m \left(\gamma', \frac{\alpha c q}{\gamma\gamma'};q\right)_n \left(\alpha, q^{-N};q\right)_{m+n}} c^{n-m} q^{mn},$$

which completes the proof of (1.24) and (1.28).

It may be mentioned that there are other double series representations for $F_{m,n}(x,y)$ that one could use instead of (5.2) in the derivation of the biorthogonality relation (1.24) which do not contain the factor $1/\left(q^{-N};q\right)_{x+y}$ that cancels out the $\left(q^{-N};q\right)_{x+y}$ factor in the weight function, but the subsequent computations turn out to be quite tedious, while the final result is, of course, the same.

References

Gasper, G. and Rahman, M. (1990). *Basic Hypergeometric Series.* Cambridge University Press, Cambridge.

Gasper, G. and Rahman, M. (2003). Some systems of multivariable orthogonal Askey-Wilson polynomials. This Proceedings.

Granovskiĭ Y. I. and Zhedanov, A. S. (1992). 'Twisted' Clebsch-Gordan coefficients for $su_q(2)$. *J. Phys. A*, 25:L1029–L1032.

Gustafson, R. A. (1987). A Whipple's transformation for hypergeometric series in $u(n)$ and multivariable hypergeometric orthogonal polynomials. *SIAM J. Math. Anal.*, 18:495–530.

Koelink, H. T. and Van der Jeugt, J. (1998). Convolutions for orthogonal polynomials from Lie and quantum algebra representations. *SIAM J. Math. Anal.*, 29:794–822.

Rahman, M. (1981). Discrete orthogonal systems corresponding to Dirichlet distribution. *Utilitas Mathematica*, 20:261–272.

Rosengren, H. (2001). Multivariable q-Hahn polynomials as coupling coefficients for quantum algebra representations. *Int. J. Math. Sci.*, 28:331–358.

Tratnik, M. V. (1991a). Some multivariable orthogonal polynomials of the Askey tableau—continuous families. *J. Math. Phys.*, 32:2065–2073.

Tratnik, M. V. (1991b). Some multivariable orthogonal polynomials of the Askey tableau—discrete families. *J. Math. Phys.*, 32:2337–2342.

van Diejen, J. F. and Stokman, J. V. (1998). Multivariable q-Racah polynomials. *Duke Math. J.*, 91:89–136.

Wilson, J. A. (1980). Some hypergeometric orthogonal polynomials. *SIAM J. Math. Anal.*, 11:690–701.

SOME SYSTEMS OF MULTIVARIABLE ORTHOGONAL ASKEY-WILSON POLYNOMIALS

George Gasper
Department of Mathematics
Northwestern University
Evanston, IL 60208

Mizan Rahman*
School of Mathematics and Statistics
Carleton University
Ottawa, Ontario
CANADA K1S 5B6

Abstract In 1991 Tratnik derived two systems of multivariable orthogonal Wilson polynomials and considered their limit cases. q-Analogues of these systems are derived, yielding systems of multivariable orthogonal Askey-Wilson polynomials and their special and limit cases.

1. Introduction

In (Tratnik, 1991a) the Wilson (Wilson, 1980) polynomials

$$w_n(x; a, b, c, d) = (a+b)_n (a+c)_n (a+d)_n \times {}_4F_3\left[\begin{matrix} -n, n+a+b+c+d-1, a+ix, a-ix \\ a+b, a+c, a+d \end{matrix}; 1\right] \quad (1.1)$$

*Supported, in part, by an NSERC grant #A6197.

were extended to the multivariable Wilson polynomials (in a different notation)

$$\begin{aligned} W_{\mathbf{n}}(\mathbf{x}) &= W_{\mathbf{n}}\left(\mathbf{x}; a, b, c, d, a_2, a_3, \ldots, a_s\right) \\ &= \left[\prod_{k=1}^{s-1} w_{n_k}\left(x_k; a + \alpha_{2,k} + N_{k-1}, b + \alpha_{2,k} + N_{k-1},\right.\right. \\ &\qquad \left.\left. a_{k+1} + ix_{k+1}, a_{k+1} - ix_{k+1}\right)\right] \\ &\quad \times w_{n_s}\left(x_s; a + \alpha_{2,s} + N_{s-1}, b + \alpha_{2,s} + N_{s-1}, c, d\right), \end{aligned} \tag{1.2}$$

where, as elsewhere,

$$\mathbf{x} = \left(x_1, \ldots, x_s\right), \; \mathbf{n} = \left(n_1, \ldots, n_s\right), \; \alpha_{j,k} = \sum_{i=j}^{k} a_i, \; \alpha_k = \alpha_{1,k}, \tag{1.3}$$

$$N_{j,k} = \sum_{i=j}^{k} n_i, \; N_k = N_{1,k}, \; \alpha_{k+1,k} = N_{k+1,k} = 0, \quad 1 \le j \le k \le s.$$

These polynomials are of total degree N_s in the variables $y_1, \ldots, y_s$ with $y_k = x_k^2$, $k = 1, 2, \ldots, s$, and they form a complete set for polynomials in these variables.

In (Askey and Wilson, 1979) and (Askey and Wilson, 1985) the notations $W_n\left(x^2; a, b, c, d\right)$ and $p_n\left(-x^2\right)$ are used for the polynomials in (1.1) and their orthogonality relation is given. Tratnik (Tratnik, 1991a, (2.5)) proved that the $W_{\mathbf{n}}(\mathbf{x})$ polynomials satisfy the orthogonality relation

$$\int_{-\infty}^{\infty} \cdots \int_{-\infty}^{\infty} W_{\mathbf{n}}\left(\mathbf{x}\right) W_{\mathbf{m}}\left(\mathbf{x}\right) \rho\left(\mathbf{x}\right) dx_1 \cdots dx_s = \lambda_{\mathbf{n}}\, \delta_{\mathbf{n},\mathbf{m}} \tag{1.4}$$

for $\operatorname{Re}\left(a, b, c, d, a_2, \ldots, a_s\right) > 0$ with

$$\begin{aligned} \rho(\mathbf{x}) &= \Gamma\left(a + ix_1\right) \Gamma\left(a - ix_1\right) \Gamma\left(b + ix_1\right) \Gamma\left(b - ix_1\right) \\ &\quad \times \left[\prod_{k=1}^{s-1} \frac{\Gamma\left(a_{k+1} + ix_{k+1} + ix_k\right) \Gamma\left(a_{k+1} - ix_{k+1} - ix_k\right)}{\Gamma\left(2ix_k\right)}\right. \\ &\quad \left. \times \frac{\Gamma\left(a_{k+1} + ix_{k+1} - ix_k\right) \Gamma\left(a_{k+1} - ix_{k+1} + ix_k\right)}{\Gamma\left(-2ix_k\right)}\right] \\ &\quad \times \frac{\Gamma\left(c + ix_s\right) \Gamma\left(c - ix_s\right) \Gamma\left(d + ix_s\right) \Gamma\left(d - ix_s\right)}{\Gamma\left(2ix_s\right) \Gamma\left(-2ix_s\right)}, \end{aligned} \tag{1.5}$$

$$\begin{aligned}\lambda_{\mathbf{n}} = (4\pi)^s \Bigg[\prod_{k=1}^{s} n_k!\,(N_k + N_{k-1} + 2\alpha_{k+1} - 1)_{n_k} \\ \times \frac{\Gamma(N_k + N_{k-1} + 2\alpha_k)\,\Gamma(n_k + 2a_{k+1})}{\Gamma(2N_k + 2\alpha_{k+1})}\Bigg] \\ \times \Gamma(a + c + \alpha_{2,s} + N_s)\,\Gamma(a + d + \alpha_{2,s} + N_s) \\ \times \Gamma(b + c + \alpha_{2,s} + N_s)\,\Gamma(b + d + \alpha_{2,s} + N_s),\end{aligned} \tag{1.6}$$

and $2a_1 = a + b$, $2a_{s+1} = c + d$.

Tratnik showed that these polynomials contain multivariable Jacobi, Meixner-Pollaczek, Laguerre, continuous Charlier, and Hermite polynomials as limit cases, and he used a permutation of the parameters and variables in (1.2) and (1.4) to show that the polynomials

$$\begin{aligned}\tilde{W}_{\mathbf{n}}(\mathbf{x}) &= \tilde{W}_{\mathbf{n}}(\mathbf{x}; a, b, c, d, a_2, a_3, \ldots, a_s) \\ &= w_{n_1}(x_1; c + \alpha_{2,s} + N_{2,s}, d + \alpha_{2,s} + N_{2,s}, a, b) \\ &\times \prod_{k=2}^{s} w_{n_k}(x_k; c + \alpha_{k+1,s} + N_{k+1,s}, d + \alpha_{k+1,s} + N_{k+1,s}, \\ &\quad a_k + ix_{k-1}, a_k - ix_{k-1})\end{aligned} \tag{1.7}$$

also form a complete system of multivariable polynomials of total degree N_s in the variables $y_k = x_k^2$, $k = 1, \ldots, s$, that is orthogonal with respect to the weight function $\rho(\mathbf{x})$ in (1.5), and with the normalization constant

$$\begin{aligned}\tilde{\lambda}_{\mathbf{n}} = (4\pi)^s \Bigg[\prod_{k=1}^{s} n_k!\,(N_{k,s} + N_{k+1,s} + 2\alpha_{k,s+1} - 1)_{n_k} \\ \times \frac{\Gamma(N_{k,s} + N_{k+1,s} + 2\alpha_{k+1,s+1})\,\Gamma(n_k + 2a_k)}{\Gamma(2N_{k,s} + 2\alpha_{k,s+1})}\Bigg] \\ \times \Gamma(a + c + \alpha_{2,s} + N_s)\,\Gamma(a + d + \alpha_{2,s} + N_s) \\ \times \Gamma(b + c + \alpha_{2,s} + N_s)\,\Gamma(b + d + \alpha_{2,s} + N_s).\end{aligned} \tag{1.8}$$

The Askey-Wilson polynomials defined as in (Askey and Wilson, 1979) and (Gasper and Rahman, 1990a) by

$$\begin{aligned}p_n(x \mid q) &= p_n(x; a, b, c, d \mid q) \\ &= a^{-n}(ab, ac, ad; q)_n \, {}_4\varphi_3\left[\begin{matrix} q^{-n}, abcdq^{n-1}, ae^{i\theta}, ae^{-i\theta} \\ ab, ac, ad \end{matrix}; q, q\right],\end{aligned} \tag{1.9}$$

where $x = \cos\theta$, are a q-analogue of the Wilson polynomials (for the definition of the q-shifted factorials and the basic hypergeometric series

${}_4\varphi_3$ see (Gasper and Rahman, 1990a). These polynomials satisfy the orthogonality relation

$$\int_{-1}^{1} p_n(x \mid q) p_m(x \mid q) \rho(x \mid q) dx = \lambda_n(q) \delta_{n,m} \tag{1.10}$$

with $\max(|q|, |a|, |b|, |c|, |d|) < 1$,

$$\begin{aligned} \rho(x \mid q) &= \rho(x; a, b, c, d \mid q) \\ &= \frac{\left(e^{2i\theta}, e^{-2i\theta}; q\right)_\infty \left(1 - x^2\right)^{-1/2}}{\left(ae^{i\theta}, ae^{-i\theta}, be^{i\theta}, be^{-i\theta}, ce^{i\theta}, ce^{-i\theta}, de^{i\theta}, de^{-i\theta}; q\right)_\infty} \end{aligned} \tag{1.11}$$

and

$$\begin{aligned} \lambda_n(q) &= \lambda_n(a, b, c, d \mid q) \\ &= \frac{2\pi (abcd; q)_\infty}{(q, ab, ac, ad, bc, bd, cd; q)_\infty} \\ &\times \frac{(q, ab, ac, ad, bc, bd, cd; q)_n \left(1 - abcdq^{-1}\right)}{\left(abcdq^{-1}; q\right)_n \left(1 - abcdq^{2n-1}\right)}. \end{aligned} \tag{1.12}$$

In this paper we extend Tratnik's systems of multivariable Wilson polynomials to systems of multivariable Askey-Wilson polynomials and consider their special cases. Some q-extensions of Tratnik's (Tratnik, 1989) multivariable biorthogonal generalization of the Wilson polynomials are considered in (Gasper and Rahman, 1990b). q-Extensions of Tratnik's (Tratnik, 1991b) system of multivariable orthogonal Racah polynomials and their special cases will be considered in a subsequent paper.

2. Multivariable Askey-Wilson polynomials

In terms of the Askey-Wilson polynomials a q-analogue of the multivariable Wilson polynomials can be defined by

$$\begin{aligned} &P_{\mathbf{n}}(\mathbf{x} \mid q) = P_{\mathbf{n}}\left(\mathbf{x}; a, b, c, d, a_2, a_3, \ldots, a_s \mid q\right) \\ &= \left[\prod_{k=1}^{s-1} p_{n_k}\left(x_k; aA_{2,k}q^{N_{k-1}}, bA_{2,k}q^{N_{k-1}}, a_{k+1}e^{i\theta_{k+1}}, a_{k+1}e^{-i\theta_{k+1}} \mid q\right)\right] \\ &\qquad \times p_{n_s}\left(x_s; aA_{2,s}q^{N_{s-1}}, bA_{2,s}q^{N_{s-1}}, c, d \mid q\right), \end{aligned} \tag{2.1}$$

where $x_k = \cos\theta_k$, $A_{j,k} = \prod_{i=j}^{k} a_i$, $A_{k+1,k} = 1$, $A_k = A_{1,k}$, $1 \le j \le k \le s$. Our main aim in this section is to show that these polynomials satisfy the orthogonality relation

$$\int_{-1}^{1}\cdots\int_{-1}^{1} P_{\mathbf{n}}(\mathbf{x}\,|\,q)P_{\mathbf{m}}(\mathbf{x}\,|\,q)\rho(\mathbf{x}\,|\,q)dx_1\cdots dx_s = \lambda_{\mathbf{n}}(q)\,\delta_{\mathbf{n},\mathbf{m}} \tag{2.2}$$

with $\max\left(|q|,|a|,|b|,|c|,|d|,|a_2|,\ldots,|a_s|\right)<1$,

$$\begin{aligned}\rho(\mathbf{x}\,|\,q) &= \rho\left(\mathbf{x};a,b,c,d,a_2,a_3,\ldots,a_s\,|\,q\right)\\ &= \left(ae^{i\theta_1},ae^{-i\theta_1},be^{i\theta_1},be^{-i\theta_1};q\right)_\infty^{-1}\end{aligned}$$

$$\times\left[\prod_{k=1}^{s-1}\frac{\left(e^{2i\theta_k},e^{-2i\theta_k};q\right)_\infty\left(1-x_k^2\right)^{-1/2}}{\left(a_{k+1}e^{i\theta_{k+1}+i\theta_k},a_{k+1}e^{i\theta_{k+1}-i\theta_k},a_{k+1}e^{i\theta_k-i\theta_{k+1}},a_{k+1}e^{-i\theta_{k+1}-i\theta_k};q\right)_\infty}\right]$$

$$\times\frac{\left(e^{2i\theta_s},e^{-2i\theta_s};q\right)_\infty\left(1-x_s^2\right)^{-1/2}}{\left(ce^{i\theta_s},ce^{-i\theta_s},de^{i\theta_s},de^{-i\theta_s};q\right)_\infty}, \tag{2.3}$$

$$\begin{aligned}\lambda_{\mathbf{n}}(q) &= \lambda_{\mathbf{n}}\left(a,b,c,d,a_2,a_3,\ldots,a_s\,|\,q\right)\\ &= (2\pi)^s\left[\prod_{k=1}^{s}\frac{\left(q,A_{k+1}^2q^{N_k+N_{k-1}-1};q\right)_{n_k}\left(A_{k+1}^2q^{2N_k};q\right)_\infty}{\left(q,A_k^2q^{N_k+N_{k-1}},a_{k+1}^2q^{n_k};q\right)_\infty}\right]\\ &\quad\times\left(acA_{2,s}q^{N_s},adA_{2,s}q^{N_s},bcA_{2,s}q^{N_s},bdA_{2,s}q^{N_s};q\right)_\infty^{-1},\end{aligned} \tag{2.4}$$

where $a_1^2=ab$ and $a_{s+1}^2=cd$. The two-dimensional case was considered in (Koelink and Van der Jeugt, 1998), but they did not give the value of the norm. First observe that by (1.10)–(1.12) the integration over x_1 in (2.2) can be evaluated to obtain that

$$\begin{aligned}&\int_{-1}^{1} p_{n_1}\left(x_1;a,b,a_2e^{i\theta_2},a_2e^{-i\theta_2}\,|\,q\right)p_{m_1}\left(x_1;a,b,a_2e^{i\theta_2},a_2e^{-i\theta_2}\,|\,q\right)\\ &\qquad\times\rho\left(x_1;a,b,a_2e^{i\theta_2},a_2e^{-i\theta_2}\,|\,q\right)dx_1\end{aligned} \tag{2.5}$$

$$\begin{aligned}&=\delta_{n_1,m_1}\frac{2\pi\left(q,aba_2^2q^{n_1-1};q\right)_{n_1}\left(aba_2^2q^{2n_1};q\right)_\infty}{\left(q,abq^{n_1},a_2^2q^{n_1};q\right)_\infty}\\ &\quad\times\left(aa_2q^{n_1}e^{i\theta_2},aa_2q^{n_1}e^{-i\theta_2},ba_2q^{n_1}e^{i\theta_2},ba_2q^{n_1}e^{-i\theta_2};q\right)_\infty^{-1}.\end{aligned}$$

After doing the integrations over $x_1, x_2, \ldots, x_j$ for a few j one is led to conjecture that

$$\int_{-1}^{1} \cdots \int_{-1}^{1} P_{\mathbf{n}}^{(j)}(\mathbf{x}\,|\,q) P_{\mathbf{m}}^{(j)}(\mathbf{x}\,|\,q) \rho^{(j)}(\mathbf{x}\,|\,q) dx_1 \cdots dx_j$$

$$= (2\pi)^j \left[\prod_{k=1}^{j} \delta_{n_k,m_k} \frac{\left(q, A_{k+1}^2 q^{N_k+N_{k-1}-1}; q\right)_{n_k} \left(A_{k+1}^2 q^{2N_k}; q\right)_\infty}{\left(q, A_k^2 q^{N_k+N_{k-1}}, a_{k+1}^2 q^{n_k}; q\right)_\infty} \right] \quad (2.6)$$

$$\times \Big(aA_{2,j+1} q^{N_j} e^{i\theta_{j+1}}, aA_{2,j+1} q^{N_j} e^{-i\theta_{j+1}},$$

$$bA_{2,j+1} q^{N_j} e^{i\theta_{j+1}}, bA_{2,j+1} q^{N_j} e^{-i\theta_{j+1}}; q \Big)_\infty^{-1},$$

where

$$P_{\mathbf{n}}^{(j)}(\mathbf{x}\,|\,q) = \prod_{k=1}^{j} p_{n_k} \Big(x_k; aA_{2,k} q^{N_{k-1}}, bA_{2,k} q^{N_{k-1}},$$

$$a_{k+1} e^{i\theta_{k+1}}, a_{k+1} e^{-i\theta_{k+1}} \,|\, q \Big),$$

$$\rho^{(j)}(\mathbf{x}\,|\,q) = \left(ae^{i\theta_1}, ae^{-i\theta_1}, be^{i\theta_1}, be^{-i\theta_1}; q \right)_\infty^{-1}$$

$$\times \prod_{k=1}^{j} \frac{\left(e^{2i\theta_k}, e^{-2i\theta_k}; q\right)_\infty \left(1 - x_k^2\right)^{-1/2}}{\left(a_{k+1} e^{i\theta_{k+1}+i\theta_k}, a_{k+1} e^{i\theta_{k+1}-i\theta_k}, a_{k+1} e^{i\theta_k-i\theta_{k+1}}, a_{k+1} e^{-i\theta_{k+1}-i\theta_k}; q\right)_\infty}$$

for $j = 1, 2, \ldots, s-1$. To prove this by induction on j, suppose that $j < s-1$, multiply (2.6) by the x_{j+1}-dependent parts of the weight function and orthogonal polynomials, and then integrate with respect to x_{j+1} to get

$$(2\pi)^j \left[\prod_{k=1}^{j} \delta_{n_k,m_k} \frac{\left(q, A_{k+1}^2 q^{N_k+N_{k-1}-1}; q\right)_{n_k} \left(A_{k+1}^2 q^{2N_k}; q\right)_\infty}{\left(q, A_k^2 q^{N_k+N_{k-1}}, a_{k+1}^2 q^{n_k}; q\right)_\infty} \right]$$

$$\times \int_{-1}^{1} p_{n_{j+1}} \left(x_{j+1}; aA_{2,j+1} q^{N_j}, bA_{2,j+1} q^{N_j}, a_{j+2} e^{i\theta_{j+2}}, a_{j+2} e^{-i\theta_{j+2}} \,|\, q \right)$$

$$\times p_{m_{j+1}} \left(x_{j+1}; aA_{2,j+1} q^{N_j}, bA_{2,j+1} q^{N_j}, a_{j+2} e^{i\theta_{j+2}}, a_{j+2} e^{-i\theta_{j+2}} \,|\, q \right)$$

$$\times \rho \left(x_{j+1}; aA_{2,j+1} q^{N_j}, bA_{2,j+1} q^{N_j}, a_{j+2} e^{i\theta_{j+2}}, a_{j+2} e^{-i\theta_{j-2}} \,|\, q \right) dx_{j+1}$$

$$= (2\pi)^{j+1}\left[\prod_{k=1}^{j+1}\delta_{n_k,m_k}\frac{\left(q,A_{k+1}^2q^{N_k+N_{k-1}-1};q\right)_{n_k}\left(A_{k+1}^2q^{2N_k};q\right)_\infty}{\left(q,A_k^2q^{N_k+N_{k-1}},a_{k+1}^2q^{n_k};q\right)_\infty}\right]$$
$$\times\left(aA_{2,j+2}q^{N_{j+1}}e^{i\theta_{j+2}},aA_{2,j+2}q^{N_{j+1}}e^{-i\theta_{j+2}},\right.$$
$$\left.bA_{2,j+2}q^{N_{j+1}}e^{i\theta_{j+2}},bA_{2,j+2}q^{N_{j+1}}e^{-i\theta_{j+2}};q\right)_\infty^{-1}, \tag{2.7}$$

which is the $j\to j+1$ case of (2.6), completing the induction proof.

Now set $j=s-1$ in (2.6) and use it and (2.5) to find that

$$\int_{-1}^{1}\cdots\int_{-1}^{1}P_{\mathbf{n}}(\mathbf{x}\,|\,q)P_{\mathbf{m}}(\mathbf{x}\,|\,q)\rho(\mathbf{x}\,|\,q)dx_1\cdots dx_s$$
$$= (2\pi)^{s-1}\left[\prod_{k=1}^{s-1}\delta_{n_k,m_k}\frac{\left(q,A_{k+1}^2q^{N_k+N_{k-1}-1};q\right)_{n_k}\left(A_{k+1}^2q^{2N_k};q\right)_\infty}{\left(q,A_k^2q^{N_k+N_{k-1}},a_{k+1}^2q^{n_k};q\right)_\infty}\right]$$
$$\times\int_{-1}^{1}p_{n_s}\left(x_s;aA_{2,s}q^{N_{s-1}},bA_{2,s}q^{N_{s-1}},c,d\,|\,q\right)$$
$$\times\, p_{m_s}\left(x_s;aA_{2,s}q^{N_{s-1}},bA_{2,s}q^{N_{s-1}},c,d\,|\,q\right)$$
$$\times\frac{\left(e^{2i\theta_s},e^{-2i\theta_s};q\right)_\infty\left(1-x_s^2\right)^{-1/2}}{\left(ce^{i\theta_s},ce^{-i\theta_s},de^{i\theta_s},de^{-i\theta_s};q\right)_\infty}$$
$$\times\left(aA_{2,s}q^{N_{s-1}}e^{i\theta_s},aA_{2,s}q^{N_{s-1}}e^{-i\theta_s},\right.$$
$$\left.b\,A_{2,s}q^{N_{s-1}}e^{i\theta_s},bA_{2,s}q^{N_{s-1}}e^{-i\theta_s};q\right)_\infty^{-1}dx_s \tag{2.8}$$
$$=\lambda_{\mathbf{n}}(q)\delta_{\mathbf{n},\mathbf{m}},$$

where $\lambda_{\mathbf{n}}(q)$ is given by (2.4). This completes the proof of (2.2).

Note that the integration region and weight function in (2.2) and (2.3) are invariant under the permutation of variables and parameters

$$a\leftrightarrow c,\ b\leftrightarrow d,\ a_{k+1}\leftrightarrow a_{s-k+1},\quad k=1,2,\ldots,s-1, \tag{2.9}$$
$$\theta_k\leftrightarrow\theta_{s-k+1},\quad k=1,2,\ldots,s.$$

Hence, when these permutations are applied to (2.2) and (2.3) the transformed polynomials also form an orthogonal system with the same weight function. Since the polynomials $P_{\mathbf{n}}(\mathbf{x}\,|\,q)$ in (2.1) are not invariant under (2.9), we obtain a second system of multivariable orthogonal Askey-Wilson polynomials, which is a q-analogue of Tratnik's second system

(1.7) of multivariable Wilson polynomials. After doing the permutation $n_k \leftrightarrow n_{s-k+1}$, $k = 1, \ldots, s$, the transformed polynomials and the normalization constant are given by

$$\begin{aligned}\tilde{P}_{\mathbf{n}}(\mathbf{x} \mid q) &= \tilde{P}_{\mathbf{n}}(\mathbf{x}; a, b, c, d, a_2, a_3, \ldots, a_s \mid q) \\ &= p_{n_1}\left(x_1; cA_{2,s}q^{N_{2,s}}, dA_{2,s}q^{N_{2,s}}, a, b \mid q\right) \\ &\times \left[\prod_{k=2}^{s} p_{n_k}\left(x_k; cA_{k+1,s}q^{N_{k+1,s}}, dA_{k+1,s}q^{N_{k+1,s}},\right.\right. \\ &\qquad \left.\left. ae^{i\theta_{k-1}}, ae^{-i\theta_{k-1}} \mid q\right)\right],\end{aligned} \tag{2.10}$$

$$\begin{aligned}\tilde{\lambda}_{\mathbf{n}}(q) &= \tilde{\lambda}_{\mathbf{n}}(a, b, c, d, a_2, a_3, \ldots, a_s \mid q) \\ &= (2\pi)^s \left[\prod_{k=1}^{s} \frac{\left(q, A_{k,s+1}^2 q^{N_{k,s}+N_{k+1,s}-1}; q\right)_{n_k} \left(A_{k,s+1}^2 q^{2N_{k,s}}; q\right)_\infty}{\left(q, A_{k+1,s+1}^2 q^{N_{k,s}+N_{k+1,s}}, a_k^2 q^{n_k}; q\right)_\infty}\right] \\ &\times \left(acA_{2,s}q^{N_s}, adA_{2,s}q^{N_s}, bcA_{2,s}q^{N_s}, bdA_{2,s}q^{N_s}; q\right)_\infty^{-1},\end{aligned} \tag{2.11}$$

with $a_1^2 = ab$, $a_{s+1}^2 = cd$, and $\max\left(|q|, |a|, |b|, |c|, |d|, |a_2|, |a_3|, \ldots, |a_s|\right) < 1$. These polynomials are of total degree N_s in the variables $x_1, \ldots, x_s$ and they form a complete set.

A five-parameter system of multivariable Askey-Wilson polynomials which is associated with a root system of type BC was introduced in (Koornwinder, 1995) and studied with four of the parameters generally complex in (Stokman, 1999).

3. Special Cases of (2.2)

First observe that the continuous dual q-Hahn polynomial defined by

$$d_n(x; a, b, c \mid q) = a^{-n} (ab, ac; q)_n \; {}_3\varphi_2 \left[\begin{matrix} q^{-n}, ae^{i\theta}, ae^{-i\theta} \\ ab, ac \end{matrix}; q, q\right] \tag{3.1}$$

is obtained by taking $d = 0$ in (1.9) and $x = \cos\theta$. Since $d_n(x; a, b, c \mid q)$ is symmetric in its parameters by (Gasper and Rahman, 1990a, (3.2.3)), we may define the multivariable dual q-Hahn polynomials by

$$\begin{aligned} D_{\mathbf{n}}(\mathbf{x}\,|\,q) &= D_{\mathbf{n}}\left(\mathbf{x}; a, b, c, a_2, a_3, \ldots, a_s \,|\, q\right) \\ &= \left[\prod_{k=1}^{s-1} d_{n_k}\left(x_k; a_{k+1}e^{i\theta_{k+1}}, a_{k+1}e^{-i\theta_{k+1}}, aA_{2,k}q^{N_{k-1}} \,|\, q\right)\right] \\ &\quad \times d_{n_s}\left(x_s; b, c, aA_{2,s}q^{N_{s-1}} \,|\, q\right), \end{aligned} \tag{3.2}$$

with $x_k = \cos\theta_k$ for $k = 1, 2, \ldots, s$. It follows from the $b = 0$ case of (2.2)–(2.4) that the orthogonality relation for these polynomials is

$$\int_{-1}^{1} \cdots \int_{-1}^{1} D_{\mathbf{n}}(\mathbf{x}\,|\,q) D_{\mathbf{m}}(\mathbf{x}\,|\,q) \rho(\mathbf{x}\,|\,q) dx_1 \cdots dx_s = \lambda_{\mathbf{n}}(q)\, \delta_{\mathbf{n},\mathbf{m}} \tag{3.3}$$

with

$$\begin{aligned} \rho(\mathbf{x}\,|\,q) &= \rho\left(\mathbf{x}; a, b, c, a_2, a_3, \ldots, a_s \,|\, q\right) \\ &= \left(ae^{i\theta_1}, ae^{-i\theta_1}; q\right)_\infty^{-1} \\ &\times \left[\prod_{k=1}^{s-1} \frac{\left(e^{2i\theta_k}, e^{-2i\theta_k}; q\right)_\infty \left(1 - x_k^2\right)^{-1/2}}{\left(a_{k+1}e^{i\theta_{k+1}+i\theta_k}, a_{k+1}e^{i\theta_{k+1}-i\theta_k}, a_{k+1}e^{i\theta_k-i\theta_{k+1}}, a_{k+1}e^{-i\theta_{k+1}-i\theta_k}; q\right)_\infty}\right] \\ &\times \frac{\left(e^{2i\theta_s}, e^{-2i\theta_s}; q\right)_\infty \left(1 - x_s^2\right)^{-1/2}}{\left(be^{i\theta_s}, be^{-i\theta_s}, ce^{i\theta_s}, ce^{-i\theta_s}; q\right)_\infty}, \end{aligned} \tag{3.4}$$

$$\begin{aligned} \lambda_{\mathbf{n}}(q) &= \lambda_{\mathbf{n}}\left(a, b, c, a_2, a_3, \ldots, a_s \,|\, q\right) \\ &= (2\pi)^s \left[\prod_{k=1}^{s} \left(q^{n_k+1}, a_{k+1}^2 q^{n_k}; q\right)_\infty^{-1}\right] \left(abA_{2,s}q^{N_s}, acA_{2,s}q^{N_s}; q\right)_\infty^{-1}, \end{aligned} \tag{3.5}$$

where $a_{s+1}^2 = bc$ and $\max\left(|q|, |a|, |b|, |c|, |a_2|, |a_3|, \ldots, |a_s|\right) < 1$.

By taking the limit $a \to 0$ in (3.2)–(3.5) we can now deduce that the multivariable Al-Salam-Chihara polynomials defined by

$$\begin{aligned} S_{\mathbf{n}}(\mathbf{x}\,|\,q) &= S_{\mathbf{n}}\left(\mathbf{x}; b, c, a_2, a_3, \ldots, a_s \,|\, q\right) \\ &= \left[\prod_{k=1}^{s-1} p_{n_k}\left(x_k; a_{k+1}e^{i\theta_{k+1}}, a_{k+1}e^{-i\theta_{k+1}} \,|\, q\right)\right] \\ &\quad \times p_{n_s}\left(x_s; b, c \,|\, q\right) \end{aligned} \tag{3.6}$$

satisfy the orthogonality relation

$$\int_{-1}^{1}\cdots\int_{-1}^{1} S_{\mathbf{n}}(\mathbf{x}\,|\,q)S_{\mathbf{m}}(\mathbf{x}\,|\,q)\rho(\mathbf{x}\,|\,q)dx_1\cdots dx_s = \lambda_{\mathbf{n}}(q)\,\delta_{\mathbf{n},\mathbf{m}} \tag{3.7}$$

with

$$\begin{aligned}\rho(\mathbf{x}\,|\,q) &\\ = \Bigg[\prod_{k=1}^{s-1} &\frac{\left(e^{2i\theta_k}, e^{-2i\theta_k}; q\right)_\infty \left(1-x_k^2\right)^{-1/2}}{\left(a_{k+1}e^{i\theta_{k+1}+i\theta_k}, a_{k+1}e^{i\theta_{k+1}-i\theta_k}, a_{k+1}e^{i\theta_k-i\theta_{k+1}}, a_{k+1}e^{-i\theta_{k+1}-i\theta_k}; q\right)_\infty}\Bigg]\\ &\times\frac{\left(e^{2i\theta_s}, e^{-2i\theta_s}; q\right)_\infty \left(1-x_s^2\right)^{-1/2}}{\left(be^{i\theta_s}, be^{-i\theta_s}, ce^{i\theta_s}, ce^{-i\theta_s}; q\right)_\infty},\end{aligned} \tag{3.8}$$

$$\lambda_{\mathbf{n}}(q) = (2\pi)^s \prod_{k=1}^{s} \left(q^{n_k+1}, a_{k+1}^2 q^{n_k}; q\right)_\infty^{-1}, \tag{3.9}$$

where $a_{s+1}^2 = bc$, $\max\left(|q|, |b|, |c|, |a_2|, |a_3|, \ldots, |a_s|\right) < 1$, and the Al-Salam-Chihara polynomial $p_n(x; b, c\,|\,q)$ is defined by

$$p_n(x; b, c\,|\,q) = b^{-n}(bc; q)_n\, {}_3\varphi_2\left[\begin{matrix} q^{-n}, be^{i\theta}, be^{-i\theta} \\ bc, 0 \end{matrix}; q, q\right], \tag{3.10}$$

see (Koekoek and Swarttouw, 1994, (3.8.1)).

Setting

$$a = q^{(2\alpha+1)/4},\ b = q^{(2\alpha+3)/4},\ c = -q^{(2\beta+1)/4},\ d = -q^{2\beta+3)/4} \tag{3.11}$$

in (2.1) and (2.2) gives a multivariable orthogonal extension of the continuous q-Jacobi polynomials $P_n^{(\alpha,\beta)}(x\,|\,q)$ defined in (Gasper and Rahman, 1990a, (7.5.24)), while setting

$$a = q^{1/2},\ b = q^{\alpha+1/2},\ c = -q^{\beta+1/2},\ d = -q^{1/2} \tag{3.12}$$

in (2.1) and (2.2) gives a multivariable orthogonal extension of the $P_n^{(\alpha,\beta)}(x; q)$ polynomials defined in (7.5.25). Also, via (Gasper and Rahman, 1990a, (7.5.33)) and (Gasper and Rahman, 1990a, (7.5.34) with $q \to q^{1/2}$) the $\alpha = \beta = \lambda - 1/2$ substitution gives a multivariable orthogonal extension of the continuous q-ultraspherical polynomials $C_n\left(x; q^\lambda\,|\,q\right)$. By letting $\lambda \to \infty$ when we use (3.12), i.e., set $a = -d = q^{1/2}$ and $b = c = 0$, we get a multivariable orthogonal extension of the continuous q-Hermite polynomials defined in (Gasper and Rahman, 1990a, Ex. 1.28).

A multivariable orthogonal extension of the continuous q-Hahn polynomials defined by

$$p_n(\cos(\theta+\varphi); a, b \mid q) \\ = \left(a^2, ab, abe^{2i\varphi}; q\right)_n \left(ae^{i\varphi}\right)^{-n} {}_4\varphi_3 \left[\begin{matrix} q^{-n}, a^2b^2q^{n-1}, ae^{2i\varphi+i\theta}, ae^{-i\theta} \\ a^2, ab, abe^{2i\varphi} \end{matrix}; q, q\right], \tag{3.13}$$

see (Gasper and Rahman, 1990a, (7.5.43)), is obtained from (2.1)–(2.4) by replacing a, b, c, d, θ_k and $x_k = \cos\theta_k$ by $a_1e^{i\varphi}$, $a_1e^{-i\varphi}$, $a_{s+1}e^{i\varphi}$, $a_{s+1}e^{-i\varphi}$, $\theta_k+\varphi$ and $\cos(\theta_k+\varphi)$, respectively.

It is clear that similar special cases of the second system of multivariable orthogonal Askey-Wilson polynomials can be obtained by appropriate specialization of the parameters in (2.10) and (2.11). Additional systems of multivariable orthogonal polynomials will be considered elsewhere.

References

Askey, R. and Wilson, J. A. (1979). A set of orthogonal polynomials that generalize the Racah coefficients or 6-j symbols. *SIAM J. Math. Anal.*, 10:1008–1016.

Askey, R. and Wilson, J. A. (1985). Some basic hypergeometric orthogonal polynomials that generalize Jacobi polynomials. *Memoirs Amer. Math. Soc.*, 319.

Gasper, G. and Rahman, M. (1990). *Basic Hypergeometric Series.* Cambridge University Press, Cambridge.

Gasper, G. and Rahman, M. (2003). q-analogues of some multivariable biorthogonal polynomials. This Proceedings.

Koekoek, R. and Swarttouw, R. F. (1998). The Askey-scheme of hypergeometric orthogonal polynomials and its q-analogue. Report 98-17, Delft University of Technology.

Koelink, H. T. and Van der Jeugt, J. (1998). Convolutions for orthogonal polynomials from Lie and quantum algebra representations. *SIAM J. Math. Anal.*, 29:794–822.

Koornwinder, T. H. (1992). Askey-Wilson polynomials for root systems of type *BC*. In *Hypergeometric functions on domains of positivity, Jack polynomials, and applications (Tampa, FL, 1991)*, volume 138 of *Contemp. Math.*, pages 189–204. Amer. Math. Soc., Providence, RI.

Stokman, J. V. (1999). On *BC* type basic hypergeometric orthogonal polynomials. *Trans. Amer. Math. Soc.*, 352:1527–1579.

Tratnik, M. V. (1989). Multivariable Wilson polynomials. *J. Math. Phys.*, 30:2001–2011.

Tratnik, M. V. (1991a). Some multivariable orthogonal polynomials of the Askey tableau—continuous families. *J. Math. Phys.*, 32:2065–2073.

Tratnik, M. V. (1991b). Some multivariable orthogonal polynomials of the Askey tableau—discrete families. *J. Math. Phys.*, 32:2337–2342.

Wilson, J. A. (1980). Some hypergeometric orthogonal polynomials. *SIAM J. Math. Anal.*, 11:690–701.

CONTINUOUS HAHN FUNCTIONS AS CLEBSCH-GORDAN COEFFICIENTS

Wolter Groenevelt and Erik Koelink
Technische Universiteit Delft
EWI-TWA
Postbus 5031
2600 GA Delft
THE NETHERLANDS
W.G.M.Groenevelt@ewi.tudelft.nl
H.T.Koelink@ewi.tudelft.nl

Hjalmar Rosengren
Department of Mathematics
Chalmers University of Technology and Göteborg University
SE-412 96 Göteborg
SWEDEN
hjalmar@math.chalmers.se

Abstract An explicit bilinear generating function for Meixner-Pollaczek polynomials is proved. This formula involves continuous dual Hahn polynomials, Meixner-Pollaczek functions, and non-polynomial ${}_3F_2$-hypergeometric functions that we consider as continuous Hahn functions. An integral transform pair with continuous Hahn functions as kernels is also proved. These results have an interpretation for the tensor product decomposition of a positive and a negative discrete series representation of $\mathfrak{su}(1,1)$ with respect to hyperbolic bases, where the Clebsch-Gordan coefficients are continuous Hahn functions.

1. Introduction

The results and techniques in this paper are mainly analytic in nature, but they are motivated by a Lie algebraic problem. As is well known, many polynomials in the Askey-scheme of orthogonal polynomials of hypergeometric type, see (Koekoek and Swarttouw, 1998), have an interpretation in the representation theory of Lie groups and Lie algebras, see, e.g., Vilenkin and Klimyk (Vilenkin and Klimyk, 1991) and

Koornwinder (Koornwinder, 1988). The Askey-scheme can be extended to families of unitary integral transforms with a hypergeometric kernel. Many of these kernels also admit group theoretic interpretations. For example the Jacobi functions, which can be considered as a non-polynomial extension of the Jacobi polynomials and are given explicitly by a certain ${}_2F_1$-hypergeometric function, have an interpretation as matrix elements for irreducible representations of the Lie group $SU(1,1)$. The Jacobi function is the kernel in the Jacobi integral transform, which can be found by spectral analysis of the hypergeometric differential operator. For an overview of Jacobi functions in representation theory, we refer to the survey paper (Koornwinder, 1995) by Koornwinder.

In this paper we give a generalization of the Jacobi functions. We consider the tensor product of a positive and a negative discrete series representation of the Lie algebra $\mathfrak{su}(1,1)$. The Clebsch-Gordan coefficients for the hyperbolic basisvectors turn out to be a certain type of non-polynomial ${}_3F_2$-hypergeometric functions, which we call continuous Hahn functions. We show that the continuous Hahn functions are the kernel in an integral transform, that generalizes the Jacobi function transform. We emphasize that the main part (Sections 11.3 and 11.5) of this paper is analytic in nature, and that the Lie algebraic interpretation is mainly restricted to Section 11.4.

The Lie algebra $\mathfrak{su}(1,1)$ is generated by the three elements H, B and C. There are four classes of irreducible unitary representations for $\mathfrak{su}(1,1)$: discrete series, i.e., the positive and the negative discrete series representations, and continuous series, i.e., the principal unitary series and the complementary series representations. There are three kinds of basis elements on which the various representations can act: the elliptic, the parabolic and the hyperbolic basis elements. These three elements are related to conjugacy classes of the group $SU(1,1)$. We consider the tensor product of a positive and a negative discrete series representation, which decomposes into a direct integral over the principal unitary series representations. Under certain condition discrete terms can appear. The Clebsch-Gordan coefficients for the standard (elliptic) basis vectors are continuous dual Hahn polynomials. We compute the Clebsch-Gordan coefficients for the hyperbolic basis vectors, which are non-polynomial extensions of the continuous (dual) Hahn polynomials, and are therefore called continuous Hahn functions. For the Clebsch-Gordan coefficients for the elliptic and parabolic basis, we refer to (Groenevelt and Koelink, 2002), respectively (Basu and Wolf, 1983), (Groenevelt, 2003).

The explicit expressions for the Clebsch-Gordan coefficients as ${}_3F_2$-series are not new, they are found by Mukunda and Radhakrishnan in (Mukunda and Radhakrishnan, 1974). However not much seems to be known about the generalized orthogonality properties of the continuous Hahn functions, i.e., they form the kernel in a unitary integral tranform (the continuous Hahn transform). Using the Lie algebraic interpretation of the continuous Hahn functions, we can compute formally the inverse of the continuous Hahn integral transform. In Section 11.5 we give an analytic proof for the integral transform pair.

The method we use to compute the Clebsch-Gordan coefficients is based on an idea by Granovskii and Zhedanov (Granovskii and Zhedanov, 1993). The idea is to consider a self-adjoint Lie algebra element $X_a = -aH + B - C$, $a \in \mathbb{R}$. The action of X_a in an irreducible representation gives a difference equation, for which the (generalized) eigenvectors can be expressed in terms of special functions and the standard basis vectors. The Clebsch-Gordan coefficients for the eigenvectors can be calculated using properties of the special functions. In (Van der Jeugt, 1997) and (Koelink and Van der Jeugt, 1998) Van der Jeugt and the second author considered the action of X_a in tensor products of positive discrete series representations of $\mathfrak{su}(1,1)$ to find convolution formulas for orthogonal polynomials. In (Groenevelt and Koelink, 2002) the action of X_a in the tensor product of a positive and a negative discrete series representation is investigated for $|a| > 1$ (the elliptic case). This leads to a bilinear summation formula for Meixner polynomials (Groenevelt and Koelink, 2002, Thm. 3.6). In this paper we consider the case $|a| < 1$ (the hyperbolic case).

The plan of the paper is as follows. In Section 11.2 we introduce the special functions we need in this paper, and give some properties of these functions.

In Section 11.3 we prove a bilinear summation formula for Meixner-Pollaczek polynomials by series manipulations. As a result we find a certain type of ${}_3F_2$-functions, which are the continuous Hahn functions. The summation formula is used in Section 11.4.2 to compute the Clebsch-Gordan coefficients for the hyperbolic bases.

In Section 11.4 we consider the tensor product of a positive and a negative discrete series representation of the Lie algebra $\mathfrak{su}(1,1)$. First we recall the basic properties of $\mathfrak{su}(1,1)$ and its irreducible unitary representations in Section 11.4.1. Then in Section 11.4.2 we diagonalize X_a, $|a| < 1$, in the various irreducible representations. This leads to generalized eigenvectors of X_a, which can be considered as hyperbolic basis vectors.

For the discrete series representations, the overlap coefficients for the eigenvectors and the standard (elliptic) basisvectors are Meixner-Pollaczek polynomials, cf. (Koornwinder, 1988, §7). For the continuous series, the overlap coefficients are Meixner-Pollaczek functions. This follows from the spectral analysis of a doubly infinite Jacobi operator, which is carried out by Masson and Repka (Masson and Repka, 1991, §3.3) and Koelink (Koelink, , §4.4.11). It turns out that the spectral projection of the Jacobi operator is on a 2-dimensional space of generalized eigenvectors. So the eigenvectors of X_a, $|a| < 1$, in the continuous series representations are 2-dimensional, and we find two linearly independent Meixner-Pollaczek functions as overlap coefficients. To determine the Clebsch-Gordan coefficients for the hyperbolic bases, we use the bilinear summation formula from Section 11.3. This leads to a pair of continuous Hahn functions as Clebsch-Gordan coefficients. By formal calculations we find an integral transform pair, with a pair of continuous Hahn functions as a kernel. To give a rigorous proof of the integral transform pair, we show that the continuous Hahn functions are eigenfunctions of a difference operator Λ. To find this operator Λ we realize H, B and C as difference operators acting on polynomials, using the difference equation for the Meixner-Pollaczek polynomials. Then Λ is a restriction of the Casimir operator in the tensor product.

The spectral analyis of this difference operator is carried out in Section 11.5. A main problem with spectral analysis of a difference operator is finding the right eigenfunctions. This is because an eigenfunction multiplied by a periodic function is again an eigenfunction. Our choice of the periodic function is mainly motivated by the Lie algebraic interpretation of the eigenfunctions. Using asymptotic methods, we find a spectral measure for the difference operator. This leads to an integral transform with a pair of continuous Hahn functions as a kernel. We call this the continuous Hahn integral transform.

Notations. We denote for a function $f : \mathbb{C} \to \mathbb{C}$

$$f^*(x) = \overline{f(\overline{x})}.$$

If $d\mu(x)$ is a positive measure, we use the notation $d\mu^{\frac{1}{2}}(x)$ for the positive measure with the property

$$d\left(\mu^{\frac{1}{2}} \times \mu^{\frac{1}{2}}\right)(x,x) = d\mu(x).$$

The hypergeometric series is defined by

$${}_pF_q\left(\begin{matrix} a_1, \ldots, a_p \\ b_1, \ldots, b_q \end{matrix} ; z\right) = \sum_{n=0}^{\infty} \frac{(a_1)_n \cdots (a_p)_n}{(b_1)_n \cdots (b_q)_n} \frac{z^n}{n!},$$

where $(a)_n$ denotes the Pochhammer symbol, defined by

$$(a)_n = \frac{\Gamma(a+n)}{\Gamma(a)} = a(a+1)(a+2)\ldots(a+n-1), \qquad n \in \mathbb{Z}_{\geq 0}.$$

Acknowledgments

We thank Ben de Pagter for useful discussions.

Dedication

We gladly dedicate this paper to Mizan Rahman who, with his unsurpassed mastery in dealing with (q)-series and his insight in the structures of formulas, has pushed the subject of (q)-special functions much further. We are also grateful to Mizan Rahman for his interest in our work, and for his willingness to help others in solving problems in this field.

2. Orthogonal polynomials and functions

In this section we recall some properties of the orthogonal polynomials and functions which we need in this paper.

Continuous dual Hahn polynomials. The Wilson polynomials, see Wilson (Wilson, 1980) or (Andrews et al., 1999, §3.8), are ${}_4F_3$-hypergeometric polynomials on top of the Askey-scheme of hypergeometric polynomials, see Koekoek and Swarttouw (Koekoek and Swarttouw, 1998). The continuous dual Hahn polynomials are a three-parameter subclass of the Wilson polynomials, and are defined by

$$s_n(y;a,b,c) = (a+b)_n(a+c)_n\,{}_3F_2\left(\begin{matrix}-n, a+ix, a-ix\\ a+b, a+c\end{matrix};1\right), \qquad x^2 = y, \quad (n \in \mathbb{Z}_{\geq 0}). \tag{2.1}$$

For real parameters a, b, c, with $a+b$, $a+c$, $b+c$ positive, the continuous dual Hahn polynomials are orthogonal with respect to a positive measure, supported on a subset of $\mathbb{R}$. The orthonormal continuous dual Hahn polynomials are defined by

$$S_n(y;a,b,c) = \frac{(-1)^n s_n(y;a,b,c)}{\sqrt{n!(a+b)_n(a+c)_n(b+c)_n}}.$$

By Kummer's transformation, see, e.g., (Andrews et al., 1999, Cor. 3.3.5), the polynomials s_n and S_n are symmetric in a, b and c. Without loss of generality we assume that a is the smallest of the real parameters

a, b and c. Let $d\mu(\cdot; a, b, c)$ be the measure defined by

$$\int_{\mathbb{R}} f(y)d\mu(y; a, b, c) =$$

$$\frac{1}{2\pi}\int_0^\infty \left|\frac{\Gamma(a+ix)\Gamma(b+ix)\Gamma(c+ix)}{\Gamma(2ix)}\right|^2 \frac{f(x^2)}{\Gamma(a+b)\Gamma(a+c)\Gamma(b+c)}\,dx$$

$$+\frac{\Gamma(b-a)\Gamma(c-a)}{\Gamma(-2a)\Gamma(b+c)}\sum_{k=0}^{K}(-1)^k\frac{(2a)_k(a+1)_k(a+b)_k(a+c)_k}{(a)_k(a-b+1)_k(a-c+1)_k k!}f(-(a+k)^2),$$

where K is the largest non-negative integer such that $a+K<0$. In particular, the measure $d\mu(\cdot; a, b, c)$ is absolutely continuous if $a \geq 0$. The measure is positive under the conditions $a+b>0$, $a+c>0$ and $b+c>0$. Then the polynomials $S_n(y; a, b, c)$ are orthonormal with respect to the measure $d\mu(y; a, b, c)$.

Meixner-Pollaczek polynomials. The Meixner-Pollaczek polynomials, see (Koekoek and Swarttouw, 1998), (Andrews et al., 1999, Ex. 6.37), are a two-parameter subclass of the Wilson polynomials, and are defined by

$$p_n^{(\lambda)}(x;\varphi) = \frac{(2\lambda)_n}{n!}e^{in\varphi}\,{}_2F_1\left(\begin{matrix}-n, \lambda+ix\\ 2\lambda\end{matrix};1-e^{-2i\varphi}\right), \quad (n \in \mathbb{Z}_{\geq 0}). \tag{2.2}$$

For $\lambda > 0$ and $0 < \varphi < \pi$, these are orthogonal polynomials with respect to a positive measure on $\mathbb{R}$. The orthonormal Meixner-Pollaczek polynomials

$$P_n(x) = P_n^{(\lambda)}(x;\varphi) = \sqrt{\frac{n!}{(2\lambda)_n}}\,p_n^{(\lambda)}(x;\varphi)$$

satisfy the following three-term recurrence relation

$$2x\sin\varphi\, P_n(x) = \alpha_n P_{n+1}(x) - 2(n+\lambda)\cos\varphi\, P_n(x) + \alpha_{n-1}P_{n-1}(x), \tag{2.3}$$

where

$$\alpha_n = \sqrt{(n+1)(n+2\lambda)}.$$

The Meixner-Pollaczek polynomials also satisfy the difference equation

$$\begin{aligned} e^{i\varphi}(\lambda-ix)y(x+i) &+ 2i[x\cos\varphi-(n+\lambda)\sin\varphi]y(x)\\ &-e^{-i\varphi}(\lambda+ix)y(x-i) = 0, \\ &y(x) = P_n^{(\lambda)}(x;\varphi). \end{aligned} \tag{2.4}$$

Define

$$w^{(\lambda)}(x;\varphi)=\frac{1}{2\pi}(2\sin\varphi)^{2\lambda}e^{(2\varphi-\pi)x}\frac{|\Gamma(\lambda+ix)|^2}{\Gamma(2\lambda)},$$

then the orthonormality relation reads

$$\int_{\mathbb{R}} P_m^{(\lambda)}(x;\varphi)P_n^{(\lambda)}(x;\varphi)w^{(\lambda)}(x;\varphi)dx=\delta_{mn}.$$

Meixner-Pollaczek functions. The Meixner-Pollaczek functions u_n, $n\in\mathbb{Z}$, can be considered as non-polynomial extensions of the Meixner-Pollaczek polynomials, see Masson and Repka (Masson and Repka, 1991) and Koelink (Koelink, , §4.4). The Meixner-Pollaczek functions are defined by

$$\begin{aligned}u_n(x;\lambda,\varepsilon,\varphi)&=(2i\sin\varphi)^{-n}\frac{\sqrt{\Gamma(n+1+\varepsilon+\lambda)\Gamma(n+\varepsilon-\lambda)}}{\Gamma(n+1+\varepsilon-ix)}\\&\times{}_2F_1\left(\begin{matrix}n+1+\varepsilon+\lambda, n+\varepsilon-\lambda\\ n+1+\varepsilon-ix\end{matrix};\frac{1}{1-e^{-2i\varphi}}\right),\quad(n\in\mathbb{Z}).\end{aligned}\tag{2.5}$$

The parameters φ and ε satisfy the conditions $0<\varphi<\pi$, $0\le\varepsilon<1$, and λ satisfies one of the following conditions: $-\frac{1}{2}<\lambda<-\varepsilon$, $-\frac{1}{2}<\lambda<\varepsilon-1$, or $\lambda=-\frac{1}{2}+i\rho$, $\rho\in\mathbb{R}$. In the last case u_n is symmetric in ρ and $-\rho$, so without loss of generality we assume $\rho\ge 0$. For $0<\varphi\le\frac{1}{6}\pi$ and $\frac{5}{6}\pi\le\varphi<\pi$ we use the unique analytic continuation of the ${}_2F_1$-function to $\mathbb{C}\setminus[1,\infty)$. Note that the Meixner-Pollaczek function is well defined for all $n\in\mathbb{Z}$, since $\Gamma(c)^{-1}{}_2F_1(a,b;c;z)$ is analytic in a, b and c.

The functions u_n and u_n^* satisfy the recurrence relation

$$2x\sin\varphi\, u_n(x)=\alpha_n u_{n+1}(x)+\beta_n u_n(x)+\alpha_{n-1}u_{n-1}(x),\tag{2.6}$$

where

$$\alpha_n=\sqrt{(n+\varepsilon+\lambda+1)(n+\varepsilon-\lambda)},\qquad \beta_n=2(n+\varepsilon)\cos\varphi.$$

Let L be the corresponding doubly infinite Jacobi operator acting on $\ell^2(\mathbb{Z})$,

$$L: e_n\mapsto \alpha_n e_{n+1}+\beta_n e_n+\alpha_{n-1}e_{n-1}.$$

L is initially defined on finite linear combinations of the basis vectors of $\ell^2(\mathbb{Z})$, and then L is essentially self-adjoint. The spectral measure of L is described by

$$\begin{aligned}\langle u,v\rangle=\frac{1}{2\pi}\int_{\mathbb{R}}&\big(\langle u,U(x)\rangle\langle U(x),v\rangle+\langle u,U^*(x)\rangle\langle U^*(x),v\rangle\\&-w_1(x)\langle u,U(x)\rangle\langle U^*(x),v\rangle-w_1^*(x)\langle u,U^*(x)\rangle\langle U(x),v\rangle\big)w_0(x)dx,\end{aligned}$$

where

$$\begin{aligned}
w_1(x) &= \frac{\Gamma(\lambda+1+ix)\Gamma(-\lambda+ix)}{\Gamma(ix-\varepsilon)\Gamma(1+\varepsilon-ix)}, \\
w_0(x) &= \frac{|\Gamma(\lambda+1-ix)\Gamma(-\lambda-ix)\Gamma(ix-\varepsilon)\Gamma(1+\varepsilon-ix)|^2}{|\Gamma(ix-\varepsilon)\Gamma(1+\varepsilon-ix)|^2-|\Gamma(\lambda+1-ix)\Gamma(-\lambda-ix)|^2} \\
&\times \frac{(2\sin\varphi)^{-2\varepsilon}e^{2x(\varphi-\frac{\pi}{2})}}{\Gamma(-\varepsilon-\lambda)\Gamma(1+\lambda-\varepsilon)\Gamma(1+\lambda+\varepsilon)\Gamma(\varepsilon-\lambda)} \\
&= (2\sin\varphi)^{-2\varepsilon}e^{2x(\varphi-\frac{\pi}{2})}, \\
U(x) &= \sum_{n=-\infty}^{\infty} u_n(x;\lambda,\varepsilon,\varphi)e_n, \quad U^*(x) = \sum_{n=-\infty}^{\infty} u_n^*(x;\lambda,\varepsilon,\varphi)e_n.
\end{aligned}$$

The spectral measure for L can be obtained from (Koelink, , §4.4.11), using the connection formulas given there. The expression for $w_0(x)$ is found using Euler's reflection formula, elementary trigonometric identities and the conditions on λ.

Let $\mathcal{H} = \mathcal{H}(\lambda,\varepsilon,\varphi)$ be the Hilbert space consisting of functions

$$\mathbf{f}:\mathbb{R}\to\mathbb{C}^2, \quad x\mapsto \begin{pmatrix} f_1(x) \\ f_2(x) \end{pmatrix},$$

with inner product defined by

$$\langle \mathbf{f},\mathbf{g}\rangle = \frac{1}{2\pi}\int_{\mathbb{R}} \begin{pmatrix} g_1(x) \\ g_2(x) \end{pmatrix}^* \begin{pmatrix} 1 & -w_1(x) \\ -w_1^*(x) & 1 \end{pmatrix} \begin{pmatrix} f_1(x) \\ f_2(x) \end{pmatrix} w_0(x)dx.$$

Observe that the square matrix inside the integral is positive definite and self adjoint.

Proposition 2.1. *The functions*

$$\mathbf{u_n} = \begin{pmatrix} u_n \\ u_n^* \end{pmatrix}$$

form an orthonormal basis of the Hilbert space $\mathcal{H}$.

The orthonormality follows by choosing u and v above as standard basis vectors e_n and e_m. The completeness of the Meixner-Pollaczek functions follows from the uniqueness of the spectral measure. An alternative, group-theoretic approach, can be found in (Vilenkin and Klimyk, 1991). It is only worked out for the smaller range of parameters corresponding to SU(1, 1) (rather than its Lie algebra or universal covering group). We will briefly describe how the analytic arguments that lie behind the approach of (Vilenkin and Klimyk, 1991) extend to the present situation. We only discuss the case $\lambda = -\frac{1}{2}+i\rho$, $\rho \in \mathbb{R}$, which is all that we need later.

Consider the space $L^2(\mathbb{T})$ on the unit circle with respect to normalized measure $|dz|/2\pi$. It has the orthonormal basis $e_n(z) = z^n$, $n \in \mathbb{Z}$. If $\lambda = -\frac{1}{2} + i\rho$, $\rho \in \mathbb{R}$ and $\varepsilon \in \mathbb{R}$, it is easily checked that

$$(Uf)(x) = \frac{1}{\sqrt{\pi}}\,(x+i)^{\lambda-\varepsilon}(x-i)^{\lambda+\varepsilon} f\left(\frac{x-i}{x+i}\right)$$

defines an isometry $U : L^2(\mathbb{T}) \to L^2(\mathbb{R})$. Next we recall that the Mellin transform, defined by

$$f(x) \mapsto \frac{1}{\sqrt{2\pi}} \int_0^\infty f(y) y^{ix-\frac{1}{2}}\, dy,$$

gives an isometry $L^2(\mathbb{R}_+) \to L^2(\mathbb{R})$. Thus, we may define a "double" Mellin transform as the isometry

$$V : L^2(\mathbb{R}) \to L^2(\mathbb{R}) \oplus L^2(\mathbb{R})$$

given by

$$(Vf)(x) = \frac{1}{\sqrt{2\pi}} \begin{pmatrix} \int_0^\infty f(y) y^{ix-\frac{1}{2}}\, dy \\ \int_0^\infty f(-y) y^{ix-\frac{1}{2}}\, dy \end{pmatrix}.$$

If we now let T_t, $t \in \mathbb{R}$, denote the translation operator $(T_t f)(x) = f(x+t)$, we may compose the above isometries to obtain the orthonormal basis

$$\mathbf{f_n} = \begin{pmatrix} f_n^+ \\ f_n^- \end{pmatrix} = (V \circ T_t \circ U)(e_n)$$

of $L^2(\mathbb{R}) \oplus L^2(\mathbb{R})$. Explicitly, we have

$$f_n^\pm(x) = \frac{1}{\sqrt{2}\,\pi} \int_0^\infty (t+i\pm y)^{\lambda-n-\varepsilon}(t-i\pm y)^{\lambda+n+\varepsilon} y^{ix-\frac{1}{2}}\, dy.$$

These integrals may be expressed in terms of Gauss's hypergeometric function; cf. (Vilenkin and Klimyk, 1991, §7.7.3). Using Kummer's identities (Erdélyi et al., 1953, §2.9), one may then express $f_n^\pm$ in terms of $u_n(x+\rho)$ and $u_n^*(x+\rho)$, where $e^{2i\varphi} = (t+i)/(t-i)$. Rewriting the identity $\langle \mathbf{f_n}, \mathbf{f_m} \rangle = \delta_{nm}$ in terms of Meixner-Pollaczek functions and making a final change of variables $x \mapsto x - \rho$, one recovers the orthonormal basis $\mathbf{u}_n$.

The group-theoretic interpretation of this proof is the following. The space $L^2(\mathbb{T})$ is a natural representation space for the principal unitary series (e_n is proportional to the e_n in (4.5) below). The operator L gives the action of a hyperbolic Lie algebra element; cf. also §4.2. It generates a one-parameter subgroup of the universal covering group of SU(1, 1), which locally may be identified with the group of linear fractional transformations of the circle that have two common fix-points. Thus, $L^2(\mathbb{T})$ splits into two invariant subspaces. The map $T_t \circ U$ corresponds to mapping the fix-points to $\{0, \infty\}$, and the one-parameter subgroup to dilations of $\mathbb{R}$. Finally, the operator V is the Fourier transform with respect to these dilations. In particular, the appearance of double eigenvalues in Proposition 2.1 has a natural geometric explanation: it corresponds to the fact that a circle falls into two pieces when removing two points.

3. A bilinear summation formula

In this section we prove a bilinear summation formula for Meixner-Pollaczeck polynomials. This summation is related to the tensor product of a positive and a negative discrete series representation of the Lie algebra $\mathfrak{su}(1,1)$, which will be explained in Section 11.4. In the summation a certain type of non-polynomial ${}_3F_2$-functions appear. These functions will be investigated in Section 11.5.

Theorem 3.1. *For*

$$\rho^2 \in \operatorname{supp} d\mu\left(\cdot; k_2 - k_1 + \frac{1}{2}, k_1 + k_2 - \frac{1}{2}, k_1 - k_2 + p + \frac{1}{2}\right),$$

$p \in \mathbb{Z}$, $x_1, x_2 \in \mathbb{R}$, and $k_1, k_2 > 0$, the Meixner-Pollaczek polynomials and the continuous dual Hahn polymials satisfy the following summation formula

$$\frac{1}{C}\sum_{n=0}^{\infty}\frac{1}{(2k_1)_{p+n}\,(2k_2)_n}s_n\left(\rho^2;k_2-k_1+\frac{1}{2},k_1+k_2-\frac{1}{2},k_1-k_2+p+\frac{1}{2}\right)$$
$$\times p_{n+p}^{(k_1)}(x_1;\varphi)\,p_n^{(k_2)}(x_2;\varphi)$$
$$=(-1)^p\frac{\Gamma\left(i(x_1-x_2)+\frac{1}{2}+i\rho\right)\Gamma\left(i(x_1-x_2)+\frac{1}{2}-i\rho\right)}{\Gamma(k_2-k_1+i(x_1-x_2)+1)\,\Gamma(k_1-k_2+i(x_1-x_2)+1+p)}$$
$$\times\,{}_3F_2\left(\begin{matrix}k_2-ix_2,k_2-k_1+\frac{1}{2}+i\rho,k_2-k_1+\frac{1}{2}-i\rho\\ 2k_2,k_2-k_1+i(x_1-x_2)+1\end{matrix};1\right)$$
$$\times\,{}_2F_1\left(\begin{matrix}p+k_1-k_2+\frac{1}{2}+i\rho,p+k_1-k_2+\frac{1}{2}-i\rho\\ k_1-k_2+i(x_1-x_2)+p+1\end{matrix};\frac{1}{1-e^{-2i\varphi}}\right)$$
$$+\frac{\Gamma\left(i(x_2-x_1)+\frac{1}{2}+i\rho\right)\Gamma\left(i(x_2-x_1)+\frac{1}{2}-i\rho\right)}{\Gamma(k_2-k_1+i(x_2-x_1)+1)\,\Gamma(k_1-k_2+i(x_2-x_1)+1+p)}$$
$$\times\,{}_3F_2\left(\begin{matrix}k_2+ix_2,k_2-k_1+\frac{1}{2}+i\rho,k_2-k_1+\frac{1}{2}-i\rho\\ 2k_2,k_2-k_1+i(x_2-x_1)+1\end{matrix};1\right)$$
$$\times\,{}_2F_1\left(\begin{matrix}p+k_1-k_2+\frac{1}{2}+i\rho,p+k_1-k_2+\frac{1}{2}-i\rho\\ k_1-k_2+i(x_2-x_1)+p+1\end{matrix};\frac{1}{1-e^{2i\varphi}}\right),$$

where

$$C=\frac{e^{-2x_1(\varphi-\frac{\pi}{2})}\,\Gamma(2k_1)}{i^p\,(2\sin\varphi)^{2k_1+p}\Gamma(k_1+ix_1)\,\Gamma(k_1-ix_1)}.$$

Here we use the convention that if p_n is a polynomial of degree n, then $p_{-n}=0$ for $n\in\mathbb{N}$.

Remark 3.2.

(i) *The* ${}_2F_1$*-series on the right hand side are Meixner-Pollaczek functions, as defined by* (2.5)*. The* ${}_3F_2$*-series on the right hand side can be considered as continuous Hahn functions, see Section* 11.5.

(ii) *Theorem* 3.1 *is a non-polynomial extension of (Groenevelt and Koelink, 2002, Thm. 3.6). Replacing* k_1+ix_1, k_2+ix_2, $e^{2i\varphi}$, ρ^2 *in Theorem* 3.1 *by* x_1, x_2, c, y *respectively, and restricting the new* x_1 *and* x_2 *to negative integers makes one term on the right hand side vanish, and we obtain (Groenevelt and Koelink, 2002, Thm. 3.6).*

(iii) *There is an interesting limit case of the summation of Theorem 3.1; for $x_2 - x_1 > 0$*

$$\sum_{n=0}^{\infty} s_n(\rho^2; k_2 - k_1 + \frac{1}{2}, k_1 + k_2 - \frac{1}{2}, k_1 - k_2 + p + \frac{1}{2})$$
$$\times \frac{1}{(2k_1)_{p+n}(2k_2)_n} L_{n+p}^{(2k_1-1)}(x_1) L_n^{(2k_2-1)}(x_2)$$
$$= \frac{e^{x_1}(x_2 - x_1)^{\frac{1}{2}-k_1-k_2-i\rho}\,\Gamma(2k_2)}{|\Gamma(k_2 - k_1 + \frac{1}{2} + i\rho)|^2}$$
$$\times {}_2F_1\left(\begin{matrix} k_1 + k_2 - \frac{1}{2} + i\rho, k_1 + k_2 - \frac{1}{2} - i\rho \\ 2k_2 \end{matrix} ; \frac{x_1}{x_1 - x_2} \right)$$
$$\times U(p + k_1 - k_2 + \frac{1}{2} - i\rho; 1 - 2i\rho; x_2 - x_1),$$

where $L_n^{(\alpha)}$ is a Laguerre polynomial as defined in (Koekoek and Swarttouw, 1998), and $U(a; b; z)$ denotes the second solution of the confluent hypergeometric differential equation in the notation of Slater (Slater, 1960):

$$U(a; b; z) =$$
$$\frac{\Gamma(1-b)}{\Gamma(1+a-b)}\, {}_1F_1\left(\begin{matrix} a \\ b \end{matrix} ; z \right) + \frac{\Gamma(b-1)}{\Gamma(a)} z^{1-b}\, {}_1F_1\left(\begin{matrix} 1+a-b \\ 2-b \end{matrix} ; z \right).$$

This formula is obtained from Theorem 3.1 as follows. We replace x_i by $-x_i/2\varphi$, $i = 1, 2$, and transform the ${}_2F_1$-series on the right hand side by (Andrews et al., 1999, (2.3.12)). Then we let $\varphi \to 0$. Here Stirling's formula is used, Euler's transformation is used for the ${}_2F_1$-series which are obtained from the ${}_3F_2$-series, and Kummer's transformation (Slater, 1960, (1.4.1)) is used for the ${}_1F_1$-series which are obtained from the ${}_2F_1$-series. This limit case can also be obtained by Lie algebraic methods, see (Groenevelt, 2003, Thm. 3.10).

(iv) *Note that from*

$$s_n\left(\rho^2; k_1 - k_2 + \frac{1}{2}, k_1 + k_2 - \frac{1}{2}, k_2 - k_1 - p + \frac{1}{2}\right) =$$
$$(-1)^p \left| \left(k_1 - k_2 + \frac{1}{2} + i\rho\right)_p \right|^2 \tag{3.1}$$
$$\times s_{n-p}\left(\rho^2; k_2 - k_1 + \frac{1}{2}, k_1 + k_2 - \frac{1}{2}, k_1 - k_2 + p + \frac{1}{2}\right),$$

see (Groenevelt and Koelink, 2002, (3.13)), it follows that the sum on the left hand side of Theorem 3.1 *is invariant under* $k_1 \leftrightarrow k_2$, $x_1 \leftrightarrow x_2$, $p \leftrightarrow -p$.

(v) *It is interesting to compare Theorem 3.1 with the results of (Ismail and Stanton, 2002), where summation formulas with a similar, but simpler, structure are obtained for various orthogonal polynomials. The method used in (Ismail and Stanton, 2002) is completely different from the method we use here.*

Proof of Theorem 3.1. We start with the sum on the left hand side, with orthonormal polynomials:

$$S = \sum_{n=0}^{\infty}(-1)^n S_n\left(\rho^2; k_2 - k_1 + \frac{1}{2}, k_1 + k_2 - \frac{1}{2}, k_1 - k_2 + p + \frac{1}{2}\right) \times P_{n+p}^{(k_1)}(x_1;\varphi)\, P_n^{(k_2)}(x_2;\varphi).$$

First we show that this sum converges absolutely. Writing out the summand R_n explicitly gives

$$R_n = K\, e^{-2in\varphi}\frac{\Gamma(2k_2+n)\,\Gamma(2k_1+p+n)}{\Gamma(n+1)\Gamma(n+p+1)}\ {}_3F_2\ {}_2F_1\ {}_2F_1$$

where K is a constant independent of n. To find the asymptotic behaviour for $n \to \infty$ of the Γ-functions, we use the asymptotic formula for the ratio of two Γ-functions (Olver, 1974, §4.5)

$$\frac{\Gamma(a+z)}{\Gamma(b+z)} = z^{a-b}\left(1 + \mathcal{O}(z^{-1})\right), \quad |z| \to \infty, \quad |\arg(z)| < \pi. \tag{3.2}$$

The asymptotics for the ${}_3F_2$-function follows from transforming the function by (Bailey, 1972, p. 15(2)) and using (3.2). This gives, for $n \to \infty$,

$$\begin{aligned}{}_3F_2&\left(\begin{matrix}-n, k_1+k_2-\frac{1}{2}+i\rho, k_1+k_2-\frac{1}{2}-i\rho \\ 2k_2, 2k_1+p\end{matrix}; 1\right)\\ &= C_1\, n^{\frac{1}{2}-k_1-k_2-i\rho} + C_2\, n^{\frac{1}{2}-k_1-k_2+i\rho},\end{aligned}$$

where C_1 and C_2 are independent of n. If ρ^2 is in the discrete part of $\operatorname{supp} d\mu\left(\cdot; k_2 - k_1 + \frac{1}{2}, k_1 + k_2 - \frac{1}{2}, k_1 - k_2 + p + \frac{1}{2}\right)$, we assume without loss of generality that $\Im(\rho) > 0$. In this case the second term in the transformation (Bailey, 1972, p. 15(2)) vanishes, so $C_2 = 0$.

The asymptotic behaviour of the ${}_2F_1$-functions follows from (Erdélyi et al., 1953, 2.3.2(14)). This gives, for $0 < \varphi < \pi$ and $n \to \infty$,

$$
\begin{aligned}
&{}_2F_1\left(\begin{matrix} -(n+p), k_1 + ix_1 \\ 2k_1 \end{matrix} ; 1 - e^{-2i\varphi}\right) \\
&= \left(C_3\, n^{-k_1 - ix_1} + C_4\, n^{-k_1 + ix_1} e^{-(n+p)\left(1 - e^{-2i\varphi}\right)}\right)\left(1 + \mathcal{O}(n^{-1})\right), \\
&{}_2F_1\left(\begin{matrix} -n, k_2 - ix_2 \\ 2k_2 \end{matrix} ; 1 - e^{2i\varphi}\right) \\
&= \left(C_5\, n^{-k_2 - ix_2} + C_6\, n^{-k_2 + ix_2} e^{-n\left(1 - e^{-2i\varphi}\right)}\right)\left(1 + \mathcal{O}(n^{-1})\right),
\end{aligned}
$$

where C_i, $i = 3, \ldots, 6$, is independent of n. Since $\Re\left(1 - e^{-2i\varphi}\right) = 2\sin^2\varphi > 0$ and $x_1, x_2 \in \mathbb{R}$, we find

$$
R_n = \left(K_1\, n^{-\frac{3}{2} + i\rho} + K_2\, n^{-\frac{3}{2} - i\rho}\right)\left(1 + \mathcal{O}(n^{-1})\right), \qquad n \to \infty,
$$

and then for $\rho \in \mathbb{R}$ the sum S converges absolutely. In case $\rho \in i\mathbb{R}$, we have $K_2 = 0$ and $\Im(\rho) > 0$, and then S still converges absolutely.

Next we write out the polynomials in S as hypergeometric series, using (2.2) and (2.1), and then we transform the ${}_3F_2$-series, using the first formula on page 142 of (Andrews et al., 1999);

$$
\begin{aligned}
&{}_3F_2\left(\begin{matrix} -n, k_2 - k_1 + \frac{1}{2} + i\rho, k_2 - k_1 + \frac{1}{2} - i\rho \\ 2k_2, p+1 \end{matrix} ; 1\right) = \\
&\frac{\left(p + k_1 - k_2 + \frac{1}{2} - i\rho\right)_n}{(p+1)_n}\, {}_3F_2\left(\begin{matrix} -n, k_2 - k_1 + \frac{1}{2} + i\rho, k_1 + k_2 - \frac{1}{2} + i\rho \\ 2k_2, k_2 - k_1 + \frac{1}{2} + i\rho - n - p \end{matrix} ; 1\right).
\end{aligned}
$$

By (Erdélyi et al., 1953, §2.9(27)) with

$$
(a, b, c, z) = \left(-n - p, k_1 + ix_1, 2k_1, 1 - e^{-2i\varphi}\right)
$$

the ${}_2F_1$-series for $P^{(k_1)}_{n+p}(x_1;\varphi)$ is written as a sum of two ${}_2F_1$-series

$$
\begin{aligned}
&{}_2F_1\left(\begin{matrix}-n-p, k_1+ix_1\\ 2k_1\end{matrix};1-e^{-2i\varphi}\right)\\
&=\frac{\left(1-e^{-2i\varphi}\right)^{-ix_1-k_1}\Gamma(2k_1)\Gamma(n+p+1)}{\Gamma(k_1-ix_1)\Gamma(n+p+1+k_1+ix_1)}\\
&\times{}_2F_1\left(\begin{matrix}k_1+ix_1, 1-k_1+ix_1\\ n+p+1+k_1+ix_1\end{matrix};\frac{1}{1-e^{-2i\varphi}}\right)\\
&+(e^{-2i\varphi})^{n+p}\frac{\left(1-e^{2i\varphi}\right)^{ix_1-k_1}\Gamma(2k_1)\Gamma(n+p+1)}{\Gamma\left(k_1+ix_1\right)\Gamma\left(n+p+1+k_1-ix_1\right)}\\
&\times{}_2F_1\left(\begin{matrix}k_1-ix_1, 1-k_1-ix_1\\ n+p+1+k_1-ix_1\end{matrix};\frac{1}{1-e^{2i\varphi}}\right).
\end{aligned}
\tag{3.3}
$$

Now the sum S splits according to this: $S = S_1 + S_2$.

First we focus on S_1. Reversing the order of summation in the ${}_2F_1$-series of the second Meixner-Pollaczek polynomial and using Euler's transformation (Andrews et al., 1999, (2.2.7)), gives

$$
\begin{aligned}
&{}_2F_1\left(\begin{matrix}-n, k_2+ix_2\\ 2k_2\end{matrix};1-e^{-2i\varphi}\right)=\\
&\qquad e^{-2in\varphi}\left(1-e^{2i\varphi}\right)^{ix_2-k_2}\frac{(k_2+ix_2)_n}{(2k_2)_n}\\
&\qquad\times\ {}_2F_1\left(\begin{matrix}k_2-ix_2, 1-2k_2-ix_2\\ 1-k_2-n-ix_2\end{matrix};\frac{1}{1-e^{-2i\varphi}}\right).
\end{aligned}
\tag{3.4}
$$

Writing out the hypergeometric series as a sum, we get

$$
\begin{aligned}
S_1 &= e^{ip\varphi}\frac{\left(1-e^{-2i\varphi}\right)^{-k_1-ix_1}\left(1-e^{2i\varphi}\right)^{ix_2-k_2}}{\Gamma\left(k_1-ix_1\right)\Gamma\left(p+1+k_1+ix_1\right)}\Gamma\left(2k_1\right)\sqrt{p!\left(2k_1\right)_p}\\
&\times\sum_{n=0}^{\infty}\sum_{j=0}^{n}\sum_{l,m=0}^{\infty}\frac{(k_2+ix_2)_n\left(p+k_1-k_2+\frac{1}{2}-i\rho\right)_n}{n!\left(p+1+k_1+ix\right)_n}\\
&\times\frac{(-n)_j\left(k_2-k_1+\frac{1}{2}+i\rho\right)_j\left(k_1+k_2-\frac{1}{2}+i\rho\right)_j}{j!\left(2k_2\right)_j\left(k_2-k_1+\frac{1}{2}+i\rho-n-p\right)_j}\left(1-e^{-2i\varphi}\right)^{-l-m}\\
&\times\frac{(k_1+ix_1)_l\left(1-k_1+ix_1\right)_l\left(k_2-ix_2\right)_m\left(1-k_2-ix_2\right)_m}{l!\left(n+p+1+k_1+ix_1\right)_l\,m!\left(1-k_2-n-ix_2\right)_m}.
\end{aligned}
$$

Next we interchange summations

$$
\sum_{n=0}^{\infty}\sum_{j=0}^{n}=\sum_{j=0}^{\infty}\sum_{n=j}^{\infty},
$$

then the sum over n becomes

$$\Sigma_1 = \sum_{n=j}^{\infty} \frac{\left(k_1 - k_2 + \frac{1}{2} - i\rho + p - j\right)_n (k_2 + ix_2 - m)_n}{(n-j)!\,(p+1+l+k_1+ix_1)_n}.$$

We substitute $k \mapsto n - j$, then we find by Stirling's formula for the summand R of S_1, for large j and k,

$$R \sim C\, j^{k_1-k_2+p} k^{-\frac{1}{2}} (j+k)^{-1-l-m},$$

where C is independent of k and j. So we see that the sum S_1 converges absolutely for $k_1 - k_2 + p < 0$, and $\frac{1}{6}\pi < \varphi < \frac{5}{6}\pi$. Now the sum Σ_1 is a multiple of a ${}_2F_1$-series with unit argument, which is summable by Gauss' summation formula (Erdélyi et al., 1953, (46), p. 104);

$$\Sigma_1 = \frac{\left(k_2 - k_1 + \frac{1}{2} - p + i\rho\right)_j (k_2 + ix_2 - m)_j}{\Gamma\left(l + j + \frac{1}{2} + k_2 + ix_1 + i\rho\right)}$$
$$\times \frac{\Gamma(p+l+1+k_1+ix_1)\,\Gamma\left(l+m+\frac{1}{2}+ix_1 - ix_2 + i\rho\right)}{\Gamma(p+l+m+k_1-k_2+ix_1-ix_2)}.$$

This gives

$$S_1 = e^{ip\varphi} \frac{\left(1-e^{-2i\varphi}\right)^{-k_1-ix_1} \left(1-e^{2i\varphi}\right)^{ix_2-k_2}}{\Gamma(k_1 - ix_1)} \Gamma(2k_1)\sqrt{p!\,(2k_1)_p}$$
$$\times \sum_{l,m=0}^{\infty} \frac{\Gamma\left(l+m+i(x_1-x_2)+\frac{1}{2}+i\rho\right)(k_2-ix_2)_m}{\Gamma(p+l+m+1+k_1-k_2+i(x_1-x_2))}$$
$$\times \frac{(k_1+ix_1)_l\,(1-k_1+ix_1)_l}{\Gamma\left(l+k_2+\frac{1}{2}+ix_1+i\rho\right) m!\,l!} \left(1-e^{-2i\varphi}\right)^{-l-m}$$
$$\times \sum_{j=0}^{\infty} \frac{\left(k_2-k_1+\frac{1}{2}+i\rho\right)_j \left(k_1+k_2-\frac{1}{2}+i\rho\right)_j (k_2+ix_2-m)_j}{j!\,(2k_2)_j \left(l+\frac{1}{2}+k_2+ix_1+i\rho\right)_j}$$

The sum over j is a ${}_3F_2$-series, which by Kummer's transformation (Andrews et al., 1999, Cor. 3.3.5) becomes

$$\frac{\Gamma\left(l+\frac{1}{2}+k_2+ix_1+i\rho\right)\Gamma\left(l+m+\frac{1}{2}+i(x_1-x_2)-i\rho\right)}{\Gamma(k_2-k_1+1+i(x_1-x_2)+l+m)\,\Gamma(l+k_1+ix_1)}$$
$$\times\ {}_3F_2\left(\begin{matrix} k_2-k_1+\frac{1}{2}+i\rho,\, k_2-k_1+\frac{1}{2}-i\rho,\, k_2-ix_2+m \\ 2k_2,\, k_2-k_1+1+i(x_1-x_2)+l+m \end{matrix};1\right).$$

Here we need the condition $k_1 > 0$ for absolute convergence. We write out the ${}_3F_2$-series explicitly as a sum over j and interchange summations

$$S_1 = \sum_{l,m,j=0}^{\infty} = \sum_{n=0}^{\infty}\sum_{j=0}^{\infty}\sum_{l+m=n}.$$

For the sum over $l+m=n$ we find

$$\sum_{m+l=n} \frac{(1-k_1+ix_1)_l\,(k_2-ix_2+j)_m}{l!\;m!} = \frac{(1+k_2-k_1+i\,(x_1-x_2)+j)_n}{n!}.$$

Now S_1 reduces to a double sum, which splits as a product of two sums, and we obtain

$$\begin{aligned}
S_1 &= e^{ip\varphi} \frac{\left(1-e^{-2i\varphi}\right)^{-k_1-ix_1}\left(1-e^{2i\varphi}\right)^{ix_2-k_2}}{\Gamma(k_1-ix_1)\,\Gamma(k_1+ix_1)} \\
&\times \frac{\Gamma\left(\frac{1}{2}+i\,(x_1-x_2)+i\rho\right)\Gamma\left(\frac{1}{2}+i\,(x_1-x_2)-i\rho\right)}{\Gamma(p+k_1-k_2+i\,(x_1-x_2)+1)\,\Gamma(k_2-k_1+i\,(x_1-x_2)+1)} \\
&\times \Gamma(2k_1)\sqrt{p!\,(2k_1)_p} \\
&\times {}_2F_1\left(\begin{matrix}\frac{1}{2}+i\,(x_1-x_2)+i\rho,\frac{1}{2}+i\,(x_1-x_2)-i\rho \\ p+k_1-k_2+i\,(x_1-x_2)+1\end{matrix};\frac{1}{1-e^{-2i\varphi}}\right) \\
&\times {}_3F_2\left(\begin{matrix}k_2-k_1+\frac{1}{2}+i\rho,k_2-k_1+\frac{1}{2}-i\rho,k_2-ix_2 \\ 2k_2,k_2-k_1+i(x_1-x_2)+1\end{matrix};1\right).
\end{aligned}$$

S_2 is calculated in the same way as S_1, only for the first step (Erdélyi et al., 1953, §2.10(4)) is used instead of (3.4). Then we obtain

$$\begin{aligned}
S_2 &= e^{-ip\varphi} \frac{\left(1-e^{-2i\varphi}\right)^{-k_2-ix_2}\left(1-e^{2i\varphi}\right)^{ix_1-k_1}}{\Gamma(k_1+ix_1)\,\Gamma(k_1-ix_1)} \Gamma(2k_1)\sqrt{p!\,(2k_1)_p} \\
&\times \frac{\Gamma\left(i\,(x_2-x_1)+\frac{1}{2}+i\rho\right)\Gamma\left(i\,(x_2-x_1)+\frac{1}{2}-i\rho\right)}{\Gamma(k_2-k_1+i\,(x_2-x_1)+1)\,\Gamma(k_1-k_2+i\,(x_2-x_1)+1+p)} \\
&\times {}_3F_2\left(\begin{matrix}k_2+ix_2,k_2-k_1+\frac{1}{2}+i\rho,k_2-k_1+\frac{1}{2}-i\rho \\ 2k_2,k_2-k_1+i\,(x_2-x_1)+1\end{matrix};1\right) \\
&\times {}_2F_1\left(\begin{matrix}\frac{1}{2}+i\,(x_2-x_1)+i\rho,\frac{1}{2}+i\,(x_2-x_1)-i\rho \\ k_1-k_2+i\,(x_2-x_1)+p+1\end{matrix};\frac{1}{1-e^{2i\varphi}}\right).
\end{aligned}$$

Finally using Euler's transformation for the ${}_2F_1$-series, using

$$\begin{aligned}
e^{ip\varphi}\left(1-e^{2i\varphi}\right)^{-p} &= (-1)^p(2i\sin\varphi)^{-p}, \\
\left(1-e^{-2i\varphi}\right)^{-ix_1}\left(1-e^{2i\varphi}\right)^{ix_1} &= e^{-2x_1\left(\varphi-\frac{\pi}{2}\right)}, \\
\left(1-e^{-2i\varphi}\right)^{-k_1}\left(1-e^{2i\varphi}\right)^{-k_1} &= (2\sin\varphi)^{-2k_1},
\end{aligned}$$

and writing the polynomials in the normalization given by (2.2) and (2.1), the theorem is proved.

Let us remark that all the series used in the proof are absolutely convergent under the conditions

$$\rho^2 \in \operatorname{supp} d\mu\left(\cdot; k_2 - k_1 + \frac{1}{2}, k_1 + k_2 - \frac{1}{2}, k_1 - k_2 + p + \frac{1}{2}\right),$$

$x_1, x_2 \in \mathbb{R}$, $k_1 > 0$, $\frac{1}{6}\pi < \varphi < \frac{5}{6}\pi$ and $k_1 - k_2 + p < 0$. The last condition can be removed using the symmetry $(k_1, k_2, p) \leftrightarrow (k_2, k_1, -p)$ and continuity in k_1 and k_2. Using the analytic continuation of the hypergeometric function, we see that the result remains valid for $0 < \varphi < \pi$. □

4. Clebsch-Gordan coefficients for hyperbolic basis vectors of $\mathfrak{su}(1,1)$

In this section we consider the tensor product of a positive and a negative discrete series representation. We diagonalize a certain self-adjoint element of $\mathfrak{su}(1,1)$ using (doubly infinite) Jacobi operators. We also give generalized eigenvectors, which can be considered as hyperbolic basis vectors. Using the summation formula from the previous section, we show that the Clebsch-Gordan coefficients for the eigenvectors are continuous Hahn functions. We find the corresponding integral transform pair by formal computations. In order to give a rigorous proof for the continuous Hahn integral transform, we realize the generators of $\mathfrak{su}(1,1)$ in the discrete series as difference operators acting on polynomials. Using these realizations, the Casimir element in the tensor product is realized as a difference operator. Spectral analysis of this difference operator is carried out in Section 11.5.

4.1 The Lie algebra $\mathfrak{su}(1,1)$

The Lie algebra $\mathfrak{su}(1,1)$ is generated by the elements H, B and C, satisfying the commutation relations

$$[H, B] = 2B, \quad [H, C] = -2C, \quad [B, C] = H. \tag{4.1}$$

There is a $*$-structure defined by $H^* = H$ and $B^* = -C$. The center of $\mathcal{U}(\mathfrak{su}(1,1))$ is generated by the Casimir element Ω, which is given by

$$\Omega = -\frac{1}{4}\left(H^2 + 2H + 4CB\right) = -\frac{1}{4}\left(H^2 - 2H + 4BC\right). \tag{4.2}$$

There are four classes of irreducible unitary representations of $\mathfrak{su}(1,1)$, see (Vilenkin and Klimyk, 1991, §6.4):

The positive discrete series representations π_k^+ are representations labelled by $k > 0$. The representation space is $\ell^2(\mathbb{Z}_{\geq 0})$ with orthonormal basis $\{e_n\}_{n\in\mathbb{Z}_{\geq 0}}$. The action is given by

$$\begin{aligned}
\pi_k^+(H)\, e_n &= 2(k+n)\, e_n,\\
\pi_k^+(B)\, e_n &= \sqrt{(n+1)(2k+n)}\, e_{n+1},\\
\pi_k^+(C)\, e_n &= -\sqrt{n(2k+n-1)}\, e_{n-1},\\
\pi_k^+(\Omega)\, e_n &= k(1-k)\, e_n.
\end{aligned} \tag{4.3}$$

The negative discrete series representations π_k^- are labelled by $k>0$. The representation space is $\ell^2(\mathbb{Z}_{\geq 0})$ with orthonormal basis $\{e_n\}_{n\in\mathbb{Z}_{\geq 0}}$. The action is given by

$$\begin{aligned}
\pi_k^-(H)\, e_n &= -2(k+n)\, e_n,\\
\pi_k^-(B)\, e_n &= -\sqrt{n(2k+n-1)}\, e_{n-1},\\
\pi_k^-(C)\, e_n &= \sqrt{(n+1)(2k+n)}\, e_{n+1},\\
\pi_k^-(\Omega)\, e_n &= k(1-k)\, e_n.
\end{aligned} \tag{4.4}$$

The principal series representations $\pi^{\rho,\varepsilon}$ are labelled by $\varepsilon\in[0,1)$ and $\rho\geq 0$, where $(\rho,\varepsilon)\neq\left(0,\frac{1}{2}\right)$. The representation space is $\ell^2(\mathbb{Z})$ with orthonormal basis $\{e_n\}_{n\in\mathbb{Z}}$. The action is given by

$$\begin{aligned}
\pi^{\rho,\varepsilon}(H)\, e_n &= 2(\varepsilon+n)\, e_n,\\
\pi^{\rho,\varepsilon}(B)\, e_n &= \sqrt{\left(n+\varepsilon+\frac{1}{2}-i\rho\right)\left(n+\varepsilon+\frac{1}{2}+i\rho\right)}\, e_{n+1},\\
\pi^{\rho,\varepsilon}(C)\, e_n &= -\sqrt{\left(n+\varepsilon-\frac{1}{2}-i\rho\right)\left(n+\varepsilon-\frac{1}{2}+i\rho\right)}\, e_{n-1},\\
\pi^{\rho,\varepsilon}(\Omega)\, e_n &= \left(\rho^2+\frac{1}{4}\right) e_n.
\end{aligned} \tag{4.5}$$

For $(\rho,\varepsilon)=\left(0,\frac{1}{2}\right)$ the representation $\pi^{0,\frac{1}{2}}$ splits into a direct sum of a positive and a negative discrete series representation: $\pi^{0,\frac{1}{2}}=\pi_{1/2}^+\oplus\pi_{1/2}^-$. The representation space splits into two invariant subspaces: $\{e_n\,|\,n<0\}\oplus\{e_n\,|\,n\geq 0\}$.

The complementary series representations $\pi^{\lambda,\varepsilon}$ are labelled by ε and λ, where $\varepsilon\in\left[0,\frac{1}{2}\right)$ and $\lambda\in\left(-\frac{1}{2},-\varepsilon\right)$ or $\varepsilon\in\left(\frac{1}{2},1\right)$ and $\lambda\in\left(-\frac{1}{2},\varepsilon-1\right)$. The representation space is $\ell^2(\mathbb{Z})$ with orthonormal basis $\{e_n\}_{n\in\mathbb{Z}}$. The action is given by

$$\begin{aligned}
\pi^{\lambda,\varepsilon}(H)\, e_n &= 2(\varepsilon+n)\, e_n, \\
\pi^{\lambda,\varepsilon}(B)\, e_n &= \sqrt{(n+\varepsilon+1+\lambda)(n+\varepsilon-\lambda)}\, e_{n+1}, \\
\pi^{\lambda,\varepsilon}(C)\, e_n &= -\sqrt{(n+\varepsilon+\lambda)(n+\varepsilon-\lambda-1)}\, e_{n-1}, \\
\pi^{\lambda,\varepsilon}(\Omega)\, e_n &= -\lambda(1+\lambda)\, e_n.
\end{aligned} \tag{4.6}$$

Note that if we formally write $\lambda = -\frac{1}{2} + i\rho$ the actions in the principal series and in the complementary series are the same.

We remark that the operators (4.3)–(4.6) are unbounded, with domain the set of finite linear combinations of the basis vectors. The representations are $*$-representations in the sense of Schmüdgen (Schmüdgen, 1990, Ch. 8).

The decomposition of the tensor product of a positive and a negative discrete series representation of $\mathfrak{su}(1,1)$ is determined in full generality in (Groenevelt and Koelink, 2002, Thm. 2.2).

Theorem 4.1. *For $k_1 \le k_2$ the decomposition of the tensor product of positive and negative discrete series representations of $\mathfrak{su}(1,1)$ is*

$$\pi^+_{k_1} \otimes \pi^-_{k_2} \cong \int_0^{\infty \, \oplus} \pi^{\rho,\varepsilon} d\rho, \qquad k_1 - k_2 \ge -\frac{1}{2}, k_1 + k_2 \ge \frac{1}{2},$$

$$\pi^+_{k_1} \otimes \pi^-_{k_2} \cong \int_0^{\infty \, \oplus} \pi^{\rho,\varepsilon} d\rho \oplus \pi^{\lambda,\varepsilon}, \qquad k_1 + k_2 < \frac{1}{2},$$

$$\pi^+_{k_1} \otimes \pi^-_{k_2} \cong \int_0^{\infty \, \oplus} \pi^{\rho,\varepsilon} d\rho \oplus \bigoplus_{\substack{j\in\mathbb{Z}_{\ge 0} \\ k_2-k_1-\frac{1}{2}-j>0}} \pi^-_{k_2-k_1-j}, \quad k_1 - k_2 < -\frac{1}{2},$$

where $\varepsilon = k_1 - k_2 + L$, L is the unique integer such that $\varepsilon \in [0,1)$, and $\lambda = -k_1 - k_2$. The intertwiner J is given by

$$J\left(e_{n_1} \otimes e_{n_2}\right) = (-1)^{n_2} \int_{\mathbb{R}} S_n\left(y; n_1 - n_2\right) e_{n_1-n_2-L}\, d\mu^{\frac{1}{2}}\left(y; n_1 - n_2\right), \tag{4.7}$$

where $n = \min\{n_1, n_2\}$, $S_n(y;p)$ is an orthonormal continuous dual Hahn polynomial,

$$S_n(y;p) = \begin{cases} S_n\left(y; k_1 - k_2 + \frac{1}{2}, k_1 + k_2 - \frac{1}{2}, k_2 - k_1 - p + \frac{1}{2}\right), & p \le 0, \\ S_n\left(y; k_2 - k_1 + \frac{1}{2}, k_1 + k_2 - \frac{1}{2}, k_1 - k_2 + p + \frac{1}{2}\right), & p \ge 0, \end{cases}$$

and $d\mu(y;p)$ is the corresponding orthogonality measure

$$d\mu(y;p) = \begin{cases} d\mu\left(y; k_1 - k_2 + \frac{1}{2}, k_1 + k_2 - \frac{1}{2}, k_2 - k_1 - p + \frac{1}{2}\right), & p \leq 0, \\ d\mu\left(y; k_2 - k_1 + \frac{1}{2}, k_1 + k_2 - \frac{1}{2}, k_1 - k_2 + p + \frac{1}{2}\right), & p \geq 0. \end{cases}$$

The inversion of (4.7) can be given explicitly, e.g., for an element

$$f \otimes e_{r-L} = \int_0^\infty f(x) e_{r-L} dx \in L^2(0,\infty) \otimes \ell^2(\mathbb{Z}) \cong \int_0^{\infty \oplus} \ell^2(\mathbb{Z}) dx$$

in the representation space of the direct integral representation, we have

$$J^*\left(f \otimes e_{r-L}\right) =$$

$$\begin{cases} \sum_{p=0}^{\infty} (-1)^{p-r} \left[\int_{\mathbb{R}} S_p(y;r) f(y) d\mu^{\frac{1}{2}}(y;r) \right] e_p \otimes e_{p-r}, & r \leq 0, \\ \sum_{p=0}^{\infty} (-1)^{p} \left[\int_{\mathbb{R}} S_p(y;r) f(y) d\mu^{\frac{1}{2}}(y;r) \right] e_{p+r} \otimes e_p, & r \geq 0. \end{cases} \tag{4.8}$$

For the discrete components in Theorem 4.1 we can replace f by a Dirac delta function at the appropriate points of the discrete mass of $d\mu(\cdot;r)$. We remark that for $k_1 = k_2 < 1/4$, the occurrence of a complementary series representation in the tensor product was discovered by Neretin (Neretin, 1986). This phenomenon was investigated from the viewpoint of operator theory in (Engliš et al., 2000).

In the following subsection we assume that discrete terms do not occur in the tensor product decomposition. From the calculations it is clear how to extend the results to the general case. At the end of Section 11.5 we briefly discuss the results for the discrete terms in the decomposition.

In the Lie algebra $\mathfrak{su}(1,1)$ three types of elements can be distinguished: the elliptic, the parabolic and the hyperbolic elements. These are related to the three conjugacy classes of the group $SU(1,1)$. A basis on which an elliptic element acts diagonally is called an elliptic basis, and similarly for the parabolic and hyperbolic elements. The basisvectors e_n in (4.3)–(4.6) are elliptic basisvectors.

We consider self-adjoint elements of the form

$$-aH + B - C \in \mathfrak{su}(1,1), \qquad (a \in \mathbb{R}),$$

in the tensor product of a positive and a negative discrete series representation. For $|a| = 1$ this is a parabolic element, for $|a| < 1$ it is hyperbolic and for $|a| > 1$ it is elliptic. For the elliptic and the parabolic case we refer to (Groenevelt and Koelink, 2002), respectively (Groenevelt, 2003). We consider the case $|a| < 1$.

4.2 Hyperbolic basisvectors

We consider a self-adjoint element X_φ in $\mathfrak{su}(1,1)$, given by

$$X_\varphi = -\cos\varphi\; H + B - C, \qquad 0 < \varphi < \pi.$$

The action of X_φ in the discrete series can be identified with the three-term recurrence relation for the Meixner-Pollaczek polynomials (2.3), cf. (Koelink and Van der Jeugt, 1998, Prop. 3.1).

Proposition 4.2. *The operators $\Theta^\pm$, defined by*

$$\begin{aligned}\Theta^\pm : \ell^2\left(\mathbb{Z}_{\geq 0}\right) &\to L^2\left(\mathbb{R}, w^{(k)}(x;\varphi)dx\right)\\ e_n &\mapsto P_n^{(k)}(\cdot;\varphi),\end{aligned}$$

are unitary and intertwine $\pi_k^\pm\left(X_\varphi\right)$ with $M(\pm 2x\sin\varphi)$.

Here M denotes the multiplication operator, i.e., $M(f)g(x) = f(x)g(x)$. Proposition 4.2 states that

$$v^\pm(x) = \sum_{n=0}^\infty P_n^{(k)}(x;\varphi)\, e_n,$$

are generalized eigenvectors of $\pi_k^\pm\left(X_\varphi\right)$ for eigenvalue $\pm 2x\sin\varphi$. These eigenvectors can be considered as hyperbolic basis vectors.

The action of X_φ in the principal unitary series can be identified with the recurrence relation for the Meixner-Pollaczek functions (2.6). Then the spectral decomposition of the corresponding doubly infinite Jacobi operator gives the following.

Proposition 4.3. *The operator $\Theta^{\rho,\varepsilon}$ defined by*

$$\begin{aligned}\Theta^{\rho,\varepsilon} : \ell^2(\mathbb{Z}) &\to \mathcal{H}\left(-\frac{1}{2} + i\rho, \varepsilon, \pi - \varphi\right),\\ e_n &\mapsto \begin{pmatrix} u_n\left(\cdot; -\frac{1}{2} + i\rho, \varepsilon, \pi - \varphi\right) \\ u_n^*\left(\cdot; -\frac{1}{2} + i\rho, \varepsilon, \pi - \varphi\right)\end{pmatrix},\end{aligned}$$

is unitary and intertwines $\pi^{\rho,\varepsilon}\left(X_\varphi\right)$ with $M(2x\sin\varphi)$, and extends to a unitary equivalence.

From Propostion 4.3 we obtain that

$$\begin{pmatrix} v_{\rho,\varepsilon}(x) \\ v_{\rho,\varepsilon}^*(x)\end{pmatrix} = \sum_{n=-\infty}^{\infty} \begin{pmatrix} u_n\left(x; -\frac{1}{2} + i\rho, \varepsilon, \pi - \varphi\right) \\ u_n^*\left(x; -\frac{1}{2} + i\rho, \varepsilon, \pi - \varphi\right)\end{pmatrix} e_n$$

is a generalized eigenvector of $\pi^{\rho,\varepsilon}(X_\varphi)$ for eigenvalue $2x\sin\varphi$.

Next we consider the action of X_φ in the tensor product. Recall that in the tensor product we need the coproduct Δ, defined by $\Delta(Y) = 1\otimes Y + Y\otimes 1$ for $Y \in \mathfrak{su}(1,1)$. Then from Proposition 4.2 we find the following.

Proposition 4.4. *The operator Υ defined by*

$$\Upsilon : \ell^2(\mathbb{Z}_{\geq 0}) \otimes \ell^2(\mathbb{Z}_{\geq 0}) \to L^2\left(\mathbb{R}^2, w^{(k_1)}(x_1;\varphi)\, w^{(k_2)}(x_2;\varphi)\, dx_1 dx_2\right),$$
$$e_{n_1}\otimes e_{n_2} \mapsto P^{(k_1)}_{n_1}(x_1;\varphi)\, P^{(k_2)}_{n_2}(x_2;\varphi),$$

is unitary and intertwines $\pi^+_{k_1}\otimes\pi^-_{k_2}(\Delta(X_\varphi))$ with $M(2(x_1-x_2)\sin\varphi)$.

In terms of the generalized eigenvectors v^+ and v^-, we find from Proposition 4.4 that

$$v^+(x_1)\otimes v^-(x_2) = \sum_{n_1,n_2=0}^{\infty} P^{(k_1)}_{n_1}(x_1;\varphi)\, P^{(k_2)}_{n_2}(x_2;\varphi)\, e_{n_1}\otimes e_{n_2},$$

is a generalized eigenvector of $\pi^+_{k_1}\otimes\pi^-_{k_2}(\Delta(X_\varphi))$ for eigenvalue $2(x_1-x_2)\sin\varphi$.

To determine the action of Υ on the representation space of the direct integral representation $\int^\oplus \pi^{\rho,\varepsilon} d\rho$, we need to find the operator $\tilde{\Upsilon}$, such that $\Upsilon = \tilde{\Upsilon}\circ J$. Here J is the intertwiner defined in Theorem 4.1. For appropriate functions g_1 and g_2 we define an operator $\tilde{\Upsilon}_g$ by

$$\tilde{\Upsilon}_{\mathbf{g}} : L^2(0,\infty)\otimes\ell^2(\mathbb{Z}) \cong \int_0^{\infty\,\oplus} \ell^2(\mathbb{Z})dx \to \int_0^{\infty\,\oplus} \mathcal{H}\left(-\frac{1}{2}+i\rho,\varepsilon,\pi-\varphi\right) d\rho,$$

$$f\otimes e_n \mapsto \int_0^\infty f(\rho)\begin{pmatrix} g_1(\rho)\\ g_2(\rho)\end{pmatrix}^* \begin{pmatrix} u_n\left(t;-\frac{1}{2}+i\rho,\varepsilon,\pi-\varphi\right)\\ u_n^*\left(t;-\frac{1}{2}+i\rho,\varepsilon,\pi-\varphi\right)\end{pmatrix} d\rho.$$

From Proposition 4.3 we see that $\tilde{\Upsilon}_{\mathbf{g}}$ intertwines $\int^\oplus \pi^{\rho,\varepsilon}(X_\varphi)\,d\rho$ with $M(2t\sin\varphi)$. The functions g_1 and g_2 for which $\Upsilon = \tilde{\Upsilon}_{\mathbf{g}}\circ J$ are the Clebsch-Gordan coefficients for the hyperbolic bases. To determine the Clebsch-Gordan coefficients we use the summation formula in Theorem 3.1. Define the continuous Hahn function by

$$\begin{aligned}\varphi_\rho(x;t) &= \varphi_\rho(x;t,k_1,k_2,\varphi)\\ &= \frac{e^{-x(2\varphi-\pi)}}{|\Gamma(k_1+ix+it)|^2}\\ &\quad\times {}_3F_2\left(\begin{matrix} k_2-ix,\, k_2-k_1+\frac{1}{2}+i\rho,\, k_2-k_1+\frac{1}{2}-i\rho\\ 2k_2,\, k_2-k_1+it+1\end{matrix};1\right).\end{aligned} \tag{4.9}$$

Theorem 4.5. *Let*

$$\begin{pmatrix} g_1(\rho) \\ g_2(\rho) \end{pmatrix}^* = \begin{pmatrix} \varphi_\rho\left(x_2; x_1 - x_2\right) \\ \varphi_\rho^*\left(x_2; x_1 - x_2\right) \end{pmatrix}^* \begin{pmatrix} m(\rho) & 0 \\ 0 & \overline{m(\rho)} \end{pmatrix},$$

where $\varphi_\rho\left(x_2; x_1 - x_2\right) = \varphi_\rho\left(x_2; x_1 - x_2, k_1, k_2, \varphi\right)$ *and* $m(\rho)$ *is given by*

$$\begin{aligned} m(\rho) = (-i)^L & \frac{(2\sin\varphi)^{-2k_1-L} e^{(x_2-x_1)(2\varphi-\pi)}}{\sqrt{2\pi}} \\ & \times \frac{\Gamma\left(\frac{1}{2} + i\left(x_2 - x_1\right) + i\rho\right) \Gamma\left(\frac{1}{2} + i\left(x_2 - x_1\right) - i\rho\right)}{\Gamma\left(k_2 - k_1 + i\left(x_2 - x_1\right) + 1\right)} \\ & \times \sqrt{\frac{\Gamma\left(2k_1\right)}{\Gamma\left(2k_2\right)}} \left| \frac{\Gamma\left(k_1 + k_2 - \frac{1}{2} + i\rho\right) \Gamma\left(k_2 - k_1 + \frac{1}{2} + i\rho\right)}{\Gamma(2i\rho)} \right|, \end{aligned}$$

then we have $\Upsilon = \tilde{\Upsilon}_{\mathbf{g}} \circ J$.

Proof. To show that $\Upsilon = \tilde{\Upsilon}_{\mathbf{g}} \circ J$, we use the summation formula of Theorem 3.1 with the orthogonal polynomials written in orthonormal form. We multiply by a continuous dual Hahn polynomial of degree n_2 with the same parameters as in Theorem 3.1, with $p = n_1 - n_2 \geq 0$. Then integrating against the corresponding orthogonality measure gives an equality with the following structure

$$P_{n_1}^{(k_1)}\left(x_1; \varphi\right) P_{n_2}^{(k_2)}\left(x_2; \varphi\right) = \int_0^\infty S_{n_2}(\rho) \left({}_3F_2\, {}_2F_1 + {}_3F_2\, {}_2F_1\right) d\rho.$$

The ${}_2F_1$-functions are the Meixner-Pollaczek functions as defined by (2.5). From Proposition 4.2 we see that the left hand side is equal to $\Upsilon\left(e_{n_1} \otimes e_{n_2}\right)\left(x_1, x_2\right)$. So from Theorem 4.1, with $n_1 \geq n_2$, and Proposition 4.3 it follows that the right hand side must be equal to

$$\int_0^\infty S_{n_2}(\rho) \mathbf{g}^*(\rho) \Theta^{\rho,\varepsilon}\left(e_{n_1-n_2-L}\right)\left(x_1 - x_2\right) d\mu^{\frac{1}{2}}(\rho),$$

where $\mathbf{g}$ is a vector containing the Clebsch-Gordan coefficients for the hyperbolic bases. This gives the desired result. For $n_1 - n_2 < 0$ the theorem follows after using (3.1). ☐

Remark 4.6. *The explicit expressions of the Clebsch-Gordan coefficients as* ${}_3F_2$*-series can also be found in Mukunda and Radhakrishnan (Mukunda and Radhakrishnan, 1974). The method used in (Mukunda* *and Radhakrishnan, 1974) is completely different from the method used here.*

In terms of the generalized eigenvectors, Theorem 4.5 states that

$$J\left(v^+(x_1)\otimes v^-(x_2)\right)$$
$$=\int_0^\infty \begin{pmatrix}\varphi_\rho(x_2;x_1-x_2)\\ \varphi_\rho^*(x_2;x_1-x_2)\end{pmatrix}^* \begin{pmatrix} m(\rho) & 0\\ 0 & \overline{m(\rho)}\end{pmatrix}\begin{pmatrix} v_{\rho,\varepsilon}(x_1-x_2)\\ v_{\rho,\varepsilon}^*(x_1-x_2)\end{pmatrix} d\rho.$$

4.3 The continuous Hahn integral transform

Since the continuous Hahn functions occur as Clebsch-Gordan coefficients for hyperbolic bases, they should satisfy (generalized) orthogonality relations. We find these relations by formal computations with the generalized eigenvectors.

For an element $f\in\ell^2(\mathbb{Z}_{\geq 0})\otimes\ell^2(\mathbb{Z}_{\geq 0})$ we have the transform pair

$$\begin{cases}(\Upsilon f)(x_1,x_2)=\left\langle f, v^+(x_1)\otimes v^-(x_2)\right\rangle,\\[2ex] f=\displaystyle\iint_{\mathbb{R}^2}(\Upsilon f)(x_1,x_2)\,v^+(x_1)\otimes v^-(x_2)\,w^{(k_1)}(x_1;\varphi)\\ \quad\times w^{(k_2)}(x_2;\varphi)\,dx_1dx_2.\end{cases}\tag{4.10}$$

Similarly for $f\in\ell^2(\mathbb{Z})$ we have the transform pair

$$\begin{cases}(\Theta^{\rho,\varepsilon}f)(x)=\left\langle f,\begin{pmatrix}v_{\rho,\varepsilon}(x)\\ v_{\rho,\varepsilon}^*(x)\end{pmatrix}\right\rangle\in\mathbb{C}^2,\\[2ex] f=\dfrac{1}{2\pi}\displaystyle\int_{\mathbb{R}}(\Theta^{\rho,\varepsilon}f)(x)^*\begin{pmatrix}1 & -w_1(x)\\ -w_1^*(x) & 1\end{pmatrix}\begin{pmatrix}v_{\rho,\varepsilon}(x)\\ v_{\rho,\varepsilon}^*(x)\end{pmatrix}w_0(x)dx.\end{cases}\tag{4.11}$$

Denoting the intertwiner J in Theorem 4.1 by $J=\int_0^{\infty\,\oplus} J_\rho d\rho$, we find from Theorem 4.5

$$(\Upsilon f)(x_1,x_2)=$$
$$\int_0^\infty \begin{pmatrix}\varphi_\rho(x_2;x_1-x_2)\\ \varphi_\rho^*(x_2;x_1-x_2)\end{pmatrix}^* \begin{pmatrix} m(\rho) & 0\\ 0 & \overline{m(\rho)}\end{pmatrix}\left(\Theta^{\rho,\varepsilon}(J_\rho f)\right)(x_1-x_2)\,d\rho.$$

We want to invert this formula. From (4.10) and Theorem 4.5 we find formally

$$J_\rho f = \iint\limits_{\mathbb{R}^2} (\Upsilon f)(x_1, x_2)\, w^{(k_1)}(x_1;\varphi)\, w^{(k_2)}(x_2;\varphi)$$
$$\times \begin{pmatrix} \varphi_\rho(x_2; x_1 - x_2) \\ \varphi_\rho^*(x_2; x_1 - x_2) \end{pmatrix}^* \begin{pmatrix} m(\rho) & 0 \\ 0 & \overline{m(\rho)} \end{pmatrix} \begin{pmatrix} v_{\rho,\varepsilon}(x_1 - x_2) \\ v_{\rho,\varepsilon}^*(x_1 - x_2) \end{pmatrix} dx_1 dx_2.$$

We substitute $x_1 \mapsto x + t$, $x_2 \mapsto x$, then (4.11) with f replaced by $J_\rho f$ gives

$$w_0(t) \begin{pmatrix} 1 & -w_1^*(t) \\ -w_1(t) & 1 \end{pmatrix} \left(\Theta^{\rho,\varepsilon}(J_\rho f)\right)(t) =$$
$$2\pi \int\limits_{\mathbb{R}} (\Upsilon f)(x+t, x) \begin{pmatrix} \overline{m(\rho)} & 0 \\ 0 & m(\rho) \end{pmatrix} \begin{pmatrix} \varphi_\rho(x;t) \\ \varphi_\rho^*(x;t) \end{pmatrix}$$
$$\times w^{(k_1)}(x+t;\varphi) w^{(k_2)}(x;\varphi) dx.$$

We denote

$$\left(\Theta^{\rho,\varepsilon}(J_\rho f)\right)(t) = \frac{2\pi}{w_0(t)} \begin{pmatrix} \overline{m(\rho)} & 0 \\ 0 & m(\rho) \end{pmatrix} \begin{pmatrix} 1 & -w_1^*(t) \\ -w_1(t) & 1 \end{pmatrix}^{-1} \mathbf{g}(\rho)$$

and $(\mathcal{F}\mathbf{g})(x) = (\Upsilon f)(x+t, x)$, then we have the following integral transform pair with the continuous Hahn functions as a kernel

$$\begin{cases} (\mathcal{F}\mathbf{g})(x) = \dfrac{2\pi}{w_0(t)} \displaystyle\int_0^\infty \begin{pmatrix} \varphi_\rho(x;t) \\ \varphi_\rho^*(x;t) \end{pmatrix}^* \begin{pmatrix} 1 & w_1^*(t) \\ w_1(t) & 1 \end{pmatrix} \mathbf{g}(\rho) \dfrac{|m(\rho)|^2}{1 - |w_1(t)|^2} d\rho \\[2ex] \mathbf{g}(\rho) = \displaystyle\int\limits_{\mathbb{R}} (\mathcal{F}\mathbf{g})(x) \begin{pmatrix} \varphi_\rho^*(x;t) \\ \varphi_\rho(x;t) \end{pmatrix} w^{(k_1)}(x+t;\varphi) w^{(k_2)}(x;\varphi) dx. \end{cases}$$

In Section 11.5 we give a rigorous proof for this integral transform pair using spectral analysis of a difference operator for which the continuous Hahn functions are eigenfunctions. In the next subsection we obtain the difference operator from the action of the Casimir element in the tensor product.

4.4 A realization of the discrete series representations.

The following lemma is based on the fact that $\mathfrak{sl}(2,\mathbb{C})$ is semi-simple, so $[\mathfrak{sl}(2,\mathbb{C}), \mathfrak{sl}(2,\mathbb{C})] = \mathfrak{sl}(2,\mathbb{C})$.

Lemma 4.7.

$$B = \frac{1}{4}[H, X_\varphi] + \frac{1}{2}X_\varphi + \frac{1}{2}\cos\varphi\, H, \quad C = \frac{1}{4}[H, X_\varphi] - \frac{1}{2}X_\varphi - \frac{1}{2}\cos\varphi\, H.$$

Proof. This follows from the definition of X_φ and the commutation relations (4.1). □

This lemma shows that to find the action of the generators H, B and C on the Meixner-Pollaczek polynomials, it is enough to find the action of H, since the action of X_φ is known. The action of H follows from the difference equation (2.4) for the Meixner-Pollaczek polynomials.

Proposition 4.8. *The operator Θ^+ intertwines the actions of the generators H, B, C in the positive discrete series, with the following difference operators:*

$$\begin{aligned}\Theta^+\pi_k^+(H) &= \left[M\left(\frac{e^{i\varphi}}{i\sin\varphi}(k-ix)\right)T_i + M\left(2\frac{\cos\varphi}{\sin\varphi}x\right)\right.\\ &\quad \left.+M\left(-\frac{e^{-i\varphi}}{i\sin\varphi}(k+ix)\right)T_{-i}\right]\Theta^+,\\ \Theta^+\pi_k^+(B) &= \left[M\left(\frac{e^{2i\varphi}}{2i\sin\varphi}(k-ix)\right)T_i + M\left(\frac{x}{\sin\varphi}\right)\right.\\ &\quad \left.+M\left(-\frac{e^{-2i\varphi}}{2i\sin\varphi}(k+ix)\right)T_{-i}\right]\Theta^+,\\ \Theta^+\pi_k^+(C) &= \left[M\left(-\frac{1}{2i\sin\varphi}(k-ix)\right)T_i + M\left(-\frac{x}{\sin\varphi}\right)\right.\\ &\quad \left.+M\left(\frac{1}{2i\sin\varphi}(k+ix)\right)T_{-i}\right]\Theta^+,\end{aligned}$$

where T denotes the shift operator: $T_a f(x) = f(x+a)$. For the negative discrete series, Θ^- intertwines the actions of H, B, C, with the following difference operators:

$$\Theta^- \pi_k^-(H) = \left[M\left(-\frac{e^{i\varphi}}{i \sin\varphi}(k - ix) \right) T_i + M\left(-2\frac{\cos\varphi}{\sin\varphi} x \right) \right.$$
$$\left. + M\left(\frac{e^{-i\varphi}}{i \sin\varphi}(k + ix) \right) T_{-i} \right] \Theta^-,$$
$$\Theta^- \pi_k^-(B) = \left[M\left(-\frac{1}{2i \sin\varphi}(k - ix) \right) T_i + M\left(-\frac{x}{\sin\varphi} \right) \right.$$
$$\left. + M\left(\frac{1}{2i \sin\varphi}(k + ix) \right) T_{-i} \right] \Theta^-,$$
$$\Theta^- \pi_k^-(C) = \left[M\left(\frac{e^{2i\varphi}}{2i \sin\varphi}(k - ix) \right) T_i + M\left(\frac{x}{\sin\varphi} \right) \right.$$
$$\left. + M\left(-\frac{e^{-2i\varphi}}{2i \sin\varphi}(k + ix) \right) T_{-i} \right] \Theta^-.$$

Proof. We find the action of H from the difference equation (2.4) for the Meixner-Pollaczek polynomials;

$$\begin{aligned} \Theta^+ \pi_k^+(H)\, e_n &= \Theta^+(2n+2k)\, e_n = (2n+2k) P_n^{(k)}(x;\varphi) \\ &= \frac{e^{i\varphi}}{i \sin\varphi}(k - ix) P_n^{(k)}(x+i;\varphi) + 2x\frac{\cos\varphi}{\sin\varphi} P_n^{(k)}(x;\varphi) \\ &\quad - \frac{e^{-i\varphi}}{i \sin\varphi}(k + ix) P_n^{(k)}(x-i;\varphi). \end{aligned}$$

The action of X_φ is given in Proposition 4.2:

$$\Theta^+ \pi_k^+(X_\varphi)\, e_n = 2x \sin\varphi\, P_n^{(k)}(x;\varphi).$$

Then Lemma 4.7 proves the proposition for the positive discrete series.

We find the action in the negative discrete series in the same way, or we use the Lie-algebra isomorphism ϑ, given by

$$\vartheta(H) = -H, \quad \vartheta(B) = C, \quad \vartheta(C) = B.$$

Then $\pi_k^+(\vartheta(Y)) = \pi_k^-(Y)$ for $Y \in \mathfrak{su}(1,1)$.

A straightforward calculation shows that these operators indeed satisfy the $\mathfrak{su}(1,1)$ commutation relations. □

Remark 4.9. *To simplify notations we denote $\Theta^+ \pi^+ (\Theta^+)^*$ by π^+, and similarly for π^-.*

In the same way as in Proposition 4.8 it can be shown that $\Theta^{\rho,\varepsilon}$ intertwines the actions of H, B and C in the principal unitary series with

2×2 diagonal matrices with difference operators as elements. This is done by finding a difference equation for the Meixner-Pollaczek functions u_n and u_n^*, using contiguous relations for ${}_2F_1$-series. We do not need these realizations here, so we will not work this out.

To express $\Delta(\Omega)$ in terms of H, B and C, the coproduct Δ is extended to $\mathcal{U}(\mathfrak{su}(1,1))$ as an algebra homomorphism. Then from the definition of the Casimir element (4.2) we find

$$\Delta(\Omega) = 1 \otimes \Omega + \Omega \otimes 1 - \frac{1}{2} H \otimes H - (C \otimes B + B \otimes C) . \tag{4.12}$$

Using this expression and Proposition 4.8, we find the following.

Proposition 4.10. *In the realizations of Proposition* 4.8, *we have*

$$\begin{aligned}
&\pi^+_{k_1} \otimes \pi^-_{k_2} (\Delta(\Omega)) \Big|_{x_1 = x+t,\ x_2 = x} \\
&= M\left(-e^{-2i\varphi} (k_1 + i(t+x)) (k_2 + ix)\right) T_{-i} \\
&+ M\left(k_1 (1-k_1) + k_2 (1-k_2) - 2(x+t)x\right) \\
&+ M\left(-e^{2i\varphi} (k_1 - i(t+x)) (k_2 - ix)\right) T_i,
\end{aligned}$$

where the shift operator T acts with respect to x.

Proof. Let $F_1(x)$ and $F_2(x)$ be polynomials in x, and let $f(x,t) = F_1(x+t)F_2(x)$, then a large but straightforward computation yields

$$\begin{aligned}
\pi^+_{k_1} \otimes \pi^-_{k_2} (\Delta(\Omega)) f(x,t) = {} & [k_1 (1-k_1) + k_2 (1-k_2) - 2(x+t)x]\, f(x,t) \\
& - e^{-2i\varphi} (k_1 + i(t+x)) (k_2 + ix)\, f(x-i,t) \\
& - e^{2i\varphi} (k_1 - i(t+x)) (k_2 - ix)\, f(x+i,t).
\end{aligned}$$

□

Remark 4.11. *The action of Ω in the tensor product can also be found from*

$$\begin{aligned}
\pi^+_k(H) &= M(2ix), & \pi^-_k(H) &= M(-2ix), \\
\pi^+_k(B) &= M(k-ix)T_i, & \pi^-_k(B) &= M\left(e^{-2i\varphi}(k+ix)\right) T_{-i}, \\
\pi^+_k(C) &= M(k+ix)T_{-i}, & \pi^-_k(C) &= M\left(e^{2i\varphi}(k-ix)\right) T_i.
\end{aligned}$$

These realizations are equivalent to the realizations given in Proposition 4.8.

In the next section we show that the continuous Hahn functions $\varphi_\rho(x; t, k_1, k_2, \varphi)$ are eigenfunctions of the difference operator of Proposition 4.10, and we work out the corresponding integral transform.

5. The continuous Hahn integral transform

In this section we study a second order difference operator. This difference operator is obtained from the action of the Casimir operator on hyperbolic basis vectors in the tensor product of a positive and a negative discrete series representation of $\mathfrak{su}(1,1)$, see Section 11.4.2. The spectral analysis of this operator, leads to an integral transform pair with a certain type of ${}_3F_2$-series as a kernel. We call these ${}_3F_2$-series continuous Hahn functions, because of their similarity to continuous (dual) Hahn polynomials. The method we use is based on asymptotics, and is essentially the same method as used by Götze (Götze, 1965) and Braaksma and Meulenbeld (Braaksma and Meulenbeld, 1967) for the Jacobi function transform by approximating with the Fourier transform.

5.1 The difference operator Λ and the Wronskian

For k_1, $k_2 > 0$, $t \in \mathbb{R}$ and $0 < \varphi < \pi$ the weight function $w(x)$ is defined by

$$w(x) = \frac{1}{2\pi} e^{(2x+t)(2\varphi-\pi)} \Gamma(k_1+it+ix)\Gamma(k_1-it-ix)\Gamma(k_2+ix)\Gamma(k_2-ix). \tag{5.1}$$

The difference operator Λ is defined by

$$\Lambda : g(x) \mapsto \alpha_+(x)g(x+i) + \beta(x)g(x) + \alpha_-(x)g(x-i),$$

where

$$\begin{aligned} \alpha_\pm(x) &= -e^{\pm 2i\varphi}(k_2 \mp ix)(k_1 \mp i(t+x)), \\ \beta(x) &= k_1(1-k_1) + k_2(1-k_2) - 2(t+x)x. \end{aligned}$$

Initially Λ is defined for those $g(x) \in L^2(\mathbb{R}, w(x)dx)$ that have an analytic continuation to the strip

$$\mathcal{S}_\varepsilon = \{z \in \mathbb{C} : |\Im(z)| < 1+\varepsilon,\ \varepsilon > 0\}.$$

The difference operator Λ corresponds to the action of the Casimir operator in the tensor product of a positive and a negative discrete series representation on hyperbolic basis vectors, see Proposition 4.10.

Remark 5.1. *Observe that the functions*

$$e^{-x(2\varphi-\pi)} p_n(x; k_1+it, k_2, k_1-it, k_2),$$

where p_n is a continuous Hahn polynomial in the notation of (Koekoek and Swarttouw, 1998), form an orthogonal basis for $L^2(\mathbb{R}, w(x)dx)$, but they are not eigenfunctions of Λ. Also, the functions

$$e^{-2\varphi x} p_n\left(x; k_1 + it, k_2, k_1 - it, k_2\right)$$

are eigenfunctions of Λ, but they are not elements of $L^2(\mathbb{R}, w(x)dx)$.

We define

$$\langle f, g \rangle_{M,N} = \int_{-M}^{N} f(x) g^*(x) w(x) dx.$$

Then $\lim_{N,M\to\infty} \langle f, g\rangle_{M,N}$ is the inner product on $L^2(\mathbb{R}, w(x)dx)$, so $\langle f, g\rangle_{M,N}$ is a truncated inner product.

Definition 5.2. *For functions f and g analytic in $\mathcal{S}_\varepsilon$, the Wronskian $[f, g]$ is defined by*

$$[f,g](y) = \int_{y}^{y+i} \left\{ f(x)g^*(x-i) - f(x-i)g^*(x) \right\} \alpha_-(x) w(x) dx.$$

Proposition 5.3. *Let f and g be analytic in $\mathcal{S}_\varepsilon$, then*

$$\langle \Lambda f, g\rangle_{N,M} - \langle f, \Lambda g\rangle_{N,M} = [f,g](N) - [f,g](-M).$$

Proof. Observe that $\alpha_+^* = \alpha_-$, $\alpha_\pm(x \mp i)w(x \mp i) = \alpha_\pm^*(x)w(x)$ and $\beta^*(x) = \beta(x)$. Furthermore for $x \in \mathbb{R}$ we have $g^*(x \pm i) = \overline{g(x \mp i)}$. This gives for the Wronskian

$$
\begin{aligned}
[f,g](y) &= \int_{y}^{y+i} \{f(x)g^*(x-i) - f(x-i)g^*(x)\}\, \alpha_-(x)w(x)dx \\
&= \int_{y}^{y+i} f(x)g^*(x-i)\alpha_+^*(x)w(x)dx \\
&- \int_{y}^{y+i} f(x-i)g^*(x)\alpha_-(x)w(x)dx \\
&= \int_{y}^{y+i} f(x)g^*(x-i)\alpha_+(x-i)w(x-i)dx \\
&- \int_{y-i}^{y} f(x)g^*(x+i)\alpha_-(x+i)w(x+i)dx.
\end{aligned}
$$

For f and g analytic in $\mathcal{S}_\varepsilon$, we find

$$
\int_{-M}^{N} (\Lambda f)(x)g^*(x)w(x)dx
$$

$$
= \int_{-M}^{N} [\alpha_+(x)f(x+i) + \beta(x)f(x) + \alpha_-(x)f(x-i)]\, g^*(x)w(x)dx
$$

$$
= \int_{-M+i}^{N+i} \alpha_+(x-i)f(x)g^*(x-i)w(x-i)dx + \int_{-M}^{N} \beta(x)f(x)g^*(x)w(x)dx
$$

$$
+ \int_{-M-i}^{N-i} \alpha_-(x+i)f(x)g^*(x+i)w(x+i)dx.
$$

We choose a different path of integration

$$
\int_{-M-i}^{N-i} = \int_{-M-i}^{-M} + \int_{-M}^{N} + \int_{N}^{N-i}
$$

and similarly for $\int_{-M+i}^{N+i}$. Then we find

$$\int_{-M}^{N} (\Lambda f)(x)g^*(x)w(x)dx =$$

$$\int_{-M-i}^{-M} \alpha_-(x+i)f(x)g^*(x+i)w(x+i)dx$$

$$-\int_{N-i}^{N} \alpha_-(x+i)f(x)g^*(x+i)w(x+i)dx$$

$$+\int_{N}^{N+i} \alpha_+(x-i)f(x)g^*(x-i)w(x-i)dx$$

$$-\int_{-M}^{-M+i} \alpha_+(x-i)f(x)g^*(x-i)w(x-i)dx$$

$$+\int_{-M}^{N} f(x)\left[\alpha_-(x+i)g^*(x+i)w(x+i) + \beta(x)g^*(x)w(x)\right.$$

$$\left. + \alpha_+(x-i)g^*(x-i)w(x-i)\right] dx.$$

The last integral $\int_{-M}^{N}$ equals

$$\int_{-M}^{N} f(x)\left[\alpha_-^*(x)g^*(x+i) + \beta^*(x)g^*(x) + \alpha_+^*(x)g^*(x-i)\right] w(x)dx$$

$$= \int_{-M}^{N} f(x)(\Lambda g)^*(x)w(x)dx,$$

and in the other four integrals we recognize $[f,g](N) - [f,g](-M)$. □

Our first goal is to show that Λ is a symmetric operator on a domain which will be specified, so we are interested in the limit of the Wronskian $[f,g](y)$ for $y \to \pm\infty$. The following lemma is useful in determining these limits.

Lemma 5.4. *Let $k_1, k_2 > 1$, $x \in \mathbb{R}$, and $-1 \le y \le 1$, then the weight function $w(x+iy)$ has the following asymptotic behaviour*

$$w(x+iy) =$$

$$\begin{cases} 2\pi e^{2t(\varphi-\pi)+4iy(\varphi-\pi)} x^{2k_1+2k_2-2} e^{-4(\pi-\varphi)x} \left(1+\mathcal{O}\left(\frac{1}{x}\right)\right), & x \to \infty, \\ 2\pi e^{2\varphi t+4i\varphi y} |x|^{2k_1+2k_2-2} e^{4\varphi x} \left(1+\mathcal{O}\left(\frac{1}{x}\right)\right), & x \to -\infty. \end{cases}$$

Proof. From Stirling's asymptotic formula (Olver, 1974, (8.16)) we find for $u > 0$ and $v \in \mathbb{R}$

$$\Gamma(u+iv) = \sqrt{2\pi}\, |v|^{u+iv-\frac{1}{2}} e^{-(u+iv)}\, e^{(iu-v)\arctan(v/u)} + \mathcal{O}\left(\frac{1}{v}\right), \qquad v \to \pm\infty.$$

We use

$$\arctan\frac{v}{u} + \arctan\frac{u}{v} = \begin{cases} \frac{\pi}{2}, & v > 0, \\ -\frac{\pi}{2}, & v < 0, \end{cases}$$

to find for $v \to \pm\infty$

$$(iu - v)\arctan\frac{v}{u} = (iu - v)\left(\pm\frac{\pi}{2} - \frac{u}{v}\right)\left(1+\mathcal{O}\left(\frac{1}{v}\right)\right).$$

So we have, for $v \to \pm\infty$,

$$\Gamma(u+iv) = \sqrt{2\pi}\, |v|^{u+iv-\frac{1}{2}} e^{-\pi|v|/2 - iv \pm \pi i u/2}\left(1 + +\mathcal{O}\left(\frac{1}{v}\right)\right).$$

Applying this formula to the four Γ-functions in (5.1) gives the asymptotic behaviour of the weight function $w(x+iy)$. □

In general, if for some $\varepsilon > 0$

$$\int_{\mathbb{R}} e^{\varepsilon|x|} d\mu(x) < \infty,$$

then the moment problem for the measure $d\mu$ is determinate, see, e.g., (de Jeu, 2003) and references therein. Using this criterion with $0 < \varepsilon < \min\{4\varphi, 4(\pi - \varphi)\}$, we find from Lemma 5.4 that the moment problem for the measure $w(x)dx$ is determinate. In particular this shows that the polynomials are dense in $L^2(\mathbb{R}, w(x)dx)$.

Let $\mathcal{D}$ be the space of polynomials on $\mathbb{R}$, then $\mathcal{D}$ is a dense subspace of $L^2(\mathbb{R}, w(x)dx)$. Since

$$\alpha_-(x+iy) = e^{-2i\varphi} x^2 \left(1+\mathcal{O}\left(\frac{1}{x}\right)\right), \qquad x \to \pm\infty, \tag{5.2}$$

it follows from Definition 5.2 and Lemma 5.4 that $\lim_{N\to\pm\infty}[f,g](N)=0$ for f, g polynomials. Hence by Proposition 5.3 we find

Proposition 5.5. *The operator $(\Lambda,\mathcal{D})$ is a densely defined symmetric operator on $L^2(\mathbb{R},w(x)dx)$.*

Remark 5.6. *The operator $(\Lambda,\mathcal{D})$ is also densely defined and symmetric on the space spanned by $e^{-x(2\varphi-\pi)}p_n(x)$, where p_n is a polynomial, cf. Remark* 5.1.

5.2 Eigenfunctions of Λ

We determine eigenfunctions of Λ, using contiguous relations for ${}_3F_2$-functions. First note that for a monic polynomial of degree n, $p_n(x)=x^n+\cdots$, we have

$$(\Lambda p_n)(x)=[\alpha_+(x)+\beta(x)+\alpha_-(x)]\,x^n+\text{lower order terms.}$$

Since $\alpha_+(x)+\beta(x)+\alpha_-(x)$ is a polynomial of degree 2, Λ raises the degree of a polynomial by 2. Therefore Λ cannot have polynomial eigenfunctions.

Let $p(x)$ be the i-periodic function

$$p(x)=\frac{1}{\pi}e^{\pi x}\sin\left(\pi\left(k_1-it-ix\right)\right), \tag{5.3}$$

and let $\varphi_\rho(x)=\varphi_\rho\left(x;t,k_1,k_2,\varphi\right)$ and $\Phi_\rho(x)=\Phi_\rho\left(x;t,k_1,k_2,\varphi\right)$ denote the functions

$$\begin{aligned}\varphi_\rho(x)&=e^{-2\varphi x}p(x)\frac{\Gamma\left(1-k_1+it+ix\right)}{\Gamma\left(k_1+ix+it\right)}\\&\quad\times{}_3F_2\left(\begin{matrix}k_2-ix,k_2-k_1+\frac12+i\rho,k_2-k_1+\frac12-i\rho\\2k_2,k_2-k_1+it+1\end{matrix};1\right),\end{aligned} \tag{5.4}$$

$$\begin{aligned}\Phi_\rho(x)&=e^{-2\varphi x}p(x)\frac{\Gamma\left(1-k_1+it+ix\right)\Gamma\left(1-k_2+ix\right)}{\Gamma\left(k_1+ix+it\right)\Gamma\left(\frac32-k_1+ix+i\rho\right)}\\&\quad\times{}_3F_2\left(\begin{matrix}k_2-k_1+\frac12+i\rho,\frac32-k_1-k_2+i\rho,\frac12-it+i\rho\\1+2i\rho,\frac32-k_1+ix+i\rho\end{matrix};1\right).\end{aligned} \tag{5.5}$$

Both ${}_3F_2$-series are absolutely convergent for $\Re\left(k_1+it+ix\right)>0$. Note that the expression for $\varphi_\rho(x)$ is the same as (4.9) after applying Euler's reflection formula.

Proposition 5.7. *For $k_1>1$, the functions $\varphi_\rho(x)$ and $\Phi_\rho(x)$ are eigenfunctions of Λ for eigenvalue $\left(\rho^2+\frac14\right)$.*

Remark 5.8.

(i) *Observe that due to the i-periodic function* $p(x)$, $\varphi_\rho(x)$ *is an entire function. Denote* $\tilde{\varphi}_\rho(x) = \varphi_\rho(x)/p(x)$, *then* $p(x)$ *cancels the poles of* $\tilde{\varphi}_\rho(x)$. *There are more choices of i-periodic functions that cancel the poles of* $\tilde{\varphi}_\rho(x)$. *One of the reasons for this particular choice, is that it appears in the Lie-algebraic interpretation of the function* $\varphi_\rho(x)$, *see Theorem* 4.5. *We come back to the choice of the i-periodic function in Remark* 5.15.

(ii) *Observe that* $\varphi^*_\rho(x)$ *is obtained from* $\varphi_\rho(x)$ *by the substitutions* $(x, t, \varphi) \mapsto (-x, -t, \pi - \varphi)$. *Since the difference operator* Λ *and the weight function* $w(x)$ *are invariant under these substitutions, it follows from Proposition* 5.7 *that, for* $k_1 > 1$, $\varphi^*_\rho(x)$ *is also an eigenfunction of* Λ *for eigenvalue* $\left(\rho^2 + \frac{1}{4}\right)$. *A similar argument shows that* $\Phi^*_\rho(x)$ *is an eigenfunction of* Λ *for eigenvalue* $\left(\rho^2 + \frac{1}{4}\right)$. *Obviously* $\Phi_{-\rho}(x)$ *and* $\Phi^*_{-\rho}(x)$ *are also eigenfunctions of* Λ *for eigenvalue* $\left(\rho^2 + \frac{1}{4}\right)$.

Proof. Combining the contiguous relations (Andrews et al., 1999, (3.7.9), (3.7.10), (3.7.13)), gives

$$
\begin{aligned}
(d-a)(e-a)\left[F(a-) - F\right] + a(a+b+c-d-e+1)\left[F(a+) - F\right] \\
= -bcF,
\end{aligned}
$$

$$
F = {}_3F_2\left(\begin{matrix} a, b, c \\ d, e \end{matrix}; 1\right), \quad F(a\pm) = {}_3F_2\left(\begin{matrix} a \pm 1, b, c \\ d, e \end{matrix}; 1\right).
$$

From this relation we find that

$$
\begin{aligned}
& e^{-2\varphi x} \frac{\Gamma(1 - k_1 + it + ix)}{\Gamma(k_1 + ix + it)} \\
& \times {}_3F_2\left(\begin{matrix} k_2 - ix, k_2 - k_1 + \frac{1}{2} + i\rho, k_2 - k_1 + \frac{1}{2} - i\rho \\ 2k_2, k_2 - k_1 + it + 1 \end{matrix}; 1\right)
\end{aligned}
$$

is an eigenfunction of Λ for eigenvalue $\rho^2 + \frac{1}{4}$. And then, since $p(x)$ is i-periodic, $\varphi_\rho(x)$ is also an eigenfunction for eigenvalue $\rho^2 + \frac{1}{4}$. Since $\varphi_\rho(x)$ must be analytic in $\mathcal{S}_\varepsilon$, the condition $k_1 > 1$ is needed for absolute convergence of the ${}_3F_2$-series at the point $x + i$.

Denote

$$
F_\pm(a\mp) = {}_3F_2\left(\begin{matrix} a, b \pm 1, c \pm 1 \\ d \pm 1, e \pm 1 \end{matrix}; 1\right).
$$

From contiguous relations (Andrews et al., 1999, (3.7.9), (3.7.10), (3.7.13)) we find

$$[(d-1)(e-1)F_-(a+) - (b-1)(c-1)F] - (a-1)(d+e-a-b-c)F \\ + \left[\frac{bc(d-a)(e-a)}{de}F_+(a-) - (d-a)(e-a)F\right] = 0.$$

From this we see that

$$\Phi_\rho(x) = \frac{\Gamma(1-k_2+ix)\Gamma(1+2i\rho)\Gamma(1-k_1+ix+it)}{\Gamma\left(\frac{1}{2}+it+i\rho\right)\Gamma\left(k_1+\frac{1}{2}+i\rho+ix\right)\Gamma\left(\frac{3}{2}-k_1+ix+i\rho\right)} \\ \times e^{-2\varphi x}p(x)\,{}_3F_2\left(\begin{matrix}\frac{1}{2}-it+i\rho, 1-k_2+ix, k_2+ix \\ k_1+\frac{1}{2}+ix+i\rho, \frac{3}{2}-k_1+i\rho+ix\end{matrix};1\right), \quad (5.6)$$

is another eigenfunction of Λ with eigenvalue $\rho^2+\frac{1}{4}$. The ${}_3F_2$ series converges absolutely if $\Re\left(\frac{1}{2}+i\rho+it\right)>0$, so absolute convergence does not depend on k_1. Using (Andrews et al., 1999, Cor. 3.3.5) this can be written as (5.5), and for this expression the condition $k_1>1$ is needed. □

The function $\varphi_\rho(x)$ can be expanded in terms of $\Phi_\rho(x)$ and $\Phi_{-\rho}(x)$.

Proposition 5.9.

$$\varphi_\rho(x) = c(\rho)\Phi_\rho(x) + c(-\rho)\Phi_{-\rho}(x),$$

where

$$c(\rho) = \frac{\Gamma(2k_2)\Gamma(k_2-k_1+it+1)\Gamma(-2i\rho)}{\Gamma\left(k_1+k_2-\frac{1}{2}-i\rho\right)\Gamma\left(\frac{1}{2}+it-i\rho\right)\Gamma\left(k_2-k_1+\frac{1}{2}-i\rho\right)}.$$

Proof. This follows from (Bailey, 1972, p. 15(2))

$${}_3F_2\left(\begin{matrix}a,b,c \\ d,e\end{matrix};1\right) = \frac{\Gamma(1-a)\Gamma(d)\Gamma(e)\Gamma(c-b)}{\Gamma(d-b)\Gamma(e-b)\Gamma(1+b-a)\Gamma(c)} \\ \times {}_3F_2\left(\begin{matrix}b, b-d+1, b-e+1 \\ 1+b-c, 1+b-a\end{matrix};1\right) \\ + \mathrm{idem}(b;c).$$

Here idem$(b;c)$ after an expression means that the expression is repeated with b and c interchanged. □

In the next subsection we consider the Wronskians $[\varphi_\rho,\varphi_\sigma](y)$ and $[\varphi_\rho^*,\varphi_\sigma](y)$ for $y\to\pm\infty$, so we need the asymptotic behaviour of φ_ρ. We find this from Proposition 5.9 and the asymptotic behaviour of Φ_ρ.

Lemma 5.10. *Let $k_1 > 1$, $\rho \in \mathbb{C}$ and $-1 \le y \le 1$. For $x \to \pm\infty$*

$$\Phi_\rho(x+iy) = p(x+iy)e^{-2\varphi(x+iy)}(ix)^{\frac{1}{2}-k_1-k_2-i\rho} \times \left\{1 + \frac{1}{2ix}\left(A(\rho) + yB(\rho)\right) + \mathcal{O}\left(\frac{1}{x^2}\right)\right\},$$

$$\Phi^*_\rho(x+iy) = p^*(x+iy)e^{-2\varphi(x+iy)}(-ix)^{\frac{1}{2}-k_1-k_2+i\overline{\rho}} \times \left\{1 + \frac{1}{-2ix}\left(\overline{A(\rho)} - y\overline{B(\rho)}\right) + \mathcal{O}\left(\frac{1}{x^2}\right)\right\},$$

where

$$A(\rho) = 2it\,(1-2k_1) + \left(k_1 - k_2 - \frac{1}{2} - i\rho\right)\left(\frac{3}{2} - k_1 - k_2 + i\rho\right) + \frac{\left(k_2 - k_1 + \frac{1}{2} + i\rho\right)\left(\frac{3}{2} - k_1 - k_2 + i\rho\right)\left(\frac{1}{2} - it + i\rho\right)}{\frac{1}{2} + i\rho},$$

$$B(\rho) = 2k_1 + 2k_2 - 1 + 2i\rho.$$

Proof. This follows from (5.5) and the asymptotic formula for the ratio of two Γ-functions (Olver, 1974, §4.5)

$$\frac{\Gamma(a+z)}{\Gamma(b+z)} = z^{a-b}\left(1 + \frac{1}{2z}(a-b)(a+b-1) + \mathcal{O}\left(\frac{1}{z^2}\right)\right), \qquad |z| \to \infty, \quad |\arg(z)| < \pi. \tag{5.7}$$

The first part of the expression for $A(\rho)$ is obtained from

$$(a_1 - b_1)(a_1 + b_1 - 1) + (a_2 - b_2)(a_2 + b_2 - 1),$$

the second part comes from the second term in the hypergeometric series. □

The asymptotic behaviour of the i-periodic function $p(x)$ is also needed;

$$p(x+iy) = \frac{e^{i\pi(k_1+2y)}}{2\pi i}e^{\pi(2x+t)} - \frac{e^{-i\pi k_1}}{2\pi i}e^{-\pi t} = \begin{cases} \dfrac{e^{i\pi(k_1+2y)}}{2\pi i}e^{\pi(2x+t)} + \mathcal{O}(1), & x \to \infty, \\[2ex] -\dfrac{e^{-i\pi k_1}}{2\pi i}e^{-\pi t} + \mathcal{O}\left(e^{2\pi x}\right), & x \to -\infty, \end{cases}$$

$$p^*(x+iy) = \frac{e^{i\pi k_1}}{2\pi i} e^{-\pi t} - \frac{e^{-i\pi(k_1-2y)}}{2\pi i} e^{\pi(2x+t)}$$
$$= \begin{cases} -\dfrac{e^{-i\pi(k_1-2y)}}{2\pi i} e^{\pi(2x+t)} + \mathcal{O}(1), & x \to \infty, \\ \dfrac{e^{i\pi k_1}}{2\pi i} e^{-\pi t} + \mathcal{O}\left(e^{2\pi x}\right), & x \to -\infty. \end{cases}$$

5.3 Continuous spectrum

We determine the spectrum of the difference operator Λ. In this subsection we consider the case where the spectrum only consists of a continuous part.

Since

$$\left|(ix)^{\frac{1}{2}-k_1-k_2-i\rho}\right|^2 = \begin{cases} |x|^{1-2k_1-2k_2+2\Im(\rho)} e^{\pi\Re(\rho)}, & x>0, \\ |x|^{1-2k_1-2k_2+2\Im(\rho)} e^{-\pi\Re(\rho)}, & x<0, \end{cases}$$

we find from Proposition 5.10 that

$$|\Phi_\rho(x)|^2 w(x) = \mathcal{O}\left(|x|^{2\Im(\rho)-1}\right), \qquad x \to \pm\infty. \tag{5.8}$$

So $\Phi_\rho(x)$ is an element of $L^2(\mathbb{R}, w(x)dx)$ for $\Im(\rho) < 0$. This shows that it is possible to give eigenfunctions of Λ for complex eigenvalues. We only consider eigenfunctions which are even in ρ, and in that case all eigenvalues of Λ are real.

First we consider the continuous spectrum of Λ. We show that $\left[\frac{1}{4}, \infty\right)$ is contained in the continuous spectrum. Assume that ρ is real and that the c-function in Proposition 5.9 does not have zeros, or, equivalently, assume that $k_1 + k_2 \geq \frac{1}{2}$ and $k_2 - k_1 > -\frac{1}{2}$. Since we only consider even functions in ρ, we may assume $\rho \geq 0$. We use Proposition 5.3 to calculate the truncated inner product of two eigenfunctions. This gives for $k_1 > 1$ and $\rho \neq \sigma$

$$\langle \varphi_\rho, \varphi_\sigma \rangle_{M,N} = \frac{[\varphi_\rho, \varphi_\sigma](N) - [\varphi_\rho, \varphi_\sigma](-M)}{\rho^2 - \sigma^2}. \tag{5.9}$$

Multiplying both sides with an arbitrary function $f(\rho)$ and integrating over ρ from 0 to ∞, gives

$$\int_0^\infty f(\rho) \langle \varphi_\rho, \varphi_\sigma \rangle_{M,N}\, d\rho = \int_0^\infty f(\rho) \frac{[\varphi_\rho, \varphi_\sigma](N) - [\varphi_\rho, \varphi_\sigma](-M)}{\rho^2 - \sigma^2} d\rho. \tag{5.10}$$

The function $f(\rho)$ must satisfy some conditions, which we shall determine later on. We take limits $N, M \to \infty$ on both sides. To determine the limits of the Wronskians, the following lemma is used.

Lemma 5.11. *Let $k_1, k_2 > 1$ and $\rho, \sigma \geq 0$. For $x \to \pm\infty$*

$$[\Phi_\rho, \Phi_\sigma](x) = \pm D^\pm(\rho,\sigma)(\rho+\sigma)|x|^{i(\sigma-\rho)}\left(1+\mathcal{O}\left(\frac{1}{x}\right)\right),$$

where

$$D^\pm(\rho,\sigma) = \frac{i}{2\pi}\, e^{t(2\varphi-\pi)} e^{\pm\frac{1}{2}\pi(\rho+\sigma+2t)}.$$

Proof. From Lemma 5.10 we find for $0 \leq y \leq 1$ and $x \to \infty$

$$\begin{aligned}
&\Phi_\rho(x+iy)\Phi_\sigma^*(x+iy-i) - \Phi_\rho(x+iy-i)\Phi_\sigma^*(x+iy)\\
&= e^{2i\varphi-4i\varphi y+4i\pi y-4\varphi x}|x|^{1-2k_1-2k_2}|p(x)|^2(ix)^{-i\rho}(-ix)^{i\sigma}\\
&\quad\times\frac{B(\rho)-\overline{B(\sigma)}}{2ix}\left(1+\mathcal{O}\left(\frac{1}{x}\right)\right)\\
&= e^{2i\varphi-4i\varphi y+4i\pi y-4\varphi x}|x|^{1-2k_1-2k_2}|p(x)|^2 e^{\pm\frac{1}{2}\pi(\rho+\sigma)}|x|^{i(\sigma-\rho)}\\
&\quad\times\frac{\rho+\sigma}{x}\left(1+\mathcal{O}\left(\frac{1}{x}\right)\right).
\end{aligned}$$

Using the asymptotic behaviour of $p(x)$ and Lemma 5.4, we find the asymptotic behaviour of the integrand of the Wronskian. Note that this is independent of y. In a similar way we find the same asymptotic behaviour of the integrand for $x \to -\infty$. Now the lemma follows from writing the Wronskian as

$$\int_x^{x+i} f(z)dz = i\int_0^1 f(x+is)ds,$$

and applying dominated convergence. □

Proposition 5.12. *Let $\sigma \geq 0$, and let f be a continuous function satisfying*

$$f(\rho) = \begin{cases} \mathcal{O}\left(e^{-\pi\rho}\rho^{2k_2-\frac{1}{2}-\varepsilon}\right), & \rho\to\infty, \quad \varepsilon>0,\\ \mathcal{O}\left(\rho^\delta\right), & \rho\to 0, \quad \delta>0,\end{cases}$$

then

$$\lim_{N\to\infty}\int_0^\infty f(\rho)\left\langle\varphi_\rho,\varphi_\sigma\right\rangle_{N,N} d\rho = W_0^{-1}(\sigma)f(\sigma),$$

where

$$W_0(\sigma) = \frac{1}{2\pi} e^{-t(2\varphi-\pi)} \, |\Gamma(\frac{1}{2} + it\sigma)\Gamma(\frac{1}{2} - it + i\sigma)|^2$$
$$\times \left| \frac{\Gamma(k_2 - k_1 + \frac{1}{2} + i\sigma)\Gamma(k_1 + k_2 - \frac{1}{2} + i\sigma)}{\Gamma(2k_2)\Gamma(k_2 - k_1 + it + 1)\Gamma(2i\sigma)} \right|^2.$$

Proof. Let $k_1, k_2 > 1$. We use (5.10) to calculate

$$\lim_{N\to\infty} \int_0^\infty f(\rho) \, \langle \varphi_\rho, \varphi_\sigma \rangle_{N,N} \; d\rho.$$

From the c-function expansion, see Proposition 5.9, we obtain

$$[\varphi_\rho, \varphi_\sigma](x) = \sum_{\epsilon,\xi\in\{-1,1\}} c(\epsilon\rho)\overline{c(\xi\sigma)} \, [\Phi_{\epsilon\rho}, \Phi_{\xi\sigma}](x).$$

Then (5.9) and Lemma 5.11 give, for $N \to \infty$,

$$\langle \varphi_\rho, \varphi_\sigma \rangle_{N,N} =$$
$$\sum_{\epsilon,\xi\in\{-1,1\}} \frac{N^{i(\xi\sigma-\epsilon\rho)}}{\epsilon\rho - \xi\sigma} c(\epsilon\rho)\overline{c(\xi\sigma)} \left[D^+(\epsilon\rho, \xi\sigma) + D^-(\epsilon\rho, \xi\sigma)\right] \left(1 + \mathcal{O}\left(\frac{1}{N}\right)\right).$$

From (5.10) we find

$$\lim_{N\to\infty} \int_0^\infty f(\rho) \, \langle \varphi_\rho, \varphi_\sigma \rangle_{N,N} \; d\rho$$
$$= \lim_{N\to\infty} \int_0^\infty f(\rho) \Big\{ \psi_1(\rho) \cos([\rho + \sigma] \ln N) + \psi_2(\rho) \sin([\rho + \sigma] \ln N)$$
$$+ \psi_3(\rho) \cos([\rho - \sigma] \ln N) + \psi_4(\rho) \frac{\sin([\rho - \sigma] \ln N)}{\rho - \sigma} \Big\} \, d\rho,$$

where

$$\psi_1(\rho) = \frac{1}{\rho + \sigma} \Big[c(\rho)\overline{c(-\sigma)} \left(D^+(\rho, -\sigma) + D^-(\rho, -\sigma) \right)$$
$$- c(-\rho)\overline{c(\sigma)} \left(D^+(-\rho, \sigma) + D^-(-\rho, \sigma) \right) \Big],$$

$$\psi_2(\rho) = \frac{-i}{\rho + \sigma} \Big[c(\rho)\overline{c(-\sigma)} \left(D^+(\rho, -\sigma) + D^-(\rho, -\sigma) \right)$$
$$+ c(-\rho)\overline{c(\sigma)} \left(D^+(-\rho, \sigma) + D^-(-\rho, \sigma) \right) \Big],$$

$$\begin{aligned}\psi_3(\rho) = \frac{1}{\rho-\sigma}\Big[&c(\rho)\overline{c(\sigma)}\left(D^+(\rho,\sigma)+D^-(\rho,\sigma)\right)\\ &-c(-\rho)\overline{c(-\sigma)}\left(D^+(-\rho,-\sigma)+D^-(-\rho,-\sigma)\right)\Big],\\ \psi_4(\rho) = -i\Big[&c(\rho)\overline{c(\sigma)}\left(D^+(\rho,\sigma)+D^-(\rho,\sigma)\right)\\ &+c(-\rho)\overline{c(-\sigma)}\left(D^+(-\rho,-\sigma)+D^-(-\rho,-\sigma)\right)\Big].\end{aligned}$$

Writing out explicitly the terms between square brackets for ψ_3 gives, for $\rho=\sigma$,

$$\begin{aligned}&e^{t(2\varphi-\pi)}\left|\frac{\Gamma(2k_2)\,\Gamma(k_2-k_1+it+1)\,\Gamma(2i\rho)}{\Gamma\left(k_1+k_2-\frac{1}{2}+i\rho\right)\Gamma\left(k_2-k_1+\frac{1}{2}+i\rho\right)}\right|^2\\ &\times\left(\frac{\sin\left(\pi(\frac{1}{2}+it+i\rho)\right)}{\left|\Gamma\left(\frac{1}{2}-it+i\rho\right)\right|^2}-\frac{\sin\left(\pi(\frac{1}{2}-it+i\rho)\right)}{\left|\Gamma\left(\frac{1}{2}+it+i\rho\right)\right|^2}\right).\end{aligned}$$

From Euler's reflection formula it follows that this is equal to zero. So ψ_3 has a removable singularity at the point $\rho=\sigma$.

From the Riemann-Lebesgue lemma, see, e.g., (Whittaker and Watson, 1963, §9.41), it follows that for $f\psi_i \in L^1(0,\infty)$, $i=1,2,3$, the terms with ψ_i, $i=1,2,3$, vanish. This leaves us with a Dirichlet integral, for which we have the property (see, e.g., (Whittaker and Watson, 1963, §9.7))

$$\lim_{t\to\infty}\frac{1}{\pi}\int_0^\infty g(x)\frac{\sin[t(x-y)]}{x-y}\,dx = g(y), \tag{5.11}$$

for a continuous function $g \in L^1(0,\infty)$. This gives for a continuous function f that satisfies $f\psi_4^{\pm} \in L^1(0,\infty)$,

$$\begin{aligned}&\lim_{N\to\infty}\int_0^\infty f(\rho)\,\langle\varphi_\rho,\varphi_\sigma\rangle_{N,N}\,d\rho = \pi f(\sigma)\psi_4(\sigma)\\ &=\frac{1}{2}e^{t(2\varphi-\pi)}\left\{\left(e^{-\pi(\sigma+t)}+e^{\pi(\sigma+t)}\right)c(\sigma)\overline{c(\sigma)}\right.\\ &\quad\left.+\left(e^{-\pi(\sigma-t)}+e^{\pi(\sigma-t)}\right)c(-\sigma)\overline{c(-\sigma)}\right\}f(\sigma)\\ &=e^{t(2\varphi-\pi)}\left|\frac{\Gamma(2k_2)\,\Gamma(k_2-k_1+it+1)\,\Gamma(2i\sigma)}{\Gamma\left(k_2-k_1+\frac{1}{2}+i\sigma\right)\Gamma\left(k_1+k_2-\frac{1}{2}+i\sigma\right)}\right|^2\\ &\quad\times\left(\frac{\sin\left(\pi\left(\frac{1}{2}+it+i\sigma\right)\right)}{\left|\Gamma\left(\frac{1}{2}-it+i\sigma\right)\right|^2}+\frac{\sin\left(\pi\left(\frac{1}{2}-it+i\sigma\right)\right)}{\left|\Gamma\left(\frac{1}{2}+it+i\sigma\right)\right|^2}\right)f(\sigma)\\ &=W_0^{-1}(\sigma)f(\sigma).\end{aligned}$$

In the last step Euler's reflection formula is used. The conditions $k_1, k_2 > 1$ can be removed by analytic continuation. Since

$$W_0^{-1}(\sigma) = \begin{cases} \mathcal{O}\left(\sigma^{1-4k_2}e^{2\pi\sigma}\right), & \sigma \to \infty, \\ \mathcal{O}\left(\sigma^{-1}\right), & \sigma \to 0, \end{cases}$$

we find that the proposition is valid for the conditions of f as stated. □

Next we consider the truncated inner product $\langle \varphi_\rho^*, \varphi_\sigma \rangle_{M,N}$ and the corresponding Wronskians. We need to find the analogues of Lemma 5.11 and Proposition 5.12.

Lemma 5.13. *Let $k_1 > 1$ and $\rho, \sigma \geq 0$. For $x \to \pm\infty$*

$$\left[\Phi_\rho^*, \Phi_\sigma\right](x) = E^\pm(\rho,\sigma)(\rho - \sigma)|x|^{i(\rho+\sigma)}\left(1 + \mathcal{O}\left(\frac{1}{x}\right)\right),$$

where

$$E^\pm(\rho,\sigma) = \frac{1}{2\pi} e^{t(2\varphi-\pi)} e^{\mp\frac{1}{2}\pi i(2k_1 - 2k_2 + i\rho + i\sigma + 2it)}.$$

Proof. The proof is similar to the proof of Lemma 5.11. □

Proposition 5.14. *Let $\sigma \geq 0$, and let f be an even continuous function satisfying*

$$f(\rho) = \begin{cases} \mathcal{O}\left(e^{-\pi\rho}\rho^{2k_2 - \frac{1}{2} - \varepsilon}\right), & \rho \to \infty, \quad \varepsilon > 0, \\ \mathcal{O}\left(\rho^\delta\right), & \rho \to 0, \quad \delta > 0, \end{cases}$$

then

$$\lim_{N\to\infty} \int_0^\infty f(\rho) \left\langle \varphi_\rho^*, \varphi_\sigma \right\rangle_{N,N} d\rho = W_1^{-1}(\sigma) f(\sigma),$$

where

$$W_1(\sigma) = \frac{1}{2\pi} e^{-t(2\varphi-\pi)} \frac{\Gamma(k_1 - k_2 + it)\Gamma\left(\frac{1}{2} - it - i\sigma\right)\Gamma\left(\frac{1}{2} - it + i\sigma\right)}{\Gamma(k_2 - k_1 - it + 1)\Gamma(2k_2)^2} \times \left|\frac{\Gamma\left(k_2 - k_1 + \frac{1}{2} + i\sigma\right)\Gamma\left(k_1 + k_2 - \frac{1}{2} + i\sigma\right)}{\Gamma(2i\sigma)}\right|^2.$$

Proof. The proof runs along the same lines as the proof of Proposition 5.12, therefore we leave out the details.

As in the proof of Proposition 5.12 we find from Lemma 5.13

$$\lim_{N\to\infty} \int_0^\infty f(\rho) \langle \varphi_\rho^*, \varphi_\sigma \rangle_{N,N} \, d\rho =$$

$$\lim_{N\to\infty} \int_0^\infty f(\rho) \Big\{ \psi_1(\rho) \cos([\rho+\sigma]\ln N) + \psi_2(\rho) \sin([\rho+\sigma]\ln N)$$
$$+ \psi_3(\rho) \cos([\rho-\sigma]\ln N) + \psi_4(\rho) \frac{\sin([\rho-\sigma]\ln N)}{\rho-\sigma} \Big\} \, d\rho,$$

where

$$\psi_1(\rho) = \frac{1}{\rho+\sigma} \Big[\overline{c(\rho)c(\sigma)} \left(E^+(\rho,\sigma) - E^-(\rho,\sigma)\right)$$
$$- \overline{c(-\rho)c(-\sigma)} \left(E^+(-\rho,-\sigma) - E^-(-\rho,-\sigma)\right) \Big],$$

$$\psi_2(\rho) = \frac{i}{\rho+\sigma} \Big[\overline{c(\rho)c(\sigma)} \left(E^+(\rho,\sigma) - E^-(\rho,\sigma)\right)$$
$$+ \overline{c(-\rho)c(-\sigma)} \left(E^+(-\rho,-\sigma) - E^-(-\rho,-\sigma)\right) \Big],$$

$$\psi_3(\rho) = \frac{-1}{\rho-\sigma} \Big[\overline{c(-\rho)c(\sigma)} \left(E^+(-\rho,\sigma) - E^-(-\rho,\sigma)\right)$$
$$- \overline{c(\rho)c(-\sigma)} \left(E^+(\rho,-\sigma) - E^-(\rho,-\sigma)\right) \Big];$$

$$\psi_4(\rho) = i \Big[\overline{c(-\rho)c(\sigma)} \left(E^+(-\rho,\sigma) - E^-(-\rho,\sigma)\right)$$
$$+ \overline{c(\rho)c(-\sigma)} \left(E^+(\rho,-\sigma) - E^-(\rho,-\sigma)\right) \Big].$$

For $\rho = \sigma$ the term between the square brackets for ψ_3 is equal to zero, so ψ_3 has a removable singularity at the point $\rho = \sigma$. So, for $f\psi_i \in L^1(0,\infty)$, $i = 1,2,3$, the terms with ψ_i, $i = 1,2,3$, vanish by the Riemann-Lebesgue lemma. This leaves us with a Dirichlet integral. Then, after applying (5.11), we find for $f\psi_4 \in L^1(0,\infty)$

$$\lim_{N\to\infty} \int_0^\infty f(\rho) \langle \varphi_\rho^*, \varphi_\sigma \rangle_{N,N} \, d\rho = \pi \psi_4(\sigma) f(\sigma).$$

Writing out ψ_4 explicitly gives the result. □

Remark 5.15. *In Remark* 5.8 *we observed that the i-periodic function* $p(x)$ *cancels the poles of* $\tilde{\varphi}_\rho(x)$. *Other obvious choices with the same* *property would be $e^{2k\pi x}p(x)$, for $k \in \mathbb{Z}$. However for $k \neq 0$, the method we used here to find an integral transform pair would fail, since the method depends on the use of the Riemann-Lebesgue lemma and the Dirichlet kernel, which can no longer be used in case $k \neq 0$. This can, e.g., be seen from Lemma 5.11, where the terms in front of $|x|^{i(\sigma-\rho)}$ would contain a factor $e^{2k\pi x}$. So this gives a heuristic argument for the choice (5.3) of the i-periodic function.*

Let f be a continuous function satisfying,

$$f(\rho) = \begin{cases} \mathcal{O}\left(\rho^{-2k_2-\varepsilon}e^{\pi\rho}\right), & \rho \to \infty, \quad \varepsilon > 0, \\ \mathcal{O}\left(\rho^{\delta}\right), & \rho \to 0, \quad \delta > 0, \end{cases} \tag{5.12}$$

and let $\mathbf{f}$ be the vector

$$\mathbf{f}(\rho) = \begin{pmatrix} \overline{f(\rho)} \\ f(\rho) \end{pmatrix}.$$

We define, for $x \in \mathbb{R}$, an operator $\tilde{\mathcal{F}}$ by

$$\left(\tilde{\mathcal{F}}\mathbf{f}\right)(x) = \int_0^\infty \left(\varphi_\rho(x)\overline{f(\rho)} + \varphi_\rho^*(x)f(\rho)\right) W_0(\rho)d\rho. \tag{5.13}$$

To verify that this is a well-defined expression, we determine the behaviour of $\varphi_\rho(x)$ and $W_0(\rho)$ for $\rho \to \infty$ and $\rho \downarrow 0$. From Thomae's transformation (Andrews et al., 1999, Cor. 3.3.6) we find

$$\varphi_\rho(x) = \begin{cases} \mathcal{O}\left(\rho^{-2k_2}e^{\pi\rho}\right), & \rho \to \infty, \\ \mathcal{O}(1), & \rho \to 0. \end{cases}$$

And from (5.7) we obtain

$$W_0(\rho) = \begin{cases} \mathcal{O}\left(\rho^{4k_2-1}e^{-2\pi\rho}\right), & \rho \to \infty, \\ \mathcal{O}(1), & \rho \to 0. \end{cases} \tag{5.14}$$

Then we see that the integral in (5.13) converges absolutely for f satisfying (5.12).

For a continuous function g satisfying

$$g(x) = \begin{cases} \mathcal{O}\left(|x|^{\frac{1}{2}-k_1-k_2-\varepsilon}e^{2(\pi-\varphi)x}\right), & x \to \infty, \quad \varepsilon > 0, \\ \mathcal{O}\left(|x|^{\frac{1}{2}-k_1-k_2-\delta}e^{-2\varphi x}\right), & x \to -\infty, \quad \delta > 0, \end{cases} \tag{5.15}$$

we define, for $\rho \geq 0$, an operator $\mathcal{G}$ by

$$(\mathcal{G}g)(\rho) = \int_{\mathbb{R}} g(x) \begin{pmatrix} \varphi_\rho^*(x) \\ \varphi_\rho(x) \end{pmatrix} w(x)dx. \tag{5.16}$$

From the asymptotic behaviour of $\varphi^*_\rho(x)$ and $w(x)$ for $x \to \pm\infty$, see Lemmas 5.4 and 5.10, it follows that the integral in (5.16) converges absolutely.

Proposition 5.16. *If $g = \tilde{\mathcal{F}}\mathbf{f}$, and g satisfies the conditions* (5.15), *then*

$$(\mathcal{G}g)(\rho) = \begin{pmatrix} 1 & W_0(\rho)/W_1(\rho) \\ \overline{W_0(\rho)/W_1(\rho)} & 1 \end{pmatrix} \begin{pmatrix} \overline{f(\rho)} \\ f(\rho) \end{pmatrix}.$$

Proof. For a function g satisfying (5.15) we define operators $\mathcal{G}_1$ and $\mathcal{G}_2$ by

$$\begin{aligned} (\mathcal{G}_1 g)(\rho) &= \int_{\mathbb{R}} g(x)\varphi^*_\rho(x)w(x)dx, \\ (\mathcal{G}_2 g)(\rho) &= \int_{\mathbb{R}} g(x)\varphi_\rho(x)w(x)dx, \end{aligned} \tag{5.17}$$

then we have

$$(\mathcal{G}g)(\rho) = \begin{pmatrix} (\mathcal{G}_1 g)(\rho) \\ (\mathcal{G}_2 g)(\rho) \end{pmatrix}.$$

If $g(x) = (\tilde{\mathcal{F}}f)(x)$ satisfies the conditions (5.15), then the integral

$$(\mathcal{G}_1 g)(\sigma) = \int_{\mathbb{R}} (\tilde{\mathcal{F}}f)(x)\varphi^*_\sigma(x)w(x)dx$$

converges absolutely. So from (5.13) we obtain

$$(\mathcal{G}_1 g)(\sigma) = \lim_{N\to\infty} \int_{-N}^{N} \varphi^*_\sigma(x)$$

$$\times \left\{ \int_0^\infty \varphi_\rho(x)\overline{f(\rho)}W_0(\rho)d\rho + \int_0^\infty \varphi^*_\rho(x)f(\rho)W_0(\rho)d\rho \right\} w(x)dx,$$

and interchanging integration gives

$$(\mathcal{G}_1 g)(\sigma) = \lim_{N\to\infty} \int_0^\infty \overline{f(\rho)}W_0(\rho) \langle \varphi_\rho, \varphi_\sigma \rangle_{N,N} \, d\rho$$

$$+ \lim_{N\to\infty} \int_0^\infty f(\rho)W_0(\rho) \langle \varphi^*_\rho, \varphi_\sigma \rangle_{N,N} \, d\rho.$$

From (5.12) and (5.14) it follows that the functions $\overline{f(\rho)}W_0(\rho)$, $f(\rho)W_0(\rho)$ satisfy the conditions for Propositions 5.12, 5.14 respectively. Applying the propositions gives $(\mathcal{G}_1 g)(\sigma) = \overline{f(\sigma)} + f(\sigma)W_0(\sigma)/W_1(\sigma)$. In the same way $(\mathcal{G}_2 g)(\sigma)$ can be calculated. So we find $(\mathcal{G}_2 g)(\sigma) = f(\sigma) + \overline{f(\sigma)}W_0(\sigma)/\overline{W_1(\sigma)}$. □

We define an operator $\mathcal{F}$ by

$$(\mathcal{F}\mathbf{f})(x) = \left(\tilde{\mathcal{F}}\begin{pmatrix} 1 & W_0/W_1 \\ W_0/\overline{W_1} & 1 \end{pmatrix}^{-1}\mathbf{f}\right)(x)$$

$$= \int_0^\infty \begin{pmatrix} \varphi^*_\rho(x) \\ \varphi_\rho(x) \end{pmatrix}^* \begin{pmatrix} 1 & -W_0(\rho)/W_1(\rho) \\ -W_0(\rho)/\overline{W_1(\rho)} & 1 \end{pmatrix} \tag{5.18}$$

$$\times \begin{pmatrix} \overline{f(\rho)} \\ f(\rho) \end{pmatrix} \frac{W_0(\rho)}{1 - \left|\frac{W_0(\rho)}{W_1(\rho)}\right|^2} d\rho.$$

From Proposition 5.16 we find $(\mathcal{G}(\mathcal{F}\mathbf{f}))(\rho) = \mathbf{f}(\rho)$.

Remark 5.17. *From Euler's reflection formula and the identity*

$$2\sin x \sin y = \cos(x-y) - \cos(x+y),$$

we find

$$1 - \left|\frac{W_0(\rho)}{W_1(\rho)}\right|^2 = \left|\frac{\Gamma\left(\frac{1}{2}+it+i\rho\right)\Gamma\left(\frac{1}{2}-it+i\rho\right)}{\Gamma\left(k_1-k_2+\frac{1}{2}+i\rho\right)\Gamma\left(k_2-k_1+\frac{1}{2}+i\rho\right)}\right|^2.$$

We define

$$\psi_\rho(x) = \frac{e^{-x(2\varphi-\pi)}\Gamma\left(\frac{1}{2}-it-i\rho\right)\Gamma\left(\frac{1}{2}-it+i\rho\right)\Gamma(2k_2)\Gamma(k_2-k_1+it+1)}{\Gamma(k_1+ix+it)\Gamma(k_2-ix)\Gamma(k_1+k_2-it)\Gamma(k_1-k_2-it+1)}$$

$$\times {}_3F_2\left(\begin{matrix} \frac{1}{2}-it+i\rho, \frac{1}{2}-it-i\rho, k_1-ix-it \\ k_1-k_2-it+1, k_1+k_2-it \end{matrix}; 1\right),$$

then from (Bailey, 1972, §3.8(1)) we obtain

$$\varphi_\rho(x) - \frac{W_0(\rho)}{\overline{W_1(\rho)}}\varphi^*_\rho(x) = \frac{\psi_\rho(x)}{\left|\Gamma\left(k_2-k_1+\frac{1}{2}+i\rho\right)\right|^2}.$$

So the definition of $\mathcal{F}$ (5.18) *is equivalent to*

$$(\mathcal{F}\mathbf{f})(x) = \int_0^\infty \begin{pmatrix} \psi^*_\rho(x) \\ \psi_\rho(x) \end{pmatrix}^* \begin{pmatrix} \overline{f(\rho)} \\ f(\rho) \end{pmatrix} W_2(\rho)d\rho,$$

where

$$W_2(\rho) = \frac{e^{-t(2\varphi-\pi)}}{2\pi}$$
$$\times \left| \frac{\Gamma\left(k_1 - k_2 + \frac{1}{2} + i\rho\right)\Gamma\left(k_2 - k_1 + \frac{1}{2} + i\rho\right)\Gamma\left(k_1 + k_2 - \frac{1}{2} + i\rho\right)}{\Gamma\left(2k_2\right)\Gamma\left(k_2 - k_1 + it + 1\right)\Gamma(2i\rho)} \right|^2 .$$

Note that $W_2(\rho)d\rho$ is the orthogonality measure for the continuous dual Hahn polynomials.

The function $\mathcal{F}\mathbf{f}$ exists for all functions $\mathbf{f}$ for which the integral (5.18) converges. We want to find a domain on which $\mathcal{F}$ is injective and isometric. We look for a set $\mathcal{S}$ of functions for which $\mathcal{FS}$ is a dense subspace of $L^2(\mathbb{R}, w(x)dx)$. Recall that the set of polynomials is a dense subspace of $L^2(\mathbb{R}, w(x)dx)$.

Lemma 5.18. *Let $p_n^{(\lambda)}(\cdot;\varphi)$ denote a Meixner-Pollazcek polynomial as defined by (2.2), then*

$$\left(\mathcal{G}p_n^{(k_2)}(\cdot;\varphi)\right)(\rho) = \mathbf{q_n}(\rho) = \begin{pmatrix} q_n(\rho) \\ \overline{q_n(\rho)} \end{pmatrix},$$

with q_n given by

$$q_n(\rho) = e^{-in\varphi}\frac{e^{(t-ik_2)(2\varphi-\pi)}}{(1-e^{-2i\varphi})^{2k_2+n}} \frac{\Gamma\left(2k_2\right)\left|\left(k_2 - k_1 + \frac{1}{2} + i\rho\right)_n\right|^2}{n!\left(k_2 - k_1 + it + 1\right)_n}$$
$$\times {}_2F_1\left(\begin{matrix} n + k_2 - k_1 + \frac{1}{2} + i\rho, n + k_2 - k_1 + \frac{1}{2} - i\rho \\ n + k_2 - k_1 + it + 1 \end{matrix}; \frac{1}{1-e^{2i\varphi}}\right).$$

Proof. Let $\mathcal{G}_1$ and $\mathcal{G}_2$ be as defined by (5.17). Since the Meixner-Pollaczek polynomial p_n is real, we have $\mathcal{G}_1 p_n = \overline{\mathcal{G}_2 p_n}$. Writing out $\overline{q_n(\rho)} = \left(\mathcal{G}_1 p_n^{(k_2)}(\cdot;\varphi)\right)(\rho)$ explicitly, gives

$$\overline{q_n(\rho)} = e^{t(2\varphi-\pi)} e^{-in\varphi} \frac{(2k_2)_n}{n!}$$
$$\times \frac{1}{2\pi} \int_{\mathbb{R}} e^{x(2\varphi-\pi)} \left|\Gamma\left(k_2+ix\right)\right|^2 {}_2F_1\left(\begin{matrix} -n, k_2-ix \\ 2k_2 \end{matrix}; 1-e^{2i\varphi}\right)$$
$$\times {}_3F_2\left(\begin{matrix} k_2+ix, k_2-k_1+\frac{1}{2}+i\rho, k_2-k_1+\frac{1}{2}-i\rho \\ 2k_2, k_2-k_1-it+1 \end{matrix}; 1\right) dx$$
$$= e^{t(2\varphi-\pi)} e^{-in\varphi} \frac{(2k_2)_n}{n!}$$
$$\times \sum_{m=0}^{\infty} \sum_{l=0}^{n} \frac{\left(k_2-k_1+\frac{1}{2}+i\rho\right)_m \left(k_2-k_1+\frac{1}{2}-i\rho\right)_m}{m!\,(2k_2)_m \left(k_2-k_1-it+1\right)_m} \frac{(-n)_l}{l!\,(2k_2)_l} \left(1-e^{2i\varphi}\right)^l$$
$$\times \frac{1}{2\pi} \int_{\mathbb{R}} e^{x(2\varphi-\pi)} \Gamma\left(k_2+ix+m\right) \Gamma\left(k_2-ix+l\right) dx.$$

The inner integral can be evaluated by (Paris and Kaminski, 2001, (3.3.9)),

$$\frac{1}{2\pi i} \int_{k-i\infty}^{k+i\infty} \Gamma(s)\Gamma(a-s) y^{-s} ds = \frac{\Gamma(a)}{(1+y)^a},$$
$$0 < \Re(s) < \Re(a), \quad |\arg(y)| < \pi,$$

with $s = k_2+m+ix$ and $a = 2k_2+m+l$. Now the sum over l becomes a terminating ${}_2F_1$-series, which can be evaluated by the Chu-Vandermonde identity (Andrews et al., 1999, Cor. 2.2.3)

$${}_2F_1\left(\begin{matrix} -n, 2k_2+m \\ 2k_2 \end{matrix}; 1\right) = \begin{cases} 0, & m < n, \\ \dfrac{(-m)_n}{(2k_2)_n}, & m \geq n. \end{cases}$$

Then $\overline{q_n(\rho)}$ reduces to a single sum, starting at $m = n$. Shifting the summation index gives the result. □

Proposition 5.19. *For a continuous function $g \in L^2(\mathbb{R}, w(x)dx)$, we have $\mathcal{F}(\mathcal{G}g) = g$.*

Proof. Let q_n be as in Lemma 5.18. We show that $(\mathcal{F}\mathbf{q_n})(x) = p_n^{(k_2)}(x;\varphi)$. From this the proposition follows, since the polynomials are dense in $L^2(\mathbb{R}, w(x)dx)$.

Transforming the ${}_2F_1$-series of q_n in Lemma 5.18 by (Andrews et al., 1999, (2.3.12)) and using the asymptotic behaviour (Erdélyi et al., 1953,

2.3.2(16)), we find

$$q_n(\rho) = \mathcal{O}\left(\rho^{k_1-k_2-\frac{1}{2}+n}\right), \qquad \rho \to \infty.$$

From Stirling's formula it follows that $|W_0(\rho)/W_1(\rho)| = \mathcal{O}\left(e^{-\pi\rho}\right)$ for $\rho \to \infty$. So by (5.14) and (5.18) we see that $\mathcal{F}\mathbf{q_n}$ exists.

We calculate

$$I(x) = \int_0^\infty \psi_\rho(x)\overline{q_n(\rho)}W_2(\rho)d\rho,$$

then according to Remark 5.17 we have $\mathcal{F}\mathbf{q_n} = I + \overline{I}$. Writing $\psi_\rho(x)$ and $\overline{q_n(\rho)}$ as a sum, and interchanging summation and integration, gives

$$I(x) = C\sum_{l=0}^\infty \sum_{m=0}^\infty \frac{\left(1-e^{-2i\varphi}\right)^{-l}}{l!\,(n+k_2-k_1-it+1)_l} \times \frac{(k_1-ix-it)_m}{m!\,(k_1-k_2-it+1)_m\,(k_1+k_2-it)_m} I_{l,m},$$

where

$$I_{l,m} = \frac{1}{2\pi}\int_0^\infty \Gamma\left(\frac{1}{2}-it+i\rho+m\right)\Gamma\left(\frac{1}{2}-it-i\rho+m\right) \times \left|\Gamma\left(k_1-k_2+\frac{1}{2}+i\rho\right)\right|^2 \times \left|\frac{\Gamma\left(n+k_2-k_1+\frac{1}{2}+i\rho+l\right)\Gamma\left(k_1+k_2-\frac{1}{2}+i\rho\right)}{\Gamma(2i\rho)}\right|^2 d\rho,$$

$$C = e^{-x(2\varphi-\pi)}\left(\frac{e^{i\varphi}}{1-e^{2i\varphi}}\right)^{2k_2+n}\frac{e^{-i\pi k_2}}{n!}\frac{1}{\Gamma(k_1+ix+it)\,\Gamma(k_2-ix)} \times \frac{1}{\Gamma(k_1+k_2-it)\,\Gamma(k_2-k_1-it+n+1)\,\Gamma(k_1-k_2-it+1)}.$$

The integral $I_{l,m}$ can be evaluated by (Andrews et al., 1999, Thm. 3.6.2);

$$I_{l,m} = \frac{\Gamma(k_2-k_1-it+n+m+l+1)\,\Gamma(k_1-k_2-it+m+1)}{\Gamma(k_1+k_2-it+n+m+l+1)} \times \Gamma(n+l+1)\,\Gamma(2k_2+l+n)\,\Gamma(2k_1)\,\Gamma(k_1+k_2-it+m).$$

Now the sum over m is a ${}_2F_1$-series which can be summed by Gauss's theorem;

$$\begin{aligned} &{}_2F_1\left(\begin{matrix} k_1 - ix - it, k_2 - k_1 - it + n + l + 1 \\ k_1 + k_2 - it + n + l + 1 \end{matrix}; 1\right) \\ &= \frac{\Gamma(k_1 + ix + it)\,\Gamma(k_1 + k_2 - it + n + l + 1)}{\Gamma(k_2 + ix + l + n + 1)\,\Gamma(2k_1)}. \end{aligned}$$

Then the sum over l becomes a ${}_2F_1$-series, and after applying Euler's transformation we obtain

$$\begin{aligned} I(x) &= \frac{e^{in\varphi}}{(1 - e^{-2i\varphi})^{ix+k_2}} \frac{\Gamma(2k_2 + n)}{\Gamma(k_2 + ix + n + 1)\,\Gamma(k_2 - ix)} \\ &\times {}_2F_1\left(\begin{matrix} k_2 + ix, 1 - k_2 + ix \\ k_2 + ix + n + 1 \end{matrix}; \frac{1}{1 - e^{-2i\varphi}}\right). \end{aligned}$$

As in (3.3) we find from this

$$\begin{aligned} &(\mathcal{F}\mathbf{q_n})(x) = I(x) + \overline{I(x)} \\ &= \frac{(2k_2)_n}{n!} e^{in\varphi}\, {}_2F_1\left(\begin{matrix} -n, k_2 + ix \\ 2k_2 \end{matrix}; 1 - e^{-2i\varphi}\right) = p_n^{(k_2)}(x; \varphi). \end{aligned}$$

Note that the condition $\frac{\pi}{6} < \varphi < \frac{5\pi}{6}$ is needed for absolute convergence of the ${}_2F_1$-series of q_n. This condition can be removed by analytic continuation. □

We define the Hilbert space $\mathcal{M}$ by

$$\mathcal{M} = \overline{\operatorname{span}\{\mathbf{q_n} \mid n \in \mathbb{Z}_{\geq 0}\}},$$

then $\mathcal{M}$ consists of functions of the form

$$\mathbf{f}(\rho) = \begin{pmatrix} f_1(\rho) \\ f_2(\rho) \end{pmatrix} \in \mathbb{C}^2.$$

The inner product on $\mathcal{M}$ is given by

$$\begin{aligned} \langle \mathbf{f}, \mathbf{g} \rangle_{\mathcal{M}} = \int_0^\infty \begin{pmatrix} g_1(\rho) \\ g_2(\rho) \end{pmatrix}^* \begin{pmatrix} 1 & -W_0(\rho)/W_1(\rho) \\ -W_0(\rho)/\overline{W_1(\rho)} & 1 \end{pmatrix} \begin{pmatrix} f_1(\rho) \\ f_2(\rho) \end{pmatrix} \\ \times \frac{W_0(\rho)}{1 - \left|\frac{W_0(\rho)}{W_1(\rho)}\right|^2} d\rho. \end{aligned} \tag{5.19}$$

Proposition 5.20. *The operator $\mathcal{F} : \mathcal{M} \to L^2(\mathbb{R}, w(x)dx)$ is an isometry.*

Proof. Let

$$\mathbf{f}(\rho) = \begin{pmatrix} f_1(\rho) \\ f_2(\rho) \end{pmatrix}, \quad \mathbf{g}(\rho) = \begin{pmatrix} g_1(\rho) \\ g_2(\rho) \end{pmatrix}.$$

We write out $\langle \mathcal{F}\mathbf{f}, \mathcal{F}\mathbf{g} \rangle_{L^2(\mathbb{R}, w(x)dx)}$ explicitly. We define

$$I_\rho(\mathbf{f}) = \varphi_\rho(x) f_1(\rho) - \varphi_\rho^*(x) f_1(\rho) \frac{\overline{W_0(\rho)}}{\overline{W_1(\rho)}} - \varphi_\rho(x) f_2(\rho) \frac{W_0(\rho)}{W_1(\rho)} + \varphi_\rho^*(x) f_2(\rho)$$

and

$$W(\rho) = \frac{W_0(\rho)}{1 - \left| \frac{W_0(\rho)}{W_1(\rho)} \right|^2},$$

then

$$
\begin{aligned}
&\langle \mathcal{F}\mathbf{f}, \mathcal{F}\mathbf{g}\rangle_{L^2(\mathbb{R},w(x)dx)} \\
&= \lim_{N\to\infty} \int_{-N}^{N}\int_{0}^{\infty} W(\rho)\, I_\rho(\mathbf{f})d\rho \,\overline{\int_0^\infty W(\sigma) I_\sigma(\mathbf{g})d\sigma}\; w(x)dx \\
&= \lim_{N\to\infty} \int_0^\infty\int_0^\infty W(\rho)\,W(\sigma) \\
&\quad\times\Bigg\{\Bigg[f_1(\rho)\overline{g_1(\sigma)} - f_1(\rho)\overline{g_2(\sigma)}\overline{\frac{W_0(\sigma)}{W_1(\sigma)}} - f_2(\rho)\overline{g_1(\sigma)}\frac{W_0(\rho)}{W_1(\rho)} \\
&\qquad + f_2(\rho)\overline{g_2(\sigma)}\frac{W_0(\rho)\overline{W_0(\sigma)}}{W_1(\rho)\overline{W_1(\sigma)}}\Bigg]\langle\varphi_\rho,\varphi_\sigma\rangle_{N,N} \\
&\quad+\Bigg[f_2(\rho)\overline{g_1(\sigma)} - f_2(\rho)\overline{g_2(\sigma)}\overline{\frac{W_0(\sigma)}{W_1(\sigma)}} - f_1(\rho)\overline{g_1(\sigma)}\overline{\frac{W_0(\rho)}{W_1(\rho)}} \\
&\qquad + f_1(\rho)\overline{g_2(\sigma)}\overline{\frac{W_0(\rho)W_0(\sigma)}{W_1(\rho)W_1(\sigma)}}\Bigg]\langle\varphi_\rho^*,\varphi_\sigma\rangle_{N,N} \\
&\quad+\Bigg[f_1(\rho)\overline{g_2(\sigma)} - f_1(\rho)\overline{g_1(\sigma)}\frac{W_0(\sigma)}{W_1(\sigma)} - f_2(\rho)\overline{g_2(\sigma)}\frac{W_0(\rho)}{W_1(\rho)} \\
&\qquad + f_2(\rho)\overline{g_1(\sigma)}\frac{W_0(\rho)W_0(\sigma)}{W_1(\rho)W_1(\sigma)}\Bigg]\langle\varphi_\rho,\varphi_\sigma^*\rangle_{N,N} \\
&\quad+\Bigg[f_2(\rho)\overline{g_2(\sigma)} - f_2(\rho)\overline{g_1(\sigma)}\frac{W_0(\sigma)}{W_1(\sigma)} - f_1(\rho)\overline{g_2(\sigma)}\overline{\frac{W_0(\rho)}{W_1(\rho)}} \\
&\qquad + f_1(\rho)\overline{g_1(\sigma)}\overline{\frac{W_0(\rho)W_0(\sigma)}{W_1(\rho)W_1(\sigma)}}\Bigg]\langle\varphi_\rho^*,\varphi_\sigma^*\rangle_{N,N}\Bigg\}\,d\rho d\sigma.
\end{aligned}
$$

Then from Propositions 5.12, 5.14 and (5.19) we obtain

$$\langle \mathcal{F}\mathbf{f}, \mathcal{F}\mathbf{g}\rangle_{L^2(\mathbb{R},w(x)dx)} = \langle \mathbf{f},\mathbf{g}\rangle_{\mathcal{M}}$$

by a straightforward calculation. □

So far we only considered the integral transform $\mathcal{F}$ in the case that $\rho^2+\frac{1}{4}$ is in the continuous spectrum of the difference operator Λ. In the next subsection we consider the discrete spectrum of Λ.

5.4 Discrete spectrum

From (5.8) it follows that for $\Im(\rho)<0$ we have $\Phi_\rho(x)\in L^2(\mathbb{R},w(x)dx)$. So if $c(-\rho)=0$ and $\Im(\rho)<0$, we find from Proposition 5.9 that $\varphi_\rho(x)=c(\rho)\Phi_\rho(x)$, and therefore $\varphi_\rho(x)\in L^2(\mathbb{R},w(x)dx)$.

There are two possible cases for $c(-\rho)=0$, cf. Theorem 4.1:

1. $k_2-k_1+\frac{1}{2}<0$, then $\rho=i\left(k_2-k_1+\frac{1}{2}+n\right)$, $n=0,\ldots,n_0$, where n_0 is the largest nonnegative integer such that $k_2-k_1+\frac{1}{2}+n_0<0$,

2. $k_1+k_2-\frac{1}{2}<0$, then $\rho=i\left(k_1+k_2-\frac{1}{2}\right)$.

Case (ii) does not occur for $k_1>1$, which is needed for convergence of the ${}_3F_2$-series of $\varphi_\rho(x+i)$. However for $k_1\le 1$ we use expression (5.6) for $\Phi_\rho(x)$ (which still converges if $k_1\le 1$) and $\varphi_\rho(x)=c(\rho)\Phi_\rho(x)$. We see that the ${}_3F_2$-series becomes a ${}_2F_1$-series of unit argument, and then, with Gauss's summation formula, we find that in case (ii) we have $\Phi_\rho(x)=e^{-2\varphi x}p(x)$. Observe that case (i) and case (ii) exclude each other.

First we consider case (i). For $\rho_n=i\left(k_2-k_1+\frac{1}{2}+n\right)$, $0\le n\le n_0$, we denote $\varphi_\rho(x)$ by $\varphi_{\rho_n}(x)$. We show that $\varphi_{\rho_n}(x)$ is orthogonal to $\varphi_{\rho_m}(x)$ and $\varphi^*_{\rho_m}(x)$ for $n\ne m$. Note that φ_{ρ_n} is given by a terminating series, cf. (5.4).

Proposition 5.21. *For* $m,n=0,\ldots,n_0$

$$\begin{aligned}\langle\varphi_{\rho_n},\varphi_{\rho_m}\rangle &= \delta_{nm}\left(2\pi i\operatorname*{Res}_{\rho=\rho_n}W_0(\rho)\right)^{-1}\\ &= \delta_{nm}\,\frac{e^{2t(\varphi-\frac{\pi}{2})}\,\Gamma\left(2k_2\right)\Gamma\left(2k_1-2k_2-1\right)}{\Gamma\left(k_1-k_2+it\right)\Gamma\left(k_1-k_2-it\right)\Gamma\left(2k_1-1\right)}\\ &\times\frac{n!\,(2k_2-2k_1+n+1)_n\,(2-2k_1)_n}{(2k_2)_n\,(2k_2-2k_1+2)_{2n}}.\end{aligned}$$

Proof. Writing out the explicit expressions (5.4) for $\varphi_{\rho_n}(x)$ and $\varphi_{\rho_m}(x)$ gives

$$\begin{aligned}
&\int_{\mathbb{R}} \varphi_{\rho_n}(x)\varphi^*_{\rho_m}(x)w(x)dx \\
&= e^{2t(\varphi-\frac{\pi}{2})}\frac{1}{2\pi}\int_{-\infty}^{\infty} {}_3F_2\left(\begin{matrix}-n, 2k_2-2k_1+n+1, k_2-ix \\ 2k_2, k_2-k_1+it+1\end{matrix};1\right) \\
&\times {}_3F_2\left(\begin{matrix}-m, 2k_2-2k_1+m+1, k_2+ix \\ 2k_2, k_2-k_1-it+1\end{matrix};1\right)\left|\frac{\Gamma(k_2+ix)}{\Gamma(k_1+it+ix)}\right|^2 dx \\
&= \sum_{k=0}^{n}\sum_{l=0}^{m} \frac{(-n)_k(2k_2-2k_1+n+1)_k(-m)_l(2k_2-2k_1+m+1)_l}{k!(2k_2)_k(k_2-k_1+it+1)_k\, l!(2k_2)_l(k_2-k_1-it+1)_l} \\
&\qquad \times e^{2t(\varphi-\frac{\pi}{2})}\frac{1}{2\pi}\int_{-\infty}^{\infty}\frac{\Gamma(k_2-ix+k)\Gamma(k_2+ix+l)}{\Gamma(k_1-it-ix)\Gamma(k_1+it+ix)}dx.
\end{aligned}$$

The integral inside the sum can be evaluated by (Paris and Kaminski, 2001, §3.3.4)

$$\frac{1}{2\pi i}\int_{-i\infty}^{i\infty}\frac{\Gamma(a+s)\Gamma(c-s)}{\Gamma(b+s)\Gamma(d-s)}ds = \frac{\Gamma(a+c)\Gamma(b+d-a-c-1)}{\Gamma(b-a)\Gamma(d-c)\Gamma(b+d-1)},$$

where $\Re(a+c-b-d) < 1$ and the path of integration separates the poles of $\Gamma(a+s)$ from the poles of $\Gamma(c-s)$. Note that the convergence condition $2k_2-2k_1+k+l<1$ is satisfied in case (i) and $k,l \le n_0$. Now we find for the double sum for $n \le m$

$$\begin{aligned}
&e^{2t(\varphi-\frac{\pi}{2})}\frac{\Gamma(2k_2)\Gamma(2k_1-2k_2-1)}{\Gamma(k_1-k_2+it)\Gamma(k_1-k_2-it)\Gamma(2k_1-1)} \\
&\qquad\times\sum_{k=0}^{n}\frac{(-n)_k(2k_2-2k_1+n+1)_k}{k!(2k_2-2k_1+2)_k} \\
&\qquad\qquad\times\sum_{l=0}^{m}\frac{(-m)_l(2k_2-2k_1+m+1)_l(2k_2+k)_l}{l!(2k_2)_l(2k_2-2k_1+k+2)_l}.
\end{aligned}$$

The sum over l is a terminating ${}_3F_2$-series, which can be evaluated by the Pfaff-Saalschütz theorem

$$
{}_3F_2\left(\begin{matrix}-m, 2k_2-2k_1+m+1, 2k_2+k\\ 2k_2, 2k_2-2k_1+k+2\end{matrix};1\right)
= \begin{cases} 0, & k<m,\\ \dfrac{(-1)^m m!\,(2-2k_1)_m}{(2k_2)_m\,(2k_2-2k_1+m+2)_m}, & k=m.\end{cases}
$$

So we find for $n \le m$

$$
\int_{\mathbb{R}} \varphi_{\rho_n}(x)\varphi^*_{\rho_m}(x)w(x)dx
$$

$$
= \delta_{nm}\,\frac{e^{2t(\varphi-\frac{\pi}{2})}\,\Gamma(2k_2)\,\Gamma(2k_1-2k_2-1)}{\Gamma(k_1-k_2+it)\,\Gamma(k_1-k_2-it)\,\Gamma(2k_1-1)}
\times \frac{n!\,(2k_2-2k_1+n+1)_n\,(2-2k_1)_n}{(2k_2)_n\,(2k_2-2k_1+n+2)_n\,(2k_2-2k_1+2)_n}.
$$

Note that from the condition $k_2-k_1+\frac{1}{2}+n<0$ follows that this expression is positive in case $n=m$. For $n \ge m$ we find the same result by interchanging the summations over k and l. A straightforward calculation shows that the expression found is equal to

$$
\delta_{nm}\left(2\pi i \operatorname*{Res}_{\rho=\rho_n} W_0(\rho)\right)^{-1}.
$$

□

Proposition 5.22. *For $m,n=0,\dots,n_0$*

$$
\langle \varphi^*_{\rho_n}, \varphi_{\rho_m}\rangle = \delta_{nm}\left(2\pi i \operatorname*{Res}_{\rho=\rho_n} W_1(\rho)\right)^{-1}
$$

$$
= \delta_{nm}\,\frac{(-1)^n n!\,e^{2t(\varphi-\frac{\pi}{2})}\,\Gamma(2k_2)\,\Gamma(2k_1-2k_2-1)}{\Gamma(k_1-k_2+it)\,\Gamma(k_1-k_2-it)\,\Gamma(2k_1-1)}
\times \frac{(k_2-k_1+it+1)_n\,(2k_2-2k_1+n+1)_n\,(2-2k_1)_n}{(k_2-k_1-it+1)_n\,(2k_2)_n\,(2k_2-2k_1+2)_{2n}}.
$$

Proof. From the first formula on page 142 in (Andrews et al., 1999) we find

$$
\varphi_{\rho_n}(x) = (-1)^n\frac{(k_2-k_1-it+1)_n}{(k_2-k_1+it+1)_n}\,\varphi^*_{\rho_n}(x).
$$

Then the explicit expression follows from Proposition 5.21. A straightforward calculation shows that the explicit expression is equal to the residue at $\rho = \rho_n$ of $W_1(\rho)$. □

A similar calculation is used for case (ii). Recall from the beginning of this subsection that in this case $\varphi_{\rho_c}(x) = e^{-2\varphi x} c(\rho_c)\, p(x)$.

Proposition 5.23. *Let* $\rho_c = i(k_1 + k_2 - \frac{1}{2})$, *then*

$$\begin{aligned}\langle \varphi_{\rho_c}, \varphi_{\rho_c}\rangle &= \left(2\pi i \operatorname*{Res}_{\rho=\rho_c} W_0(\rho)\right)^{-1}\\ &= e^{2t(\varphi-\frac{\pi}{2})}\frac{\Gamma(2k_2)\,\Gamma(1-2k_1-2k_2)}{\Gamma(1-2k_1)}\\ &\times\left|\frac{\Gamma(k_2-k_1+it+1)}{\Gamma(k_1+k_2+it)\,\Gamma(1-k_1-k_2+it)}\right|^2.\end{aligned}$$

Proof. The proof is similar to the proof of Proposition 5.21. □

Proposition 5.24. *Let* $\rho_c = i(k_1 + k_2 - \frac{1}{2})$, *then*

$$\langle \varphi^*_{\rho_c}, \varphi_{\rho_c}\rangle = \left(2\pi i \operatorname*{Res}_{\rho=\rho_c} W_1(\rho)\right)^{-1}$$

$$= \frac{e^{2t(\varphi-\frac{\pi}{2})}\Gamma(2k_2)\,\Gamma(1-2k_1-2k_2)\,\Gamma(k_2-k_1-it+1)}{\Gamma(1-2k_1)\,\Gamma(k_1+k_2-it)\,\Gamma(1-k_1-k_2-it)\,\Gamma(k_1-k_2+it)}.$$

Proof. We use Euler's reflection formula to write $p(x)$ in terms of Γ-functions, then we have

$$\langle \varphi^*_{\rho_c}, \varphi_{\rho_c}\rangle = e^{2t(\varphi-\frac{\pi}{2})}\overline{c(\rho_c)}^2$$

$$\times\frac{1}{2\pi}\int_{\mathbb{R}}\frac{\Gamma(k_2+ix)\,\Gamma(k_2-ix)\,\Gamma(k_1-it-ix)}{\Gamma(k_1+it+ix)\,\Gamma(1-k_1-it-ix)\,\Gamma(1-k_1-it-ix)}dx.$$

We use a special case of (Slater, 1966, (4.5.1.2)) to evaluate the integral;

$$\frac{1}{2\pi i}\int_{-i\infty}^{i\infty}\frac{\Gamma(a+s)\Gamma(b-s)\Gamma(c-s)}{\Gamma(d+s)\Gamma(e-s)\Gamma(f-s)}ds$$

$$= \frac{\Gamma(a+b)\Gamma(a+c)}{\Gamma(d-a)\Gamma(a+e)\Gamma(a+f)}\,{}_3F_2\left(\begin{matrix}a+b, a+c, 1+a-d\\ a+e, a+f\end{matrix};1\right),$$

where $\Re(d+e+f-a-b-c) > 0$ and the path of integration separates the poles of $\Gamma(a+s)$ from the poles of $\Gamma(b-s)$ and $\Gamma(c-s)$. We put

$$s = -k_2 + ix, \quad a = 2k_2, \quad b = 0, \quad c = k_1 - k_2 - it,$$
$$d = k_1 + k_2 + it, \quad e = 1 - k_1 - k_2 - it, \quad f = 1 - k_1 - k_2 - it,$$

then we find that the ${}_3F_2$-series reduces to a ${}_2F_1$-series, which can be evaluated by Gauss's summation formula. From this the result follows. □

Next we show that if $\rho^2 + \frac{1}{4}$ is in the discrete spectrum of Λ, $\varphi_\rho(x)$ is orthogonal to $\varphi_\sigma(x)$ and $\varphi_\sigma^*(x)$, if σ is real.

Proposition 5.25. *For $\sigma \in [0, \infty)$ and $\rho = \rho_n$, or $\rho = \rho_c$, we have*

$$\langle \varphi_\rho, \varphi_\sigma \rangle = 0, \qquad \langle \varphi_\rho, \varphi_\sigma^* \rangle = 0.$$

Proof. By Propositions 5.3 and 5.7

$$\begin{aligned} &\lim_{N,M\to\infty} \int_{-M}^{N} \varphi_\rho(x)\varphi_\sigma^*(x)w(x)dx \\ &= \lim_{N,M\to\infty} \frac{[\varphi_\rho, \varphi_\sigma](N) - [\varphi_\rho, \varphi_\sigma](-M)}{\rho^2 - \sigma^2}. \end{aligned}$$

We show that the limit of each Wronskian is zero. Let

$$f(x) = \left(\varphi_\rho(x)\varphi_\sigma^*(x-i) - \varphi_\rho(x-i)\varphi_\sigma^*(x)\right)\alpha_-(x)w(x).$$

First we consider the asymptotic behaviour of $f(-M)$ and $f(N)$.

For $\rho = \rho_n$ and $0 \le y \le 1$ we find from Lemmas 5.4, 5.10 and Proposition 5.9

$$f(x+iy) = \mathcal{O}\left(|x|^{\frac{1}{2}-k_1+k_2+n}\right), \qquad x \to \pm\infty.$$

For $\rho = \rho_c$ we find

$$f(x+iy) = \mathcal{O}\left(|x|^{2k_1+2k_2-1}\right), \qquad x \to \pm\infty.$$

Here the implied constants do not depend on y. Then dominated convergence gives the result.

The proof for $\langle \varphi_\rho, \varphi_\sigma^* \rangle = 0$ runs along the same lines. □

Remark 5.26. *The explicit calculations in this subsection can be carried out because of the choice (5.3) of the i-periodic function $p(x)$. It is not likely that with another choice for the function $p(x)$ all the calculations can be done explicitly. This gives another (heuristic) argument for the choice of $p(x)$.*

5.5 The continuous Hahn integral transform

We combine the results of Subsection 11.5.3 with the results of Subsection 11.5.4.

Let $k_1, k_2 > 0$, $0 < \varphi < \pi$ and $t \in \mathbb{R}$. Let $\varphi_\rho(x)$ be the function given by

$$\varphi_\rho(x) = \frac{e^{-x(2\varphi-\pi)}}{\Gamma(k_1+ix+it)\,\Gamma(k_1-ix-it)} \times {}_3F_2\left(\begin{matrix} k_2-ix, k_2-k_1+\frac{1}{2}+i\rho, k_2-k_1+\frac{1}{2}-i\rho \\ 2k_2, k_2-k_1+it+1 \end{matrix};1\right).$$

We denote

$$h(\rho) = \frac{W_0(\rho)}{W_1(\rho)} = \frac{\Gamma\left(\frac{1}{2}+it+i\rho\right)\Gamma\left(\frac{1}{2}+it-i\rho\right)}{\Gamma(k_1-k_2+it)\,\Gamma(k_2-k_1+it+1)},$$

$$W(\rho) = \frac{W_0(\rho)}{1-|h(\rho)|^2} = \frac{1}{2\pi}e^{-t(2\varphi-\pi)} \times \left|\frac{\Gamma\left(k_2-k_1+\frac{1}{2}+i\rho\right)^2\Gamma\left(k_1-k_2+\frac{1}{2}+i\rho\right)\Gamma\left(k_1+k_2-\frac{1}{2}+i\rho\right)}{\Gamma(2k_2)\,\Gamma(k_2-k_1+it+1)\,\Gamma(2i\rho)}\right|^2,$$

$$w(x) = \frac{1}{2\pi}e^{(2x+t)(2\varphi-\pi)}\left|\Gamma(k_1+it+ix)\,\Gamma(k_2+ix)\right|^2.$$

Let $\mathcal{M}$ be the Hilbert space given by

$$\mathcal{M} = \overline{\operatorname{span}\{\mathbf{q_m} \mid m \in \mathbb{Z}_{\geq 0}\}},$$

where $\mathbf{q_m}$ is as in Lemma 5.18. For functions $\mathbf{f} = \begin{pmatrix} f_1 \\ f_2 \end{pmatrix}$, the inner product on $\mathcal{M}$ is given by:

(i) For $k_1+k_1-\frac{1}{2} \geq 0$, $k_2-k_1+\frac{1}{2} \geq 0$

$$\langle \mathbf{f}, \mathbf{g} \rangle_{\mathcal{M}} = \int_0^\infty \begin{pmatrix} g_1(\rho) \\ g_2(\rho) \end{pmatrix}^* \begin{pmatrix} 1 & -h(\rho) \\ -\overline{h(\rho)} & 1 \end{pmatrix} \begin{pmatrix} f_1(\rho) \\ f_2(\rho) \end{pmatrix} W(\rho)d\rho.$$

(ii) For $k_1+k_2-\frac{1}{2} < 0$, $\rho_c = i\left(k_1+k_2-\frac{1}{2}\right)$,

$$\langle \mathbf{f}, \mathbf{g} \rangle_{\mathcal{M}} = \int_0^\infty \begin{pmatrix} g_1(\rho) \\ g_2(\rho) \end{pmatrix}^* \begin{pmatrix} 1 & -h(\rho) \\ -\overline{h(\rho)} & 1 \end{pmatrix} \begin{pmatrix} f_1(\rho) \\ f_2(\rho) \end{pmatrix} W(\rho)d\rho + \begin{pmatrix} g_1(\rho_c) \\ g_2(\rho_c) \end{pmatrix}^* \begin{pmatrix} 1 & -h(\rho_c) \\ -\overline{h(\rho_c)} & 1 \end{pmatrix} \begin{pmatrix} f_1(\rho_c) \\ f_2(\rho_c) \end{pmatrix} 2\pi i \operatorname*{Res}_{\rho=\rho_c} W(\rho).$$

(iii) For $k_2 - k_1 + \frac{1}{2} < 0$, $\rho_n = i\left(k_2 - k_1 + \frac{1}{2} + n\right)$, $n = 0, \ldots, n_0$, where n_0 is the largest integer such that $-i\rho_{n_0} < 0$,

$$\langle \mathbf{f}, \mathbf{g} \rangle_{\mathcal{M}} = \int_0^\infty \begin{pmatrix} g_1(\rho) \\ g_2(\rho) \end{pmatrix}^* \begin{pmatrix} 1 & -h(\rho) \\ -\overline{h(\rho)} & 1 \end{pmatrix} \begin{pmatrix} f_1(\rho) \\ f_2(\rho) \end{pmatrix} W(\rho) d\rho$$
$$+ \sum_{n=0}^{n_0} \begin{pmatrix} g_1(\rho_n) \\ g_2(\rho_n) \end{pmatrix}^* \begin{pmatrix} f_1(\rho_n) \\ f_2(\rho_n) \end{pmatrix} \pi i \operatorname*{Res}_{\rho=\rho_n} W(\rho) \left(1 - |h(\rho)|^2\right).$$

Observe that for $m > n$ we have $\mathbf{q_m}(\rho_n) = \mathbf{0}$. For $m \leq n$ it follows from the way q_m is calculated in Proposition 5.18, that $q_m(\rho_n) = \overline{h(\rho_n) q_m(\rho_n)}$.

For a continuous function $\mathbf{f} \in \mathcal{M}$ we define the linear operator $\mathcal{F}$ $\mathcal{M} \to L^2(\mathbb{R}, w(x)dx)$ by

$$(\mathcal{F}\mathbf{f})(x) = \left\langle \mathbf{f}, \begin{pmatrix} \varphi_\rho^*(x) \\ \varphi_\rho(x) \end{pmatrix} \right\rangle_{\mathcal{M}}.$$

We call $\mathcal{F}$ the continuous Hahn integral transform.

Theorem 5.27. *The continuous Hahn integral transfrom $\mathcal{F} : \mathcal{M} \to L^2(\mathbb{R}, w(x)dx)$ is unitary and its inverse is given by*

$$\left(\mathcal{F}^{-1} g\right)(\rho) = \int_{\mathbb{R}} g(x) \begin{pmatrix} \varphi_\rho^*(x) \\ \varphi_\rho(x) \end{pmatrix} w(x) dx.$$

Proof. For case (i) this follows from Propositions 5.16, 5.19 and 5.20.

For case (ii) we only have to check that Propositions 5.12 and 5.14 still hold with the discrete mass point in $\rho = \rho_c$ added to the integral. From Propositions 5.23, 5.24 and 5.25 we find

$$f(\rho_c) \langle \varphi_{\rho_c}, \varphi_\sigma \rangle = \begin{cases} 0, & \sigma \in [0, \infty), \\ \left(2\pi i \operatorname*{Res}_{\sigma=\rho_c} W_0(\sigma)\right)^{-1} f(\rho_c), & \sigma = \rho_c, \end{cases}$$

$$f(\rho_c) \langle \varphi_{\rho_c}^*, \varphi_\sigma \rangle = \begin{cases} 0, & \sigma \in [0, \infty), \\ \left(2\pi i \operatorname*{Res}_{\sigma=\rho_c} W_1(\sigma)\right)^{-1} f(\rho_c), & \sigma = \rho_c. \end{cases}$$

Now the proof for case (ii) is completely analogous to the proof of case (i).

For case (iii) injectivity and surjectivity of $\mathcal{F}$ can be proved in the same way as case (i). We check that $\mathcal{F}$ is an isometry. The continuous part follows from Proposition 5.20, so we only have to check for the discrete part. We write out $\langle \mathcal{F}\mathbf{q_k}, \mathcal{F}\mathbf{q_l}\rangle_{L^2(\mathbb{R},w(x)dx)}$, $k,l \in \mathbb{Z}_{\geq 0}$, for the discrete part of $\mathcal{F}\mathbf{q_k}$ and $\mathcal{F}\mathbf{q_l}$. From Propositions 5.21, 5.22 and $q_k(\rho_n) = \overline{h(\rho_n) q_k(\rho_n)}$ we find, for $n, m = 0, \ldots, n_0$,

$$\begin{aligned}
&\int_{\mathbb{R}} \sum_{n=0}^{n_0} \Big(\varphi_{\rho_n}(x)\overline{q_k(\rho_n)} + \varphi^*_{\rho_n}(x) q_k(\rho_n)\Big)\ \pi i \operatorname*{Res}_{\rho=\rho_n} W_0(\rho) \\
&\qquad \times \sum_{m=0}^{n_0} \Big(\varphi_{\rho_m}(x)\overline{q_l(\rho_m)} + \varphi^*_{\rho_m}(x) q_l(\rho_m)\Big)\ \pi i \operatorname*{Res}_{\rho=\rho_m} W_0(\rho)\, w(x)dx \\
&= \sum_{n,m=0}^{n_0} \Big(\overline{q_k(\rho_n)\, q_l(\rho_m)}\, \langle \varphi_{\rho_n}, \varphi^*_{\rho_m}\rangle + \overline{q_k(\rho_n)} q_l(\rho_m)\, \langle \varphi_{\rho_n}, \varphi_{\rho_m}\rangle \\
&\qquad + q_k(\rho_n)\, \overline{q_l(\rho_m)}\, \langle \varphi^*_{\rho_n}, \varphi^*_{\rho_m}\rangle \\
&\qquad + q_k(\rho_n)\, q_l(\rho_m)\, \langle \varphi^*_{\rho_n}, \varphi_{\rho_m}\rangle \Big) (\pi i)^2 \operatorname*{Res}_{\rho=\rho_n} W_0(\rho) \operatorname*{Res}_{\rho=\rho_m} W_0(\rho) \\
&= \sum_{n=0}^{n_0} \Big(q_k(\rho_n)\, \overline{q_l(\rho_n)} + \overline{q_k(\rho_n)} q_l(\rho_n)\Big)\ \pi i \operatorname*{Res}_{\rho=\rho_n} W_0(\rho).
\end{aligned}$$

Here we recognize the discrete part of the inner product $\langle \mathbf{q_k}, \mathbf{q_l}\rangle_{\mathcal{M}}$. Combined with Propositions 5.20 and 5.25 this shows that $\mathcal{F}$ acts isometric on the basis elements $\mathbf{q_k}$. By linearity $\mathcal{F}$ extends to an isometry. □

The continuous Hahn integral transform in case (i) corresponds exactly to the integral transform we found in §11.4.3 by formal computations.

Remark 5.28. *Let us denote the operator Λ by $\Lambda(k_1, k_2, t)$, let $w(x) = w(x; k_1, k_2, t)$, and let T_t denote the shift operator. Observe that*

$$w(x+t; k_2, k_1, -t) = w(x; k_1, k_2, t).$$

It is clear that $L^2(\mathbb{R}, w(x; k_1, k_2, t)\, dx)$ is invariant under the action of T_t. A short calculation shows that $T_{-t} \circ \Lambda(k_1, k_2, t) \circ T_t = \Lambda(k_2, k_1, -t)$, so $\varphi_\rho(x+t; -t, k_2, k_1, \varphi)$ is an eigenfunction of $\Lambda(k_1, k_2, t)$ for eigenvalue $\rho^2 + \frac{1}{4}$. Going through the whole machinery of this section again, then gives another spectral measure of $\Lambda(k_1, k_2, t)$, namely the one we found with $k_1 \mapsto k_2$, $k_2 \mapsto k_1$ and $t \mapsto -t$.

Finally we compare the spectrum of the difference operator Λ with the tensor product decomposition in Theorem 4.1. The discrete term in case (ii) in this section corresponds to one complementary series representation in the tensor product decomposition in Theorem 4.1. Case

(iii) does not occur in Theorem 4.1, since it is assumed that $k_1 \le k_2$. If we no longer assume this, the discrete terms in case (iii) correspond to a finite number of positive discrete series representations.

The finite number of negative discrete series in Theorem 4.1 can be obtained as described in Remark 5.28. In order to obtain the Clebsch-Gordan coefficients in this case from the summation formula in Theorem 3.1, we need to consider different overlap coefficients for the continuous series representations. Let $v_n(x;\lambda,\varepsilon,\varphi) := u_{-n}(x;\lambda,-\varepsilon,\pi-\varphi)$, where u_n is the Meixner-Pollaczek function as defined by (2.5). Then

$$\begin{pmatrix} \tilde{v}_{\rho,\varepsilon}(x) \\ \tilde{v}^*_{\rho,\varepsilon}(x) \end{pmatrix} = \sum_{n=-\infty}^{\infty} \begin{pmatrix} v_n\left(x; -\frac{1}{2}+i\rho, \varepsilon, \pi-\varphi\right) \\ v^*_n\left(x; -\frac{1}{2}+i\rho, \varepsilon, \pi-\varphi\right) \end{pmatrix} e_n,$$

is a generalized eigenvector of $\pi^{\rho,\varepsilon}(X_\varphi)$, see (Koelink, , §4.4.11). From Theorem 3.1, with $(k_1,k_2,x_1,x_2,p) \mapsto (k_2,k_1,x_2,x_1,-p)$ we find the Clebsch-Gordan coefficients for the eigenvector $\begin{pmatrix} \tilde{v}_{\rho,\varepsilon}(x_1-x_2) \\ \tilde{v}^*_{\rho,\varepsilon}(x_1-x_2) \end{pmatrix}$, and these Clebsch-Gordan coefficients are multiples of continuous Hahn functions

$$\varphi_\rho\left(x_1; x_2-x_1, k_2, k_1, \varphi\right)$$

and

$$\varphi^*_\rho\left(x_1; x_2-x_1, k_2, k_1, \varphi\right).$$

In this case the discrete mass points in the measure for the continuous Hahn transform correspond to one complementary series representation, or a finite number of negative discrete series representations.

References

Andrews, G. E., Askey, R., and Roy, R. (1999). *Special Functions*, volume 71 of *Encycl. Math. Appl.* Cambridge Univ. Press, Cambridge.

Bailey, W. N. (1972). *Generalized Hypergeometric Series.* Hafner, New York.

Basu, D. and Wolf, K. B. (1983). The Clebsch-Gordan coefficients of the three-dimensianal Lorentz algebra in the parabolic basis. *J. Math. Phys.*, 24:478–500.

Braaksma, B. L. J. and Meulenbeld, B. (1967). Integral transforms with generalized Legendre functions as kernels. *Compositio Math.*, 18:235–287.

de Jeu, M. (2003). Determinate multidimensional measures, the extended Carleman theorem and quasi-analytic weights. *Ann. Prob.*, 31:1205–1227. Available electronically: math.CA/0111019.

Engliš, M., Hille, S. C., Peetre, J., Rosengren, H., and Zhang, G. (2000). A new kind of Hankel-Toeplitz type operator connected with the complementary series. *Arab J. Math. Sci.*, 6:49–80.

Erdélyi, A., Magnus, W., Oberhettinger, F., and Tricomi, F. G. (1953). *Higher Transcendental Functions*, volume I, II. McGraw-Hill, New York-Toronto-London.

Götze, F. (1965). Verallgemeinerung einer integraltransformation von Mehler-Fock durch den von Kuipers und Meulenbeld eingefürten Kern $p_k^{m,n}(z)$. *Indag. Math.*, 27:396–404.

Granovskii, Y. I. and Zhedanov, A. S. (1993). New construction of $3nj$-symbols. *J. Phys. A: Math. Gen.*, 26:4339–4344.

Groenevelt, W. (2003). Laguerre functions and representations of $\mathfrak{su}(1,1)$. *Indag. Math.* To appear.

Groenevelt, W. and Koelink, E. (2002). Meixner functions and polynomials related to Lie algebra representations. *J. Phys. A: Math. Gen.*, 35:65–85.

Ismail, M. E. H. and Stanton, D. (2002). q-integral and moment representations for q-orthogonal polynomials. *Canad. J. Math.*, 54:709–735.

Koekoek, R. and Swarttouw, R. F. (1998). The Askey-scheme of hypergeometric orthogonal polynomials and its q-analogue. Technical Report 98-17, Technical University Delft, Delft.

Koelink, E. Spectral theory and special functions. In Marcellán, F., Van Assche, W., and Alvarez-Nordarse, R., editors, *Proceedings of the 2000 SIAG OP-SF Summer School on Orthogonal Polynomials and Special Functions (Laredo, Spain)*. Nova Science Publishers, Hauppauge, NY. Available electronically: math.CA/0107036.

Koelink, H. T. and Van der Jeugt, J. (1998). Convolutions for orthogonal polynomials from Lie and quantum algebra representations. *SIAM J. Math. Anal.*, 29:794–822.

Koornwinder, T. H. (1984). Jacobi functions and analysis on noncompact semisimple Lie groups. In Askey, R. A., Koornwinder, T. H., and Schempp, W., editors, *Special functions: Group Theoretical Aspects and Applications*, Math. Appl., pages 1–85. D. Reidel Publ. Comp., Dordrecht.

Koornwinder, T. H. (1988). Group theoretic interpretations of Askey's scheme of hypergeometric orthogonal polynomials. In Alfaro, et al., M., editor, *Orthogonal Polynomials and Their Applications*, volume 1329 of *Lecture Notes Math.*, pages 46–72. Springer-Verlag, Berlin.

Masson, D. R. and Repka, J. (1991). Spectral theory of Jacobi matrices in $\ell^2(\mathbb{Z})$ and the $\mathfrak{su}(1,1)$ Lie algebra. *SIAM J. Math. Anal.*, 22:1131–1146.

Mukunda, N. and Radhakrishnan, B. (1974). Clebsch-Gordan problem and coefficients for the three-dimensional Lorentz group in a continuous basis. II. *J. Math. Phys.*, 15:1332–1342.

Neretin, Y. A. (1986). Discrete occurrences of representations of the complementary series in tensor products of unitary representations. *Funktsional. Anal. i Prilozhen.*, 20:79–80. English translation: Funct. Anal. Appl. **20** (1986), 68–70.

Olver, F. W. J. (1974). *Asymptotics and Special Functions.* Academic Press, New York.

Paris, R. B. and Kaminski, D. (2001). *Asymptotics and Mellin-Barnes Integrals*, volume 85 of *Encycl. Math. Appl.* Cambridge Univ. Press, Cambridge.

Schmüdgen, K. (1990). *Unbounded Operator Algebras and Representation Theory*, volume 37 of *Operator theory: Advances and Applications.* Birkhäuser Verlag, Basel.

Slater, L. J. (1960). *Confluent Hypergeometric Functions.* Cambridge Univ. Press, Cambridge.

Slater, L. J. (1966). *Generalized Hypergeometric Functions.* Cambridge Univ. Press, Cambridge.

Van der Jeugt, J. (1997). Coupling coefficients for Lie algebra representations and addition formulas for special functions. *J. Math. Phys.*, 38:2728–2740.

Vilenkin, N. J. and Klimyk, A. U. (1991). *Representations of Lie Groups and Special Functions*, volume 1. Kluwer Academic Publishers, Dordrecht.

Whittaker, E. T. and Watson, G. N. (1963). *A Course of Modern Analysis*. Cambridge Univ. Press, Cambridge, fourth edition.

Wilson, J. A. (1980). Some hypergeometric orthogonal polynomials. *SIAM J. Math. Anal.*, 11:690–701.

NEW PROOFS OF SOME q-SERIES RESULTS*

Mourad E.H. Ismail
Department of Mathematics
University of Central Florida
Orlando, FL 32816
ismail@math.usf.edu

Ruiming Zhang
Department of Mathematics
University of South Florida
Tampa, FL 33620-5700

Abstract We use a Sheffer classification technique to give very short proofs of the addition theorem for the $\mathcal{E}_q$ function, the representation of $\mathcal{E}_q$ as a multiple of a ${}_2\varphi_1$, and a relatively new representation of $\mathcal{E}_q$. A direct proof of the evaluation of the connection coefficients of the Askey-Wilson polynomials and the Nassrallah-Rahman integral are also given. A simple proof of a characterization theorem for the continuous q-Hermite polynomials is also given.

Keywords: Sheffer classification, delta operators, polynomial bases, Askey-Wilson operators, the q-exponential function $\mathcal{E}_q$.

1. Introduction

By a polynomial sequence $\{f_n(x)\}$ we mean a sequence of polynomials such that $f_n(x)$ is a polynomial of exact degree n in x. Let T be a linear operator defined on polynomials and reduces the degree of a polynomial by 1, hence Tx is a nonzero constant. Given such operator one can inductively define a polynomial sequence $\{f_n(x)\}$ such that

$$f_0(x) = 1, \quad (Tf_n)(x) = f_{n-1}(x), n > 0. \tag{1.1}$$

*Research partially supported by NSF grant DMS 99-70865. This work was done at the University of South Florida

We say that a polynomial sequence $\{f_n(x)\}$ with $f_0(x) = 1$ belongs to an operator T if $(Tf_n)(x) = f_{n-1}(x)$. It is known that given a sequence of polynomials $\{f_n(x)\}$ with f_n of precise degree n then there exists an operator T of the form

$$T = \sum_{k=0}^{\infty} a_k(x) D^{k+1}, \qquad D = \frac{d}{dx}.$$

such that $a_k(x)$ has degree at most k and $Tf_n(x) = f_{n-1}(x)$. For details, see (Rainville, 1971, Chapter 13).

Theorem 1.1. *Assume that two polynomial sequences* $\{r_n(x)\}$ *and* $\{s_n(x)\}$ *belong to the same operator* T. *Then there is a sequence of constants* $\{a_n\}$ *with* $a_0 = 1$ *such that*

$$s_n(x) = \sum_{k=0}^{n} a_{n-k} r_k(x). \tag{1.2}$$

Conversely if (1.2) *holds with* $a_0 = 1$, *and* $\{r_n(x)\}$ *belongs to* T *then* $\{s_n(x)\}$ *also belongs to* T.

Theorem 1.1 is implicit in the Sheffer classification (Rainville, 1971), the umbral calculus (Rota et al., 1973) and the q-umbral calculus (Ismail, 2001) where additional assumptions are made but these assumptions do not enter in the proof of (1.2). In §2 we apply Theorem 1 to give few line proofs of three identities whose original proofs are lengthy.

Throughout this work we shall follow the notations and terminology in (Andrews et al., 1999) and (Gasper and Rahman, 1990) for basic hypergeometric series.

Theorem 1.1 will be applied where T is related to the the Askey-Wilson operator $\mathcal{D}_q$ which we now define. Given a function f we set $\breve{f}\left(e^{i\theta}\right) := f(x)$, $x = \cos\theta$, that is

$$\breve{f}(z) = f((z + 1/z)/2), \quad z = e^{i\theta}.$$

In other words we think of $f(\cos\theta)$ as a function of $e^{i\theta}$. The Askey-Wilson divided difference operator $\mathcal{D}_q$, (Ismail, 1995), (Askey and Wilson, 1985, p. 32), is defined by

$$(\mathcal{D}_q f)(x) = \frac{\breve{f}\left(q^{1/2}e^{i\theta}\right) - \breve{f}\left(q^{-1/2}e^{i\theta}\right)}{\left(q^{1/2} - q^{-1/2}\right) i \sin\theta}, \quad x = \cos\theta. \tag{1.3}$$

It easy to see that the action of $\mathcal{D}_q$ on Chebyshev polynomials is given by

$$\mathcal{D}_q T_n(x) = \frac{q^{n/2} - q^{-n/2}}{q^{1/2} - q^{-1/2}} U_{n-1}(x),$$

hence $\mathcal{D}_q$ reduces the degree of a polynomial by one and

$$\lim_{q\to 1} \mathcal{D}_q = \frac{d}{dx}.$$

For applications of $\mathcal{D}_q$ to deriving summation theorems see (Ismail, 1995) and (Cooper, 1996).

The polynomial bases we will be concerned with are

$$\begin{aligned}
\varphi_n(\cos\theta; a) &= \left(ae^{i\theta}, ae^{-i\theta}; q\right)_n, \\
\varphi_n(\cos\theta) &= \left(q^{1/4}e^{i\theta}, q^{1/4}e^{-i\theta}; q^{1/2}\right)_n, \\
\rho_n(\cos\theta) &= \left(1+e^{2i\theta}\right)\left(-q^{2-n}e^{2i\theta}; q^2\right)_{n-1} e^{-in\theta}, \\
u_n(\cos\theta, \cos\varphi) &= e^{-in\varphi}\left(-e^{i\varphi}q^{(1-n)/2}e^{i\theta}, -e^{i\varphi}q^{(1-n)/2}e^{-i\theta}; q\right)_n.
\end{aligned} \tag{1.4}$$

In the calculus of the Askey-Wilson operator the bases $\{\varphi_n(x;a)\}$, $\{\varphi_n(x)\}$, play the role played by $\{(1-2ax+a^2)^n\}$, and $\{(1-x)^n\}$, respectively, in the differential and integral calculus. On the other hand $\{\rho_n(x)\}$, and $\{u_n(x,y)\}$ play the role of $\{x^n\}$ and $\{(x+y)^n\}$, respectively, with $y = \cos\varphi$. For expansions of entire functions in the first three bases in (1.4) we refer the interested reader to (Ismail and Stanton, 2003b).

The q-exponential function of Ismail and Zhang (Ismail and Zhang, 1994) is

$$\begin{aligned}
\mathcal{E}_q(\cos\theta; t) &= \frac{(t^2;q^2)_\infty}{(qt^2;q^2)_\infty} \sum_{n=0}^{\infty} \frac{(-it)^n}{(q;q)_n} q^{n^2/4} \\
&\quad \times \left(-iq^{(1-n)/2}e^{i\theta}, -iq^{(1-n)/2}e^{-i\theta}; q\right)_n.
\end{aligned} \tag{1.5}$$

Ismail and Zhang (Ismail and Zhang, 1994) also defined a two variable q-exponential function by

$$\begin{aligned}
\mathcal{E}_q(\cos\theta, \cos\varphi; \alpha) &= \frac{(\alpha^2;q^2)_\infty}{(q\alpha^2;q^2)_\infty} \\
\times \sum_{n=0}^{\infty} &\left(-e^{i(\varphi+\theta)}q^{(1-n)/2}, -e^{i(\varphi-\theta)}q^{(1-n)/2}; q\right)_n \frac{\left(\alpha e^{-i\varphi}\right)^n}{(q;q)_n} q^{n^2/4}.
\end{aligned} \tag{1.6}$$

It is evident from (1.5) that the function $\mathcal{E}_q(x;t)$ is entire in x for all t, $|t| < 1$ and is analytic in t for $|t| < 1$ and all x.

In §3 we give a lemma, Lemma 3.1, which is a special case of the Nassrallah-Rahman integral (Nassrallah and Rahman, 1985) by using a change of base formula from (Ismail, 1995). Lemma 3.1 and q-integration by parts (Brown et al., 1996) are used to solve the connection coefficient problem for Askey-Wilson polynomials. The original proof of the connection coefficients is in (Askey and Wilson, 1985). In §4 we show that Lemma 3.1 and the Watson transformation imply the full Nassrallah-Rahman result.

In (Al-Salam, 1995) W. Al-Salam proved the following characterization theorem:

Theorem 1.2. *Let $\{p_n(x)\}$ be a polynomial sequence, $p_0(x) = 1$. If $\{p_n(x)\}$ are orthogonal polynomials and*

$$\sum_{n=0}^{\infty} \frac{q^{n^2/4}}{(q;q)_n} p_n(x) t^n = A(t)\mathcal{E}_q(x;t), \tag{1.7}$$

$A(t)$ is a formal power series in t, then $p_n(x)$ is a multiple of the q-Hermite polynomials, (Askey and Ismail, 1983).

In §5 we use Theorem 1.1 and Corollary 2.1 together with properties of continuous q-Hermite polynomials to give a simple proof of Theorem 1.2. We outline a derivation of every formula used, so the developments presented here are self-contained.

We hope that the elementary nature of the proofs presented here will lead to a better understanding of the subject matter.

2. Theorem 1.1 and its Applications

First note that given a sequence of polynomials $\{p_n(x)\}$ with $p_n(x)$ of degree n and $p_0(x) = 1$ we can define a linear operator T on the vector space of polynomials by $Tp_0(x) = 0$ and $Tp_n(x) = p_{n-1}$ for $n > 0$. Hence T always exists. For representation of T as a differential operator, possibly of infinite order, see Theorem 70 of §122 in Rainville (Rainville, 1971).

Proof of Theorem 1.1. Let

$$s_n(x) = \sum_{k=0}^{n} a_{n,k} r_k(x).$$

Thus

$$\sum_{k=0}^{n-1} a_{n-1,k} r_k(x) = s_{n-1}(x) = T s_n(x)$$

$$= \sum_{k=1}^{n} a_{n,k} T r_k(x) = \sum_{k=1}^{n} a_{n,k} r_{k-1}(x).$$

Hence $a_{n,k+1} = a_{n-1,k}$, or $a_{n,k+1} = a_{n-k-1,0}$ and (1.2) follows. Conversely given (1.2) construct the operator T to which $\{r_n(x)\}$ belongs. It follows from (1.2) that $\{s_n(x)\}$ belongs to T and the result follows. □

The following corollary is a restatement of Theorem 1.1 in the language of formal power series.

Corollary 2.1. *Two polynomial sequences* $\{r_n(x)\}$, *and* $\{s_n(x)\}$, *with* $r_0(x) = s_0(x) = 1$ *belong to the same operator* T *if and only if*

$$\sum_{n=0}^{\infty} s_n(x)t^n = A(t) \sum_{n=0}^{\infty} r_n(x)t^n, \tag{2.1}$$

where $A(t) = \sum\limits_{n=0}^{\infty} a_n t^n$ *with* $a_0 = 1$ *and* a_n *is independent of* x.

Ismail and Stanton (Ismail and Stanton, 2003a) proved

$$\mathcal{E}_q(\cos\theta; t) = \frac{(-t; q^{1/2})_\infty}{(qt^2; q^2)_\infty} \, {}_2\varphi_1 \left(\begin{matrix} q^{1/4}e^{i\theta}, q^{1/4}e^{-i\theta} \\ -q^{1/2} \end{matrix} \,\middle|\, q^{1/2}, -t \right), \tag{2.2}$$

and

$$\mathcal{E}_q(\cos\theta; t) = \sum_{k=0}^{\infty} \frac{\left(1 + e^{2i\theta}\right)\left(-e^{2i\theta}q^{-k}; q^2\right)_k}{(q;q)_k \left(1 + e^{2i\theta}q^{-k}\right)} \, q^{k^2/4} t^k e^{-ik\theta}. \tag{2.3}$$

Suslov (Suslov, 2001) proved the following addition theorem for $\mathcal{E}_q$

$$\mathcal{E}_q(x, y; t) = \mathcal{E}_q(x; t)\, \mathcal{E}_q(y; t). \tag{2.4}$$

Recall that

$$\mathcal{D}_q \varphi_n(x) = \frac{2q^{1/4}\left(1 - q^n\right)}{q - 1} \varphi_{n-1}(x), \tag{2.5}$$

$$\mathcal{D}_q u_n(x, y) = 2q^{(1-n)/2} \frac{1 - q^n}{1 - q} u_{n-1}(x, y), \tag{2.6}$$

and that (Ismail and Stanton, 2003a)

$$\mathcal{D}_q \rho_n(x) = 2q^{(1-n)/2} \frac{1 - q^n}{1 - q} \rho_{n-1}(x). \tag{2.7}$$

Proof of (2.2). It is clear from (2.5), (2.6), and (1.4) that

$$\frac{(q-1)^n q^{-n/4}}{2^n (q;q)_n} \varphi_n(x), \quad \text{and} \quad \frac{(1-q)^n q^{n(n-1)/4}}{2^n (q;q)_n} u_n(x, 0)$$

belong to the operator $\mathcal{D}_q$. Thus Corollary 2.1 and (1.5), after replacing t by $tq^{1/4}$, shows that there is a power series $B(t)$ such that

$$\mathcal{E}_q(x;t) = B(t)\,{}_2\varphi_1\left(\begin{matrix} q^{1/4}e^{i\theta}, q^{1/4}e^{-i\theta} \\ -q^{1/2} \end{matrix}\,\middle|\, q^{1/2}, -t\right). \tag{2.8}$$

Both series defining the above ${}_2\varphi_1$ and $\mathcal{E}_q(x;t)$ are analytic functions of t for $|t| < 1$ and are entire functions of x. Moreover $\mathcal{E}_q(x;t)$ and the ${}_2\varphi_1$ in (2.8) are equal to 1 at $t = 0$. Hence $B(t)$ is analytic in a neighborhood of $t = 0$. Substituting $x = \left(q^{1/4} + q^{-1/4}\right)/2$ we see that

$$\begin{aligned} B(t) &= \mathcal{E}_q\left(\left(q^{1/4} + q^{-1/4}\right)/2; t\right) \\ &= \frac{(t^2;q^2)_\infty}{(qt^2;q^2)_\infty} \sum_{n=0}^{\infty} \frac{q^{n^2/4}(-i)^n}{(q;q)_n} \left(-iq^{(1-2n)/4}; q^{1/2}\right)_{2n} t^n \\ &= \frac{(t^2;q^2)_\infty}{(qt^2;q^2)_\infty} \sum_{n=0}^{\infty} \frac{\left(-iq^{1/4}, iq^{1/4}; q^{1/2}\right)_n}{(q;q)_n} t^n \\ &= \frac{(t^2;q^2)_\infty}{(qt^2;q^2)_\infty} \sum_{n=0}^{\infty} \frac{(-q^{1/2};q)_n}{(q;q)_n} t^n = \frac{(t^2;q^2)_\infty\,(-tq^{1/2};q)_\infty}{(qt^2;q^2)_\infty\,(t;q)_\infty}, \end{aligned}$$

where we used the q-binomial theorem in the last step. A simplification completes the proof. □

Note that (2.2) implies

$$\begin{aligned} \mathcal{E}_q(0;t) &= \frac{\left(-t;q^{1/2}\right)_\infty}{(qt^2;q^2)_\infty}\,{}_1\varphi_0\left(-q^{1/2}; q, -t\right) \\ &= \frac{\left(-t;q^{1/2}\right)_\infty (tq^{1/2};q)_\infty}{(qt^2;q^2)_\infty (-t;q)_\infty}. \end{aligned}$$

Thus

$$\mathcal{E}_q(0;t) = 1. \tag{2.9}$$

Proof of (2.3) *and* (2.4). The relationships (2.6) and (2.7) imply that

$$\left\{\frac{(1-q)^n q^{n(n-1)/4}}{2^n (q;q)_n}\,\rho_n(x)\right\}, \quad \text{and} \quad \left\{\frac{(1-q)^n q^{n(n-1)/4}}{2^n (q;q)_n}\,u_n(x,y)\right\}$$

belong to $\mathcal{D}_q$ for all y. Thus there are formal power series $A(t)$ and $B(t)$ such that

$$\begin{aligned} \mathcal{E}_q(x;t) &= A(t) \sum_{n=0}^{\infty} \frac{q^{n^2/4}}{(q;q)_n}\,\rho_n(x)\,t^n, \\ \mathcal{E}_q(x,y;t) &= B(t)\,\mathcal{E}_q(x;t). \end{aligned}$$

This establishes (2.3) and (2.4) since $\rho_n(0) = \delta_{n,0}$, $\mathcal{E}_q(0, y; t) = \mathcal{E}_q(y; t)$ imply $A(t) = 1$ and $B(t) = \mathcal{E}_q(y; t)$. $\square$

3. Connection Coefficients

Ismail (Ismail, 1995) proved a q-Taylor series expansion for polynomials in $\varphi_n(x; a)$ and applied it to derive the q-analogue of the Pfaff-Saalschütz theorem in the form

$$\varphi_n(x; b) = \sum_{k=0}^{n} \frac{(q; q)_n \, (b/a)^k}{(q; q)_k \, (q; q)_{n-k}} \left(abq^k, b/a; q\right)_{n-k} \varphi_k(x; a). \tag{3.1}$$

Let

$$w(\cos\theta; \mathbf{a}) = \frac{\left(e^{2i\theta}, e^{-2i\theta}; q\right)_\infty}{\sin\theta \prod_{j=1}^{4} (a_j e^{i\theta}, a_j e^{-i\theta}; q)_\infty}, \tag{3.2}$$

where $\mathbf{a}$ stands for the vector (a_1, a_2, a_3, a_4). The Askey-Wilson integral is (Askey and Wilson, 1985)

$$\begin{aligned} I(a_1, a_2, a_3, a_4) &:= \int_0^\pi \frac{(e^{2i\theta}, e^{-2i\theta}; q)_\infty \, d\theta}{\prod_{j=1}^{4} (a_j e^{i\theta}, a_j e^{-i\theta}; q)_\infty} \\ &= \frac{2\pi \, (a_1 a_2 a_3 a_4; q)_\infty}{(q; q)_\infty \prod_{1 \le j < k \le 4} (a_j a_k; q)_\infty}. \end{aligned} \tag{3.3}$$

Lemma 3.1. *We have the evaluation*

$$\begin{aligned} &\int_0^\pi \frac{(e^{2i\theta}, e^{-2i\theta}; q)_\infty (\alpha e^{i\theta}, \alpha e^{-i\theta}; q)_n \, d\theta}{\prod_{j=1}^{4} (a_j e^{i\theta}, a_j e^{-i\theta}; q)_\infty} \\ &= \frac{2\pi \, (\alpha/a_4, \alpha a_4; q)_n \, (a_1 a_2 a_3 a_4; q)_\infty}{(q; q)_\infty \prod_{1 \le j < k \le 4} (a_j a_k; q)_\infty} \\ &\times {}_4\varphi_3\left(\begin{matrix} q^{-n}, a_1 a_4, a_2 a_4, a_3 a_4 \\ \alpha a_4, a_1 a_2 a_3 a_4, q^{1-n} a_4/\alpha \end{matrix} \,\middle|\, q, q \right) \end{aligned} \tag{3.4}$$

$$\begin{aligned} &= \frac{2\pi \, (a_1 a_2 a_3 a_4; q)_\infty \, (\alpha a_4; q)_n}{(q; q)_\infty \prod_{1 \le j < k \le 4} (a_j a_k; q)_\infty} \\ &\times \sum_{k=0}^{n} \frac{(q; q)_n}{(q; q)_k} \frac{(a_1 a_4, a_2 a_4, a_3 a_4; q)_k \, (\alpha/a_4; q)_{n-k}}{(\alpha a_4, a_1 a_2 a_3 a_4; q)_k \, (q; q)_{n-k}} \left(\frac{\alpha}{a_4}\right)^k \end{aligned} \tag{3.5}$$

Proof. Apply (3.1) with $b = \alpha$ and $a = a_4$ to see that the left-hand side of (3.4) is

$$\sum_{k=0}^{n} \frac{(q, \alpha a_4; q)_n \, (\alpha/a_4)^k}{(\alpha a_4, q; q)_k \, (q; q)_{n-k}} \, (\alpha/a_4; q)_{n-k} \, I\left(a_1, a_2, a_3, q^k a_4\right)$$

$$= \frac{2\pi \, (q, \alpha a_4; q)_n \, (a_1 a_2 a_3 a_4; q)_\infty}{(q; q)_\infty \prod\limits_{1 \le j < k \le 4} (a_j a_k; q)_\infty}$$

$$\times \sum_{k=0}^{n} \frac{(a_1 a_4, a_2 a_4, a_3 a_4; q)_k \, (\alpha/a_4; q)_{n-k}}{(q, \alpha a_4, a_1 a_2 a_3 a_4; q)_k \, (q; q)_{n-k}} \, (\alpha/a_4)^k,$$

and the results follow. □

Define the inner product (Brown et al., 1996)

$$\langle f, g \rangle := \int_{-1}^{1} \frac{f(x) \, \overline{g(x)}}{\sqrt{1-x^2}} \, dx. \tag{3.6}$$

Brown, Evans, and Ismail (Brown et al., 1996) proved that if $\breve{f}(z)$ and $\breve{g}(z)$ are analytic in the ring $q^{1/2} \le |z| \le q^{-1/2}$ then

$$\langle \mathcal{D}_q f, g \rangle = -\left\langle f, \sqrt{1-x^2} \, \mathcal{D}_q \left(g(x) \left(1-x^2\right)\right)^{-1/2} \right\rangle$$
$$+ \frac{\pi\sqrt{q}}{1-q} \left[f\left(\frac{1}{2}\left(q^{1/2} + q^{-1/2}\right)\right) \overline{g(1)} - f\left(-\frac{1}{2}\left(q^{1/2} + q^{-1/2}\right)\right) \overline{g(-1)} \right]. \tag{3.7}$$

The Askey-Wilson polynomials are defined by

$$p_n(\cos\theta; \mathbf{a}) = \frac{(a_1 a_2, a_1 a_3, a_1 a_4; q)_n}{a_1^n} \times {}_4\varphi_3\left(\begin{matrix} q^{-n}, a_1 a_2 a_3 a_4 q^{n-1}, a_1 e^{i\theta}, a_1 e^{-i\theta} \\ a_1 a_2, a_1 a_3, a_1 a_4 \end{matrix} \,\middle|\, q, q \right). \tag{3.8}$$

They have the property

$$\mathcal{D}_q p_n(x; \mathbf{a}) = 2q^{(1-n)/2} \frac{(1-q^n)\left(1 - a_1 a_2 a_3 a_4 q^{n-1}\right)}{1-q} p_{n-1}\left(x; q^{1/2}\mathbf{a}\right), \tag{3.9}$$

and satisfy the Rodrigues type formula

$$w(x; \mathbf{a}) \, p_n(x; \mathbf{a}) = \left[\frac{q-1}{2}\right]^n q^{n(n-1)/4} \mathcal{D}^n \left[w\left(x; \mathbf{a}q^{n/2}\right)\right]. \tag{3.10}$$

In the above $c\,\mathbf{a} = (c\,a_1, c\,a_2, c\,a_3, c\,a_4)$. Clearly the form of the Askey-Wilson polynomials in (3.8) makes the polynomials symmetric under $a_i \leftrightarrow a_j$, $i, j = 1, 2, 3$. The symmetry $a_1 \leftrightarrow a_2$ is the Sears transformation which can be proved using (3.1), see (Ismail, 1995). On the other hand (3.10) makes the full symmetry transparent.

The orthogonality relation of the Askey-Wilson polynomials is

$$\int_{-1}^{1} w(x; \mathbf{a}) p_m(x; \mathbf{a}) p_n(x; \mathbf{a})\, dx = h_n(\mathbf{a}) \delta_{m,n}, \tag{3.11}$$

where

$$\begin{aligned} h_n(\mathbf{a}) = &\frac{2\pi (q;q)_n\, (1 - a_1a_2a_3a_4/q)}{(a_1a_2a_3a_4/q;q)_n\, (1 - a_1a_2a_3a_4q^{2n-1})} \\ &\times \frac{(a_1a_2a_3a_4;q)_\infty}{(q;q)_\infty} \prod_{1 \le j < k \le 4} \frac{(a_ja_k;q)_n}{(a_ja_k;q)_\infty}. \end{aligned} \tag{3.12}$$

Theorem 3.2. *The Askey-Wilson polynomials satisfy the connection relation*

$$p_n(x; \mathbf{b}) = \sum_{k=0}^{n} c_{n,k}(\mathbf{a}, \mathbf{b})\, p_k(x; \mathbf{a}), \tag{3.13}$$

where

$$\begin{aligned} c_{n,k}(\mathbf{b}, \mathbf{a}) = &\frac{(q;q)_n\, b_4^{k-n}\, (b_1b_2b_3b_4q^{n-1};q)_k\, (b_1b_4, b_2b_4, b_3b_4;q)_n}{(q;q)_{n-k}\, (q, a_1a_2a_3a_4q^{k-1};q)_k\, (b_1b_4, b_2b_4, b_3b_4;q)_k} \\ &\times q^{k(k-n)} \sum_{j,l \ge 0} \frac{(q^{k-n}, b_1b_2b_3b_4q^{n+k-1}, a_4b_4q^k;q)_{j+l}\, q^{j+l}}{(b_1b_4q^k, b_2b_4q^k, b_3b_4q^k;q)_{j+l}\, (q;q)_j (q;q)_l} \\ &\times \frac{(a_1a_4q^k, a_2a_4q^k, a_3a_4q^k;q)_l\, (b_4/a_4;q)_j}{(a_4b_4q^k, a_1a_2a_3a_4q^{2k};q)_l} \left(\frac{b_4}{a_4}\right)^l. \end{aligned} \tag{3.14}$$

Proof. Clearly the coefficients $c_{n,k}$ exist and are given by

$$h_k(\mathbf{a}) c_{n,k} = \left\langle \sqrt{1 - x^2}\, p_n(x; \mathbf{b}), w(x; \mathbf{a}) p_k(x; \mathbf{a}) \right\rangle. \tag{3.15}$$

Using (3.7)–(3.10), (3.1), and (3.15) we proceed in the following steps

$$\begin{aligned}
h_k(\mathbf{a})c_{n,k} &= \left[\frac{q-1}{2}\right]^k q^{k(k-1)/4}\left\langle \sqrt{1-x^2}\,p_n(x;\mathbf{b}), \mathcal{D}_q^k w\left(x;q^{k/2}\mathbf{a}\right)\right\rangle \\
&= \left[\frac{1-q}{2}\right]^k q^{k(k-1)/4}\left\langle \mathcal{D}_q^k p_n(x;\mathbf{b}), \sqrt{1-x^2}\,w\left(x;q^{k/2}\mathbf{a}\right)\right\rangle \\
&= q^{k(k-n)/2}\left(b_1b_2b_3b_4q^{n-1};q\right)_k \frac{(q;q)_n}{(q;q)_{n-k}} \\
&\quad\times\left\langle p_{n-k}\left(x;q^{k/2}\mathbf{b}\right), \sqrt{1-x^2}\,w\left(x;q^{k/2}\mathbf{a}\right)\right\rangle \\
&= b_4^{k-n}\left(b_1b_2b_3b_4q^{n-1};q\right)_k\left(b_1b_4q^k, b_2b_4q^k, b_3b_4q^k;q\right)_{n-k} \\
&\quad\times q^{k(k-n)}\frac{(q;q)_n}{(q;q)_{n-k}}\sum_{j=0}^{n-k}\frac{\left(q^{k-n}, b_1b_2b_3b_4q^{n+k-1};q\right)_j}{\left(q, b_1b_4q^k, b_2b_4q^k, b_3b_4q^k;q\right)_j} q^j \\
&\quad\times\left\langle \varphi_j\left(x;b_4q^{k/2}\right), \sqrt{1-x^2}\,w\left(x;q^{k/2}\mathbf{a}\right)\right\rangle.
\end{aligned}$$

Using Lemma 3.1 we see that the j-sum is

$$\begin{aligned}
&\frac{2\pi\left(a_1a_2a_3a_4q^{2k};q\right)_\infty}{(q;q)_\infty \prod\limits_{1\le r<s\le 4}\left(a_ra_sq^k;q\right)_\infty} \\
&\times\sum_{j=0}^{n-k}\frac{\left(q^{k-n}, b_1b_2b_3b_4q^{n+k-1}, a_4b_4q^k;q\right)_j}{\left(b_1b_4q^k, b_2b_4q^k, b_3b_4q^k;q\right)_j} \\
&\times\sum_{l=0}^{j}\frac{\left(a_1a_4q^k, a_2a_4q^k, a_3a_4q^k;q\right)_l}{\left(q, a_4b_4q^k, a_1a_2a_3a_4q^{2k};q\right)_l}\frac{(b_4/a_4;q)_{j-l}}{(q;q)_{j-l}}\left(\frac{b_4}{a_4}\right)^l,
\end{aligned}$$

and some manipulations and the use of (3.12) one completes the proof. □

Theorem 3.2 seems to be new. Although Askey and Wilson (Askey and Wilson, 1985) only considered the case when $a_4 = b_4$, they were aware that the connection coefficients are double sums, as Askey kindly pointed out in a private conversation. To get the Askey-Wilson result set $a_4 = b_4$ in (3.13) to obtain

$$c_{n,k}\left(b_1, b_2, b_3, a_4; a_1, a_2, a_3, a_4\right)$$
$$= \left(b_1 b_2 b_3 a_4 q^{n-1}; q\right)_k \frac{q^{k(k-n)}(q;q)_n \left(b_1 a_4 q^k, b_2 a_4 q^k, b_3 a_4 q^k; q\right)_{n-k}}{a_4^{n-k}(q;q)_{n-k}\left(q, a_1 a_2 a_3 a_4 q^{k-1}; q\right)_k}$$
$$\times {}_5\varphi_4\left(\left.\begin{matrix} q^{k-n}, b_1 b_2 b_3 a_4 q^{n+k-1}, a_1 a_4 q^k, a_2 a_4 q^k, a_3 a_4 q^k \\ b_1 a_4 q^k, \quad b_2 a_4 q^k, \quad b_3 a_4 q^k, \quad a_1 a_2 a_3 a_4 q^{2k} \end{matrix}\,\right|\, q, q\right). \tag{3.16}$$

Askey and Wilson (Askey and Wilson, 1985) also pointed out that if in addition to $a_4 = b_4$, we also have $b_j = a_j$ for $j = 2, 3$ then the ${}_5\varphi_4$ becomes a ${}_3\varphi_2$ which can be summed by the q-analogue of the Pfaff-Saalschütz theorem. This is evident from (3.14).

Now define an $(N+1) \times (N+1)$ lower triangular matrix $C(\mathbf{a}, \mathbf{b})$ whose n, k element is $c_{n,k}(\mathbf{a}, \mathbf{b})$, with $0 \le k \le n \le N$. Thus (3.13) is

$$X(\mathbf{b}) = C(\mathbf{b}, \mathbf{a}) X(\mathbf{a}), \tag{3.17}$$

where $X(\mathbf{a})$ is a column vector whose jth component is $p_j(x; \mathbf{a})$, $0 \le j \le N$. Therefore the family of matrices $C(\mathbf{a}, \mathbf{b})$ has the property

$$C(\mathbf{c}, \mathbf{b}) C(\mathbf{b}, \mathbf{a}) = C(\mathbf{c}, \mathbf{a}). \tag{3.18}$$

Furthermore

$$[C(\mathbf{b}, \mathbf{a})]^{-1} = C(\mathbf{a}, \mathbf{b}). \tag{3.19}$$

The implications of the orthogonality relation $[C(\mathbf{b}, \mathbf{a})]^{-1} C(\mathbf{a}, \mathbf{b}) = I$, I being the identity matrix will be explored in a future work.

4. The Nassrallah-Rahman Integral

We now state and prove the Nassrallah-Rahman integral, (Gasper and Rahman, 1990, 6.3.7)).

Theorem 4.1. *We have for* $|a_j| < 1$; $1 \le j \le 5$, *the evaluation*

$$\int_0^\pi \frac{\left(e^{2i\theta}, e^{-2i\theta}; q\right)_\infty \left(\alpha e^{i\theta}, \alpha e^{-i\theta}; q\right)_\infty d\theta}{\prod_{j=1}^{5} \left(a_j e^{i\theta}, a_j e^{-i\theta}; q\right)_\infty}$$
$$= \frac{2\pi \left(\alpha/a_4, \alpha a_4, a_1 a_2 a_3 a_4, a_1 a_3 a_4 a_5, a_2 a_3 a_4 a_5; q\right)_\infty}{\left(q, a_1 a_2 a_3 a_4^2 a_5; q\right)_\infty \prod_{1 \le j < k \le 5} \left(a_j a_k; q\right)_\infty}$$
$$\times {}_8W_7\left(a_1 a_2 a_3 a_4^2 a_5/q; a_1 a_4, a_2 a_4, a_3 a_4, a_4 a_5, a_1 a_2 a_3 a_4 a_5/\alpha; q, \alpha/a_4\right). \tag{4.1}$$

Proof. Let $a_5 = \alpha q^n$ and apply Lemma 3.1. Next apply (III.20) in (Gasper and Rahman, 1990) to the ${}_4\varphi_3$ in Lemma 3.1 with the choices:

$$a = a_1 a_4, \quad b = a_2 a_4, \quad c = a_3 a_4, \quad e = a_1 a_2 a_3 a_4,$$
$$d = \alpha a_4, \quad f = q^{1-n} a_4/\alpha, \quad \mu = a_1 a_2 a_3 a_4^2 \alpha q^{n-1}.$$

This establishes the theorem when $a_5 = \alpha q^n$. Since both sides of (4.1) are analytic functions of α the identity theorem for analytic functions establishes the result. □

5. A Characterization Theorem

Recall the generating function of the continuous q-Hermite polynomials (Askey and Ismail, 1983)

$$\sum_{n=0}^{\infty} \frac{H_n(\cos\theta \mid q)}{(q;q)_n} t^n = \frac{1}{(te^{i\theta}, te^{-i\theta}; q)_\infty}, \tag{5.1}$$

which implies

$$\mathcal{D}_q H_n(x \mid q) = \frac{2q^{(1-n)/2}}{1-q} (1-q^n) H_{n-1}(x \mid q), \tag{5.2}$$

and

$$H_{2n+1}(0 \mid q) = 0, \qquad \frac{H_{2n}(0 \mid q)}{(q;q)_{2n}} = \frac{(-1)^n}{(q^2;q^2)_n}. \tag{5.3}$$

Before proving Al-Salam's theorem we record the generating function (Ismail and Zhang, 1994)

$$\left(qt^2;q^2\right)_\infty \mathcal{E}_q(x;t) = \sum_{n=0}^{\infty} \frac{q^{n^2/4}}{(q;q)_n} H_n(x \mid q) t^n. \tag{5.4}$$

To prove formula (5.4) simply note that both

$$\frac{(1-q)^n q^{n(n-1)/4}}{2^n (q;q)_n} H_n(x \mid q) \quad \text{and} \quad \frac{(1-q)^n q^{n(n-1)/4}}{2^n (q;q)_n} u_n(x,0)$$

belong to the operator $\mathcal{D}$. Therefore

$$\sum_{n=0}^{\infty} \frac{q^{n^2/4}}{(q;q)_n} H_n(x \mid q) t^n = A(t) \mathcal{E}_q(x;t),$$

and

$$A(t) = \sum_{n=0}^{\infty} \frac{q^{n^2/4}}{(q;q)_n} H_n(0 \mid q) t^n = \sum_{n=0}^{\infty} \frac{q^{n^2}(-1)^n}{(q^2;q^2)_n} t^{2n} = \left(qt^2;q^2\right)_\infty,$$

and (5.4) follows.

Recall the three term recurrence relation (Askey and Ismail, 1983)

$$2xH_n(x \mid q) = H_{n+1}(x \mid q) + (1 - q^n)\, H_{n-1}(x \mid q), \tag{5.5}$$

hence $r_n(x) = H_n(x \mid q)/(q;q)_n$ satisfies

$$2xr_n(x) = \left(1 - q^{n+1}\right) r_{n+1}(x) + r_{n-1}(x). \tag{5.6}$$

Proof of Theorem 1.2. Let $\{p_n(x)\}$ have the generating function (1.7). Thus $s_n(x) = p_n/(q;q)_n$ is related to the r_n's via

$$s_n(x) = \sum_{k=0}^{n} a_k\, q^{-nk/2} r_{n-k}(x). \tag{5.7}$$

Let the three term recurrence relation of the s_n's be

$$2xs_n(x) = \left(1 - q^{n+1}\right) s_{n+1}(x) + c_n s_n(x) + d_n s_{n-1}(x). \tag{5.8}$$

Substitute from (5.7) into (5.8) and use (5.6) to obtain

$$\begin{aligned}
&\sum_{k=0}^{n} a_k q^{-nk/2} \left[\left(1 - q^{n-k+1}\right) r_{n-k+1} + r_{n-k-1}\right] \\
&= c_n \sum_{k=0}^{n} a_k\, q^{-nk/2} r_{n-k}(x) + d_n \sum_{k=0}^{n-1} a_k q^{-k(n-1)/2} r_{n-k-1}(x) \\
&\quad +(1 - q^{n+1}) \sum_{k=0}^{n+1} a_k q^{-k(n+1)/2} r_{n-k+1}(x).
\end{aligned} \tag{5.9}$$

By equating the coefficients of $r_n(x)$ and $r_{n-1}(x)$ in (5.9) we find

$$\begin{aligned}
c_n &= a_1 q^{-(n+1)/2} \left(q^{1/2} - 1\right) \left(1 + q^{n+1/2}\right), \\
d_n &= 1 + a_2 q^{-n-1} (q - 1) \left(1 + q^n\right) \\
&\quad -a_1^2 q^{-n-1/2} \left(q^{1/2} - 1\right) \left(1 + q^{n+1/2}\right).
\end{aligned} \tag{5.10}$$

For $k \geq 2$ equating coefficients of $r_{n-k}(x)$ in (5.9) then replacing c_n and d_n by their values from (5.10) give

$$\begin{aligned}
&a_{k+1} \left(1 - q^{n-k}\right) + a_{k-1} q^n = a_{k+1} \left(1 - q^{n+1}\right) q^{-(k+1)/2} \\
&\quad +a_k a_1 q^{-1/2} \left(q^{1/2} - 1\right) \left(1 + q^{n+1/2}\right) + a_{k-1} q^{(k-1)/2} \\
&\times \left[q^n + a_2 \left(1 - q^{-1}\right) \left(1 + q^n\right) - a_1^2 \left(1 - q^{-1/2}\right) \left(1 + q^{n+1/2}\right)\right].
\end{aligned} \tag{5.11}$$

Since the above equation must hold for all $n > 3$, the coefficients of q^n and the constant terms on both sides must be identical. This gives two recurrence relations for a_k, which must be compatible. If $a_1 = a_2 = 0$ a calculation proves $a_k = 0$ for all $k \geq 3$ and we conclude that $p_n(x) = H_n(x \mid q)$. When $a_1 = 0$, but $a_2 \neq 0$ then (5.11) shows that $a_{2k+1} = 0$ for $k \geq 0$ and

$$a_{2k} = \frac{(1-q)}{1-q^k} q^{2k-2} a_2 \, a_{2k-2}. \tag{5.12}$$

The consistence condition implies $a_2 \left(1 - q^{-1}\right) = -1$. Hence $p_n(x) = H_n(x \mid 1/q)$. Through a change of variable the polynomials are orthogonal when $q > 1$ on the imaginary axis (Askey, 1989). If $a_1 a_2 \neq 0$ the compatibility condition leads to a contradiction, so this case does not arise. □

Dedication

It is very appropriate to dedicate this paper to our good friend Mizan Rahman for his influence and seminal contributions to the subject of q-series and for always being a friend who has always been there for us.

Acknowledgments

The authors gratefully acknowledge the page full of remarks made by each of two referees who read the paper very carefully. The referees' comments improved both the presentation and accuracy of the final version.

References

Al-Salam, W. A. (1995). Characterization of the Rogers q-Hermite polynomials. *Internat. J. Math. Math. Sci.*, 18:641–647.

Andrews, G. E., Askey, R. A., and Roy, R. (1999). *Special Functions.* Cambridge University Press, Cambridge.

Askey, R. A. (1989). Continuous q-Hermite polynomials when $q > 1$. In Stanton, D., editor, *q-Series and Partitions*, IMA Volumes in Mathematics and Its Applications, pages 151–158. Springer-Verlag, New York.

Askey, R. A. and Ismail, M. E. H. (1983). A generalization of ultraspherical polynomials. In Erdős, P., editor, *Studies in Pure Mathematics*, pages 55–78. Birkhauser, Basel.

Askey, R. A. and Wilson, J. A. (1985). Some basic hypergeometric orthogonal polynomials that generalize Jacobi polynomials. *Memoirs Amer. Math. Soc.*, 54(319):iv+55.

Brown, B. M., Evans, W. D., and Ismail, M. E. H. (1996). The Askey-Wilson polynomials and q-Sturm-Liouville problems. *Math. Proc. Cambridge Phil. Soc.*, 119:1–16.

Cooper, S. (1996). The Askey-Wilson operator and the ${}_6\varphi_5$ summation formula. Preprint.

Gasper, G. and Rahman, M. (1990). *Basic Hypergeometric Series.* Cambridge University Press, Cambridge.

Ismail, M. E. H. (1995). The Askey-Wilson operator and summation theorems. In Ismail, M., Nashed, M. Z., Zayed, A., and Ghaleb, A., editors, *Mathematical Analysis, Wavelets, and Signal Processing*, volume 190 of *Contemporary Mathematics*, pages 171–178. American Mathematical Society, Providence.

Ismail, M. E. H. (2001). An operator calculus for the Askey-Wilson operator. *Annals of Combinatorics*, 5:333–348.

Ismail, M. E. H. and Stanton, D. (2003a). Applications of q-Taylor theorems. *J. Comp. Appl. Math.*, 153:259–272.

Ismail, M. E. H. and Stanton, D. (2003b). q-Taylor theorems, polynomial expansions, and interpolation of entire functions. *J. Approx. Theory*, 123:125–146.

Ismail, M. E. H. and Zhang, R. (1994). Diagonalization of certain integral operators. *Advances in Math.*, 109:1–33.

Nassrallah, B. and Rahman, M. (1985). Projection formulas, a reproducing kernel and a generating function for q-Wilson polynomials. *SIAM J. Math. Anal.*, 16:186–197.

Rainville, E. D. (1971). *Special Functions.* Chelsea Publishing Co., Bronx, NY.

Rota, G.-C., Kahaner, D., and Odlyzko, A. (1973). On the foundations of combinatorial theory. VIII. Finite operator calculus. *J. Math. Anal. Appl.*, 42:684–760.

Suslov, S. K. (1997). Addition theorems for some q-exponential and trigonometric functions. *Methods and Applications of Analysis*, 4:11–32.

THE LITTLE q-JACOBI FUNCTIONS OF COMPLEX ORDER

Kevin W. J. Kadell
Department of Mathematics and Statistics
Arizona State University
Tempe, AZ 85287-1804

Abstract We use Ismail's argument and an elementary combinatorial identity to prove the q-binomial theorem, the symmetry of the Rogers-Fine function, Ramanujan's ${}_1\psi_1$ sum, and Heine's q-Gauss sum and give many other proofs of these results. We prove a special case of Heine's ${}_2\varphi_1$ transformation and write Ramanujan's ${}_1\psi_1$ sum as the nonterminating q-Chu-Vandermonde sum. We show that the q-Saalschütz and q-Chu-Vandermonde sums are equivalent to the evaluations of certain moments and to the orthogonality of the little q-Jacobi polynomials; hence the q-Chu-Vandermonde sum implies the q-Saalschütz sum. We extend the little q-Jacobi polynomials naturally to the little q-Jacobi functions of complex order. We show that the nonterminating q-Saalschütz and q-Chu-Vandermonde sums are equivalent to the evaluations of certain moments and, using the Liouville-Ismail argument, to two orthogonality relations. We show that the nonterminating q-Chu-Vandermonde sum implies the nonterminating q-Saalschütz sum.

Keywords: little q-Jacobi polynomials, basic hypergeometric functions, bilateral basic hypergeometric functions, q-binomial theorem, Rogers-Fine symmetric function, Ramanujan's ${}_1\psi_1$ sum, Heine's q-Gauss sum and ${}_2\varphi_1$ transformation, q-Chu-Vandermonde sum, q-Saalschütz sum, nonterminating sum, and Macdonald, Bidenharn-Louck, Heckman-Opdam, Koornwinder and Sahi-Knop polynomials

1. Introduction and summary

The celebrated Askey tableau of orthogonal polynomials (see (Koekoek and Swarttouw, 1994)) is founded on the Jacobi and little q-Jacobi polynomials (see (Hahn, 1949)). It is naturally related to the standard results for ordinary and basic hypergeometric functions surveyed by Gasper and Rahman (Gasper and Rahman, 1990). See also (Andrews, 1965; Andrews, 1969; Andrews, 1972; Andrews, 1998), (Andrews and Askey, 1977; Andrews and Askey, 1978; Andrews and Askey, 1981), (Askey,

1978), (Askey and Ismail, 1979), (Askey and Wilson, 1985), (Fine, 1988), (Gasper, 1997), (Heine, 1847; Heine, 1878), (Ismail, 1977), (Ismail and Rahman, 1995), (Joichi and Stanton, 1989), (Kadell, 1987a; Kadell, 1987b), (Rogers, 1983; Rogers, 1916), (Rahman and Suslov, 1994b; Rahman and Suslov, 1994a; Rahman and Suslov, 1996; Rahman and Suslov, 1998a; Rahman and Suslov, 1998b), (Suslov, 1998) by Andrews, Askey, Fine, Gasper, Heine, Ismail, Joichi, Kadell, Rahman, Rogers, Stanton, Suslov and Wilson.

Askey (Askey, 1980) conjectured a number of extensions of Selberg's integral (Selberg, 1944) which use the beta-type distributions of the Askey tableau. Macdonald (Macdonald, 1982) and Morris (Morris, II, 1982) conjectured certain constant term identities associated with root systems. Anderson (Anderson, 1991) and Aomoto (Aomoto, 1987) developed alternative proofs of Selberg's integral. See (Evans, 1994), (Garvan, 1989a; Garvan, 1989b), (Garvan and Gonnet, 1991), (Gustafson, 1990), (Habsieger, 1988), (Habsieger, 1986), (Kadell, 1988a; Kadell, 1994a; Kadell, 1994b; Kadell, 1998), (Mehta, 1967), (Stembridge, 1988), (Zeilberger, 1988; Zeilberger, 8990) by Evans, Garvan, Gonnet, Gustafson, Habsieger, Kadell, Mehta, Stembridge and Zeilberger for proofs of these conjectures and closely related results.

Heckman and Opdam (Heckman, 1987; Heckman and Opdam, 1987; Opdam, 1988a; Opdam, 1988b) introduced Jacobi polynomials and basic hypergeometric functions associated with root systems. Opdam (Opdam, 1989) gave a norm evaluation which eclipsed the Macdonald-Morris constant term q-conjecture. Baker and Forrester (Baker and Forrester, 1999), Kadell (Kadell, 1988b; Kadell, 1993; Kadell, 1997; Kadell, 2000a; Kadell, 2000b), Kaneko (Kaneko, 1993; Kaneko, 1996; Kaneko, 1998), Knop and Sahi (Knop and Sahi, 1997), Macdonald (Macdonald, 1995 Chap. VI), Opdam (Opdam, 1995), Richards (Richards, 1989), Stanley (Stanley, 1989), and Vinet and Lapointe (Vinet and Lapointe, 1995) developed further aspects of the theory associated with the root system A_{n-1}. Sahi and Knop (Sahi, 1994; Sahi and Knop, 1996) introduced polynomials related to the hypergeometric distribution which generalized the Biedenharn-Louck (Biedenharn and Louck, 1989) polynomials. Koornwinder (Koornwinder, 1995) introduced polynomials associated with the root system BC_n. See also Kadell (Kadell, 2003) and Macdonald (Macdonald, 1998).

In (Kadell, 2000b), we used the classical ratio of alternants to extend the Schur functions to partitions with complex parts. Following Proctor (Proctor, 1989), we gave a combinatorial representation as an alternating sum over the symmetric group, and established the Pieri formula and two Selberg q-integrals which extend results of Hua (Hua, 1963) and Kadell (Kadell, 1993; Kadell, 1997).

Our goal is to lay the groundwork for extending these marvelous polynomials naturally to functions of complex arguments, to give q-integrals which serve as orthogonality relations, and to generalize other basic properties. While (Kadell, 2000b) treats the case $k=1$ (see also Macdonald (Macdonald, 1992)), we treat the case $n=2$ with a focus on q-series identities and Ismail's argument (Ismail, 1977).

Let $|q|<1$, let $n\geq 0$ be a nonnegative integer, and let $(a;q)_n = \prod_{i=1}^{n}\left(1-aq^{i-1}\right)$. The basic hypergeometric function ${}_{r+1}\varphi_r\left[a_1,\ldots,a_{r+1}; b_1,\ldots,b_r;q,t\right]$ is given by

$$ {}_{r+1}\varphi_r\left[\begin{matrix}a_1,\ldots,a_{r+1}\\ b_1,\ldots,b_r\end{matrix};q,t\right] = \sum_{i=0}^{\infty}\frac{(a_1;q)_i\cdots(a_{r+1};q)_i}{(q;q)_i\,(b_1;q)_i\cdots(b_r;q)_i}\,t^i, \tag{1.1} $$

where we assume throughout that $|t|<1$ and that there are no poles in the denominator.

The q-Saalschütz sum (Gasper and Rahman, 1990, (1.7.2)), (Andrews, 1998, (3.3.12)) is given by

$$ {}_3\varphi_2\left[\begin{matrix}q^{-n},a,b\\ c,abq^{1-n}/c\end{matrix};q,q\right] = \frac{(c/a;q)_n\,(c/b;q)_n}{(c;q)_n\,(c/ab;q)_n}. \tag{1.2} $$

If we set $b=0$ in the q-Saalschütz sum (1.2) and relabel the parameters, we obtain the q-Chu-Vandermonde sum (Gasper and Rahman, 1990, (1.5.2)), (Andrews, 1998, (3.3.10))

$$ {}_2\varphi_1\left[\begin{matrix}q^{-n},a\\ b\end{matrix};q,bq^n/a\right] = \frac{(b/a;q)_n}{(b;q)_n}. \tag{1.3} $$

If we let n tend to infinity in the q-Saalschütz sum (1.2), we obtain Heine's (Heine, 1878) q-Gauss sum (Gasper and Rahman, 1990, (1.5.1)), (Andrews, 1998, Corollary 2.4)

$$ {}_2\varphi_1\left[\begin{matrix}a,b\\ c\end{matrix};q,c/ab\right] = \frac{(c/a;q)_\infty\,(c/b;q)_\infty}{(c;q)_\infty\,(c/ab;q)_\infty}. \tag{1.4} $$

Observe that the q-Chu-Vandermonde sum (1.3) is the case $a=q^{-n}$ of (1.4) with the parameters relabeled.

Letting b and c in Heine's q-Gauss sum (1.4) tend to zero with $c/b = at$, we obtain the q-binomial theorem (Gasper and Rahman, 1990, (1.3.2)), (Andrews, 1998, (2.2.1))

$$ {}_1\varphi_0\left[\begin{matrix}a\\ \end{matrix};q,t\right] = \frac{(at;q)_\infty}{(t;q)_\infty}. \tag{1.5} $$

Let $q \neq 0$ and let n be an integer. We may extend the q-Pockhammer symbol to integers by

$$(a;q)_n = \frac{(a;q)_\infty}{(aq^n;q)_\infty}. \tag{1.6}$$

The basic bilateral hypergeometric function ${}_r\psi_r\,[a_1,\ldots,a_r;b_1,\ldots,b_r;q,t]$ is given by

$${}_r\psi_r\left[\begin{matrix}a_1,\ldots,a_r\\ b_1,\ldots,b_r\end{matrix};q,t\right] = \sum_{i=-\infty}^{\infty} \frac{(a_1;q)_i\cdots(a_r;q)_i}{(b_1;q)_i\cdots(b_r;q)_i}\,t^i, \tag{1.7}$$

where we assume throughout that $|a_1\cdots a_r/b_1\cdots b_r| < |t| < 1$ and that there are no poles in the denominator.

Andrews (Andrews, 1969) used a limit of Heine's q-Gauss sum (1.4) to obtain Ramanujan's ${}_1\psi_1$ sum (Gasper and Rahman, 1990, (5.2.1))

$${}_1\psi_1\left[\begin{matrix}a\\ b\end{matrix};q,t\right] = \frac{(q;q)_\infty\,(b/a;q)_\infty\,(at;q)_\infty\,(q/at;q)_\infty}{(b;q)_\infty\,(q/a;q)_\infty\,(t;q)_\infty\,(b/at;q)_\infty}, \tag{1.8}$$

where $|b/a| < |t| < 1$. Ismail (Ismail, 1977) used the q-binomial theorem (1.5) and analysis to give a short and elegant proof of (1.8); he also gives references to other proofs of (1.8). We now call his argument Ismail's argument. See Kadell (Kadell, 1987b) for a probabilistic proof of (1.8).

Rogers (Rogers, 1983; Rogers, 1916) and Fine (Fine, 1988) studied the function

$$F(a,b;t) = \sum_{i=0}^{\infty} \frac{(aq;q)_i}{(bq;q)_i}\,t^i. \tag{1.9}$$

Many of their results are related to the fact that the function

$$\frac{1}{(1-b)}\,F(ab,b;t) = \frac{1}{(1-t)}\,F(at,t;b) \tag{1.10}$$

is symmetric in b and t. See Starcher (Starcher, 1930) for some applications to number theory.

All of these q-series identities are consequences of Heine's (Heine, 1847; Heine, 1878) ${}_2\varphi_1$ transformation (Gasper and Rahman, 1990, (1.4.1)), (Andrews, 1998, Corollary 2.3)

$${}_2\varphi_1\left[\begin{matrix}a,b\\ c\end{matrix};q,t\right] = \frac{(b;q)_\infty\,(at;q)_\infty}{(c;q)_\infty\,(t;q)_\infty}\,{}_2\varphi_1\left[\begin{matrix}c/b,t\\ at\end{matrix};q,b\right]. \tag{1.11}$$

Gasper and Rahman (Gasper and Rahman, 1990, Section 2.10) used the case $t = q$ of (1.11) to write Ramanujan's ${}_1\psi_1$ sum (1.8) as the nonterminating q-Chu-Vandermonde sum (Gasper and Rahman, 1990, (2.10.13))

$$ {}_2\varphi_1\left[\begin{matrix} a,b \\ c \end{matrix};q,q\right] - \frac{q}{c}\,\frac{(q^2/c;q)_\infty\,(a;q)_\infty\,(b;q)_\infty}{(c;q)_\infty\,(aq/c;q)_\infty\,(bq/c;q)_\infty}\,{}_2\varphi_1\left[\begin{matrix} aq/c,bq/c \\ q^2/c \end{matrix};q,q\right] = \frac{(q/c;q)_\infty\,(abq/c;q)_\infty}{(aq/c;q)_\infty\,(bq/c;q)_\infty}. \tag{1.12} $$

Observe that (1.12) follows by setting $c=0$ and replacing e by c in the nonterminating q-Saalschütz sum (Gasper and Rahman, 1990, (2.10.12))

$$ \begin{aligned} &{}_3\varphi_2\left[\begin{matrix} a,b,c \\ e,abcq/e \end{matrix};q,q\right] \\ &\quad -\frac{q}{e}\,\frac{(q^2/e;q)_\infty\,(a;q)_\infty\,(b;q)_\infty\,(c;q)_\infty\,(abcq^2/e^2;q)_\infty}{(e;q)_\infty\,(aq/e;q)_\infty\,(bq/e;q)_\infty\,(cq/e;q)_\infty\,(abcq/e;q)_\infty} \\ &\qquad\qquad \times\,{}_3\varphi_2\left[\begin{matrix} aq/e,bq/e,cq/e \\ q^2/e,abcq^2/e^2 \end{matrix};q,q\right] \\ &= \frac{(q/e;q)_\infty\,(abq/e;q)_\infty\,(acq/e;q)_\infty\,(bcq/e;q)_\infty}{(aq/e;q)_\infty\,(bq/e;q)_\infty\,(cq/e;q)_\infty\,(abcq/e;q)_\infty}. \end{aligned} \tag{1.13} $$

See Sears (Sears, 1951) for another proof of (1.13).

Let α, β have positive real part and let μ be complex. Let $q \neq 0$ with a fixed natural logarithm and

$$ q^\mu = e^{\mu ln(q)}. \tag{1.14} $$

Following Jackson (Jackson, 1910), we have the q-integral

$$ \int_0^1 f(t)\,d_qt = (1-q)\sum_{i=0}^{\infty} q^i\,f(q^i) \tag{1.15} $$

and the q-gamma function

$$ \Gamma_q(\alpha) = (1-q)^{1-\alpha}\,\frac{(q;q)_\infty}{(q^\alpha;q)_\infty}. \tag{1.16} $$

Gasper and Rahman (Gasper and Rahman, 1990, Section 1.10) showed that the q-beta integral (Gasper and Rahman, 1990, (1.11.7))

$$ \int_0^1 t^{\alpha-1}\,\frac{(tq;q)_\infty}{(tq^\beta;q)_\infty}\,d_qt = \frac{\Gamma_q(\alpha)\,\Gamma_q(\beta)}{\Gamma_q(\alpha+\beta)} \tag{1.17} $$

follows using the q-binomial theorem (1.5); we may view (1.17) as a q-integral formulation of (1.5). See Kadell (Kadell, 1987b) for a probabilistic proof of (1.17).

Hahn (Hahn, 1949) showed that the little q-Jacobi polynomials (Gasper and Rahman, 1990, (7.3.1))

$$p_n^{\alpha,\beta}(t) = {}_2\varphi_1\left[\begin{matrix} q^{-n}, q^{n+\alpha+\beta-1} \\ q^{\alpha} \end{matrix}; q, tq\right] \tag{1.18}$$

are orthogonal with respect to the q-beta integral (1.17). Andrews and Askey (Andrews and Askey, 1977) solved the connection coefficient problem and deduced the Rogers-Ramanujan identities.

Let $m \geq 0$, $\ell \geq 0$ be nonnegative integers, let $\alpha+x$, $\beta+y$ have positive real part, and let

$$I_n^{\alpha,\beta}(x,y) = \int_0^1 t^{\alpha+x-1} \frac{(tq;q)_\infty}{(tq^{\beta+y};q)_\infty} p_n^{\alpha,\beta}(t)\, d_q t \tag{1.19}$$

denote the moment of $t^x \left(tq^\beta;q\right)_\infty / \left(tq^{\beta+y};q\right)_\infty$ with respect to the weight function (1.17) times the little q-Jacobi polynomial $p_n^{\alpha,\beta}(t)$. The q-Saalschütz sum (1.2), the evaluation of the moment $I_n^{\alpha,\beta}(x,0)$, and the orthogonality of the little q-Jacobi polynomials are equivalent.

Similarly, the q-Chu-Vandermonde sum (1.3), the evaluation of the moment $I_n^{\alpha,\beta}(x, n-x-1)$, and the orthogonality of the little q-Jacobi polynomials are equivalent.

Our main result is to extend the little q-Jacobi polynomials naturally to the little q-Jacobi functions of complex order. We show that the nonterminating q-Saalschütz (1.13) and q-Chu-Vandermonde (1.12) sums are equivalent to the evaluations of the moments $I_n^{\alpha,\beta}(x,0)$ and $I_n^{\alpha,\beta}(x, n-x-1)$, respectively, and, using Ismail's argument (Ismail, 1977), that (1.12) implies (1.13).

In Section 2, we use an elementary combinatorial identity and Ismail's argument (Ismail, 1977) to establish the q-binomial theorem (1.5) and Ramanujan's ${}_1\psi_1$ sum (1.8).

In Section 3, we recall (Kadell, 1987a, Section 7) the path function proof of the q-binomial theorem (1.5). We extend the path function to the plane and establish the symmetry (1.10) of the Rogers-Fine function. We show that (1.10) also follows using Ismail's argument (Ismail, 1977).

In Section 4, we recall (Kadell, 1987a, Section 7) the path function proof of Heine's q-Gauss sum (1.4) which implies Ramanujan's ${}_1\psi_1$ sum (1.8) which implies the symmetry (1.10) of the Rogers-Fine function. We use Ismail's argument (Ismail, 1977) to establish Heine's q-Gauss sum

(1.4) using Ramanujan's ${}_1\psi_1$ sum (1.8) and the elementary combinatorial identity of Section 2.

In Section 5, we give a path function proof of the case $t = q$ of Heine's ${}_2\varphi_1$ transformation (1.11) which implies the symmetry (1.10) of the Rogers-Fine function. Following Gasper and Rahman (Gasper and Rahman, 1990, Section 2.10), we write Ramanujan's ${}_1\psi_1$ sum (1.8) as the nonterminating q-Chu-Vandermonde sum (1.12).

In Section 6, we show following Andrews and Askey (Andrews and Askey, 1977) that the q-Saalschütz sum (1.2) and the q-Chu-Vandermonde sum (1.3) are equivalent to the evaluations of the moments $I_n^{\alpha,\beta}(x,0)$ and $I_n^{\alpha,\beta}(x, n-x-1)$, respectively, and that each of these is equivalent to the orthogonality of the little q-Jacobi polynomials. Hence the q-Chu-Vandermonde sum (1.3) implies the q-Saalschütz sum (1.2).

In Section 7, we extend the little q-Jacobi polynomials naturally to the little q-Jacobi functions of complex order. We show that the nonterminating q-Saalschütz (1.13) and q-Chu-Vandermonde (1.12) sums are equivalent to the evaluations of the moments $I_n^{\alpha,\beta}(x,0)$ and $I_n^{\alpha,\beta}(x, n-x-1)$, respectively, and, using the Liouville-Ismail argument (Hille, 1973, Theorem 8.2.2) (Ismail, 1977), to two orthogonality relations. We show that (1.12) implies (1.13).

In Section 8, we conclude with some thoughts on the Hankel determinant, the Rodriguez formula, the slinky rule for the Schur functions, the q-Dyson polynomials, q-series identities, and the Vinet operator.

2. The q-binomial theorem (1.5) and Ramanujan's ${}_1\psi_1$ sum (1.8)

In this section, we use an elementary combinatorial identity and Ismail's argument (Ismail, 1977) to establish the q-binomial theorem (1.5) and Ramanujan's ${}_1\psi_1$ sum (1.8).

Let $\pi \in S_n$ be a permutation of the integers from one through n. We define the inversion number of π by

$$\operatorname{inv}(\pi) = \Big|\{(i,j) \mid 1 \le i < j \le n \text{ and } \pi(i) > \pi(j)\}\Big|. \tag{2.1}$$

Let π' denote the permutation obtained by removing n from π. We then have

$$\operatorname{inv}(\pi) = n - \pi^{-1}(n) + \operatorname{inv}(\pi'). \tag{2.2}$$

The sum of a finite geometric series is given by

$$\sum_{i=0}^{n-1} q^{n-i} = \frac{(1-q^n)}{(1-q)}. \tag{2.3}$$

Using induction on n and taking $i = \pi^{-1}(n)$ in (2.3), we have the generating function

$$\sum_{\pi \in S_n} q^{\mathrm{inv}(\pi)} = \frac{(q;q)_n}{(1-q)^n}. \tag{2.4}$$

Let $0 \le m \le n$ and let $[a,b] = \{i \mid a \le i \le b\}$ denote the interval from a to b. Let $M \subseteq [1,n]$ with $|M| = m$. We define the inversion number of M by

$$\mathrm{inv}(M) = \sum_{\substack{1 \le i < j \le n \\ i \in M,\ j \notin M}} 1. \tag{2.5}$$

Let $M \subseteq [1,n]$ be defined by $i \in M \iff \pi(i) \le m$ and let π_1 and π_2 denote the permutations obtained by listing the values of π in $[1,m]$ and $[m+1,n]$, respectively. We then have

$$\sum_{i \in M} i - 1 = \binom{m}{2} + \mathrm{inv}(M), \tag{2.6}$$

where $\sum\limits_{i=1}^{m} i - 1 = \binom{m}{2} = m(m-1)/2$ and

$$\mathrm{inv}(\pi) = \mathrm{inv}\,(\pi_1) + \mathrm{inv}\,(\pi_2) + \mathrm{inv}(M). \tag{2.7}$$

Using (2.4), (2.6) and (2.7), we have

$$\frac{(q;q)_m}{(1-q)^m}\,\frac{(q;q)_{n-m}}{(1-q)^{n-m}} \sum_{\substack{M \subseteq [1,n] \\ |M| = m}} q^{\sum_{i \in M} i-1} = q^{\binom{m}{2}}\,\frac{(q;q)_n}{(1-q)^n}. \tag{2.8}$$

Since the powers of $(1-q)$ cancel, we may rearrange (2.8) as the generating function

$$\sum_{\substack{M \subseteq [1,n] \\ |M| = m}} q^{\sum_{i \in M} i-1} = q^{\binom{m}{2}}\,\frac{(q;q)_n}{(q;q)_m\,(q;q)_{n-m}} = q^{\binom{m}{2}}\,\frac{\left(q^{n-m+1};q\right)_m}{(q;q)_m}. \tag{2.9}$$

See Kendall and Stuart (Kendall and Stuart, 1973) for the natural statistical version of this argument.

Observe that (2.9) gives the Laurent expansion

$$(x;q)_n = \sum_{i=0}^{n} (-x)^i\, q^{\binom{i}{2}}\,\frac{\left(q^{n-i+1};q\right)_i}{(q;q)_i} \tag{2.10}$$

of a polynomial whose zeroes form a geometric series. Using the identity (Gasper and Rahman, 1990, (I.7))

$$(a;q)_n = (-a)^n q^{\binom{n}{2}} \left(q^{1-n}/a;q\right)_n \tag{2.11}$$

for reversing the q-Pockhammer symbol, we have

$$\begin{aligned}\frac{(q^{-n};q)_i}{(q;q)_i}(xq^n)^i &= (-q^{-n})^i q^{\binom{i}{2}} \frac{\left(q^{n-m+1};q\right)_i}{(q;q)_i}(xq^n)^i \\ &= (-x)^i q^{\binom{i}{2}} \frac{\left(q^{n-i+1};q\right)_i}{(q;q)_i}.\end{aligned} \tag{2.12}$$

Hence we may write (2.10) as

$${}_1\varphi_0\left[\begin{matrix}q^{-n} \\ \end{matrix};q,xq^n\right] = (x;q)_n, \tag{2.13}$$

which is the case $a = q^{-n}$, $t = xq^n$ of the q-binomial theorem (1.5).

Replacing a and t by $1/a$ and ax, respectively, in the q-binomial theorem (1.5), we have the Laurent expansion

$${}_1\varphi_0\left[\begin{matrix}1/a \\ \end{matrix};q,ax\right] = \frac{(x;q)_\infty}{(ax;q)_\infty} \tag{2.14}$$

in the disc $|x| < 1/|a|$.

Recall the uniform convergence theorem Hille (Hille, 1973, Theorem 7.10.3) that a sum of functions which are analytic on the domain $\mathcal{D}$ and converges uniformly on compact subsets of $\mathcal{D}$ converges to a function which is analytic on $\mathcal{D}$ and we may differentiate term by term.

Observe that

$$(1/a;q)_i\, a^i = (a-1)\cdots\left(a-q^{i-1}\right) \tag{2.15}$$

is a polynomial and hence is an entire function of a. Using comparison with the sum (2.3) of a geometric series for the left side and an analysis of the partial products on the right side, we see that the functions on both sides of (2.14) are analytic in a in the disc $|a| < 1/|x|$.

Recall the identity theorem Hille (Hille, 1973, Section 8.1) that two functions which are analytic in the domain $\mathcal{D}$ and agree at infinitely many points which include an accumulation point in $\mathcal{D}$ agree throughout $\mathcal{D}$.

Observe that (2.14) holds when $a = q^n$ since in that case it reduces to (2.13). Hence we see that (2.14) holds for a in the disc $|a| < 1/|x|$.

We call this argument, see Ismail (Ismail, 1977) and Askey and Ismail (Askey and Ismail, 1979), Ismail's argument.

Let $\pi = (\pi_1, \pi_2, \ldots)$ be a partition; thus $\pi_1 \geq \pi_2 \geq \cdots \geq 0$ and the norm of π, denoted by $|\pi| = \sum_{i=1}^{\infty} \pi_i$, is finite. Let $\ell(\pi)$ denote the number of non zero parts of π.

Observe that

$$M = \{\pi_1 + m,\ \pi_2 + m - 1,\ \ldots,\ \pi_m + 1\} \tag{2.16}$$

gives a bijecton between partitions π with $\pi_1 \leq n-1$, $\ell(\pi) \leq m$ and subsets $M \subseteq [1, n+m-1]$ with $|M| = m$ such that

$$\sum_{i \in M} i - 1 = \binom{m}{2} + |\pi|. \tag{2.17}$$

We see that (2.9) becomes the generating function

$$\sum_{\pi_1 \leq n-1,\ \ell(\pi) \leq m} q^{|\pi|} = \frac{(q^n; q)_m}{(q; q)_m} \tag{2.18}$$

for partitions with at most m parts which are less than or equal to $n-1$.

Observe (see Andrews (Andrews, 1998, Chapter 1)) that (2.18) gives the Laurent expansion

$${}_1\varphi_0 \left[\begin{matrix} q^n \\ \ \end{matrix} ; q, t \right] = \frac{1}{(t; q)_n} \tag{2.19}$$

in the disc $|t| < 1$. Since (2.19) is the case $a = q^n$ of the q-binomial theorem (1.5), we see that (1.5) follows by applying Ismail's argument [34] to the parameter a.

Let $m \geq 0$, $n \geq 0$ be nonnegative integers. Using (2.11) and the fact that $m + \binom{m}{2} = \binom{m+1}{2}$, we have

$$\begin{aligned} (x; q)_n \, (q/x; q)_m &= (x; q)_n \, (-q/x)^m \, q^{\binom{m}{2}} \, (x/q^m; q)_m \\ &= (-1/x)^m \, q^{\binom{m+1}{2}} \, (x/q^m; q)_{m+n} \, . \end{aligned} \tag{2.20}$$

Observe that the zeroes of (2.20) form a finite geometric series. Using the substitution $i = s + m$ to shift the index of summation and then replacing s by i, we see by (2.11) that

$$(x/q^m; q)_{m+n} = \sum_{i=-m}^{n} (-x/q^m)^{i+m} \, q^{\binom{i+m}{2}} \, \frac{(q^{n-i+1}; q)_{i+m}}{(q; q)_{i+m}}. \tag{2.21}$$

Substituting (2.21) into (2.20) and using the fact that $\binom{m+1}{2} - m(i + m) + \binom{i+m}{2} = \binom{i}{2}$, we obtain

$$(x; q)_n \, (q/x; q)_m = \sum_{i=-m}^{n} (-x)^i \, q^{\binom{i}{2}} \, \frac{(q^{n-i+1}; q)_{i+m}}{(q; q)_{i+m}}, \tag{2.22}$$

which is an equivalent formulation of (2.10).

Observe that if we let m and n tend to infinity in (2.22) and multiply by $(q;q)_\infty$, we obtain the Jacobi triple product identity (Gasper and Rahman, 1990, (1.6.1))

$$(q;q)_\infty\,(x;q)_\infty\,(q/x;q)_\infty = \sum_{i=-\infty}^{\infty} (-x)^i\, q^{\binom{i}{2}}. \tag{2.23}$$

The reader may compare this with the proof of (2.23) given by Andrews (Andrews, 1965, Theorem 2.8).

We now use a variant of Ismail's argument (Ismail, 1977) to establish Ramanujan's ${}_1\psi_1$ sum (1.8) using the formulation (2.22) of the q-binomial theorem (1.5).

Making the substitution $x = at$, $A = 1/a$, $B = b$ in Ramanujan's ${}_1\psi_1$ sum (1.8) and rearranging the result, we have the equivalent Laurent expansion

$$\frac{(x;q)_\infty\,(q/x;q)_\infty}{(Ax;q)_\infty\,(B/x;q)_\infty} = \frac{(Aq;q)_\infty\,(B;q)_\infty}{(q;q)_\infty\,(AB;q)_\infty}\;{}_1\psi_1\left[\begin{matrix}1/A\\ B\end{matrix};q,Ax\right] \tag{2.24}$$

in the annulus $|B| < |x| < 1/|A|$. Observe that we require $|AB| < 1$ in order that the annulus is not empty.

Setting $A = q^n$, $B = q^{m+1}$ in (2.24), we have

$$(x;q)_n\,(q/x;q)_m = \frac{\left(q^{n+1};q\right)_m}{(q;q)_m}\;{}_1\psi_1\left[\begin{matrix}q^{-n}\\ q^{m+1}\end{matrix};q,xq^n\right]. \tag{2.25}$$

Observe by (1.6) that the ${}_1\psi_1$ on the right side of (2.25) is a sum from $-m$ to n. Using (2.11), we have

$$\left(q^{-n};q\right)_i = \left(-q^{-n}\right)^i q^{\binom{i}{2}} \left(q^{n-i+1};q\right)_i. \tag{2.26}$$

Substituting (2.26) into (2.25), we readily obtain (2.22).

Following Ismail (Ismail, 1977), we observe that the functions on both sides of (2.24) are analytic in A and B in the discs $|A| < 1/|x|$ and $|B| < |x|$, respectively. Since by (2.22) we have that (2.24) holds when $A = q^n$, $B = q^{m+1}$, we see that (2.24) follows by two successive applications of Ismail's argument (Ismail, 1977) to the parameters A and B. That is, we fix $B = q^{m+1}$ and establish (2.24) when A is in the disc $|A| < 1/|x|$. Then, we fix $A = q^n$ and establish (2.24) when B is in the disc $|B| < |x|$.

3. The symmetry (1.10) of the Rogers-Fine function

In this section, we recall (Kadell, 1987a, Section 7) the path function proof of the q-binomial theorem (1.5). We extend the path function to the plane and establish the symmetry (1.10) of the Rogers-Fine function. We show that (1.10) also follows using Ismail's argument (Ismail, 1977).

Suppressing the parameters throughout, we let η be an invertible linear operator and use the convention

$$t_0 = 1, \quad a_0 = 0. \tag{3.1}$$

Recall (Kadell, 1987a, Section 6) that the separation $\{a_n\}_{n\geq 1}$ of $\{t_n\}_{n\geq 1}$ with respect to η, which is recursively defined by (3.1) and (Kadell, 1987a, (6.11))

$$t_n + a_{n+1} = a_n + \eta(t_n), \quad n \geq 1, \tag{3.2}$$

provides a measure (Kadell, 1987a, (6.13))

$$a_n = \eta\left(1 + t_1 + \cdots + t_{n-1}\right) - \left(1 + t_1 + \cdots + t_{n-1}\right), \qquad n \geq 1, \tag{3.3}$$

of how close η comes to fixing the partial sums $1 + \sum_{i=1}^{n-1} t_i$.

Taking the limit of (3.3), we see that

$$\Pi = 1 + \sum_{i=1}^{\infty} t_i \tag{3.4}$$

satisfies the functional equation

$$\Pi = \eta(\Pi) - \lim_{n\to\infty} a_n. \tag{3.5}$$

Since (3.2) reduces to

$$(1-t) + t\left(1 - aq^n\right) = (1 - q^n) + q^n\left(1 - at\right), \tag{3.6}$$

we have (Kadell, 1987a, Section 7) the separation

$$a_n = \frac{1}{(1-t)} \frac{(a;q)_n}{(q;q)_{n-1}} t^n, \; t_n = \frac{(a;q)_n}{(q;q)_n} t^n, \qquad n \geq 1, \tag{3.7}$$

with respect to the operator η given by

$$\eta(\Pi(a;t)) = \frac{(1-at)}{(1-t)} \Pi(a;tq). \tag{3.8}$$

Since

$$\lim_{n\to\infty} a_n = 0, \tag{3.9}$$

we see by (3.5) that

$$\Pi(a;t) = \sum_{i=0}^{\infty} \frac{(a;q)_i}{(q;q)_i} t^i \tag{3.10}$$

satisfies the functional equation

$$\Pi(a;t) = \frac{(1-at)}{(1-t)}\Pi(a;tq). \tag{3.11}$$

The q-binomial theorem (1.5) follows by repeated application of (3.11).

We define a path function (Kadell, 1987a, (7.6)) by taking

$$\begin{aligned} H_{i,j} &= \eta^{j-1}(a_i) = \frac{(at;q)_{j-1}}{(t;q)_j}\frac{(a;q)_i}{(q;q)_{i-1}}\, t^i\, q^{i(j-1)}, \\ V_{i,j} &= \eta^{j-1}(t_i) = \frac{(at;q)_{j-1}}{(t;q)_{j-1}}\frac{(a;q)_i}{(q;q)_i}\, t^i\, q^{i(j-1)}. \end{aligned} \tag{3.12}$$

to be the integrals over paths starting at the point (i,j) and moving one unit to the right or downward, respectively, and extending by linearity. Observe by (1.6) that

$$i \text{ is a negative integer} \implies \frac{1}{(q;q)_i} = \frac{(q^{i+1};q)_\infty}{(q;q)_\infty} = 0, \tag{3.13}$$

so that the path function (3.12) is zero over the upper half plane.

Observe that η moves the underlying path one unit to the right. Applying η^{j-1} to (3.2), we obtain

$$V_{i,j} + H_{i+1,j} = H_{i,j} + V_{i,j+1}, \tag{3.14}$$

which gives the Cauchy property that the integral over a path depends only on the endpoints. Observe that (3.5) follows by integrating from $(0,1)$ to $(n,2)$ going through $(0,2)$ or $(n,1)$.

We may view (3.2) as saying that $\{a_n\}_{n\geq 1}$ and $\{t_n\}_{n\geq 1}$ represent elements of a group which correspond to translations to the right and downward and which commute with each other. As Ismail's argument (Ismail, 1977) suggests, we may extend the path function (3.12) by replacing $(q;q)_{n-1}$ and $(q;q)_n$ by $(b;q)_{n-1}$ and $(b;q)_n$, respectively. However, we obtain an equivalent extension naturally using the symmetry (1.10) of the Rogers-Fine function.

Let $|b|<1$, $|t|<1$. Using (1.9), the symmetry (1.10) of the Rogers-Fine function becomes

$$\sum_{i=0}^{\infty}\frac{(abq;q)_i}{(b;q)_{i+1}}\, t^i = \sum_{i=0}^{\infty}\frac{(atq;q)_i}{(t;q)_{i+1}}\, b^i. \tag{3.15}$$

Placing no restriction on i and j, we set

$$H_{0,j} = \frac{(abq;q)_j}{(b;q)_{j+1}}\, t^j,\; V_{i,0} = \frac{(atq;q)_i}{(t;q)_{i+1}}\, b^i, \tag{3.16}$$

so that (3.15) becomes

$$\sum_{j=0}^{\infty} H_{0,j} = \sum_{i=0}^{\infty} V_{i,0}. \tag{3.17}$$

Observe that the operators

$$\begin{aligned} \eta_{\mathrm{H}}(\Pi(a,b;t)) &= t\,\frac{(1-abq)}{(1-b)}\,\Pi(a,bq;t), \\ \eta_{\mathrm{V}}(\Pi(a,b;t)) &= b\,\frac{(1-atq)}{(1-t)}\,\Pi(a,b;tq), \end{aligned} \tag{3.18}$$

are invertible and commute with each other, and that we have

$$H_{0,j} = \eta_{\mathrm{H}}^{j}(H_{0,0}),\; V_{i,0} = \eta_{\mathrm{V}}^{i}(V_{0,0}), \tag{3.19}$$

in agreement with (3.16). We see using (3.19) that the path function

$$\begin{aligned} H_{i,j} &= \eta_{\mathrm{V}}^{i}\left(H_{0,j}\right) = \frac{(atq;q)_i}{(t;q)_i}\,\frac{(abq;q)_j}{(b;q)_{j+1}}\,b^i t^j\,q^{ij}, \\ V_{i,j} &= \eta_{\mathrm{H}}^{j}\left(V_{i,0}\right) = \frac{(atq;q)_i}{(t;q)_{i+1}}\,\frac{(abq;q)_j}{(b;q)_j}\,b^i t^j\,q^{ij}, \end{aligned} \tag{3.20}$$

satisfies the translation property

$$\begin{aligned} H_{i+m,j} &= \eta_{\mathrm{V}}^{m}\left(H_{i,j}\right), \; H_{i,j+m} = \eta_{\mathrm{H}}^{m}\left(H_{i,j}\right), \\ V_{i+m,j} &= \eta_{\mathrm{V}}^{m}\left(V_{i,j}\right), \;\; V_{i,j+m} = \eta_{\mathrm{H}}^{m}\left(V_{i,j}\right). \end{aligned} \tag{3.21}$$

We easily check that

$$H_{0,0} + V_{0,1} = \frac{1}{(1-b)} + \frac{(1-abq)}{(1-b)\,(1-t)}\,t = \frac{(1-abtq)}{(1-b)\,(1-t)} \tag{3.22}$$

and, interchanging b and t,

$$V_{0,0} + H_{1,0} = \frac{1}{(1-t)} + \frac{(1-atq)}{(1-b)\,(1-t)}\,b = \frac{(1-abtq)}{(1-b)\,(1-t)}. \tag{3.23}$$

Since these are equal, we see that the Cauchy integral formula (3.14) holds when $i = j = 0$. Using (3.21), we may extend (3.14) to the plane.

The formulation (3.17) of the symmetry (1.10) of the Rogers-Fine function now follows using the Cauchy integral formula (3.14) and the fact that

$$0 = \lim_{n\to\infty} \sum_{i=0}^{n} V_{i,n} = \lim_{n\to\infty} \sum_{j=0}^{n} V_{n,j}. \tag{3.24}$$

Andrews and Askey (Andrews and Askey, 1978) gave a simple proof of Ramanujan's ${}_1\psi_1$ sum (1.8) which essentially used the q-binomial theorem (1.5) and the extended path function (3.19).

We see by the Cauchy integral formula (3.14) that the integrals

$$\sum_{j=1}^{\infty} H_{n,j} = \sum_{i=n}^{\infty} V_{i,1} \tag{3.25}$$

of the path function (3.12) from $(n, 1)$ to infinity along horizontal and vertical paths are equal. The reader may verify that (3.25) equals $t^n\,(a;q)_n/(q;q)_{n-1}$ times the case $b = q^n$ of the symmetry (1.10) of the Rogers-Fine function. Hence (1.10) also follows by applying Ismail's argument (Ismail, 1977) to the parameter b.

4. Heine's q-Gauss sum (1.4)

In this section, we recall (Kadell, 1987a, Section 7) the path function proof of Heine's q-Gauss sum (1.4) which implies Ramanujan's ${}_1\psi_1$ sum (1.8) which implies the symmetry (1.10) of the Rogers-Fine function. We use Ismail's argument (Ismail, 1977) to establish Heine's q-Gauss sum (1.4) using Ramanujan's ${}_1\psi_1$ sum (1.8) and the elementary combinatorial identity of Section 2.

Since (3.2) reduces to

$$\begin{aligned}(1-b)\,(1-cq^n) - (c/a)\,(1-aq^n)\,(1-bq^n)\\ = -b\,(1-q^n)\,(1-cq^n) + (1-c/a)\,(1-bq^n),\end{aligned} \tag{4.1}$$

we have the separation

$$\begin{aligned} a_n &= -b\,\frac{(a;q)_n\,(bq;q)_{n-1}}{(q;q)_{n-1}\,(c;q)_n}\,(c/ab)^n,\\ t_n &= \frac{(a;q)_n\,(b;q)_n}{(q;q)_n\,(c;q)_n}\,(c/ab)^n,\\ &\quad n \geq 1,\end{aligned} \tag{4.2}$$

with respect to the operator η given by

$$\eta(\Pi(a,b;c)) = \frac{(1-c/a)}{(1-c)}\,\Pi(a,bq;cq). \tag{4.3}$$

Since (3.9) holds, we see by (3.5) that

$$\Pi(a,b;c) = \sum_{i=0}^{\infty} \frac{(a;q)_i\,(b;q)_i}{(q;q)_i\,(c;q)_i}\,(c/ab)^i \tag{4.4}$$

satisfies the functional equation

$$\Pi(a,b;c) = \frac{(1-c/a)}{(1-c)}\,\Pi(a,bq;cq). \tag{4.5}$$

Heine's q-Gauss sum (1.4) follows by repeated application of (4.5) and the q-binomial theorem (1.5).

Observe by (1.6) that

$$(a;q)_{-n} = \frac{(a;q)_\infty}{(aq^{-n};q)_\infty} = \frac{1}{(aq^{-n};q)_n}. \tag{4.6}$$

Using (2.11), (4.6) and the fact that $n^2 - \binom{n}{2} = \binom{n+1}{2}$, we have

$$(a;q)_{-n} = \frac{1}{(-aq^{-n})^n\, q^{\binom{n}{2}}\,(q/a;q)_n} = (-1/a)^n\, q^{\binom{n+1}{2}}\,\frac{1}{(q/a;q)_n} \tag{4.7}$$

and hence

$$\begin{aligned}\sum_{i=-\infty}^{-1} \frac{(a;q)_i}{(b;q)_i}\,t^i &= \sum_{i=1}^{\infty} \frac{(a;q)_{-i}}{(b;q)_{-i}}\,t^{-i} = \sum_{i=1}^{\infty} \frac{(q/b;q)_i}{(q/a;q)_i}\,(b/at)^i \\ &= \frac{1}{t}\,\frac{(b-q)}{(a-q)} \sum_{i=0}^{\infty} \frac{(q^2/b;q)_i}{(q^2/a;q)_i}\,(b/at)^i.\end{aligned} \tag{4.8}$$

Substituting (4.8) into Ramanujan's ${}_1\psi_1$ sum (1.8), replacing a, b by abq, bq, respectively, and dividing by $(1-b)$, we have the Laurent expansion

$$\begin{aligned}&\frac{1}{(1-b)} \sum_{i=0}^{\infty} \frac{(abq;q)_i}{(bq;q)_i}\,t^i + \frac{1}{t}\,\frac{1}{(1-ab)} \sum_{i=0}^{\infty} \frac{(q/b;q)_i}{(q/ab;q)_i}\,(1/at)^i \\ &= \frac{(q;q)_\infty\,(1/a;q)_\infty\,(abtq;q)_\infty\,(1/abt;q)_\infty}{(b;q)_\infty\,(1/ab;q)_\infty\,(t;q)_\infty\,(1/at;q)_\infty}\end{aligned} \tag{4.9}$$

in the annulus $1/|a| < |t| < 1$.

Observe that the function on the right side of (4.9) is symmetric in b and t. Recall that the Laurent expansion of a function which is analytic in a given annulus is unique. Hence the sum of the terms in (4.9) with nonnegative powers of t is symmetric in b and t, which is the symmetry (1.10) of the Rogers-Fine function.

Let $[\omega]\,f$ denote the coefficient of the monomial ω in the Laurent expansion of the function f in a given annulus.

Let $h \geq 0$ be a nonnegative integer and let $|AB| < 1$. Using the q-binomial theorem (1.5) to extract the coefficient of x^h in the non empty annulus $|B| < |x| < 1/|A|$, we obtain

$$\begin{aligned}
&\left[x^h\right] \frac{(x;q)_\infty\,(q/x;q)_\infty}{(Ax;q)_\infty\,(B/x;q)_\infty} \\
&\quad= \left[x^h\right] \left(\sum_{i=0}^{\infty} \frac{(1/A;q)_i}{(q;q)_i}\,(Ax)^i\right) \left(\sum_{i=0}^{\infty} \frac{(q/B;q)_i}{(q;q)_i}\,(B/x)^i\right) \\
&\quad= \sum_{i=0}^{\infty} \frac{(1/A;q)_{i+h}}{(q;q)_{i+h}}\,A^{i+h}\,\frac{(q/B;q)_i}{(q;q)_i}\,B^i \\
&\quad= A^h\,\frac{(1/A;q)_h}{(q;q)_h}\,{}_2\varphi_1\left[\begin{matrix} q^h/A,\, q/B \\ q^{h+1} \end{matrix};q, AB\right].
\end{aligned} \tag{4.10}$$

Using the formulation (2.24) of Ramanujan's ${}_1\psi_1$ sum, we have

$$\left[x^h\right] \frac{(x;q)_\infty\,(q/x;q)_\infty}{(Ax;q)_\infty\,(B/x;q)_\infty} = \frac{(Aq;q)_\infty\,(B;p)_\infty}{(q;q)_\infty\,(AB;q)_\infty}\,\frac{(1/A;q)_h}{(B;q)_h}\,A^h. \tag{4.11}$$

Equating (4.10) and (4.11) and solving for the ${}_2\varphi_1$, we obtain

$${}_2\varphi_1\left[\begin{matrix} q^h/A,\, q/B \\ q^{h+1} \end{matrix};q, AB\right] = \frac{(Aq;q)_\infty\,(Bq^h;q)_\infty}{(q^{h+1};q)_\infty\,(AB;q)_\infty}. \tag{4.12}$$

Taking $A = q^h/a$ and $B = q/b$ in (4.12), we see that Heine's q-Gauss sum (1.4) holds when $c = q^{h+1}$. Observe that $AB = c/ab$ and that the functions on both sides of (1.4) are analytic in c in the disc $|c| < |ab|$, which is $|AB| < 1$. Hence (1.4) follows by applying Ismail's argument (Ismail, 1977) to the parameter c.

We may use the q-binomial theorem (1.5) and Heine's q-Gauss sum (1.4) to compute the coefficients of x^{-h} in the function on the left side of (2.24). Alternatively, we may observe that replacing x, A and B by q/x, B/q and Aq, respectively, fixes the function on the left side of (2.24) and the annulus $|B| < |x| < 1/|A|$ and, by (4.7), reverses the ${}_1\psi_1$ sum on the right side of (2.24). Thus we have another proof of Ramanujan's ${}_1\psi_1$ sum (1.8).

The reader may compare this with Andrews' proof (Andrews, 1969) of Ramanujan's ${}_1\psi_1$ sum (1.8) using Heine's q-Gauss sum (1.4).

Let $m \geq 0$, $n \geq 0$ and $h \geq 0$ be nonnegative integers and let $0 \leq h \leq m + n$. Observe that

$$(x;q)_{m+n} = (x;q)_m\,(xq^m;q)_n\,, \tag{4.13}$$

which holds by (1.6) for all integers m, n. Using (2.11), the formulation (2.13) of the q-binomial theorem (1.5), and the fact that $2\binom{h}{2} + (m +$

$n-h+1)h=(m+n)h$, we have

$$\left[x^h\right](x;q)_{m+n}=(-1)^h\,q^{\binom{h}{2}}\,\frac{(q^{m+n-h+1};q)_h}{(q;q)_h}=q^{(m+n)h}\,\frac{(q^{-m-n};q)_h}{(q;q)_h}. \tag{4.14}$$

Using (4.14) and the formulation (2.13) of the q-binomial theorem (1.5), we have

$$\begin{aligned}\left[x^h\right](x;q)_{m+n}&=\left[x^h\right]\left(\sum_{i=0}^{m}\frac{(q^{-m};q)_i}{(q;q)_i}\,x^i\,q^{mi}\right)\left(\sum_{i=0}^{n}\frac{(q^{-n};q)_i}{(q;q)_i}\,x^i\,q^{(m+n)i}\right)\\&=q^{mh}\sum_{i=0}^{h}\frac{(q^{-m};q)_{h-i}}{(q;q)_{h-i}}\,\frac{(q^{-n};q)_i}{(q;q)_i}\,q^{ni}.\end{aligned} \tag{4.15}$$

Using (4.7) and (4.13), we have

$$\begin{aligned}(q^{-m};q)_{h-i}&=(q^{-m};q)_h\left(q^{h-m};q\right)_{-i}\\&=(q^{-m};q)_h\,(-1)^i\,q^{(m-h)i}\,q^{\binom{i+1}{2}}\,\frac{1}{(q^{m-h+1};q)_i},\end{aligned} \tag{4.16}$$

$$(q;q)_{h-i}=(q;q)_h\left(q^{h+1};q\right)_{-i}=(q;q)_h\,(-1)^i\,q^{-(h+1)i}\,q^{\binom{i+1}{2}}\,\frac{1}{(q^{-h};q)_i}. \tag{4.17}$$

Substituting (4.16) and (4.17) into (4.15) gives

$$\left[x^h\right](x;q)_m\,(xq^m;q)_n=q^{mh}\,\frac{(q^{-m};q)_h}{(q;q)_h}\,{}_2\varphi_1\left[\begin{matrix}q^{-h},q^{-n}\\q^{m-h+1}\end{matrix};q,q^{m+n+1}\right]. \tag{4.18}$$

Equating (4.14) and (4.18), solving for the ${}_2\varphi_1$, and using (2.11), we obtain

$${}_2\varphi_1\left[\begin{matrix}q^{-h},q^{-n}\\q^{m-h+1}\end{matrix};q,q^{m+n+1}\right]=q^{nh}\,\frac{(q^{m+n-h+1};q)_h}{(q^{-m};q)_h}=\frac{(q^{-m-n};q)_h}{(q^{m-h+1};q)_h}. \tag{4.19}$$

Replacing a and b by $1/a$ and $1/b$, respectively, in Heine's q-Gauss sum (1.4), we have

$${}_2\varphi_1\left[\begin{matrix}1/a,1/b\\c\end{matrix};q,abc\right]=\frac{(ac;q)_\infty\,(bc;q)_\infty}{(c;q)_\infty\,(abc;q)_\infty}. \tag{4.20}$$

Using (2.12) and (2.12) with a replaced by b, we see that the functions on both sides of (4.20) are analytic in a, b and c in the disc $|abc|<1$. Since by (4.19) we have that (4.20) holds when $a=q^h$, $b=q^n$ and

$c = q^{m-h+1}$, we see that (4.20) follows by three successive applications of Ismail's argument (Ismail, 1977) to the parameters c, b and a.

Observe that the first two applications of Ismail's argument (Ismail, 1977) gives the q-Chu-Vandermonde sum (1.3). We may avoid the first two applications of Ismail's argument and establish (1.3) directly by considering $\left[x^h\right](x;q)_\infty/(x/ab;q)_\infty$.

5. The nonterminating q-Chu-Vandermonde sum (1.12)

In this section, we give a path function proof of the case $t = q$ of Heine's ${}_2\varphi_1$ transformation (1.11) which implies the symmetry (1.10) of the Rogers-Fine function. Following Gasper and Rahman (Gasper and Rahman, 1990, Section 2.10), we write Ramanujan's ${}_1\psi_1$ sum (1.8) as the nonterminating q-Chu-Vandermonde sum (1.12).

Since (3.2) reduces to

$$\begin{aligned} &q^n\left(1-cq^n\right)\left(1-b\right)-\left(1-aq^n\right)\left(1-bq^n\right)\\ &\quad=-\left(1-q^n\right)\left(1-cq^n\right)+q^n\left(a-c\right)\left(1-bq^n\right), \end{aligned} \tag{5.1}$$

we have the separation

$$a_n = -\frac{(a;q)_n\,(bq;q)_{n-1}}{(q;q)_{n-1}\,(c;q)_n},\quad t_n = \frac{(a;q)_n\,(b;q)_n}{(q;q)_n\,(c;q)_n}\,q^n, \qquad n\geq 1, \tag{5.2}$$

with respect to the operator η given by

$$\eta(\Pi(a,b;c)) = \frac{(a-c)}{(1-c)}\,\Pi(a,bq;cq). \tag{5.3}$$

Comparing (4.1) and (5.1), we see that there must be a relation between the separations (4.2) and (5.2). Applying η^{-1} to (3.2) and rearranging the result, we have

$$t_n - \eta^{-1}\left(a_{n+1}\right) = -\eta^{-1}\left(a_n\right)+\eta^{-1}\left(t_n\right), \qquad n\geq 1. \tag{5.4}$$

Hence the separation $\{\alpha_n\}_{n\geq 1}$ of $\{t_n\}_{n\geq 1}$ with respect to η^{-1} is given by (Kadell, 1987a, (6.28))

$$\alpha_n = -\eta^{-1}\left(a_n\right), \qquad n\geq 1. \tag{5.5}$$

The reader may check that the separation (5.2) is obtained by replacing q by $1/q$ in the separation (5.5) obtained from (4.2) and (4.3).

Observe that

$$\lim_{n\to\infty} a_n = -\frac{(a;q)_\infty\,(b;q)_\infty}{(q;q)_\infty\,(c;q)_\infty}. \tag{5.6}$$

Observe by (3.5) that the function

$$\Pi(a,b;c)=\sum_{i=0}^{\infty}\frac{(a;q)_i\,(b;q)_i}{(q;q)_i\,(c;q)_i}\,q^i \tag{5.7}$$

satisfies the q-difference equation

$$\Pi(a,b;c)=\frac{(a-c)}{(1-c)}\,\Pi(a,bq;cq)+\frac{(a;q)_\infty\,(bq;q)_\infty}{(q;q)_\infty\,(c;q)_\infty}. \tag{5.8}$$

Repeated application of (5.8) and multiplication and division by $(1-b)$ yields

$${}_2\varphi_1\left[\begin{matrix}a,b\\c\end{matrix};q,q\right]=\frac{(a;q)_\infty\,(b;q)_\infty}{(q;q)_\infty\,(c;q)_\infty}\sum_{i=0}^{\infty}\frac{(c/a;q)_i}{(b;q)_{i+1}}\,a^i, \tag{5.9}$$

which is the case $t=q$ of Heine's ${}_2\varphi_1$ transformation (1.11) with a and b interchanged.

Replacing a, b and c by t, b and $abtq$, respectively, in (5.9) and rearranging the result, we have

$$\frac{1}{(1-b)}\,F(ab,b;t)=\sum_{i=0}^{\infty}\frac{(abq;q)_i}{(b;q)_{i+1}}\,t^i={}_2\varphi_1\left[\begin{matrix}b,t\\abtq\end{matrix};q,q\right]\frac{(q;q)_\infty\,(abtq;q)_\infty}{(b;q)_\infty\,(t;q)_\infty}. \tag{5.10}$$

Since the function on the right side of (5.10) is symmetric in b and t, we have another proof of the symmetry (1.10) of the Rogers-Fine function.

Replacing a, b and c by t, b/q and at, respectively, in (5.9), multiplying by $(1-b/q)$, and rearranging the result, we have

$$\sum_{i=0}^{\infty}\frac{(a;q)_i}{(b;q)_i}\,t^i=\frac{(q;q)_\infty\,(at;q)_\infty}{(b;q)_\infty\,(t;q)_\infty}\,{}_2\varphi_1\left[\begin{matrix}b/q,t\\at\end{matrix};q,q\right]. \tag{5.11}$$

Replacing a, b and t by q^2/b, q^2/a and b/at, respectively, in (5.11), we have

$$\sum_{i=0}^{\infty}\frac{\left(q^2/b;q\right)_i}{\left(q^2/a;q\right)_i}\,(b/at)^i=\frac{(q;q)_\infty\,\left(q^2/at;q\right)_\infty}{\left(q^2/a;q\right)_\infty\,(b/at;q)_\infty}\,{}_2\varphi_1\left[\begin{matrix}q/a,b/at\\q^2/at\end{matrix};q,q\right]. \tag{5.12}$$

Let $|b/a|<|t|<1$. Observe using (4.8) that we may write Ramanujan's ${}_1\psi_1$ sum (1.8) as

$$\begin{aligned}\sum_{i=0}^{\infty}\frac{(a;q)_i}{(b;q)_i}\,t^i+\frac{1}{t}\frac{(b-q)}{(a-q)}\sum_{i=0}^{\infty}\frac{\left(q^2/b;q\right)_i}{\left(q^2/a;q\right)_i}\,(b/at)^i\\=\frac{(q;q)_\infty\,(b/a;q)_\infty\,(at;q)_\infty\,(q/at;q)_\infty}{(b;q)_\infty\,(q/a;q)_\infty\,(t;q)_\infty\,(b/at;q)_\infty}.\end{aligned} \tag{5.13}$$

Substituting (5.11) and (5.12) into (5.13), multiplying by $(b;q)_\infty (t;q)_\infty / (q;q)_\infty (at;q)_\infty$, and using the facts that $(b-q) = -q(1-b/q)$ and $(a-q)\left(q^2/a;q\right)_\infty = a(q/a;q)_\infty$, we obtain

$$\begin{aligned}
&{}_2\varphi_1\left[\begin{matrix} b/q, t \\ at \end{matrix}; q, q\right] - \frac{q}{at}\frac{(b/q;q)_\infty (t;q)_\infty \left(q^2/at;q\right)_\infty}{(at;q)_\infty (q/a;q)_\infty (b/at;q)_\infty}\, {}_2\varphi_1\left[\begin{matrix} q/a, b/at \\ q^2/at \end{matrix}; q, q\right] \\
&\quad = \frac{(b/a;q)_\infty (q/at;q)_\infty}{(q/a;q)_\infty (b/at;q)_\infty}.
\end{aligned} \tag{5.14}$$

Observe that if we replace a, b and t by c/b, aq and b, respectively, in (5.14), then we obtain the nonterminating q-Chu-Vandermonde sum (1.12).

6. The little q-Jacobi polynomials

In this section, we show following Andrews and Askey (Andrews and Askey, 1977) that the q-Saalschütz sum (1.2) and the q-Chu-Vandermonde sum (1.3) are equivalent to the evaluations of the moments $I_n^{\alpha,\beta}(x,0)$ and $I_n^{\alpha,\beta}(x,n-x-1)$, respectively, and that each of these is equivalent to the orthogonality of the little q-Jacobi polynomials. Hence the q-Chu-Vandermonde sum (1.3) implies the q-Saalschütz sum (1.2).

Let $m \geq 0$, $\ell \geq 0$ be nonnegative integers and let $0 \neq |q| < 1$ with a fixed natural logarithm of q. Using (1.14), (1.16) and the functional equation

$$\Gamma_q(\alpha+1) = \frac{(1-q^\alpha)}{(1-q)}\,\Gamma_q(\alpha) \tag{6.1}$$

for the q-gamma function, we obtain

$$\begin{aligned}
I_n^{\alpha,\beta}(x,y) &= \sum_{i=0}^{n} \frac{(q^{-n};q)_i \left(q^{n+\alpha+\beta-1};q\right)_i}{(q;q)_i (q^\alpha;q)_i}\, q^i \int_0^1 t^{\alpha+x+i-1} \frac{(tq;q)_\infty}{(tq^{\beta+y};q)_\infty}\, d_qt \\
&= \sum_{i=0}^{n} \frac{(q^{-n};q)_i \left(q^{n+\alpha+\beta-1};q\right)_i}{(q;q)_i (q^\alpha;q)_i} \frac{\Gamma_q(\alpha+x+i)\,\Gamma_q(\beta+y)}{\Gamma_q(\alpha+\beta+x+y+i)}\, q^i \\
&= \frac{(q;q)_\infty \left(q^{\alpha+\beta+x+y};q\right)_\infty}{\left(q^{\alpha+x};q\right)_\infty \left(q^{\beta+y};q\right)_\infty}\, {}_3\varphi_2\left[\begin{matrix} q^{-n}, q^{n+\alpha+\beta-1}, q^{\alpha+x} \\ q^\alpha, q^{\alpha+\beta+x+y} \end{matrix}; q, q\right].
\end{aligned} \tag{6.2}$$

Setting $y=0$ in (6.2), taking

$$a = q^{n+\alpha+\beta-1},\ b = q^{\alpha+x} \text{ and } c = q^\alpha \tag{6.3}$$

in the q-Saalschütz sum (1.2), and using (2.11), Andrews and Askey (Andrews and Askey, 1977) obtained

$$\begin{aligned} I_n^{\alpha,\beta}(x,0) &= \frac{(q;q)_\infty \left(q^{\alpha+\beta+x};q\right)_\infty}{(q^{\alpha+x};q)_\infty (q^\beta;q)_\infty} {}_3\varphi_2 \left[\begin{matrix} q^{-n}, q^{n+\alpha+\beta-1}, q^{\alpha+x} \\ q^\alpha, q^{\alpha+\beta+x} \end{matrix} ; q, q \right] \\ &= \frac{(q;q)_\infty \left(q^{\alpha+\beta+x};q\right)_\infty}{(q^{\alpha+x};q)_\infty (q^\beta;q)_\infty} \frac{\left(q^{1-n-\beta};q\right)_n (q^{-x};q)_n}{(q^\alpha;q)_n \left(q^{1-n-x-\alpha-\beta};q\right)_n} \\ &= q^{n(\alpha+x)} \frac{(q;q)_\infty \left(q^{\alpha+\beta+x+n};q\right)_\infty}{(q^{\alpha+x};q)_\infty (q^{\beta+n};q)_\infty} \frac{(q^{-x};q)_n}{(q^\alpha;q)_n}. \end{aligned} \tag{6.4}$$

Using the inverse $q^\alpha = c$, $q^\beta = aq^{1-n}/c$ and $q^x = b/c$ of (6.3), we see that (6.4) is equivalent to the q-Saalschütz sum (1.2).

Since $0 \le m \le n-1$ implies that $(q^{-m};q)_n = 0$, we have the orthogonality relation

$$0 \le m \le n-1 \implies I_n^{\alpha,\beta}(m,0) = 0 \tag{6.5}$$

and hence the little q-Jacobi polynomials are orthogonal on $(0,1)$ with respect to the q-beta integral (1.17).

Observe that

$$\frac{(c;q)_n}{(c;q)_i} = \left(cq^i;q\right)_{n-i}. \tag{6.6}$$

Using (2.11), we have

$$\begin{aligned} \frac{(c/ab;q)_n}{(abq^{1-n}/c;q)_i} q^i &= \frac{1}{(-abq^{1-n}/c)^i q^{\binom{i}{2}}} \frac{(c/ab;q)_n}{(cq^{n-i}/ab;q)_i} q^i \\ &= (-c/ab)^i q^{ni-\binom{i}{2}} (c/ab;q)_{n-i}. \end{aligned} \tag{6.7}$$

Observe that

$$a^n (c/a;q)_n = (a-c)\cdots\left(a-cq^{n-1}\right). \tag{6.8}$$

Using (6.6–8) and (6.8) with a replaced by b to multiply the q-Saalschütz sum (1.2) by $(ab)^n (c;q)_n (c/ab;q)_n$, we obtain

$$\begin{aligned} S_n(a,b,c) &= \sum_{i=0}^n \frac{(q^{-n};q)_i}{(q;q)_i} q^{ni-\binom{i}{2}} (ab)^{n-i} (-c)^i \\ &\quad \times (a;q)_i (b;q)_i \left(cq^i;q\right)_{n-i} (c/ab;q)_{n-i} \\ &= (a-c)\cdots\left(a-cq^{n-1}\right)(b-c)\cdots\left(b-cq^{n-1}\right), \end{aligned} \tag{6.9}$$

which is a polynomial formulation of the q-Saalschütz sum (1.2).

Let n, a and b be fixed with $n \geq 0$, $a \neq 0$, $b \neq 0$ and a/b not an integer power of q. Since (6.3) gives $b = cq^x$, we see taking $x = m$ in the orthogonality relation (6.5) that

$$0 \leq m \leq n-1 \implies S_n\left(a, b, bq^{-m}\right) = 0. \tag{6.10}$$

Using the symmetry

$$S_n(b, a, c) = S_n(a, b, c), \tag{6.11}$$

we see that (6.10) becomes

$$0 \leq m \leq n-1 \implies S_n\left(a, b, aq^{-m}\right) = 0. \tag{6.12}$$

We have

$$S_n(a, b, 0) = (ab)^n. \tag{6.13}$$

Observe that $S_n(a, b, c)$ has degree at most $2n$ as a polynomial in c and that (6.10), (6.12) and (6.13) give the values, in agreement with (6.9), of $S_n(a, b, c)$ at $2n+1$ distinct values of c. We may remove the restrictions on a and b by successively considering $S_n(a, b, c)$ as a polynomial in a and b. Thus the orthogonality of the little q-Jacobi polynomials, which is given by (6.5), implies the polynomial formulation (6.9) of the q-Saalschütz sum (1.2). Hence the orthogonality of the little q-Jacobi polynomials is equivalent to the q-Saalschütz sum (1.2).

We may use (2.11) to reverse the q-Chu-Vandermonde sum (1.3). The result is (Gasper and Rahman, 1990, (1.5.3))

$${}_2\varphi_1\left[\begin{matrix} q^{-n}, a \\ b \end{matrix}; q, q\right] = a^n \frac{(b/a; q)_n}{(b; q)_n}. \tag{6.14}$$

Setting $y = n - x - 1$ in (6.2) and taking

$$a = q^{\alpha+x} \text{ and } b = q^{\alpha} \tag{6.15}$$

in the reversed q-Chu-Vandermonde sum (6.14), we obtain

$$\begin{aligned} I_n^{\alpha,\beta}(x, n-x-1) &= \frac{(q;q)_\infty \left(q^{\alpha+\beta+n-1}; q\right)_\infty}{\left(q^{\alpha+x}; q\right)_\infty \left(q^{\beta+n-x-1}; q\right)_\infty} {}_2\varphi_1\left[\begin{matrix} q^{-n}, q^{\alpha+x} \\ q^{\alpha} \end{matrix}; q, q\right] \\ &= q^{n(\alpha+x)} \frac{(q;q)_\infty \left(q^{\alpha+\beta+n-1}; q\right)_\infty}{\left(q^{\alpha+x}; q\right)_\infty \left(q^{\beta+n-x-1}; q\right)_\infty} \frac{(q^{-x}; q)_n}{(q^{\alpha}; q)_n}. \end{aligned} \tag{6.16}$$

Using the inverse $q^{\alpha} = b$ and $q^x = a/b$ of (6.16), we see that (6.17) is equivalent to the reversed q-Chu-Vandermonde sum (6.14) and hence is equivalent to the q-Chu-Vandermonde sum (1.3).

Since $0 \le \ell \le n-1$ implies that $(q^{-\ell};q)_n = 0$, we have the orthogonality relation

$$0 \le \ell \le n-1 \implies I_n^{\alpha,\beta}(\ell, n-\ell-1) = 0. \tag{6.17}$$

Since we may use any basis for the polynomials in t with degree less than or equal to $n-1$, we see by (6.18) that the little q-Jacobi polynomials are orthogonal on $(0,1)$ with respect to the q-beta integral (1.17).

Using (6.6) and (6.8) with c replaced by b to multiply the reversed q-Chu-Vandermonde sum (6.14) by $(b;q)_n$, we obtain

$$T_n(a,b) = \sum_{i=0}^{n} \frac{(q^{-n};q)_i}{(q;q)_i} q^i (a;q)_i \left(bq^i;q\right)_{n-i} = (a-b)\cdots\left(a-bq^{n-1}\right), \tag{6.18}$$

which is a polynomial formulation of the reversed q-Chu-Vandermonde sum (6.14).

Let n and a be fixed with $n \ge 0$, $a \ne 0$ and a/b not an integer power of q. Since (6.16) gives $a = bq^x$, we see taking $x = \ell$ in the orthogonality relation (6.18) that

$$0 \le \ell \le n-1 \implies T_n\left(a, aq^{-\ell}\right) = 0. \tag{6.19}$$

We have

$$[b^n]\, T_n(a,b) = (-1)^n q^{\binom{n}{2}}. \tag{6.20}$$

Observe that (6.20) gives the values, in agreement with (6.19), of $T_n(a,b)$ at n distinct values of b and that (6.21) gives the coefficient of the leading term of $T_n(a,b)$ as a polynomial in b. We may remove the restrictions on a by considering $T_n(a,b)$ as a polynomial in a. Thus the orthogonality of the little q-Jacobi polynomials, which is given by (6.18), implies the polynomial formulation (6.19) of the reversed q-Chu-Vandermonde sum (6.13) and hence implies the q-Chu-Vandermonde sum (1.3). Hence the orthogonality of the little q-Jacobi polynomials is equivalent to the q-Chu-Vandermonde sum (1.3).

Observe that the q-Chu-Vandermonde sum (1.3) is equivalent to and hence implies the orthogonality of the little q-Jacobi polynomials, which is equivalent to and hence implies the q-Saalschütz sum (1.2).

7. The little q-Jacobi functions

In this section, we extend the little q-Jacobi polynomials naturally to the little q-Jacobi functions of complex order. We show that the nonterminating q-Saalschütz (1.13) and q-Chu-Vandermonde (1.12) sums are equivalent to the evaluations of the moments $I_n^{\alpha,\beta}(x,0)$ and $I_n^{\alpha,\beta}(x, n-$

$x-1$), respectively, and, using the Liouville-Ismail argument (Hille, 1973, Theorem 8.2.2), (Ismail, 1977), to two orthogonality relations. We show that (1.12) implies (1.13).

Let n, μ be complex, let $x+1$ have positive real part, and let $m \geq 0$, $\ell \geq 0$ be nonnegative integers. Let $0 \neq |q| < 1$, $t \neq 0$ with fixed natural logarithms of q and t. Observe that (1.6) and (1.14) extend the q-Pockhammer symbol $(a;q)_n = (a;q)_\infty / (aq^n;q)_\infty$ to complex n and that (4.13) continues to hold. The shifted basic hypergeometric function ${}_{r+1}\varphi_r^{q^\mu}[a_1,\ldots,a_{r+1};b_1,\ldots,b_r;q,t]$ is given by

$$\begin{aligned}
&{}_{r+1}\varphi_r^{q^\mu}\left[\begin{matrix}a_1,\ldots,a_{r+1}\\ b_1,\ldots,b_r\end{matrix};q,t\right] \\
&= \sum_{i=-\infty}^{\infty} \frac{(a_1;q)_{i+\mu}\cdots(a_{r+1};q)_{i+\mu}}{(q;q)_{i+\mu}(b_1;q)_{i+\mu}\cdots(b_r;q)_{i+\mu}} t^{i+\mu} \\
&= t^\mu \frac{(a_1;q)_\infty\cdots(a_{r+1};q)_\infty}{(a_1q^\mu;q)_\infty\cdots(a_{r+1}q^\mu;q)_\infty} \frac{(q^{\mu+1};q)_\infty}{(q;q)_\infty} \\
&\quad \times \frac{(b_1q^\mu;q)_\infty\cdots(b_rq^\mu;q)_\infty}{(b_1;q)_\infty\cdots(b_r;q)_\infty} {}_{r+1}\psi_{r+1}\left[\begin{matrix}a_1q^\mu,\ldots,a_{r+1}q^\mu\\ q^{\mu+1},b_1q^\mu,\ldots,b_rq^\mu\end{matrix};q,t\right].
\end{aligned} \tag{7.1}$$

Setting $\mu = 0$ in (7.1), we see that ${}_{r+1}\varphi_r^1 = {}_{r+1}\varphi_r$ is the usual basic hypergeometric function.

In (Kadell, 2000b), we used the classical ratio of alternants and the tool

$$\sum_{i=a}^{b} f(i) = \sum_{i=a}^{\infty} f(i) - \sum_{i=b+1}^{\infty} f(i), \tag{7.2}$$

which interprets sums with complex limits, to extend the Schur functions to partitions with complex parts. Following Proctor (Proctor, 1989), we gave a combinatorial representation as an alternating sum over the symmetric group of a sum over representatives of tableaux of complex shape which may be indexed by certain sequences of nonnegative integers. Observe by (3.13) that

$${}_{r+1}\psi_{r+1}\left[\begin{matrix}a_1,\ldots,a_{r+1}\\ b_1,\ldots,b_r,q\end{matrix};q,t\right] = {}_{r+1}\varphi_r\left[\begin{matrix}a_1,\ldots,a_{r+1}\\ b_1,\ldots,b_r\end{matrix};q,t\right]. \tag{7.3}$$

Since (7.1) and (7.3) give

$$
\begin{aligned}
{}_2\varphi_1^{q^{1-\alpha}} &\left[\begin{matrix} q^{-n}, q^{n+\alpha+\beta-1} \\ q^{\alpha} \end{matrix} ; q, tq\right] \\
&= (tq)^{1-\alpha} \frac{(q^{-n};q)_\infty \left(q^{n+\alpha+\beta-1};q\right)_\infty \left(q^{2-\alpha};q\right)_\infty}{(q^{1-n-\alpha};q)_\infty (q^{n+\beta};q)_\infty (q^{\alpha};q)_\infty} \\
&\quad \times {}_2\varphi_1 \left[\begin{matrix} q^{1-n-\alpha}, q^{n+\beta} \\ q^{2-\alpha} \end{matrix} ; q, tq\right],
\end{aligned}
\tag{7.4}
$$

it is natural to extend the little q-Jacobi polynomials to functions of complex order n by

$$
\begin{aligned}
p_n^{\alpha,\beta}(t) = {}_2\varphi_1^{1} &\left[\begin{matrix} q^{-n}, q^{n+\alpha+\beta-1} \\ q^{\alpha} \end{matrix} ; q, tq\right] - {}_2\varphi_1^{q^{1-\alpha}} \left[q^{-n}, q^{n-\alpha+\beta-1} q^{\alpha}; q, tq\right] \\
&- (tq)^{1-\alpha} \frac{(q^{-n};q)_\infty \left(q^{n+\alpha+\beta-1};q\right)_\infty \left(q^{2-\alpha};q\right)_\infty}{(q^{1-n-\alpha};q)_\infty (q^{n+\beta};q)_\infty (q^{\alpha};q)_\infty} \\
&\times {}_2\varphi_1 \left[\begin{matrix} q^{1-n-\alpha}, q^{n+\beta} \\ q^{2-\alpha} \end{matrix} ; q, tq\right],
\end{aligned}
\tag{7.5}
$$

which is a difference of shifted basic hypergeometric functions. We may avoid poles by assuming that $n+\alpha-1$, $-n-\beta$ and $-\alpha$ are not non-negative integers.

Using (1.16), (1.17) and (6.1), we have

$$
\begin{aligned}
&\int_0^1 t^{x} \frac{(tq;q)_\infty}{\left(tq^{\beta+y};q\right)_\infty} {}_2\varphi_1 \left[\begin{matrix} q^{1-n-\alpha}, q^{n+\beta} \\ q^{2-\alpha} \end{matrix} ; q, tq\right] d_q t \\
&= \sum_{i=0}^{\infty} \frac{\left(q^{1-n-\alpha};q\right)_i \left(q^{n+\beta};q\right)_i}{(q;q)_i \left(q^{2-\alpha};q\right)_i} q^i \int_0^1 t^{x+i} \frac{(tq;q)_\infty}{\left(tq^{\beta+y};q\right)_\infty} d_q t \\
&= \sum_{i=0}^{\infty} \frac{\left(q^{1-n-\alpha};q\right)_i \left(q^{n+\beta};q\right)_i}{(q;q)_i \left(q^{2-\alpha};q\right)_i} \frac{\Gamma_q(x+i+1)\,\Gamma_q(\beta+y)}{\Gamma_q(\beta+x+y+i+1)} q^i \\
&= \frac{(q;q)_\infty \left(q^{\beta+x+y+1};q\right)_\infty}{\left(q^{x+1};q\right)_\infty \left(q^{\beta+y};q\right)_\infty} {}_3\varphi_2 \left[\begin{matrix} q^{1-n-\alpha}, q^{n+\beta}, q^{x+1} \\ q^{2-\alpha}, q^{\beta+x+y+1} \end{matrix} ; q, q\right].
\end{aligned}
\tag{7.6}
$$

Combining our results (6.2), (7.5) and (7.6), we obtain

$$
\begin{aligned}
I_n^{\alpha,\beta}(x,y) &= \frac{(q;q)_\infty \left(q^{\alpha+\beta+x+y};q\right)_\infty}{\left(q^{\alpha+x};q\right)_\infty \left(q^{\beta+y};q\right)_\infty} {}_3\varphi_2\left[\begin{matrix} q^{-n}, q^{n+\alpha+\beta-1}, q^{\alpha+x} \\ q^{\alpha}, q^{\alpha+\beta+x+y} \end{matrix} ;q,q\right] \\
&\quad - q^{1-\alpha} \frac{\left(q^{-n};q\right)_\infty \left(q^{n+\alpha+\beta-1};q\right)_\infty \left(q^{2-\alpha};q\right)_\infty}{\left(q^{1-n-\alpha};q\right)_\infty \left(q^{n+\beta};q\right)_\infty \left(q^{\alpha};q\right)_\infty} \\
&\quad \times \frac{(q;q)_\infty \left(q^{\beta+x+y+1};q\right)_\infty}{\left(q^{x+1};q\right)_\infty \left(q^{\beta+y};q\right)_\infty} {}_3\varphi_2\left[\begin{matrix} q^{1-n-\alpha}, q^{n+\beta}, q^{x+1} \\ q^{2-\alpha}, q^{\beta+x+y+1} \end{matrix} ;q,q\right].
\end{aligned}
\tag{7.7}
$$

Setting $y = 0$ in (7.7), we have

$$
\begin{aligned}
I_n^{\alpha,\beta}(x,0) &= \frac{(q;q)_\infty \left(q^{\alpha+\beta+x};q\right)_\infty}{\left(q^{\alpha+x};q\right)_\infty \left(q^{\beta};q\right)_\infty} {}_3\varphi_2\left[\begin{matrix} q^{-n}, q^{n+\alpha+\beta-1}, q^{\alpha+x} \\ q^{\alpha}, q^{\alpha+\beta+x} \end{matrix} ;q,q\right] \\
&\quad - q^{1-\alpha} \frac{\left(q^{-n};q\right)_\infty \left(q^{n+\alpha+\beta-1};q\right)_\infty \left(q^{2-\alpha};q\right)_\infty}{\left(q^{1-n-\alpha};q\right)_\infty \left(q^{n+\beta};q\right)_\infty \left(q^{\alpha};q\right)_\infty} \\
&\quad \times \frac{(q;q)_\infty \left(q^{\beta+x+1};q\right)_\infty}{\left(q^{x+1};q\right)_\infty \left(q^{\beta};q\right)_\infty} {}_3\varphi_2\left[\begin{matrix} q^{1-n-\alpha}, q^{n+\beta}, q^{x+1} \\ q^{2-\alpha}, q^{\beta+x+1} \end{matrix} ;q,q\right].
\end{aligned}
\tag{7.8}
$$

Taking

$$
a = q^{-n},\ b = q^{n+\alpha+\beta-1},\ c = q^{\alpha+x} \text{ and } e = q^{\alpha} \tag{7.9}
$$

in the nonterminating q-Saalschütz sum (1.13) and substituting into (7.8), we obtain

$$
I_n^{\alpha,\beta}(x,0) = \frac{(q;q)_\infty \left(q^{1-\alpha};q\right)_\infty \left(q^{1-n+x};q\right)_\infty \left(q^{n+\alpha+\beta+x};q\right)_\infty}{\left(q^{1-n-\alpha};q\right)_\infty \left(q^{n+\beta};q\right)_\infty \left(q^{x+1};q\right)_\infty \left(q^{\alpha+x};q\right)_\infty}. \tag{7.10}
$$

Using the inverse $q^{\alpha} = e$, $q^{x} = c/e$, $q^{\beta} = abq/e$ and $q^{n} = 1/a$ of (7.9), we see that (7.10) is equivalent to the nonterminating q-Saalschütz sum (1.13).

Setting $y = n - x - 1$ in (7.7), we have

$$
\begin{aligned}
&I_n^{\alpha,\beta}(x, n-x-1) \\
&= \frac{(q;q)_\infty \left(q^{\alpha+\beta+n-1};q\right)_\infty}{\left(q^{\alpha+x};q\right)_\infty \left(q^{\beta+n-x-1};q\right)_\infty} {}_2\varphi_1\left[\begin{matrix} q^{-n}, q^{\alpha+x} \\ q^{\alpha} \end{matrix} ;q,q\right] \\
&\quad - q^{1-\alpha} \frac{\left(q^{-n};q\right)_\infty \left(q^{n+\alpha+\beta-1};q\right)_\infty \left(q^{2-\alpha};q\right)_\infty (q;q)_\infty}{\left(q^{1-n-\alpha};q\right)_\infty \left(q^{\alpha};q\right)_\infty \left(q^{x+1};q\right)_\infty \left(q^{\beta+n-x-1};q\right)_\infty} \\
&\quad \times {}_2\varphi_1\left[\begin{matrix} q^{1-n-\alpha}, q^{x+1} \\ q^{2-\alpha} \end{matrix} ;q,q\right].
\end{aligned}
\tag{7.11}
$$

Taking

$$a = q^{-n}, \quad b = q^{\alpha+x} \text{ and } c = q^{\alpha} \tag{7.12}$$

in the nonterminating q-Chu-Vandermonde sum (1.12) and substituting into (7.11), we obtain

$$I_n^{\alpha,\beta}(x, n-x-1) = \frac{(q;q)_\infty \left(q^{1-\alpha};q\right)_\infty \left(q^{1-n+x};q\right)_\infty \left(q^{\alpha+\beta+n-1};q\right)_\infty}{\left(q^{1-n-\alpha};q\right)_\infty \left(q^{\beta+n-x-1};q\right)_\infty \left(q^{x+1};q\right)_\infty \left(q^{\alpha+x};q\right)_\infty}. \tag{7.13}$$

Using the inverse $q^\alpha = c$, $q^x = b/c$ and $q^n = 1/a$ of (7.12), we see that (7.13) is equivalent to the nonterminating q-Chu-Vandermonde sum (1.12).

Observe that the evaluations (7.10) and (7.13) of the moments $I_n^{\alpha,\beta}(x,0)$ and $I_n^{\alpha,\beta}(x, n-x-1)$ give the orthogonality relations

$$\begin{gathered} n-x-1,\ \alpha-1 \text{ or } -n-\alpha-\beta-x \text{ is a nonnegative integer} \\ \Longrightarrow I_n^{\alpha,\beta}(x,0) = 0 \end{gathered} \tag{7.14}$$

and

$$\begin{gathered} n-x-1,\ \alpha-1 \text{ or } -n-\alpha-\beta+1 \text{ is a nonnegative integer} \\ \Longrightarrow I_n^{\alpha,\beta}(x, n-x-1) = 0, \end{gathered} \tag{7.15}$$

respectively.

We may rearrange the nonterminating q-Saalschütz sum (1.13) as

$$\begin{aligned} \mathcal{S}(a,b,c;e) &= (aq/e;q)_\infty (bq/e;q)_\infty (cq/e;q)_\infty \\ &\quad \times \sum_{i=0}^{\infty} \left(eq^i;q\right)_\infty \left(abcq^{i+1}/e;q\right)_\infty \frac{(a;q)_i (b;q)_i (c;q)_i}{(q;q)_i} q^i \\ &\quad - \frac{q}{e} (a;q)_\infty (b;q)_\infty (c;q)_\infty \\ &\quad \times \sum_{i=0}^{\infty} \left(q^{2+i}/e;q\right)_\infty \left(abcq^{2+i}/e^2;q\right)_\infty \\ &\quad \times \frac{(aq/e;q)_i (bq/e;q)_i (cq/e;q)_i}{(q;q)_i} q^i \\ &= (e;q)_\infty (q/e;q)_\infty (abq/e;q)_\infty (acq/e;q)_\infty (bcq/e;q)_\infty. \end{aligned} \tag{7.16}$$

Since (7.9) gives $e/acq = q^{n-x-1}$, $e/q = q^{\alpha-1}$ and $e/bcq = q^{-n-\alpha-\beta-x}$, we see that the orthogonality relation (7.14) gives

$$\begin{gathered} m \geq 0 \Longrightarrow \mathcal{S}\left(a,b,c;acq^{m+1}\right) \\ = \mathcal{S}\left(a,b,c;q^{m+1}\right) = \mathcal{S}\left(a,b,c;bcq^{m+1}\right) = 0. \end{gathered} \tag{7.17}$$

Observe that the functions on both sides of (7.16) satisfy

$$\mathcal{S}\left(aq/e, bq/e, cq/e; q^2/e\right) = -\frac{e}{q}\,\mathcal{S}(a,b,c;e) \tag{7.18}$$

and they are both symmetric in a, b and c. Hence by (7.17), we have

$$m \geq 0 \implies \mathcal{S}\left(a,b,c;abq^{m+1}\right) = \mathcal{S}\left(a,b,c;q^{1-m}\right) = 0. \tag{7.19}$$

Observe that (7.17) and (7.19) express the orthogonality relation (7.14) in terms of the function $\mathcal{S}(a,b,c;e)$ and identify the factors on the right side of (7.16).

We may rearrange the nonterminating q-Chu-Vandermonde sum (1.12) as

$$\begin{aligned}\mathcal{T}(a,b;c) &= (aq/c;q)_\infty\,(bq/c;q)_\infty \sum_{i=0}^{\infty}\left(cq^i;q\right)_\infty \frac{(a;q)_i\,(b;q)_i}{(q;q)_i}\,q^i \\ &\quad - \frac{q}{c}\,(a;q)_\infty\,(b;q)_\infty \sum_{i=0}^{\infty}\left(q^{2+i}/c;q\right)_\infty \frac{(aq/c;q)_i\,bq/c;q)_i}{(q;q)_i}\,q^i \\ &= (c;q)_\infty\,(q/c;q)_\infty\,(abq/c;q)_\infty.\end{aligned} \tag{7.20}$$

Since (7.12) gives $c/abq = q^{n-x-1}$ and $c/q = q^{\alpha-1}$, we see that the orthogonality relation (7.15) gives

$$\ell \geq 0 \implies \mathcal{T}\left(a,b;abq^{\ell+1}\right) = \mathcal{T}\left(a,b;q^{\ell+1}\right) = 0. \tag{7.21}$$

Observe that the functions on both sides of (7.21) satisfy

$$\mathcal{T}\left(aq/e, bq/e; q^2/c\right) = -\frac{c}{q}\,\mathcal{T}(a,b;c). \tag{7.22}$$

Hence by (7.21), we have

$$\ell \geq 0 \implies \mathcal{T}\left(a,b;q^{1-\ell}\right) = 0. \tag{7.23}$$

Observe that (7.21) and (7.23) express the orthogonality relation (7.15) in terms of the function $\mathcal{T}(a,b;c)$ and identify the factors on the right side of (7.20).

Since

$$\mathcal{S}(a,b,0;c) = \mathcal{T}(a,b;c), \tag{7.24}$$

we see that the orthogonality relation (7.14), which is given by (7.17) and (7.19), implies the orthogonality relation (7.15), which is given by (7.21) and (7.23).

Observe that

$$t^x \frac{(tq^\beta;q)_\infty}{(tq^{\beta+y};q)_\infty} = t^x \frac{(tq^\beta;q)_\infty}{(tq^{\beta+y+1};q)_\infty} + q^{\beta+y}\, t^{x+1} \frac{(tq^\beta;q)_\infty}{(tq^{\beta+y};q)_\infty} \tag{7.25}$$

and hence

$$I_n^{\alpha,\beta}(s,y) = I_n^{\alpha,\beta}(s,y+1) + q^{\beta+y}\, I_n^{\alpha,\beta}(s+1,y). \tag{7.26}$$

We now show that the orthogonality relation (7.15) implies that

$$\begin{aligned} &n-m-x-1 \text{ and } m \text{ are nonnegative integers} \\ &\quad \Longrightarrow I_n^{\alpha,\beta}(x,n-m-x-1) = 0. \end{aligned} \tag{7.27}$$

We proceed by induction on m. The case $m = 0$ of (7.27) follows by (7.15). For $m > 0$, we set $s = x$, $y = n-m-x-1$ in (7.26) and use our induction assumption.

Setting $m = n-x-1$ in (7.27) gives the first part of the orthogonality relation (7.14), which by (7.17) is equivalent to the third part of the orthogonality relation (7.14). Letting $m \geq 0$, we observe that

$$\begin{aligned} &\sum_{i=0}^{\infty} \left(q^{m+i+1};q\right)_\infty \left(abcq^{i-m};q\right)_\infty \frac{(a;q)_i\,(b;q)_i\,(c;q)_i}{(q;q)_i}\, q^i \\ &= \sum_{i=0}^{\infty} \left(q^{i+1};q\right)_\infty \left(abcq^{i-m};q\right)_\infty \frac{(a;q)_i\,(b;q)_i\,(c;q)_i}{(q;q)_{m+i}}\, q^i \end{aligned} \tag{7.28}$$

and, replacing i by $m+i$,

$$\begin{aligned} &\sum_{i=0}^{\infty} \left(q^{i-m+1};q\right)_\infty \left(abcq^{i-2m};q\right)_\infty \frac{(a/q^m;q)_i\,(b/q^m;q)_i\,(c/q^m;q)_i}{(q;q)_i}\, q^i \\ &= \sum_{i=0}^{\infty} \left(q^{i+1};q\right)_\infty \left(abcq^{i-m};q\right)_\infty \\ &\qquad \times \frac{(a/q^m;q)_{m+i}\,(b/q^m;q)_{m+i}\,(c/q^m;q)_{m+i}}{(q;q)_{m+i}}\, q^{m+i}. \end{aligned} \tag{7.29}$$

Using (7.28) and (7.29), we obtain the second part of the orthogonality relation (7.17), which is equivalent to the second part of the orthogonality relation (7.14). Hence the orthogonality relations (7.14) and (7.15) are equivalent.

We now show that the orthogonality relation (7.14) implies the formulation (7.16) of the nonterminating q-Saalschütz sum (1.13). Let a, b and e be fixed, non zero complex numbers with $|abq| > |e|$, a/b and e are not integer powers of q, and abq/e is not the reciprocal of a nonnegative integer power of q. Let

$$\begin{aligned} f(c) &= \frac{(aq/e;q)_\infty\,(bq/e;q)_\infty\,(cq/e;q)_\infty\,(abcq/e)_\infty}{(q/e;q)_\infty\,(abq/e;q)_\infty\,(acq/e;q)_\infty\,(bcq/e;q)_\infty}\,{}_3\varphi_2\left[\begin{matrix} a,b,c \\ e, abcq/e \end{matrix};q,q\right] \\ &\quad - \frac{q}{(e-q)}\,\frac{(a;q)_\infty\,(b;q)_\infty\,(c;q)_\infty\,\left(abcq^2/e^2;q\right)_\infty}{(e;q)_\infty\,(abq/e;q)_\infty\,(acq/e;q)_\infty\,(bcq/e;q)_\infty} \\ &\qquad \times {}_3\varphi_2\left[\begin{matrix} aq/e, bq/e, cq/e \\ q^2/e, abcq^2/e^2 \end{matrix};q,q\right] \end{aligned} \tag{7.30}$$

denote the sum on the left side of (7.16) divided by the product on the right side of (7.16).

Observe that the orthogonality relation (7.14) implies (7.17) and (7.19), which identify the factors on the right side of (7.16). Since the function on the right side of (7.16) has at most simple zeroes, we see that f is an entire function of a, b and c.

Observe that $f(c)$ is analytic and hence bounded on the disc

$$\Omega_1 = \{c \mid |ac| < |e| \text{ and } |bc| < |e|\}. \tag{7.31}$$

Let $\epsilon > 0$ be given and set

$$\begin{aligned} \Omega_2 = \big\{c \notin \Omega_1 \mid &\left|acq^{1+m}/e - 1\right| > \epsilon \\ &\text{and } \left|bcq^{1+m}/e - 1\right| \text{ for all } m \geq 0\big\}. \end{aligned} \tag{7.32}$$

Let $c \in \Omega_2$. Hence $c \neq 0$ and we may write

$$cq^z = \omega \text{ where } z \text{ is an integer and } |q| \leq |\omega| < 1. \tag{7.33}$$

Using the identity (2.11) for reversing the q-Pockhammer symbol and (4.13), we see that

$$\begin{aligned} (xcq/e;q)_\infty &= (xcq/e;q)_z\,\left(xcq^{1+z}/e\right)_\infty \\ &= (-xcq/e)^z\,q^{\binom{z}{2}}\,(e/x\omega;q)_z\,(x\omega q/e)_\infty. \end{aligned} \tag{7.34}$$

Using (7.33) and (7.34), we have

$$\frac{(cq/e;q)_\infty\,(abcq/e)_\infty}{(acq/e;q)_\infty\,(bcq/e;q)_\infty} = \frac{(e/\omega;q)_z\,(\omega q/e;q)_\infty\,(e/ab\omega;q)_z\,(ab\omega q/e;q)_\infty}{(e/a\omega;q)_z\,(a\omega q/e;q)_\infty\,(e/b\omega;q)_z\,(b\omega q/e;q)_\infty}, \tag{7.35}$$

$$\begin{aligned} &\frac{(c;q)_\infty\,\left(abcq^2/e^2\right)_\infty}{(acq/e;q)_\infty\,(bcq/e;q)_\infty} \\ &\quad = \frac{(q/\omega;q)_z\,(\omega;q)_\infty\,\left(e^2/ab\omega q;q\right)_z\,\left(ab\omega q^2/e^2;q\right)_\infty}{\left(e^2/a\omega q;q\right)_z\,\left(a\omega q^2/e^2;q\right)_\infty\,\left(e^2/b\omega q;q\right)_z\,\left(b\omega q^2/e^2;q\right)_\infty}. \end{aligned} \tag{7.36}$$

Observe by (7.31–33) that z is bounded below. Using (7.35) and (7.36), we see that both of the ratios of products on the right side of (7.30) are bounded as functions of c. Using (7.33) and the identity (2.11) for reversing the q-Pockhammer symbol, we have

$$\begin{aligned} \frac{(c;q)_i}{(abcq/e;q)_i} &= (e/abq)^i \frac{\left(q^{1+z-i}/\omega;q\right)_i}{\left(q^{z-i}/ab\omega;q\right)_i}, \qquad 0 \le i < z, \\ \frac{(c;q)_i}{(abcq/e;q)_i} &= (e/abq)^z \frac{(q/\omega;q)_z}{(1/ab\omega;q)_z} \frac{(\omega;q)_{i-z}}{(abq\omega/e;q)_{i-z}}, \qquad i \ge z. \end{aligned} \tag{7.37}$$

Using (7.37), (7.37) with c replaced by cq/e, and the fact that $|e/abq| < 1$, we see that ${}_3\varphi_2[a,b,c;e,abcq/e;q,q]$ and ${}_3\varphi_2\left[aq/e, bq/e, cq/e; q^2/e, abcq^2/e^2; q, q\right]$ are bounded as functions of c. Thus $f(c)$ is bounded on Ω_2.

Recall Liouville's theorem Hille (Hille, 1973, Theorem 8.2.2) that a bounded entire function is constant. The proof uses the Laurent series representation Hille (Hille, 1973, Theorem 8.1.1)

$$f(c) = \sum_{n=0}^{\infty} a_n \left(c - c_0\right)^n, \tag{7.38}$$

where the coefficients are given by Hille (Hille, 1973, (8.1.2), (8.2.3))

$$a_n = \frac{1}{2\pi i} \int_{C_r} \frac{f(z)}{\left(z - c_0\right)^{n+1}}\, dz, \tag{7.39}$$

and the contour $C_r = \{z \mid |z - c_0| = r\}$ and c are inside a disc with center c_0 on which f is analytic. If we take $\epsilon < 1 - |q|^{1/4}$, then for each c_0 there exists a sequence $\{r_n\}_{n \ge 1}$ of radii with $\lim_{n\to\infty} r_n = \infty$ and $C_{r_n} \subseteq \Omega_1 \cup \Omega_2$ $n \ge 1$. Using (7.39), we have $a_0 = f(c_0)$ and $a_n = 0$, $n \ge 1$. Hence (7.38) gives $f(c) = f(c_0)$. Taking $c_0 = e/q$, we have $f(c) = f(e/q) = 1$ as required. We may remove the restrictions on a, b and e since f is entire and hence continuous in each of these parameters.

We call this argument the Liouville-Ismail argument.

We have come full circle since the nonterminating q-Chu-Vandermonde sum (1.12) implies in turn (7.13), (7.15), (7.14) and the formulation (7.16) of the nonterminating q-Saalschütz sum (1.13), which implies (1.12).

8. Conclusion

In this section, we conclude with some thoughts on the Hankel determinant, the Rodriguez formula, the slinky rule for the Schur functions, the q-Dyson polynomials, q-series identities, and the Vinet operator.

The q-Saalschütz sum (1.13) or the Hankel determinant ((Szegő, 1975, (2.1.5))) or Rodriguez formula (Atakishiyev, Rahman and Suslov (Atakishiyev et al., 1995)) for the little q-Jacobi polynomials together with the evaluation of the q-measure for a family of orthogonal polynomials in the Askey tableau imply the orthogonality of the polynomials. We hope to extend the Hankel determinant, the Rodriguez formula, and the little q-Jacobi functions (7.5) to the Askey tableau.

Observe that the renormalized little q-Jacobi functions

$$\begin{aligned}(tq)^{(\alpha-1)/2} & \left(q^{1-n-\alpha};q\right)_\infty \left(q^{n+\beta};q\right)_\infty (q^\alpha;q)_\infty\, p_n^{\alpha,\beta}(t) \\ &= (tq)^{(\alpha-1)/2}\left(q^{1-n-\alpha};q\right)_\infty \left(q^{n+\beta};q\right)_\infty (q^\alpha;q)_\infty \\ &\quad \times {}_2\varphi_1\left[\begin{matrix} q^{-n}, q^{n+\alpha+\beta-1} \\ q^\alpha \end{matrix};q,tq\right] \\ &- (tq)^{(1-\alpha)/2}\left(q^{-n};q\right)_\infty \left(q^{n+\alpha+\beta-1};q\right)_\infty \left(q^{2-\alpha};q\right)_\infty \\ &\quad \times {}_2\varphi_1\left[\begin{matrix} q^{1-n-\alpha}, q^{n+\beta} \\ q^{2-\alpha} \end{matrix};q,tq\right]\end{aligned} \tag{8.1}$$

are antisymmetric under

$$\alpha \to 2-\alpha,\ \beta \to \beta,\ n \to n+\alpha-1, \tag{8.2}$$

and the antisymmetry allows us to recover (7.5). The slinky rule for the Schur functions was essential in establishing (Kadell, 2000b) the combinatorial representation as an alternating sum over the symmetric group. This paper follows the success for $n = 2$ of the thesis (Kadell, 2000b) that we may extend the Macdonald polynomials to functions by reading the slinky rule from the Selberg q-integral (Kaneko (Kaneko, 1996), Macdonald (Macdonald, 1995, Section 6.9, Example 3)) and using the tool

$$\sum_{i=a}^{b} f(i) = \sum_{i=a}^{\infty} f(i) - \sum_{i=b+k}^{\infty} f(i), \tag{8.3}$$

where, following (Macdonald, 1995, Chap. VI), we have $t = q^k$. Following (Kadell, 1994a; Kadell, 1997; Kadell, 1998; Kadell, 2000a; Kadell, 2000b; Kadell, 2003), we hope that this idea will work for the Askey tableau, the root system BC_n, and the q-Dyson polynomials.

The pioneering work (Baker and Forrester, 1999), (Biedenharn and Louck, 1989), (Heckman, 1987), (Heckman and Opdam, 1987), (Kaneko, 1993; Kaneko, 1996; Kaneko, 1998), (Koornwinder, 1995), (Opdam, 1988a; Opdam, 1988b; Opdam, 1989) by Baker, Biedenharn, Koornwinder, Louck, Forrester, Heckman, Kaneko and Opdam focuses on the multivariable basic hypergeometric functions and orthogonal polynomials associated with root systems. We hope to extend the Macdonald (Macdonald, 1995, Chap. VI), Bidenharn-Louck (Biedenharn and Louck, 1989), Heckman-Opdam (Heckman, 1987), (Heckman and Opdam, 1987), (Opdam, 1988a; Opdam, 1988b), Koornwinder (Koornwinder, 1995), and Sahi-Knop (Sahi and Knop, 1996) polynomials to functions of complex argument and to extend the properties of orthogonal polynomials and q-series identities. We hope to follow Section 5 of this paper and combine the transformation formulas of Baker and Forrester (Baker and Forrester, 1999) with Kaneko's multivariable ${}_1\psi_1$ sum (Kaneko, 1998) to give a multivariable nonterminating q-Chu-Vandermonde sum and a multivariable Rogers-Fine symmetric function. We hope to use Ismail's argument (Ismail, 1977) to prove Heine's ${}_2\varphi_1$ transformation (1.11) and extend this to the multivariable setting.

We hope to give a Vinet operator (see (Vinet and Lapointe, 1995)) of complex order which extends the Rodriguez formulas for the Askey tableau.

References

Anderson, G. (1991). A short proof of Selberg's generalized beta formula. *Forum Math.*, 3:415–417.

Andrews, G. E. (1965). A simple proof of Jacobi's triple product identity. *Proc. Amer. Math. Soc.*, 16:333–334.

Andrews, G. E. (1969). On Ramanujan's summation of ${}_1\psi_1(a;b;z)$. *Proc. Amer. Math. Soc.*, 22:552–553.

Andrews, G. E. (1972). Two theorems of Gauss and allied identities proved arithmetically. *Pacific J. Math.*, 41:563–568.

Andrews, G. E. (1998). *The Theory of Partitions.* Cambridge Mathematical Library. Cambridge University Press, Cambridge. Reprint of the 1976 original.

Andrews, G. E. and Askey, R. (1977). Enumeration of partitions: the role of Eulerian series and q-orthogonal polynomials. In Aigner, M., editor, *Higher Combinatorics*, pages 3–26. Reidel, Boston, Mass.

Andrews, G. E. and Askey, R. (1978). A simple proof of Ramanujan's summation of the ${}_1\psi_1$. *Aequationes Math.*, 18:333–337.

Andrews, G. E. and Askey, R. (1981). Another q-extension of the beta function. *Proc. Amer. Math. Soc.*, 81:97–102.

Aomoto, K. (1987). Jacobi polynomials associated with Selberg's integral. *SIAM J. Math. Anal.*, 18:545–549.

Askey, R. (1978). The q-gamma and q-beta functions. *Appl. Anal.*, 8:125–141.

Askey, R. (1980). Some basic hypergeometric extensions of integrals of Selberg and Andrews. *SIAM J. Math. Anal.*, 11:938–951.

Askey, R. and Ismail, M. E.-H. (1979). The very well poised ${}_6\psi_6$. *Proc. Amer. Math. Soc.*, 77:218–222.

Askey, R. and Wilson, J. (1985). Some basic hypergeometric orthogonal polynomials that generalize Jacobi polynomials. *Mem. Amer. Math. Soc.*, 54(319):iv+55.

Atakishiyev, N. M., Rahman, M., and Suslov, S. K. (1995). On classical orthogonal polynomials. *Constr. Approx.*, 11:181–226.

Baker, T. H. and Forrester, P. J. (1999). Transformation formulas for multivariable basic hypergeometric series. *Methods Appl. Anal.*, 6(2):147–164. Dedicated to Richard A. Askey on the occasion of his 65th birthday, Part II.

Biedenharn, L. C. and Louck, J. D. (1989). A new class of symmetric polynomials defined in terms of tableaux. *Adv. in Appl. Math.*, 10:396–438.

Evans, R. (1994). Multidimensional beta and gamma integrals. *Contemp. Math.*, 166:341–357.

Fine, N. J. (1988). *Basic Hypergeometric Series and Applications*, volume 27 of *Mathematical Surveys and Monographs.* Amer. Math. Soc., Providence, RI.

Garvan, F. and Gonnet, G. (1991). Macdonald's constant term conjectures for exceptional root systems. *Bull. Amer. Math. Soc. (N.S.)*, 24:343–347.

Garvan, F. G. (1989a). A beta integral associated with the root system g_2. *SIAM J. Math. Anal.*, 19:1462–1474.

Garvan, F. G. (1989b). A proof of the Macdonald-Morris root system conjecture for f_4. *SIAM J. Math. Anal.*, 21:803–821.

Gasper, G. (1997). Elementary derivations of summation and transformation formulas for q-series. In Ismail, M. E. H., Masson, D. R., and Rahman, M., editors, *Special Functions, q-Series and Related Topics (Toronto, ON, 1995)*, volume 14 of *Fields Instit. Commun.*, pages 55–70. Amer. Math. Soc., Providence, RI.

Gasper, G. and Rahman, M. (1990). *Basic Hypergeometric Series.* Cambridge Univ. Press, Cambridge.

Gustafson, R. A. (1990). A generalization of Selberg's beta integral. *Bull. Amer. Math. Soc. (N.S.)*, 22:97–105.

Habsieger, L. (1986). La q-conjecture de Macdonald-Morris pour g_2. *C. R. Acad. Sci. Paris Sér. I Math.*, 303:211–213.

Habsieger, L. (1988). Une q-intégrale de Selberg-Askey. *SIAM J. Math. Anal.*, 19:1475–1489.

Hahn, W. (1949). Über polynome, die gleichzeitig zwei verschieclen orthogonalsystemen angehören. *Math. Nachr.*, 2:263–278.

Heckman, G. J. (1987). Root systems and hypergeometric functions II. *Compositio Math.*, 64:353–373.

Heckman, G. J. and Opdam, E. M. (1987). Root systems and hypergeometric functions I. *Compositio Math.*, 64:329–352.

Heine, E. (1847). Untersuchungen über die Riehe .... *J. reine angew. Math.*, 34:285–328.

Heine, E. (1878). *Handbuch der Kugelfunctionen, Theorie und Anwendungen*, volume 1. Reimer, Berlin.

Hille, E. (1973). *Analytic Function Theory*, volume 1. Chelsea, New York, second edition.

Hua, L. K. (1963). *Harmonic Analysis of Functions of Several Complex Variables in the Classical Domains*, volume 6 of *Transl. Math. Monogr.* Amer. Math. Soc., Providence, RI.

Ismail, M. E.-H. (1977). A simple proof of Ramanujan's ${}_1\psi_1$ summation formula. *Proc. Amer. Math. Soc.*, 63:185–186.

Ismail, M. E.-H. and Rahman, M. (1995). Some basic bilateral series and integrals. *Pacific J. Math.*, 170:497–515.

Jackson, F. H. (1910). On q-definite integrals. *Quart. J. Pure Appl. Math.*, 41:193–203.

Joichi, J. T. and Stanton, D. (1989). An involution for Jacobi's identity. *Discrete Math.*, 73:261–271.

Kadell, K. W. J. (1987a). Path functions and generalized basic hypergeometric functions. *Mem. Amer. Math. Soc.*, 65(360):iv+54.

Kadell, K. W. J. (1987b). A probabilistic proof of Ramanujan's ${}_1\psi_1$ sum. *SIAM J. Math. Anal.*, 18:1539–1548.

Kadell, K. W. J. (1988a). A proof of Askey's conjectured q-analogue of Selberg's integral and a conjecture of Morris. *SIAM J. Math. Anal.*, 19:969–986.

Kadell, K. W. J. (1988b). The q-Selberg polynomials for $n = 2$. *Trans. Amer. Math. Soc.*, 310(2):535–553.

Kadell, K. W. J. (1993). An integral for the product of two Selberg-Jack symmetric polynomials. *Compositio Math.*, 87:5–43.

Kadell, K. W. J. (1994a). A proof of the q-Macdonald-Morris conjecture for bc_n. *Mem. Amer. Math. Soc.*, 108(516):vi+80.

Kadell, K. W. J. (1994b). A simple proof of an Aomoto type extension of the q-Morris theorem. *Contemp. Math.*, 166:167–181.

Kadell, K. W. J. (1997). The Selberg-Jack symmetric functions. *Adv. Math.*, 130:33–102.

Kadell, K. W. J. (1998). A simple proof of an Aomoto-type extension of Gustafson's Askey-Wilson Selberg q-integral. *Methods Appl. Anal.*, 5:125–142.

Kadell, K. W. J. (2000a). A Dyson constant term orthogonality relation. *J. Combin. Theory Ser. A*, 89(2):291–297.

Kadell, K. W. J. (2000b). The Schur functions for partitions with complex parts. In *q-series from a contemporary perspective (South Hadley, MA, 1998)*, volume 254 of *Contemp. Math.*, pages 247–270. Amer. Math. Soc., Providence, RI.

Kadell, K. W. J. (2003). Selberg sums for the Sahi-Knop polynomials. Preprint.

Kaneko, J. (1993). Selberg integrals and hypergeometric series associated with Jack polynomials. *SIAM J. Math. Anal.*, 24:1086–1110.

Kaneko, J. (1996). q-Selberg integrals and Macdonald polynomials. *Ann. Sci. École Norm. Sup.*, 29(4):583–637.

Kaneko, J. (1998). A ${}_1\psi_1$ summation theorem for Macdonald polynomials. *Ramanujan J.*, 2:379–386.

Kendall, M. G. and Stuart, A. (1973). *The Advanced Theory of Statistics*, volume 2. Hafner, New York.

Knop, F. and Sahi, S. (1997). A recursion and a combinatorial formula for Jack polynomials. *Invent. Math.*, 128:9–22.

Koekoek, R. and Swarttouw, R. F. (1994). The Askey-scheme of hypergeometric orthogonal polynomials and its q-analogue. Faculty TWI 94–05, Delft University of Technology. Electronic version: `http://aw.twi.tudelft.nl/~koekoek/askey/`.

Koornwinder, T. (1995). Jacobi polynomials of type bc, Jack polynomials, limit transitions and $o(\infty)$. *Contemp. Math.*, 190:283–286.

Macdonald, I. G. (1982). Some conjectures for root systems and finite reflection groups. *SIAM J. Math. Anal.*, 13:988–1007.

Macdonald, I. G. (1992). Schur functions: theme and variations. In *Séminaire Lotharingien de Combinatoire, (Saint-Nabor, 1992)*, volume 498 of *Publ. Inst. Rech. Math. Av.*, pages 5–39. Univ. Louis Pasteur, Strasbourg.

Macdonald, I. G. (1995). *Symmetric Functions and Hall Polynomials.* Oxford Univ. Press, Oxford, 2nd edition.

Macdonald, I. G. (1998). *Symmetric Functions and Orthogonal Polynomials*, volume 12 of *Univ. Lecture Ser.* Amer. Math. Soc., Providence, RI.

Mehta, M. L. (1967). *Random Matrices and the Statistical Theory of Energy Levels.* Academic Press, New York.

Morris, II, W. G. (1982). *Constant term identities for finite and affine root systems: conjectures and theorems.* PhD thesis, University of Wisconsin-Madison.

Opdam, E. (1989). Some applications of hypergeometric shift operators. *Invent. Math.*, 98:1–18.

Opdam, E. (1995). Harmonic analysis for certain representations of graded Hecke algebras. *Acta Math.*, 175:75–121.

Opdam, E. M. (1988a). Root systems and hypergeometric functions III. *Compositio Math.*, 67:21–49.

Opdam, E. M. (1988b). Root systems and hypergeometric functions IV. *Compositio Math.*, 67:191–209.

Proctor, R. (1989). Equivalence of the combinatorial and the classical definitions of the Schur functions. *J. Combin. Theory Ser. A*, 51:135–137.

Rahman, M. and Suslov, S. K. (1994a). Barnes and Ramanujan-type integrals on the q-linear lattice. *SIAM J. Math. Anal.*, 25:1002–1022.

Rahman, M. and Suslov, S. K. (1994b). The Pearson equation and the beta integral. *SIAM J. Math. Anal.*, 25:646–693.

Rahman, M. and Suslov, S. K. (1996). A unified approach to the summation and integration formulas for q-hypergeometric functions, I. *J. Statist. Plan. Infer.*, 54:101–118.

Rahman, M. and Suslov, S. K. (1998a). A unified approach to the summation and integration formulas for q-hypergeometric functions, II. *Methods Appl. Anal.*, 5(4):399–412.

Rahman, M. and Suslov, S. K. (1998b). A unified approach to the summation and integration formulas for q-hypergeometric functions, III. *Methods Appl. Anal.*, 5(4):413–424.

Richards, D. S. P. (1989). Analogs and extensions of Selberg's integral. In *q-series and Partitions (Minneapolis, MN, 1988)*, volume 18 of *IMA Vol. Math. Appl.*, pages 109–137. Springer, New York.

Rogers, L. J. (1916). On two theorems of combinatory analysis and some allied identities. *Proc. London Math. Soc.*, 16(2):315–336.

Rogers, L. J. (1983). On a three-fold symmetry in the elements of Heine's series. *Proc. London Math. Soc.*, 24:171–191.

Sahi, S. (1994). The spectrum of certain invariant differential operators associated to a Hermitian symmetric space. In Brylinski, J.-L., Brylinski, R., Guillemin, V., and Kac, V., editors, *Lie Theory and Geometry in Honor of Bertram Kostant*, volume 123 of *Progr. Math.*, pages 569–576. Birkhäuser, Boston, Mass.

Sahi, S. and Knop, F. (1996). Difference equations and symmetric polynomials defined by their zeroes. *Internat. Math. Res. Notices*, 10:473–486.

Sears, D. B. (1951). Transformations of basic hypergeometric functions of special type. *Proc. London Math. Soc.*, 53(2):467–483.

Selberg, A. (1944). Bemerkninger om et multipelt integral. *Norsk Mat. Tidsskr.*, 26:71–78.

Stanley, R. P. (1989). Some combinatorial properties of Jack symmetric functions. *Adv. Math.*, 77:76–115.

Starcher, G. W. (1930). On identities arising from solutions of q-difference equations and some interpretations in number theory. *Amer. J. Math.*, 53:801–816.

Stembridge, J. (1988). A short proof of Macdonald's root-system conjecture for the root systems of type a. *Proc. Amer. Math. Soc.*, 102:777–786.

Suslov, S. K. (1998). Multiparameter Ramanujan-type q-beta integrals. *Ramanujan J.*, 2:351–369.

Szegö, G. (1939). *Orthogonal Polynomials*, volume 23 of *Colloquium Publications*. Amer. Math. Soc.

Vinet, L. and Lapointe, L. (1995). A Rodriguez formula for the Jack polynomials and the Macdonald-Stanley conjecture. *Internat. Math. Res. Notices*, 9:419–424.

Zeilberger, D. (1988). A unified approach to Macdonald's root-system conjectures. *SIAM J. Math. Anal.*, 19:987–1013.

Zeilberger, D. (1989/90). A Stembridge-Stanton style elementary proof of the Habsieger-Kadell q-Morris identity. *Discrete Math.*, 79:313–322.

A SECOND ADDITION FORMULA FOR CONTINUOUS q-ULTRASPHERICAL POLYNOMIALS

Tom H. Koornwinder
Korteweg-de Vries Institute
University of Amsterdam
Plantage Muidergracht 24
1018 TV Amsterdam
THE NETHERLANDS
thk@science.uva.nl

Abstract This paper provides the details of Remark 5.4 in the author's paper "Askey-Wilson polynomials as zonal spherical functions on the $SU(2)$ quantum group," SIAM J. Math. Anal. **24** (1993), 795–813. In formula (5.9) of the 1993 paper a two-parameter class of Askey-Wilson polynomials was expanded as a finite Fourier series with a product of two ${}_3\varphi_2$'s as Fourier coefficients. The proof given there used the quantum group interpretation. Here this identity will be generalized to a 3-parameter class of Askey-Wilson polynomials being expanded in terms of continuous q-ultraspherical polynomials with a product of two ${}_2\varphi_2$'s as coefficients, and an analytic proof will be given for it. Then Gegenbauer's addition formula for ultraspherical polynomials and Rahman's addition formula for q-Bessel functions will be obtained as limit cases. This q-analogue of Gegenbauer's addition formula is quite different from the addition formula for continuous q-ultraspherical polynomials obtained by Rahman and Verma in 1986. Furthermore, the functions occurring as factors in the expansion coefficients will be interpreted as a special case of a system of biorthogonal rational functions with respect to the Askey-Roy q-beta measure. A degenerate case of this biorthogonality are Pastro's biorthogonal polynomials associated with the Stieltjes-Wigert polynomials.

1. Introduction

Rahman and Verma (Rahman and Verma, 1986) obtained the following *addition formula for continuous q-ultraspherical polynomials*:

$$\begin{aligned} &p_n\left(\cos\theta; a, aq^{\frac{1}{2}}, -a, -aq^{\frac{1}{2}} \mid q\right) \\ &= \sum_{k=0}^{n} \frac{(q;q)_n \left(a^4q^n, a^4q^{-1}, a^2q^{\frac{1}{2}}, -a^2q^{\frac{1}{2}}, -a^2; q\right)_k a^{n-k}}{(q;q)_k (q;q)_{n-k} (a^4q^{-1};q)_{2k} \left(a^2q^{\frac{1}{2}}, -a^2q^{\frac{1}{2}}, -a^2; q\right)_n} \\ &\quad \times p_{n-k}\left(\cos\varphi; aq^{\frac{1}{2}k}, aq^{\frac{1}{2}(k+1)}, -aq^{\frac{1}{2}k}, -aq^{\frac{1}{2}(k+1)} \mid q\right) \\ &\quad \times p_{n-k}\left(\cos\psi; aq^{\frac{1}{2}k}, aq^{\frac{1}{2}(k+1)}, -aq^{\frac{1}{2}k}, -aq^{\frac{1}{2}(k+1)} \mid q\right) \\ &\quad \times p_k\left(\cos\theta; ae^{i(\varphi+\psi)}, ae^{-i(\varphi+\psi)}, ae^{i(\varphi-\psi)}, ae^{i(\psi-\varphi)} \mid q\right). \end{aligned} \tag{1.1}$$

The formula is here written in the form given in (Gasper and Rahman, 1990, Exercise 8.11). Use (Gasper and Rahman, 1990) also for notation of (q-)hypergeometric functions and (q-)shifted factorials. Throughout it is supposed that $0 < q < 1$.

Formula (1.1) is given in terms of *Askey-Wilson polynomials* (see (Askey and Wilson, 1985) or (Gasper and Rahman, 1990, §7.5)):

$$p_n(\cos\theta; a, b, c, d \mid q) := a^{-n} (ab, ac, ad; q)_n \, r_n(\cos\theta; a, b, c, d \mid q) \qquad (n \in \mathbb{Z}_{\geq 0}) \tag{1.2}$$

(symmetric in a, b, c, d), where

$$r_n(\cos\theta; a, b, c, d \mid q) := {}_4\varphi_3\left(\begin{matrix} q^{-n}, abcdq^{n-1}, ae^{i\theta}, ae^{-i\theta} \\ ab, ac, ad \end{matrix}; q, q\right). \tag{1.3}$$

The *continuous q-ultraspherical polynomials* are the special case $b = aq^{\frac{1}{2}}$, $c = -a$, $d = -aq^{\frac{1}{2}}$ of the Askey-Wilson polynomials, often notated as follows (see (Gasper and Rahman, 1990, (7.4.14))):

$$\begin{aligned} C_n\left(x; a^2 \mid q\right) &= \frac{\left(a^4;q\right)_n}{(q;q)_n a^n} r_n\left(x; a, aq^{\frac{1}{2}}, -a, -aq^{\frac{1}{2}} \mid q\right) \\ &= \frac{\left(a^2;q\right)_n}{\left(q, a^4q^n; q\right)_n} p_n\left(x; a, aq^{\frac{1}{2}}, -a, -aq^{\frac{1}{2}} \mid q\right). \end{aligned} \tag{1.4}$$

A further specialization to $a = \frac{1}{4}$, i.e., $(a, b, c, d) = \left(\frac{1}{4}, \frac{3}{4}, -\frac{1}{4}, -\frac{3}{4}\right)$, yields the *continuous q-Legendre polynomials*. For this case $a = \frac{1}{4}$ Koelink was able to give two different poofs of the addition formula (1.1) from a quantum group interpretation on $SU_q(2)$, see (Koelink, 1994) and (Koelink, 1997).

If a is replaced by $q^{\frac{1}{2}\lambda}$ in (1.1) and the limit is taken for $q \uparrow 1$, then a version of the *addition formula for ultraspherical polynomials* is obtained:

$$C_n^\lambda(\cos\theta) = \sum_{k=0}^{n} \frac{2^{2k}\,(2\lambda+2k-1)\,(n-k)!\,(\lambda)_k^2}{(2\lambda-1)_{n+k+1}} \\ \times (\sin\varphi)^k\, C_{n-k}^{\lambda+k}(\cos\varphi)\,(\sin\psi)^k\, C_{n-k}^{\lambda+k}(\cos\psi)\, C_k^{\lambda-\frac{1}{2}}\left(\frac{\cos\theta-\cos\varphi\cos\psi}{\sin\varphi\sin\psi}\right). \tag{1.5}$$

Here *ultraspherical polynomials* are defined by

$$C_n^\lambda(x) := \frac{(2\lambda)_n}{n!}\, {}_2F_1\left(\begin{matrix}-n, n+2\lambda\\ \lambda+\frac{1}{2}\end{matrix}; \frac{1}{2}(1-x)\right). \tag{1.6}$$

By elementary substitution the addition formula (1.5) transforms into the familiar addition formula for ultraspherical polynomials:

$$C_n^\lambda(\cos\varphi\cos\psi+\sin\varphi\sin\psi\cos\theta) = \sum_{k=0}^{n} \frac{2^{2k}\,(2\lambda+2k-1)\,(n-k)!\,(\lambda)_k^2}{(2\lambda-1)_{n+k+1}} \\ \times (\sin\varphi)^k\, C_{n-k}^{\lambda+k}(\cos\varphi)\,(\sin\psi)^k\, C_{n-k}^{\lambda+k}(\cos\psi)\, C_k^{\lambda-1/2}(\cos\theta), \tag{1.7}$$

see (Erdélyi et al., 1953, 10.9(34)), but watch out for the misprint 2^m which should be 2^{2m}; see also the references given in (Askey, 1975, Lecture 4). For the removable singularity at $\lambda = \frac{1}{2}$ in (1.7) observe that

$$\lim_{\lambda\to\frac{1}{2}} \frac{k}{2\lambda-1}\, C_k^{\lambda-\frac{1}{2}}(\cos\theta) = \cos(k\theta) \quad (k>0); \qquad C_0^{\lambda-\frac{1}{2}}(\cos\theta) = 1. \tag{1.8}$$

An elementary transformation comparable to the passage from (1.5) to (1.7) cannot be performed on the q-level. It is a generally observed phenomenon in q-theory that, for a classical formula involving a parameter dependent function with an argument transformed by a parameter dependent transformation, a possible q-analogue has the transformation parameters occurring in the function parameters. Compare for instance the p_k factor in (1.1) with the $C_k^{\lambda-\frac{1}{2}}$ factor in (1.5).

In (Koornwinder, 1993, (5.9)) I obtained the following formula as a spin-off of the interpretation of certain Askey-Wilson polynomials as zonal spherical functions on the quantum group $SU_q(2)$:

$$\begin{aligned} & p_n(\cos\theta; -q^{\frac12(\sigma+\tau+1)}, -q^{\frac12(-\sigma-\tau+1)}, q^{\frac12(\sigma-\tau+1)}, q^{\frac12(-\sigma+\tau+1)} \mid q) \\ &= \sum_{k=-n}^{n} \frac{(q^{n+1};q)_n (q;q)_{2n}}{(q;q)_{n+k}(q;q)_{n-k}} q^{\frac12(n-k)(n-k+\sigma+\tau)} \\ &\times {}_3\varphi_2\left(\begin{matrix} q^{-n+k}, q^{-n}, q^{-n-\sigma} \\ q^{-2n}, 0 \end{matrix}; q, q\right) {}_3\varphi_2\left(\begin{matrix} q^{-n+k}, q^{-n}, q^{-n-\tau} \\ q^{-2n}, 0 \end{matrix}; q, q\right) e^{ik\theta}. \end{aligned} \tag{1.9}$$

Here the summand, with $e^{ik\theta}$ omitted, is invariant under the transformation $k \to -k$, as can be seen by twofold application of (Gasper and Rahman, 1990, (3.2.3)). The ${}_3\varphi_2$'s were viewed in (Koornwinder, 1993) as dual q-Krawtchouk polynomials (see their definition in (Koekoek and Swarttouw, 1998, §3.17)), but here we will prefer to consider them as certain (unusual) q-analogues of ultraspherical polynomials, for which an expression in terms of a ${}_2\varphi_2$ is more suitable. For this purpose apply a ${}_3\varphi_2 \to {}_2\varphi_2$ transformation obtained from formulas (1.5.4) and (III.7) in (Gasper and Rahman, 1990). Then, after the substitutions $q^{\frac12\sigma} \to is^{-1}$, $q^{\frac12\tau} \to it$ in (1.9), we can write (1.9) equivalently as

$$\begin{aligned} & p_n\left(\cos\theta; st^{-1}q^{\frac12}, s^{-1}tq^{\frac12}, -stq^{\frac12}, -s^{-1}t^{-1}q^{\frac12} \mid q\right) \\ &= (-1)^n q^{\frac12 n^2} (q;q)_n \sum_{k=0}^{n} q^{\frac12 k} \frac{\left(q^{n+1};q\right)_k (q^{-n};q)_k}{(q;q)_k^2} \\ &\times (st)^{k-n} \left(-q^{-k+1}s^2, -q^{-k+1}t^{-2}; q\right)_{n-k} \\ &\times {}_2\varphi_2\left(\begin{matrix} q^{-n+k}, q^{n+k+1} \\ q^{k+1}, -q^{k+1}s^2 \end{matrix}; q, -qs^2\right) \\ &\times {}_2\varphi_2\left(\begin{matrix} q^{-n+k}, q^{n+k+1} \\ q^{k+1}, -q^{k+1}t^{-2} \end{matrix}; q, -qt^{-2}\right) (1+\delta_{k,0})\cos(k\theta). \end{aligned} \tag{1.10}$$

After the substitutions $q^\sigma = \tan^2 \frac12\varphi$, $q^\tau = \tan^2 \frac12\psi$ in (1.9), the limit for $q \uparrow 1$ becomes the case $\lambda = \frac12$ of the addition formula (1.7) (combined with (1.8)), i.e., the addition formula for Legendre polynomials $P_n(x) = C_n^{\frac12}(x)$.

The first main result of this paper is the following addition formula for continuous q-ultraspherical polynomials. Thus I will finally fulfill my promise of bringing out "reference [9]" of my paper (Koornwinder, 1993), which reference was mentioned there as being in preparation.

Theorem 1.1. *We have, in notation* (1.3),

$$r_n\left(\cos\theta; ast^{-1}q^{\frac{1}{2}}, ats^{-1}q^{\frac{1}{2}}, -astq^{\frac{1}{2}}, -as^{-1}t^{-1}q^{\frac{1}{2}} \mid q\right)$$
$$= (-1)^n a^{2n} q^{\frac{1}{2}n(n+1)}\left(1+a^2\right)$$
$$\times \sum_{k=0}^{n} \frac{\left(1-a^2q^k\right)\left(q^{-n}, a^4q, a^4q^{n+1}; q\right)_k \left(a^{-2}st^{-1}q^{\frac{1}{2}}\right)^k}{\left(1-a^4q^k\right)(q;q)_k\,(a^2q;q)_k^2\,(-a^2s^2q;q)_k\,(-a^2t^{-2}q;q)_k}$$
$$\times {}_2\varphi_2\left(\begin{matrix} q^{-n+k}, a^4q^{n+k+1} \\ a^2q^{k+1}, -a^2s^2q^{k+1}\end{matrix}; q, -s^2q\right)$$
$$\times {}_2\varphi_2\left(\begin{matrix} q^{-n+k}, a^4q^{n+k+1} \\ a^2q^{k+1}, -a^2t^{-2}q^{k+1}\end{matrix}; q, -t^{-2}q\right) r_k\left(\cos\theta; a, -a, aq^{\frac{1}{2}}, -aq^{\frac{1}{2}} \mid q\right), \tag{1.11}$$

or, in notation (1.2),

$$p_n\left(\cos\theta; ast^{-1}q^{\frac{1}{2}}, ats^{-1}q^{\frac{1}{2}}, -astq^{\frac{1}{2}}, -as^{-1}t^{-1}q^{\frac{1}{2}} \mid q\right)$$
$$= (-1)^n q^{\frac{1}{2}n^2} \sum_{k=0}^{n} a^{n-k}\left(a^2q^{k+1}; q\right)_{n-k}$$
$$\times q^{\frac{1}{2}k}\,\frac{\left(q^{-n}, a^4q^{n+1}; q\right)_k}{(q, a^4q^k; q)_k}\,\frac{\left(-a^2s^2q^{k+1}, -a^2t^{-2}q^{k+1}; q\right)_{n-k}}{(st^{-1})^{n-k}}$$
$$\times {}_2\varphi_2\left(\begin{matrix} q^{-n+k}, a^4q^{n+k+1} \\ a^2q^{k+1}, -a^2s^2q^{k+1}\end{matrix}; q, -s^2q\right)$$
$$\times {}_2\varphi_2\left(\begin{matrix} q^{-n+k}, a^4q^{n+k+1} \\ a^2q^{k+1}, -a^2t^{-2}q^{k+1}\end{matrix}; q, -t^{-2}q\right) p_k\left(\cos\theta; a, -a, aq^{\frac{1}{2}}, -aq^{\frac{1}{2}} \mid q\right). \tag{1.12}$$

For $a = 1$ formula (1.12) specializes to formula (1.10) (with usage of a Chebyshev case of the Askey-Wilson polynomials, see (Askey and Wilson, 1985, (4.21))). Its limit case for $q \uparrow 1$ (after the substitutions $a = q^{\frac{1}{2}\lambda-\frac{1}{4}}$, $s = \tan\left(\frac{1}{2}\varphi\right)$, $t = \tan\left(\frac{1}{2}\psi\right)$) is the addition formula (1.7) for ultraspherical polynomials of general order λ. It is also possible to obtain Rahman's addition formula (Rahman, 1988, (1.10)) for q-Bessel functions as a formal limit case of (1.12), see Remark 2.2. For precise correspondence of (1.11) with (1.1) one should replace a by $aq^{-\frac{1}{4}}$ in (1.11).

The left-hand side of (1.12) is invariant under the symmetries $(s,t) \to (t,s)$, $(s,t) \to \left(-s^{-1}, t\right)$ and $(s,t) \to \left(s, -t^{-1}\right)$. These symmetries are also visible on the right-hand side of (1.12) if we take into account that

$$\frac{(-a^2s^2q^{k+1};q)_{n-k}}{s^{n-k}}\ {}_2\varphi_2\left(\begin{matrix}q^{-n+k},a^4q^{n+k+1}\\ a^2q^{k+1},-a^2s^2q^{k+1}\end{matrix};q,-s^2q\right)$$

is invariant under the transformation $s \to s^{-1}$ up to the factor $(-1)^{n-k}$ (by (3.14)).

The proof of Theorem 1.1 (see details in §14.2) is quite similar to the proof of (1.1) in (Rahman and Verma, 1986). We consider (1.11) as a connection formula which connects Askey-Wilson polynomials of different order. There are certain choices of the orders of the Askey-Wilson polynomials in a connection formula

$$r_n(\cos\theta;\alpha,\beta,\gamma,\delta\mid q)=\sum_{k=0}^{n}A_{n,k}\,r_k(\cos\theta;a,b,c,d\mid q) \qquad (1.13)$$

for which the connection coefficients $A_{n,k}$ factorize. Formula (1.1) is one example; formula (1.11) is another example.

At the end of §2 a degenerate addition formula for continuous q-ultraspherical polynomials will be given as a limit case of the addition formula (1.11). As a further limit case we obtain a degenerate addition formula for q-Bessel functions.

The ${}_2\varphi_2$ factors on the right-hand side of (1.11) tend, after the mentioned substitutions and as $q \uparrow 1$, to ultraspherical polynomials $C_{n-k}^{\lambda+k}$ of argument $\cos\varphi$ resp. $\cos\psi$, so we might expect that these ${}_2\varphi_2$'s also satisfy some kind of (bi)orthogonality relations. This is indeed the case $\alpha=\beta$ of Theorem 1.2 below. See its proof in §14.3.

Theorem 1.2. *Let $\alpha,\beta>-1$. Define a system of rational functions in t by*

$$p_n^{(\alpha,\beta)}(t;q,\mathrm{rat}) := {}_2\varphi_2\left(\begin{matrix}q^{-n},q^{n+\alpha+\beta+1}\\ q^{\alpha+1},-tq^{\beta+1}\end{matrix};q,-tq\right) \qquad (n\in\mathbb{Z}_{\geq 0}), \quad (1.14)$$

and define, with additional parameter $c\in\mathbb{R}$, a measure on $[0,\infty)$ by

$$d\mu_{\alpha,\beta,c;q}(t) := \frac{q^{-c^2}\,\Gamma_q(c)\,\Gamma_q(1-c)\,\Gamma_q(\alpha+\beta+2)}{\Gamma(c)\,\Gamma(1-c)\,\Gamma_q(\alpha+1)\,\Gamma_q(\beta+1)} \times \frac{t^{c-1}\,(-tq^{\beta+1},-t^{-1}q^{\alpha+2};q)_\infty}{(-tq^{-c},-t^{-1}q^{1+c};q)_\infty}\,dt, \qquad (1.15)$$

where Γ_q *is defined in* (3.2). *Then*

$$\int_0^\infty p_n^{(\alpha,\beta)}(t;q,\mathrm{rat})\,p_m^{(\beta,\alpha)}(t^{-1}q;q,\mathrm{rat})\,d\mu_{\alpha,\beta,c;q}(t) = \frac{(-1)^n\left(1-q^{n+\alpha+\beta+1}\right)(q;q)_n\,\delta_{n,m}}{q^{\frac{1}{2}n(n-1)}\left(1-q^{2n+\alpha+\beta+1}\right)\left(q^{\alpha+\beta+2};q\right)_n}. \tag{1.16}$$

The case $n=m=0$ of (1.16), i.e., $\int_0^\infty d\mu_{\alpha,\beta,c;q}(t)=1$, is precisely the q-beta integral of Askey and Roy (Askey and Roy, 1986, (3.4)), which extends a q-beta integral of Ramanujan (Ramanujan, 1915, (19)). The Askey-Roy integral was independently obtained by Gasper (see (Gasper, 1984) and also (Gasper, 1987)) and by Thiruvenkatachar and Venkatachaliengar (see (Askey, 1988, p. 93)).

Note that the two ${}_2\varphi_2$'s in (1.11) (with $a=q^{\frac{1}{2}\alpha}$), can be rewritten in terms of (1.14): as $p_{n-k}^{(\alpha+k,\alpha+k)}(s^2;q,\mathrm{rat})$ and $p_{n-k}^{(\alpha+k,\alpha+k)}(t^{-2};q,\mathrm{rat})$, respectively.

The biorthogonality measure in (1.16) is evidently not unique, because of the parameter c. Further illustration of the non-uniqueness of the measure for these biorthogonality relation is provided by a q-integral variant of (1.16). In order to state this, we need the following definition of q-integral on $(0,\infty)$ (f arbitrary function on $(0,\infty)$ for which the sum absolutely converges):

$$\int_0^{s\cdot\infty} f(t)\,d_qt := (1-q)\sum_{k=-\infty}^{\infty} sq^k\,f(sq^k) \qquad (s>0). \tag{1.17}$$

Theorem 1.3. *The functions defined by* (1.14) *also satisfy the biorthogonality relations*

$$\frac{\Gamma_q(\alpha+\beta+2)}{\Gamma_q(\alpha+1)\,\Gamma_q(\beta+1)}\int_0^{s\cdot\infty} p_n^{(\alpha,\beta)}(t;q,\mathrm{rat})\,p_m^{(\beta,\alpha)}(t^{-1}q;q,\mathrm{rat}) \tag{1.18}$$
$$\times\frac{t^{-1}\left(-tq^{\beta+1},-t^{-1}q^{\alpha+2};q\right)_\infty}{(-t,-t^{-1}q;q)_\infty}\,d_qt$$
$$=\frac{(-1)^n\left(1-q^{n+\alpha+\beta+1}\right)(q;q)_n\,\delta_{n,m}}{q^{\frac{1}{2}n(n-1)}\left(1-q^{2n+\alpha+\beta+1}\right)\left(q^{\alpha+\beta+2};q\right)_n} \qquad (\alpha,\beta>-1,\ s>0).$$

The case $n=m=0$ of (1.3) is a q-beta integral first given by Gasper (Gasper, 1987), but it is essentially Ramanujan's ${}_1\psi_1$ sum; see some further discussion in §14.3. The proof of Theorem 1.3 is by a completely analogous argument as I will give in §14.3 for the proof of Theorem 1.2.

The paper concludes in §14.4 with some open questions and with some specializations of Theorem 1.2. Pastro's (Pastro, 1985) biorthogonal polynomials associated with the Stieltjes-Wigert polynomials occur as a special case.

Acknowledgments

I did the work communicated here essentially already in the beginning of the nineties. During that time, Mizan Rahman sent me very useful hints concerning the material which is now in §2, while René Swarttouw carefully checked (and corrected) my computations. One of the referees made some very good suggestions, which resulted, among others, into Remarks 2.1 and 2.4. Finally I thank Erik Koelink for stimulating me to publish this work after such a long time.

2. Proof of the new addition formula

In this section I prove the second addition formula for continuous q-ultraspherical polynomials, stated in Theorem 1.1. Let us first consider the general connection formula (1.13). We can split up this connection into three successive connections of more simple nature:

$$r_n(\cos\theta;\alpha,\beta,\gamma,\delta\mid q)=\sum_{j=0}^{n} c_{n,j}\left(\alpha e^{i\theta},\alpha e^{-i\theta};q\right)_j, \tag{2.1}$$

$$\left(\alpha e^{i\theta},\alpha e^{-i\theta};q\right)_j=\sum_{l=0}^{j} d_{j,l}\left(a e^{i\theta},a e^{-i\theta};q\right)_l, \tag{2.2}$$

$$\left(a e^{i\theta},a e^{-i\theta};q\right)_l=\sum_{k=0}^{l} e_{l,k}\, r_k(\cos\theta;a,b,c,d\mid q). \tag{2.3}$$

Then

$$A_{n,k}=\sum_{j=0}^{n-k}\sum_{l=0}^{j} c_{n,j+k}\, d_{j+k,l+k}\, e_{l+k,k} \tag{2.4}$$

and

$$A_{n,k}=\sum_{m=0}^{n-k}\sum_{i=0}^{m} c_{n,n-i}\, d_{n-i,n-m}\, e_{n-m,k}. \tag{2.5}$$

The coefficients $c_{n,j}$, $d_{j,l}$ and $e_{l,k}$ can be explicitly given as

$$c_{n,j} = \frac{\left(q^{-n}, \alpha\beta\gamma\delta q^{n-1}; q\right)_j q^j}{(\alpha\beta, \alpha\gamma, \alpha\delta, q; q)_j}, \tag{2.6}$$

$$d_{j,l} = \frac{(q, a\alpha; q)_j \left(a^{-1}\alpha; q\right)_{j-l} \alpha^l}{(q, a\alpha; q)_l (q; q)_{j-l} a^l}, \tag{2.7}$$

$$e_{l,k} = \frac{(abcd; q)_{2k} \left(q^{-l}; q\right)_k (ab, ac, ad; q)_l q^{lk}}{(abcdq^{k-1}, q; q)_k (abcd; q)_{k+l}}. \tag{2.8}$$

Here (2.6) follows from (1.3), formula (2.7) can be obtained by rewriting the q-Saalschütz formula (Gasper and Rahman, 1990, (1.7.2)), and (2.8) follows from (Askey and Wilson, 1985, (2.6), (2.5)).

It turns out that in the two double sums (2.4), (2.5) of $A_{n,k}$ the inner sum can be written as a balanced ${}_4\varphi_3$ of argument q:

$$\begin{aligned} A_{n,k} = (-1)^k q^{\frac{1}{2}k(k+1)} &\frac{\left(q^{-n}, \alpha\beta\gamma\delta q^{n-1}, ab, ac, ad; q\right)_k \alpha^k}{(q, abcdq^{k-1}, \alpha\beta, \alpha\gamma, \alpha\delta; q)_k a^k} \\ \times \sum_{j=0}^{n-k} &\frac{\left(q^{-n+k}, \alpha\beta\gamma\delta q^{n+k-1}, \alpha a q^k, \alpha a^{-1}; q\right)_j q^j}{\left(\alpha\beta q^k, \alpha\gamma q^k, \alpha\delta q^k, q; q\right)_j} \\ \times {}_4\varphi_3 &\left(\begin{matrix} q^{-j}, abq^k, acq^k, adq^k \\ a\alpha q^k, a\alpha^{-1}q^{1-j}, abcdq^{2k} \end{matrix} ; q, q \right) \end{aligned} \tag{2.9}$$

and

$$\begin{aligned} A_{n,k} = {} & (-1)^n q^{-\frac{1}{2}n(n-2k-1)} \\ & \times \frac{\left(1 - abcdq^{2k-1}\right) (q^{-n}; q)_k \left(\alpha\beta\gamma\delta q^{n-1}, ab, ac, ad; q\right)_n \alpha^n}{\left(1 - abcdq^{k-1}\right) (q; q)_k \left(abcdq^k, \alpha\beta, \alpha\gamma, \alpha\delta; q\right)_n a^n} \\ & \times \sum_{m=0}^{n-k} \frac{\left(q^{-n+k}, (abcd)^{-1}q^{1-n-k}, \alpha a^{-1}, (\alpha a)^{-1}q^{1-n}; q\right)_m q^m}{\left(q, (ab)^{-1}q^{1-n}, (ac)^{-1}q^{1-n}, (ad)^{-1}q^{1-n}; q\right)_m} \\ \times {}_4\varphi_3 & \left(\begin{matrix} q^{-m}, (\alpha\beta)^{-1}q^{-n+1}, (\alpha\gamma)^{-1}q^{-n+1}, (\alpha\delta)^{-1}q^{-n+1} \\ (\alpha\beta\gamma\delta)^{-1}q^{-2n+2}, (\alpha a)^{-1}q^{-n+1}, a\alpha^{-1}q^{-m+1} \end{matrix} ; q, q \right). \end{aligned} \tag{2.10}$$

The sums in (2.9) and (2.10) can be compared with the *Bateman type product formula* (Gasper and Rahman, 1990, (8.4.7)):

$$r_n\left(\cos\varphi; a, aq^{\frac{1}{2}}, -a, -aq^{\frac{1}{2}} \mid q\right) r_n\left(\cos\psi; a, aq^{\frac{1}{2}}, -a, -aq^{\frac{1}{2}} \mid q\right)$$
$$= \frac{1+a^2q^n}{1+a^2} (-1)^n q^{-\frac{1}{2}n} \sum_{m=0}^{n} \frac{\left(q^{-n}, a^4q^n, -e^{i\psi-i\varphi}q^{\frac{1}{2}}, -a^2e^{i\varphi-i\psi}q^{\frac{1}{2}}; q\right)_m q^m}{\left(q, a^2q^{\frac{1}{2}}, -a^2q^{\frac{1}{2}}, -a^2q; q\right)_m}$$
$$\times {}_4\varphi_3\left(\begin{matrix} q^{-m}, a^2, a^2e^{2i\varphi}, a^2e^{-2i\psi} \\ a^4, -a^2e^{i\varphi-i\psi}q^{\frac{1}{2}}, -e^{i\varphi-i\psi}q^{-m+\frac{1}{2}} \end{matrix}; q, q\right). \tag{2.11}$$

The sum in (2.9) can be matched with the right-hand side of (2.11) precisely for those values of the parameters $a, b, c, d, \alpha, \beta, \gamma, \delta$ in (1.13) which occur in the Rahman-Verma addition formula (1.1), i.e., for $ab = cd$, $\alpha^2 = cdq$, $\alpha = \beta q^{\frac{1}{2}} = -\gamma q^{\frac{1}{2}} = -\delta$. In fact, this will prove (1.1). The proof in (Rahman and Verma, 1986) is essentially along these lines.

Next we see that the sum in (2.10) can be matched with the right-hand side of (2.11) precisely for those values of the parameters $a, b, c, d, \alpha, \beta, \gamma, \delta$ in (1.13) when $\alpha\beta = \gamma\delta = a^2q$, $b = -c = aq^{\frac{1}{2}} = -dq^{\frac{1}{2}}$. Thus put $b = aq^{\frac{1}{2}}$, $c = -aq^{\frac{1}{2}}$, $d = -a$, $\alpha = ast^{-1}q^{\frac{1}{2}}$, $\beta = ats^{-1}q^{\frac{1}{2}}$, $\gamma = -astq^{\frac{1}{2}}$, $\delta = -as^{-1}t^{-1}q^{\frac{1}{2}}$ in (1.13) and (2.10). Then these two formulas specialize to:

$$r_n\left(\cos\theta; ast^{-1}q^{\frac{1}{2}}, ats^{-1}q^{\frac{1}{2}}, -astq^{\frac{1}{2}}, -as^{-1}t^{-1}q^{\frac{1}{2}} \mid q\right)$$
$$= \sum_{k=0}^{n} A_{n,k}\, r_k\left(\cos\theta; a, -a, aq^{\frac{1}{2}}, -aq^{\frac{1}{2}} \mid q\right), \tag{2.12}$$

$$A_{n,k} = (-1)^n s^n t^{-n} q^{\frac{1}{2}n(-n+2k+2)}$$
$$\times \frac{(1+a^2)(1-a^4q^{2k})\left(q^{-n}, a^4q; q\right)_k \left(a^4q^{n+1}; q\right)_n^2}{(1+a^2q^n)(1-a^4q^k)(q, a^4q^{n+1}; q)_k (a^2q; q)_n^2} \frac{1}{(-s^2a^2q, -t^{-2}a^2q; q)_n}$$
$$\times \sum_{m=0}^{n-k} \frac{\left(q^{-n+k}, a^{-4}q^{-n-k}, st^{-1}q^{\frac{1}{2}}, a^{-2}ts^{-1}q^{-n+\frac{1}{2}}; q\right)_m q^m}{\left(q, a^{-2}q^{\frac{1}{2}-n}, -a^{-2}q^{\frac{1}{2}-n}, -a^{-2}q^{1-n}; q\right)_m}$$
$$\times {}_4\varphi_3\left(\begin{matrix} q^{-m}, a^{-2}q^{-n}, -a^{-2}s^{-2}q^{-n}, -a^{-2}t^2q^{-n} \\ a^{-4}q^{-2n}, a^{-2}ts^{-1}q^{-n+\frac{1}{2}}, ts^{-1}q^{-m+\frac{1}{2}} \end{matrix}; q, q\right). \tag{2.13}$$

If we next make the successive substitutions $n \to n-k$, $a \to a^{-1}q^{-\frac{1}{2}n}$, $e^{i\varphi} \to is^{-1}$, $e^{i\psi} \to -it^{-1}$ in (2.11) and compare with (2.13) then we can write (2.12) as follows:

$$r_n\left(\cos\theta; ast^{-1}q^{\frac{1}{2}}, ats^{-1}q^{\frac{1}{2}}, -astq^{\frac{1}{2}}, -as^{-1}t^{-1}q^{\frac{1}{2}} \mid q\right)$$
$$= \frac{s^n t^{-n} q^{-\frac{1}{2}n(n-1)}}{(a^2q;q)_n^2}$$
$$\times \frac{(a^4q^{n+1};q)_n^2}{(-s^2a^2q, t^{-2}a^2q;q)_n} \sum_{k=0}^{n} (-1)^k q^{k(n+\frac{1}{2})} \frac{1-a^2q^k}{1-a^2} \frac{(q^{-n}, a^4;q)_k}{(q, a^4q^{n+1};q)_k}$$
$$\times r_{n-k}\left(\frac{s-s^{-1}}{2i}; a^{-1}q^{-\frac{1}{2}n}, a^{-1}q^{-\frac{1}{2}n+\frac{1}{2}}, -a^{-1}q^{-\frac{1}{2}n}, -a^{-1}q^{-\frac{1}{2}n+\frac{1}{2}} \mid q\right)$$
$$\times r_{n-k}\left(\frac{t^{-1}-t}{2i}; a^{-1}q^{-\frac{1}{2}n}, a^{-1}q^{-\frac{1}{2}n+\frac{1}{2}}, -a^{-1}q^{-\frac{1}{2}n}, -a^{-1}q^{-\frac{1}{2}n+\frac{1}{2}} \mid q\right)$$
$$\times r_k\left(\cos\theta; a, -a, aq^{\frac{1}{2}}, -aq^{\frac{1}{2}} \mid q\right). \tag{2.14}$$

In order to make the two r_{n-k} factors on the right-hand side above into closer q-analogues of the two $C_{n-k}^{\lambda+k}$ factors on the right-hand side of (1.7), we will use the following string of identities:

$$r_n\left(\cos\theta; a, aq^{\frac{1}{2}}, -a, -aq^{\frac{1}{2}} \mid q\right) = {}_4\varphi_3\left(\begin{matrix} q^{-n}, a^4q^{n-1}, ae^{i\theta}, ae^{-i\theta} \\ a^2q^{\frac{1}{2}}, -a^2, -a^2q^{\frac{1}{2}} \end{matrix}; q, q\right)$$
$$= \frac{(a^2;q)_n\, a^n\, e^{in\theta}}{(a^4;q)_n}\, {}_2\varphi_1\left(\begin{matrix} q^{-n}, a^2 \\ a^{-2}q^{1-n} \end{matrix}; q, a^{-2}e^{-2i\theta}q\right)$$
$$= \frac{(a^2, a^{-2}e^{-2i\theta}q^{1-n};q)_n\, a^n\, e^{in\theta}}{(a^4;q)_n}\, {}_2\varphi_2\left(\begin{matrix} q^{-n}, a^{-4}q^{1-n} \\ a^{-2}q^{1-n}, a^{-2}e^{-2i\theta}q^{1-n} \end{matrix}; q, e^{-2i\theta}q\right). \tag{2.15}$$

For the proof use successively (7.4.14), (7.4.2) and (1.5.4) in (Gasper and Rahman, 1990).

Proof of Theorem 1.1. This follows from (2.14) by twofold substitution of (2.15). Here replace n by $n-k$ and a by $a^{-1}q^{-\frac{1}{2}n}$ in (2.14), and replace $e^{i\theta}$ by is^{-1} for the first substitution and by it for the second substitution. □

Remark 2.1. *Let the divided difference operator D_q acting on a function F of argument $e^{i\theta}$ be given by*

$$D_qF\left(e^{i\theta}\right) := \frac{\delta_q F(e^{i\theta})}{\delta_q \cos\theta}, \quad \delta_q F\left(e^{i\theta}\right) := F\left(q^{\frac{1}{2}}e^{i\theta}\right) - F\left(q^{-\frac{1}{2}}e^{i\theta}\right).$$

Then D_q, acting on both sides of the addition formula (1.12), sends this to the same formula with n replaced by $n-1$ and a replaced by $q^{\frac{1}{2}}a$ (apply (Koekoek and Swarttouw, 1998, (3.1.9))). Thus, if formula (1.12) is already known for $a = 1$ (i.e., if formula (1.10) is known) then the procedure just sketched yields this formula for $a = q^{\frac{1}{2}j}$ for all $j \in \mathbb{Z}_{\geq 0}$, i.e., for infinitely many disctinct values of a. Since, for fixed n, both sides of (1.12) are rational in a, formula (1.12) will then be valid for general a. Since formula (1.9), equivalent to (1.10), can be obtained by a quantum group interpretation, we can say that it is possible to prove formula (1.12) by arguments in a quantum group setting, followed by minor analytic, but not very computational reasoning.

Proof that (1.11) *has limit* (1.7) *as* $q \uparrow 1$. (after the substitutions $a = q^{\frac{1}{2}\lambda-\frac{1}{4}}$, $s = \tan\left(\frac{1}{2}\varphi\right)$, $t = \tan\left(\frac{1}{2}\psi\right)$).

The limits for $q \uparrow 1$ of the factors $r_n(\cos\theta)$, $r_k(\cos\theta)$, ${}_2\varphi_2\left(-s^2q\right)$ and ${}_2\varphi_2\left(-t^{-2}q\right)$ in (1.11), after the above substitutions and after substitution of (1.3) for the r_n and r_k factors yields respectively:

$$
{}_2F_1\left(-n, n+2\lambda; \lambda+\frac{1}{2}; \frac{1}{2}-\frac{1}{2}(\cos\varphi\cos\psi+\sin\varphi\sin\psi\cos\theta)\right),
$$

$$
{}_2F_1\left(-k, k+2\lambda-1; \lambda; \frac{1}{2}-\frac{1}{2}\cos\theta\right),
$$

$$
{}_2F_1\left(-n+k, 2\lambda+n+k; \lambda+k+\frac{1}{2}; \frac{1}{2}-\frac{1}{2}\cos\varphi\right),
$$

$$
{}_2F_1\left(-n+k, 2\lambda+n+k; \lambda+k+\frac{1}{2}; \frac{1}{2}+\frac{1}{2}\cos\psi\right).
$$

Express these ${}_2F_1$'s as ultraspherical polynomials by (1.6). The limit of the coefficients on the right-hand side of (1.11) (after the above substitutions) is also easily computed. □

Remark 2.2. *Replace s by $sq^{\frac{1}{2}n}$ and t by $tq^{-\frac{1}{2}n}$ in (1.12) and let $n \to \infty$. Then formally we obtain Rahman's addition formula for q-Bessel functions, see (Rahman, 1988, (1.10)) (but watch out for the misprint*

$\Gamma_q(\nu+1)$ *which should be* $\Gamma_q^2(\nu+1)$):

$$
\begin{aligned}
&{}_2\varphi_1\left(\begin{matrix}-astq^{\frac{1}{2}}e^{i\theta}, -astq^{\frac{1}{2}}e^{-i\theta}\\ a^2q\end{matrix}; q, -a^{-2}t^{-2}\right)\\
&= \frac{1}{(-a^{-2}t^{-2};q)_\infty}\sum_{k=0}^{\infty}\frac{(-1)^k q^{\frac{1}{2}k^2}a^{-k}s^k t^{-k}}{(q,a^2q,a^4q^k;q)_k} \qquad (2.16)\\
&\times {}_0\varphi_1\left(\begin{matrix}-\\ a^2q^{k+1}\end{matrix}; q, -s^2q^{k+1}\right){}_0\varphi_1\left(\begin{matrix}-\\ a^2q^{k+1}\end{matrix}; q, -t^{-2}q^{k+1}\right)\\
&\times p_k\left(\cos\theta; a, -a, aq^{\frac{1}{2}}, -aq^{\frac{1}{2}} \mid q\right).
\end{aligned}
$$

According to (Rahman, 1988, (1.10)), the further conditions $0 < a < 1$, $0 < s < t^{-1}$, $\theta \in \mathbb{R}$ *should be imposed here. If we replace in* (2.16) s *by* $(1-q)s$, t *by* $(1-q)^{-1}t$, a *by* $q^{\frac{1}{2}\alpha}$, *and if we let* $q \uparrow 1$ *then we formally obtain the familiar Gegenbauer's addition formula for Bessel functions* J_α, *(see (Erdélyi et al., 1953, 7.15(32))).*

In (2.16) *the left-hand side is an Askey-Wilson q-Bessel function (see (Koelink and Stokman, 2001, §2.3)), earlier studied under the name q-Bessel function on a q-quadratic grid in (Ismail et al., 1999) and (Bustoz and Suslov, 1998). The* ${}_0\varphi_1$*'s on the right-hand side of* (2.16) *are Jackson's second q-Bessel functions, usually written (see (Gasper and Rahman, 1990, Exercise 1.24)) as*

$$
J_\alpha^{(2)}(x;q) = \frac{(q^{\alpha+1};q)_\infty}{(q;q)_\infty}(x/2)^\alpha\, {}_0\varphi_1\left(\begin{matrix}-\\ q^{\alpha+1}\end{matrix}; q, -\frac{1}{4}x^2q^{\alpha+1}\right). \qquad (2.17)
$$

Koelink (Koelink, 1991, (3.6.18)) gave an interpretation of the case $a = 1$ *of* (2.16) *on the quantum group of plane motions. This interpretation is similar to the interpretation given in (Koornwinder, 1993, (5.9)) for equation* (1.9).

Remark 2.3. *If we multiply both sides of the addition formula* (1.11) *with* $\left(-a^2t^{-2}q;q\right)_n$ *and if we next let* $t \to iaq^{\frac{1}{2}n}$ *in* (1.11) *then we obtain* *a degenerate form of the addition formula* (1.11):

$$\frac{\left(-ise^{i\theta}q^{\frac{1}{2}(1-n)},-ise^{-i\theta}q^{\frac{1}{2}(1-n)};q\right)_n}{(-a^2s^2q;q)_n}$$
$$=\sum_{k=0}^{n}\left(-ia^{-1}q^{\frac{1}{2}(n+1)}\right)^k\frac{\left(q^{-n},a^4;q\right)_k}{(q,a^2;q)_k}\frac{s^k}{(-a^2s^2q;q)_k}$$
$$\times\,{}_2\varphi_2\left(\begin{matrix}q^{-n+k},a^4q^{n+k+1}\\a^2q^{k+1},-a^2s^2q^{k+1}\end{matrix};q,-s^2q\right)r_k\left(\cos\theta;a,aq^{\frac{1}{2}},-aq^{\frac{1}{2}},-a\mid q\right). \tag{2.18}$$

Integrated forms of (1.11) *and* (2.18) *can be obtained by integrating both sides of these formulas with respect to the measure* $\left|\frac{\left(e^{2i\theta};q\right)_\infty}{\left(a^2e^{2i\theta};q\right)_\infty}\right|^2 d\theta$ *on* $[0,\pi]$, *i.e., with respect to the orthogonality measure for the continuous q-ultraspherical polynomials* $r_k\left(\cos\theta;a,q^{\frac{1}{2}}a,-q^{\frac{1}{2}}a,-a\mid q\right)$ *(see (Gasper and Rahman, 1990, §7.4)). This will yield a product formula and an integral representation, respectively, for the functions* $p_n^{(\alpha,\alpha)}(s^2;q,\mathrm{rat})$ *(with* $a=q^{\frac{1}{2}\alpha}$*).*

Remark 2.4. *In* (2.18) *replace* s *by* $sq^{\frac{1}{2}n}$ *and let* $n\to\infty$. *Then we formally obtain a degenerate addition formula for q-Bessel functions:*

$$\left(-isq^{\frac{1}{2}}e^{i\theta},-isq^{\frac{1}{2}}e^{-i\theta};q\right)_\infty=\sum_{k=0}^{\infty}i^ka^{-k}s^kq^{\frac{1}{2}k^2}\frac{\left(a^4;q\right)_k}{(q,a^2;q)_k}$$
$$\times\,{}_0\varphi_1\left(\begin{matrix}-\\a^2q^{k+1}\end{matrix};q,-s^2q^{k+1}\right)r_k\left(\cos\theta;a,aq^{\frac{1}{2}},-aq^{\frac{1}{2}},-a\mid q\right) \tag{2.19}$$

If we replace in (2.19) a *by* $q^{\frac{1}{2}\alpha}$, *and if we substitute* (2.17) *and* (1.4) *then we can rewrite* (2.19) *as*

$$\left(-isq^{\frac{1}{2}}e^{i\theta},-isq^{\frac{1}{2}}e^{-i\theta};q\right)_\infty=\frac{q^{\frac{1}{2}\alpha^2}}{s^\alpha}\frac{(q;q)_\infty}{\left(q^{\alpha+1};q\right)_\infty}\sum_{k=0}^{\infty}i^kq^{\frac{1}{2}k^2+\frac{1}{2}k\alpha}\frac{1-q^{\alpha+k}}{1-q^\alpha}$$
$$\times J_{\alpha+k}^{(2)}\left(2sq^{-\frac{1}{2}\alpha};q\right)C_k\left(\cos\theta;q^\alpha\mid q\right). \tag{2.20}$$

It is interesting to compare formula (2.20) *with Ismail and Zhang (Ismail and Zhang, 1994, (3.32)). They expand there a generalized q-exponential function* $\mathcal{E}_q(z;-i,b/2)$ *in terms of the* $C_k\left(z;q^\alpha\mid q\right)$ *and they obtain almost the same expansion coefficients as in* (2.20), *including Jackson's second q-Bessel functions, but they have a factor* $q^{\frac{1}{4}k^2}$, *where* (2.20) *has a factor* $q^{\frac{1}{2}k^2+\frac{1}{2}k\alpha}$.

3. Rational biorthogonal functions for the Askey-Roy q-beta measure

The *Askey-Roy q-beta integral* (see (Askey and Roy, 1986, (3.4)), (Gasper, 1987), (Askey, 1988, pp. 92, 93) (Gasper and Rahman, 1990, Exercise 6.17(ii))) is as follows:

$$\int_0^\infty t^{c-1} \frac{(-tq^{\beta+1}, -t^{-1}q^{\alpha+2}; q)_\infty}{(-tq^{-c}, -t^{-1}q^{1+c}; q)_\infty}\, dt$$
$$= \frac{q^{c^2}\,\Gamma(c)\,\Gamma(1-c)\,\Gamma_q(\alpha+1)\,\Gamma_q(\beta+1)}{\Gamma_q(c)\,\Gamma_q(1-c)\,\Gamma_q(\alpha+\beta+2)}, \operatorname{Re}\alpha, \operatorname{Re}\beta > -1,\ c \in \mathbb{R}. \quad (3.1)$$

Here the *q-gamma function* is defined by

$$\Gamma_q(z) := \frac{(q;q)_\infty}{(q^z;q)_\infty}\,(1-q)^{1-z}. \quad (3.2)$$

The special case $c = \alpha+1$ of (3.1) goes back (without proof) to Ramanujan in Chapter 16 of his second notebook (see (Berndt, 1991, p. 29, Entry 14)), and later in his paper (Ramanujan, 1915, (19)), with subsequent proof by Hardy (Hardy, 1915).

When we let $q \uparrow 1$ in (3.1) then we formally obtain the *beta integral* on $(0,\infty)$:

$$\int_0^\infty t^\alpha\,(1+t)^{-\alpha-\beta-2}\,dt = \frac{\Gamma(\alpha+1)\,\Gamma(\beta+1)}{\Gamma(\alpha+\beta+2)} \qquad (\operatorname{Re}\alpha, \operatorname{Re}\beta > -1). \quad (3.3)$$

When we move the orthogonality relations

$$\int_{-1}^1 P_n^{(\alpha,\beta)}(x)\,P_m^{(\alpha,\beta)}(x)\,(1-x)^\alpha\,(1+x)^\beta\,dx$$
$$= \frac{2^{\alpha+\beta+1}\,\Gamma(n+\alpha+1)\,\Gamma(n+\beta+1)}{(2n+\alpha+\beta+1)\,n!\,\Gamma(n+\alpha+\beta+1)}\,\delta_{n,m} \qquad (\alpha,\beta > -1) \quad (3.4)$$

of the *Jacobi polynomials*

$$P_n^{(\alpha,\beta)}(x) := \frac{(\alpha+1)_n}{n!}\,{}_2F_1\left(\begin{matrix}-n, n+\alpha+\beta+1\\ \alpha+1\end{matrix}; \frac{1}{2}(1-x)\right) \quad (3.5)$$

(see (Erdélyi et al., 1953, §10.8)) to $[0,\infty)$ by the substitution $x = (1-t)/(1+t)$, then we obtain orthogonality relations

$$\int_0^\infty P_n^{(\alpha,\beta)}\left(\frac{1-t}{1+t}\right) P_m^{(\alpha,\beta)}\left(\frac{1-t}{1+t}\right) t^\alpha (1+t)^{-\alpha-\beta-2}\,dt$$
$$= \frac{\Gamma(n+\alpha+1)\,\Gamma(n+\beta+1)}{(2n+\alpha+\beta+1)\,n!\,\Gamma(n+\alpha+\beta+1)}\,\delta_{n,m} \qquad (\alpha,\beta>-1). \quad (3.6)$$

Thus the rational functions $t \mapsto P_n^{(\alpha,\beta)}((1-t)/(1+t))$, $n \in \mathbb{Z}_{\geq 0}$, are orthogonal with respect to the beta measure on $[0,\infty)$ of which the total mass is given in (3.3). We would like to find q-analogues of these orthogonal rational functions such that the orthogonality measure is the q-beta measure in (3.1).

From (1.15) we have

$$d\mu_{\alpha,\beta,c;q}(t) = \frac{w_{\alpha,\beta,c;q}(t)\,dt}{\int_0^\infty w_{\alpha,\beta,c;q}(s)\,ds}, \quad \text{where}$$
$$w_{\alpha,\beta,c;q}(t) := t^{c-1}\,\frac{(-tq^{\beta+1}, -t^{-1}q^{\alpha+2};q)_\infty}{(-tq^{-c}, -t^{-1}q^{1+c};q)_\infty}. \quad (3.7)$$

Observe from (3.1) that, for $k,l \in \mathbb{Z}_{\geq 0}$,

$$\int_0^\infty \frac{t^k}{(-tq^{\beta+1};q)_k}\,\frac{t^{-l}}{(-t^{-1}q^{\alpha+2};q)_l}\,w_{\alpha,\beta,c;q}(t)\,dt$$
$$= q^{-(k-l)(c+k-l)} \int_0^\infty w_{\alpha+k,\beta+l,c+k-l;q}(t)\,dt$$
$$= \frac{q^{c(c+k-l)}\,\Gamma(c+k-l)\,\Gamma(1-c-k+l)\,\Gamma_q(\alpha+k+1)\,\Gamma_q(\beta+l+1)}{\Gamma_q(c+k-l)\,\Gamma_q(1-c-k+l)\,\Gamma_q(\alpha+\beta+k+l+2)}$$
$$= \frac{(q^{\alpha+1};q)_k\,(q^{\beta+1};q)_l}{q^{\frac12(k-l)(k-l-1)}\,(q^{\alpha+\beta+2};q)_{k+l}}\;\frac{q^{c^2}\,\Gamma(c)\,\Gamma(1-c)\,\Gamma_q(\alpha+1)\,\Gamma_q(\beta+1)}{\Gamma_q(c)\,\Gamma_q(1-c)\,\Gamma_q(\alpha+\beta+2)}.$$

Thus

$$\int_0^\infty \frac{q^{\frac12 k(k-1)}\,t^k}{(q^{\alpha+1}, -tq^{\beta+1};q)_k}\,\frac{q^{\frac12 l(l-1)}\,(t^{-1}q)^l}{(q^{\beta+1}, -t^{-1}q^{\alpha+2};q)_l}\,d\mu_{\alpha,\beta,c;q}(t) = \frac{q^{kl}}{(q^{\alpha+\beta+2};q)_{k+l}}. \quad (3.8)$$

Proof of Theorem 1.2. Multiply both sides of (3.8) with the factor $\dfrac{(q^{-n}, q^{n+\alpha+\beta+1}; q)_k \, q^k}{(q;q)_k}$ and sum from $k = 0$ to n. Then the right-hand side becomes

$$\frac{1}{(q^{\alpha+\beta+2};q)_l}\, {}_2\varphi_1\left(\begin{matrix} q^{-n}, q^{n+\alpha+\beta+1} \\ q^{\alpha+\beta+l+2} \end{matrix}; q, q^{l+1}\right) = \frac{(q^{l-n+1};q)_n}{(q^{\alpha+\beta+2};q)_{l+n}} = \frac{(q;q)_n\, \delta_{n,l}}{(q^{\alpha+\beta+2};q)_{2n}} \quad (l = 0, 1, \ldots, n)$$

by (Gasper and Rahman, 1990, (1.5.2)). Thus

$$\int_0^\infty {}_2\varphi_2\left(\begin{matrix} q^{-n}, q^{n+\alpha+\beta+1} \\ q^{\alpha+1}, -tq^{\beta+1} \end{matrix}; q, -tq\right) \frac{q^{l(l-1)/2}\,(tq^{-1})^l}{(q^{\beta+1}, -t^{-1}q^{\alpha+2};q)_l}\, d\mu_{\alpha,\beta,c;q}(t) = \frac{(q;q)_n\, \delta_{n,l}}{(q^{\alpha+\beta+2};q)_{2n}}, \quad l = 0, 1, \ldots, n, \tag{3.9}$$

$$\int_0^\infty \frac{q^{k(k-1)/2}\, t^k}{(q^{\alpha+1}, -tq^{\beta+1};q)_k}\, {}_2\varphi_2\left(\begin{matrix} q^{-m}, q^{m+\alpha+\beta+1} \\ q^{\beta+1}, -t^{-1}q^{\alpha+2} \end{matrix}; q, -t^{-1}q^2\right) d\mu_{\alpha,\beta,c;q}(t) = \frac{(q;q)_m\, \delta_{m,k}}{(q^{\alpha+\beta+2};q)_{2m}}, \quad k = 0, 1, \ldots, m, \tag{3.10}$$

where (3.10) is obtained by a similar argument as (3.9). Then (3.9) and (3.10) together with (1.14) imply the biorthogonality relations (1.16). □

The q-integral version of the Askey-Roy integral (3.1) is

$$\frac{(-sq^{-c}, -s^{-1}q^{1+c};q)_\infty}{s^c\,(-s, -s^{-1}q;q)_\infty} \int_0^{s\cdot\infty} t^{c-1} \times \frac{(-tq^{\beta+1}, -t^{-1}q^{\alpha+2};q)_\infty}{(-tq^{-c}, -t^{-1}q^{1+c};q)_\infty}\, d_qt = \frac{\Gamma_q(\alpha+1)\,\Gamma_q(\beta+1)}{\Gamma_q(\alpha+\beta+2)}. \tag{3.11}$$

Here $s > 0$, $\operatorname{Re}\alpha, \operatorname{Re}\beta > -1$, $c \in \mathbb{R}$, and the q-integral is defined by (1.17). The case $s = q^c$ of (3.11) (which is no real restriction) was given in (Gasper, 1987) (see also (Askey, 1988, (2.27))) and in (Gasper and Rahman, 1990, Exercise 6.17(i))). Another approach to (3.11) is presented in (De Sole and Kac, 2003).

For f any function on $(0,\infty)$ for which the sum below converges absolutely, we have:

$$\frac{(-sq^{-c},-s^{-1}q^{1+c};q)_\infty}{s^c\,(-s,-s^{-1}q;q)_\infty}\int_0^{s\cdot\infty} f(t)\,t^{c-1}\,\frac{(-tq^{\beta+1},-t^{-1}q^{\alpha+2};q)_\infty}{(-tq^{-c},-t^{-1}q^{1+c};q)_\infty}\,d_qt$$
$$=\frac{(1-q)\,(-sq^{\beta+1},-s^{-1}q^{\alpha+2};q)_\infty}{(-s,-s^{-1}q;q)_\infty}\sum_{k=-\infty}^{\infty} f(sq^k)\,\frac{(-sq^{-\alpha-1};q)_k}{(-sq^{\beta+1};q)_k}\,q^{k(\alpha+1)}. \tag{3.12}$$

The right-hand side, and thus the left-hand side of (3.12) is independent of c. Henceforth we will take $c=0$ without loss of information. Then (3.11) together with (3.12) takes the form

$$\int_0^{s\cdot\infty} t^{-1}\,\frac{(-tq^{\beta+1},-t^{-1}q^{\alpha+2};q)_\infty}{(-t,-t^{-1}q;q)_\infty}\,d_qt=\frac{(1-q)\,(-sq^{\beta+1},-s^{-1}q^{\alpha+2};q)_\infty}{(-s,-s^{-1}q;q)_\infty}$$
$$\times\sum_{k=-\infty}^{\infty}\frac{(-sq^{-\alpha-1};q)_k}{(-sq^{\beta+1};q)_k}\,q^{k(\alpha+1)}=\frac{\Gamma_q(\alpha+1)\,\Gamma_q(\beta+1)}{\Gamma_q(\alpha+\beta+2)}. \tag{3.13}$$

The second equality in (3.13) is Ramanujan's ${}_1\psi_1$ sum (Gasper and Rahman, 1990, (5.2.1)). This observation is the usual way to prove (3.11). With a completely analogous argument as used for the proof of Theorem 1.2, we can next prove Theorem 1.3. Details are omitted.

In completion of this section, observe the following symmetry of the functions $p_n^{(\alpha,\beta)}(t;q,\mathrm{rat})$:

$$\begin{aligned}
p_n^{(\alpha,\beta)}(t;q,\mathrm{rat}) &= {}_2\varphi_2\left(\begin{matrix}q^{-n},q^{n+\alpha+\beta+1}\\ q^{\alpha+1},-tq^{\beta+1}\end{matrix};q,-tq\right)\\
&=\frac{1}{(-tq^{\beta+1};q)_n}\,{}_2\varphi_1\left(\begin{matrix}q^{-n},q^{-n-\beta}\\ q^{\alpha+1}\end{matrix};q,-tq^{\beta+n+1}\right)\\
&=\frac{(-1)^n\,(q^{\beta+1};q)_n\,t^n}{(q^{\alpha+1},-tq^{\beta+1};q)_n}\,{}_2\varphi_1\left(\begin{matrix}q^{-n},q^{-n-\alpha}\\ q^{\beta+1}\end{matrix};q,-t^{-1}q^{\alpha+n+1}\right)\\
&=\frac{(-1)^n\,(q^{\beta+1};q)_n}{(q^{\alpha+1};q)_n}\,\frac{t^n\,(-t^{-1}q^{\alpha+1};q)_n}{(-tq^{\beta+1};q)_n}\,{}_2\varphi_2\left(\begin{matrix}q^{-n},q^{n+\alpha+\beta+1}\\ q^{\beta+1},-t^{-1}q^{\alpha+1}\end{matrix};q,-t^{-1}q\right)\\
&=\frac{(-1)^n\,(q^{\beta+1};q)_n}{(q^{\alpha+1};q)_n}\,\frac{t^n\,(-t^{-1}q^{\alpha+1};q)_n}{(-tq^{\beta+1};q)_n}\,p_n^{(\beta,\alpha)}(t^{-1};q,\mathrm{rat}).
\end{aligned} \tag{3.14}$$

Here the first and last equality are just (1.14). We use (Gasper and Rahman, 1990, (1.5.4)) for the second and fourth equality, while the third equality is obtained by reversion of summation order in a terminating q-hypergeometric series (see (Gasper and Rahman, 1990, Exercise 1.4.(ii))).

4. Concluding remarks

The results of this paper lead to several interesting questions. I formulate two of these questions here. I also discuss specializations of Theorem 1.2.

As I already mentioned in §1, the new addition formula in the case of the continuous q-Legendre polynomials (formula (1.9)) was first obtained in a quantum group context, where a two-parameter family of Askey-Wilson polynomials, including the continuous q-Legendre polynomials, was interpreted as spherical functions on the $SU_q(2)$ quantum group. Here the left and right invariance of the spherical functions was no longer with respect to the diagonal quantum subgroup, but infinitesimally with respect to twisted primitive elements in the corresponding quantized universal enveloping algebra. The σ and τ variables in the addition formula (1.9) are parameters for the twisted primitive elements occurring in the left respectively right invariance property. On the other hand, the ${}_3\varphi_2$ factors on the right-hand side of (1.9), involving σ resp. τ, can be rewritten as the functions $p_{n-k}^{(k,k)}(\,.\,;q,\mathrm{rat})$ of argument $-q^{-\sigma}$ resp. $-q^{-\tau}$. So I wonder whether an interpretation of these last functions and of their biorthogonality (discussed in §14.3) can be given in the context of $SU_q(2)$.

A second question is whether the biorthogonality relations for the functions $p_n^{(\alpha,\beta)}(t;q,\mathrm{rat})$ (Theorems 1.2 and 1.3) fit into a more general class of biorthogonal rational functions. In fact, several papers have appeared during the last 10 or 20 years which discuss explicit systems of biorthogonal rational functions depending on many parameters and expressed as q-hypergeometric functions, see Rahman (Rahman, 1986), Wilson (Wilson, 1991), Ismail and Masson (Ismail and Masson, 1995) and Spiridonov and Zhedanov (Spiridonov and Zhedanov, 2000). However, I did not see how the functions $p_n^{(\alpha,\beta)}(t;q,\mathrm{rat})$ and their biorthogonality relations can be obtained as special or limit case of families discussed in these references. If the functions $p_n^{(\alpha,\beta)}(t;q,\mathrm{rat})$ are indeed unrelated to the functions discussed in these references, then it is a natural question how to generalize the system of biorthogonal functions $p_n^{(\alpha,\beta)}(t;q,\mathrm{rat})$.

On the other hand, our functions $p_n^{(\alpha,\beta)}(t;q,\mathrm{rat})$ have some interesting limit cases, one of which has occurred earlier in literature. When we take limits for $\alpha \to \infty$ and/or $\beta \to \infty$ in (1.14), then we obtain:

$$p_n^{(\alpha,\infty)}(t;q,\mathrm{rat}) := {}_1\varphi_1\left(q^{-n};q^{\alpha+1};q,-qt\right), \tag{4.1}$$

$$p_n^{(\infty,\beta)}(t;q,\mathrm{rat}) := {}_1\varphi_1\left(q^{-n};-tq^{\beta+1};q,-qt\right), \tag{4.2}$$

$$p_n^{(\infty,\infty)}(t;q,\mathrm{rat}) := {}_1\varphi_1\left(q^{-n};0;q,-qt\right). \tag{4.3}$$

In (4.1) and (4.3) we have polynomials of degree n in t, rather than rational functions in t. The limit case $\beta \to \infty$ of the biorthogonality relations (1.16) then becomes:

$$\begin{aligned}\int_0^\infty p_n^{(\alpha,\infty)}(t;q,\mathrm{rat})\, p_m^{(\infty,\alpha)}(t^{-1}q;q,\mathrm{rat})\, d\mu_{\alpha,\infty,c;q}(t)\\ = (-1)^n\, q^{-\frac{1}{2}n(n-1)}\,(q;q)_n\,\delta_{n,m},\end{aligned} \tag{4.4}$$

where

$$\begin{aligned}d\mu_{\alpha,\infty,c;q}(t) := \frac{q^{-c^2}\,\Gamma_q(c)\,\Gamma_q(1-c)\,\left(q^{\alpha+1};q\right)_\infty}{\Gamma(c)\,\Gamma(1-c)\,(1-q)\,(q;q)_\infty}\\ \times \frac{t^{c-1}\left(-t^{-1}q^{\alpha+2};q\right)_\infty}{\left(-tq^{-c},-t^{-1}q^{1+c};q\right)_\infty}\,dt \qquad (\alpha > -1).\end{aligned} \tag{4.5}$$

The further limit case $\alpha \to \infty$ of the biorthogonality relations (4.4) then becomes

$$\begin{aligned}\frac{q^{-c^2}\,\Gamma_q(c)\,\Gamma_q(1-c)}{\Gamma(c)\,\Gamma(1-c)\,(1-q)\,(q;q)_\infty} \int_0^\infty p_n^{(\infty,\infty)}(t;q,\mathrm{rat})\\ \times p_m^{(\infty,\infty)}(t^{-1}q;q,\mathrm{rat})\,\frac{t^{c-1}\,dt}{\left(-tq^{-c},-t^{-1}q^{1+c};q\right)_\infty}\\ = (-1)^n\, q^{-\frac{1}{2}n(n-1)}\,(q;q)_n\,\delta_{n,m}.\end{aligned} \tag{4.6}$$

Similar limit cases can be considered for the biorthogonality relations (1.3). The biorthogonality relations (4.6) are essentially the ones observed by Pastro (Pastro, 1985, pp. 532, 533). He also points out that biorthogonality relations of the form $\int P_n(t) Q_m(t^{-1})\, d\mu(t) = h_n\,\delta_{n,m}$ with P_n and Q_n polynomials of degree n can be rewritten as orthogonality relations on the real line for Laurent polynomials. This is indeed the case in (4.6). Pastro also observes that the biorthogonality measure occurring in (4.6) is a (non-unique) orthogonality measure for the Stieltjes-Wigert polynomials (see (Koekoek and Swarttouw, 1998, §3.27)).

Note

For further discussion and references concerning the Ismail-Zhang expansion of a generalized q-exponential function, mentioned in Remark 2.4, see §4.5 in S. K. Suslov, *An Introduction to Basic Fourier Series*, Kluwer, 2003.

References

Askey, R. (1975). *Orthogonal polynomials and special functions*, volume 21 of *Regional Conference Series in Applied Math.* SIAM, Philadelphia, PA.

Askey, R. (1988). Beta integrals and q-extensions. In Balakrishnan, R., Padmanabhan, K. S., and Thangaraj, V., editors, *Proceedings of the Ramanujan Centennial International Conference (Annamalainagar, 1987)*, volume 1 of *RMS Publ.*, pages 85–102, Annamalainagar. Ramanujan Mathematical Society.

Askey, R. and Roy, R. (1986). More q-beta integrals. *Rocky Mountain J. Math.*, 16:365–372.

Askey, R. and Wilson, J. (1985). Some basic hypergeometric orthogonal polynomials that generalize Jacobi polynomials. *Mem. Amer. Math. Soc.*, 54(319):iv+55.

Berndt, B. C., editor (1991). *Ramanujan's notebooks, Part III.* Springer-Verlag, New York.

Bustoz, J. and Suslov, S. K. (1998). Basic analog of Fourier series on a q-quadratic grid. *Methods Appl. Anal.*, 5:1–38.

De Sole, A. and Kac, V. (2003). On integral representations of q-gamma and q-beta functions. Preprint: `arXiv:math.QA/0302032`.

Erdélyi, A., Magnus, W., Oberhettinger, F., and Tricomi, F. G. (1953). *Higher transcendental functions, Vol. II.* McGraw-Hill.

Gasper, G. (1984). Letter to R. Askey dated July 23, 1984. Personal communication.

Gasper, G. (1987). Solution to problem #6497 (q-analogues of a gamma function identity, by R. Askey). *Amer. Math. Monthly*, 94:199–201.

Gasper, G. and Rahman, M. (1990). *Basic hypergeometric series*, volume 35 of *Encyclopedia of Mathematics and its Applications.* Cambridge University Press, Cambridge.

Hardy, G. H. (1915). Proof of a formula of Mr. Ramanujan. *Messenger Math.*, 44:18–21. Reprinted in *Collected Papers of G. H. Hardy*, Vol. 5, Oxford, 1972, pp. 594–597.

Ismail, M. E. H. and Masson, D. R. (1995). Generalized orthogonality and continued fractions. *J. Approx. Theory*, 83:1–40.

Ismail, M. E. H., Masson, D. R., and Suslov, S. K. (1999). The q-Bessel function on a q-quadratic grid. In van Diejen, J. F. and Vinet, L., editors, *Algebraic methods and q-special functions (Montréal, QC, 1996)*, volume 22 of *CRM Proc. Lecture Notes*, pages 183–200. Amer. Math. Soc., Providence, RI.

Ismail, M. E. H. and Zhang, R. (1994). Diagonalization of certain integral operators. *Adv. Math.*, 109:1–33.

Koekoek, R. and Swarttouw, R. F. (1998). The Askey-scheme of hypergeometric orthogonal polynomials and its q-analogue. Report 98-17, Faculty of Technical Mathematics and Informatics, Delft University of Technology. Electronic version: `http://aw.twi.tudelft.nl/~koekoek/askey/`.

Koelink, E. (1997). Addition formulas for q-special functions. In Ismail, M. E. H., Masson, D. R., and Rahman, M., editors, *Special functions, q-series and related topics (Toronto, ON, 1995)*, volume 14 of *Fields Institute Communications*, pages 109–129. Amer. Math. Soc., Providence, RI. Electronic version: `arXiv:math.QA/9506216`.

Koelink, E. and Stokman, J. V. (2001). The Askey-Wilson function transform scheme. In Bustoz, J., Ismail, M. E. H., and Suslov, S. K., editors, *Special Functions 2000: Current perspective and future directions (Tempe, AZ, 2000)*, volume 30 of *NATO Science Series II*, pages 221–241. Kluwer Academic Publishers, Dordrecht. Electronic version: `arXiv:math.CA/9912140`.

Koelink, H. T. (1991). *On quantum groups and q-special functions*. PhD thesis, University of Leiden.

Koelink, H. T. (1994). The addition formula for continuous q-Legendre polynomials and associated spherical elements on the $SU(2)$ quantum group related to Askey-Wilson polynomials. *SIAM J. Math. Anal.*, 25:197–217.

Koornwinder, T. H. (1993). Askey-Wilson polynomials as zonal spherical functions on the $SU(2)$ quantum group. *SIAM J. Math. Anal.*, 24:795–813.

Pastro, P. I. (1985). Orthogonal polynomials and some q-beta integrals of Ramanujan. *J. Math. Anal. Appl.*, 112:517–540.

Rahman, M. (1986). An integral representation of a ${}_{10}\varphi_9$ and continuous bi-orthogonal ${}_{10}\varphi_9$ rational functions. *Canad. J. Math.*, 38:605–618.

Rahman, M. (1988). An addition theorem and some product formulas for q-Bessel functions. *Canad. J. Math.*, 45:1203–1221.

Rahman, M. and Verma, A. (1986). Product and addition formula for the continuous q-ultraspherical polynomials. *SIAM J. Math. Anal.*, 17:1461–1474.

Ramanujan, S. (1915). Some definite integrals. *Messenger Math.*, 44:10–18. Reprinted in Collected papers of Srinivasa Ramanujan, Cambridge University Press, 1927; Chelsea, New York, 1962.

Spiridonov, V. and Zhedanov, A. (2000). Spectral transformation chains and some new biorthogonal rational functions. *Comm. Math. Phys.*, 210:49–83.

Wilson, J. A. (1991). Orthogonal functions from Gram determinants. *SIAM J. Math. Anal.*, 22:1147–1155.

A BILATERAL SERIES INVOLVING BASIC HYPERGEOMETRIC FUNCTIONS

Hjalmar Rosengren
Department of Mathematics
Chalmers University of Technology and Göteborg University
SE-412 96 Göteborg
SWEDEN
hjalmar@math.chalmers.se

Abstract We prove a summation formula for a bilateral series whose terms are products of two basic hypergeometric functions. In special cases, series of this type arise as matrix elements of quantum group representations.

1. Introduction

The object of the present paper is to study bilateral series of a type that first appeared in the work of Koelink and Stokman (Koelink and Stokman, 2001), in the context of harmonic analysis on the SU(1, 1) quantum group. They needed to compute sums of the form

$$\sum_{n=-\infty}^{\infty} {}_2\varphi_1\left(\begin{matrix} a, b \\ c \end{matrix}; q, xq^n\right) {}_2\varphi_1\left(\begin{matrix} d, e \\ f \end{matrix}; q, yq^n\right) t^n, \tag{1.1}$$

which appeared as matrix elements of quantum SU(1, 1) representations. Several special cases were considered, with different, very technical, proofs (some of these proofs were found by Mizan Rahman, see the appendix to (Koelink and Stokman, 2001)).

In joint work with Koelink (Koelink and Rosengren, 2002), we gave an extension of the summation formulas from (Koelink and Stokman, 2001), with a unified proof. Namely, we showed that under natural conditions of convergence and the assumption

$$abde = cf, \qquad fx = dey \tag{1.2}$$

(and thus also $cy = abx$), the sum in (1.1) can be expressed as the sum of two ${}_8W_7$ series, or, alternatively, as the sum of three balanced ${}_4\varphi_3$

series. From the quantum algebraic viewpoint, which was only briefly mentioned in (Koelink and Rosengren, 2002), this computes a general class of matrix elements for the strange, the complementary and the principal unitary series of quantum SU(1,1). The proof in (Koelink and Rosengren, 2002) is similar to, though not an extension of, one of Rahman's proofs in (Koelink and Stokman, 2001). In particular, it is quite technical and involves rather non-obvious applications of q-series transformations.

In a related paper (Stokman, 2003), Stokman considered matrix elements for the principal unitary series and gave a very simple proof for their expression as ${}_8W_7$ series. The idea is simple but powerful: the representations are realized on $L^2(\mathbb{T})$, where one wants to compute the scalar product of two particular functions. Expanding both functions as Fourier series gives a special case of the sum (1.1). Stokman's observation, which is our starting point, is that the integral defining the scalar product can be computed very easily using residue calculus.

Our aim is to use Stokman's method not only to give a simple proof of the summation formula in (Koelink and Rosengren, 2002), but also to give a far-reaching extension of this identity.

Dedication

This paper would not have been written without the pioneering contributions of Mizan Rahman. I'm happy to know him, not only through his work as a master of identities, but also as a genuinely kind and helpful person. It is a pleasure to dedicate this paper to him as a small token of my appreciation.

2. A bilateral summation

We will now state the main result of the paper. Throughout, we use the standard notation of (Gasper and Rahman, 1990). The base q is assumed to satisfy $0 < |q| < 1$. We denote by ${}_{r+1}\varphi_r$ not only the convergent basic hypergeometric series, but also its analytic continuation to $\mathbb{C} \setminus \mathbb{R}_{\geq 1}$; cf. (Gasper and Rahman, 1990, §4.5).

Theorem 2.1. *Let x, $y \in \mathbb{C} \setminus \mathbb{R}_{\geq 0}$, and let the other parameters below satisfy*

$$\max_{ij}\left(|a_i c_j|\right) < |t| < 1, \tag{2.1}$$

$$|qyb_1 \cdots b_k| < |xa_1 \cdots a_{k+1}|. \tag{2.2}$$

Then the following identity holds:

$$\sum_{n=-\infty}^{\infty} {}_{k+1}\varphi_k\left(\begin{matrix} a_1,\dots,a_{k+1} \\ b_1,\dots,b_k \end{matrix};q,xq^n\right) {}_{l+1}\varphi_l\left(\begin{matrix} c_1,\dots,c_{l+1} \\ d_1,\dots,d_l \end{matrix};q,yq^n\right) t^n$$
$$= \frac{(q,xt,q/xt,a_1,\dots,a_{k+1},b_1/t,\dots,b_k/t;q)_\infty}{(t,x,q/x,a_1/t,\dots,a_{k+1}/t,b_1,\dots,b_k;q)_\infty}$$
$$\times {}_{k+l+2}\varphi_{k+l+1}\left(\begin{matrix} t,qt/b_1,\dots,qt/b_k,c_1,\dots,c_{l+1} \\ qt/a_1,\dots,qt/a_{k+1},d_1,\dots,d_l \end{matrix};q,\frac{qyb_1\cdots b_k}{xa_1\cdots a_{k+1}}\right)$$
$$+ \frac{(q,a_1x,q/a_1x,yt/a_1,qa_1/yt,a_2,\dots,a_{k+1},b_1/a_1,\dots,b_k/a_1;q)_\infty}{(x,q/x,y,q/y,t/a_1,a_2/a_1,\dots,a_{k+1}/a_1,b_1,\dots,b_k;q)_\infty}$$
$$\times \frac{(c_1,\dots,c_{l+1},a_1d_1/t,\dots,a_1d_1/t;q)_\infty}{(a_1c_1/t,\dots,a_1c_{l+1}/t,d_1,\dots,d_l;q)_\infty}$$
$$\times {}_{k+l+2}\varphi_{k+l+1}\left(\begin{matrix} a_1,qa_1/b_1,\dots,qa_1/b_k,a_1c_1/t,\dots,a_1c_{l+1}/t \\ qa_1/t,qa_1/a_2,\dots,qa_1/a_{k+1},a_1d_1/t,\dots,a_1d_l/t \end{matrix};\right.$$
$$\left. q,\frac{qyb_1\cdots b_k}{xa_1\cdots a_{k+1}}\right) + \text{idem}\,(a_1;a_2,\dots,a_{k+1})\,. \tag{2.3}$$

Here, idem $(a_1; a_2, \dots, a_{k+1})$ denotes the sum of the k terms obtained by interchanging a_1 in the second term with each of $a_2, \dots, a_{k+1}$ (thus, we have in total $k + 2$ terms on the right-hand side of (2.3)). As is customary, we implicitly assume that the parameters are such that one never divides by zero. Note also that interchanging the roles of ${}_{k+1}\varphi_k$ and ${}_{l+1}\varphi_l$ gives an alternative expression for the series, which is valid when (2.2) is replaced by the condition $|qxd_1 \cdots d_l| < |yc_1 \cdots c_{l+1}|$. These two expressions are related by (Gasper and Rahman, 1990, (4.10.10)).

In (Koelink and Rosengren, 2002), we computed the sum when $k = l = 1$ and, crucially to the methods used there, (1.2) holds. The latter condition means exactly that the three ${}_4\varphi_3$ series on the right-hand side of (2.3) are balanced. In this case Theorem 2.1 reduces to (Koelink and Rosengren, 2002, (3.13)), rather than the alternative expression as a sum of two ${}_8W_7$ series given in (Koelink and Rosengren, 2002, Proposition 3.1).

Our main tool is the following lemma from (Koelink and Rosengren, 2002), where it was used only in the case $k = 5$. Here we need the general case.

Lemma 2.2. *For*

$$\max_i\,(|a_i|) < |t| < 1 \tag{2.4}$$

and $x \in \mathbb{C} \setminus \mathbb{R}_{\geq 0}$ *one has*

$$\sum_{n=-\infty}^{\infty} {}_{k+1}\varphi_k \left(\begin{matrix} a_1, \dots, a_{k+1} \\ b_1, \dots, b_k \end{matrix} ; q, xq^n \right) t^n$$

$$= \frac{(q, a_1, \dots, a_{k+1}, xt, q/xt, b_1/t, \dots, b_k/t; q)_\infty}{(x, q/x, b_1, \dots, b_k, t, a_1/t, \dots, a_{k+1}/t; q)_\infty}. \qquad (2.5)$$

In (Koelink and Rosengren, 2002), this was proved using the explicit formula for analytic continuation of ${}_{k+1}\varphi_k$ series. It is possible to give a much simpler proof using integral representations. Namely, the coefficient of t^n in the Laurent expansion of the right-hand side in the annulus (2.4) is given by

$$\frac{(q, a_1, \dots, a_{k+1}; q)_\infty}{(x, q/x, b_1, \dots, b_k; q)_\infty} \frac{1}{2\pi i} \int \frac{(xt, q/xt, b_1/t, \dots, b_k/t; q)_\infty}{(t, a_1/t, \dots, a_{k+1}/t; q)_\infty} t^{-n-1} \, dt, \qquad (2.6)$$

where the integral is over a positively oriented contour encircling the origin inside the annulus. This is an integral of the form (Gasper and Rahman, 1990, (4.9.4)), which is computed there using residue calculus. If $|xq^n| < 1$, its value is given by (Gasper and Rahman, 1990, (4.10.9)) as the ${}_{k+1}\varphi_k$ series in (2.5). Since (2.6) is analytic in x, this also holds for $|xq^n| \geq 1$ in the sense of analytic continuation.

Remark 2.3. *More generally, (Gasper and Rahman, 1990, (4.10.9)) may be used to express the Laurent coefficients of*

$$P(z) = \prod_{i=1}^{m} \frac{(\alpha_i z; q)_\infty}{(\gamma_i z; q)_\infty} \prod_{i=1}^{n} \frac{(\beta_i/z; q)_\infty}{(\delta_i/z; q)_\infty} \qquad (2.7)$$

in the annulus

$$\max_i |\delta_i| < |z| < \min_i (1/|\gamma_i|) \qquad (2.8)$$

as sums of analytically continued basic hypergeometric series.

Proof of Theorem 2.1. Similarly as in (Koelink and Rosengren, 2002), it is easy to check that (2.1) is the natural condition for absolute convergence of the left-hand side of (2.3). We first rewrite this series as an integral. For this we make the preliminary assumption that t is real with

$$\max_{ij} (|a_i|, |c_j|) < t^{1/2} < 1. \qquad (2.9)$$

Writing $f_k(t; a, b, x)$ for either side of (2.5), we consider the integral

$$\frac{1}{2\pi i} \int_{|z|=t^{1/2}} f_k(z; a, b, x) \, f_l(\bar{z}; c, d, y) \, \frac{dz}{z}.$$

Using the expression on the left-hand side of (2.5) together with orthogonality of the monomials, we find that it equals the sum in (2.1).

To compute the integral, we plug in the expressions from the right-hand side of (2.5). Up to a constant, this gives an integral of the form

$$\frac{1}{2\pi i}\int P(z)\,\frac{dz}{z},$$

where P is as in (2.7), with

$$\alpha = (x, q/yt, d_1/t, \dots, d_l/t)\,, \qquad \beta = (q/x, yt, b_1, \dots, b_k)\,,$$
$$\gamma = (1, c_1/t, \dots, c_{l+1}/t)\,, \qquad \delta = (t, a_1, \dots, a_{k+1})\,.$$

Note that (2.8) and (2.9) are equivalent. Thus, we again have an integral of the form (Gasper and Rahman, 1990, (4.9.4)). The condition (2.2) is precisely (Gasper and Rahman, 1990, (4.10.2)), which ensures that the singularity at $z = 0$ does not contribute to the integral. The value of the integral is then given by (Gasper and Rahman, 1990, (4.10.8)) as a sum of $k+2$ terms, each being a ${}_{k+l+4}\varphi_{k+l+3}$ series. However, because of the condition $\alpha_1\beta_1 = \alpha_2\beta_2 = q$, they all reduce to type ${}_{k+l+2}\varphi_{k+l+1}$. This proves Theorem 2.1 under the assumption (2.9). By analytic continuation in t, this may be replaced with the weaker condition (2.1). □

Remark 2.4. *Using instead of Lemma* 2.2 *the Laurent expansion of the general product* (2.7) *gives a generalization of Theorem* 2.1, *where the two basic hypergeometric series on the left-hand side of* (2.3) *are replaced by finite sums of such series.*

In the case $l = 0$, one may use the q-binomial theorem to sum the series ${}_1\varphi_0$ in Theorem 2.1. The condition $y \in \mathbb{C} \setminus \mathbb{R}_{\geq 0}$ is then superfluous. We find this special case interesting enough to write out explicitly. Compared to Theorem 2.1 we have made the change of variables $y \mapsto c$, $c_1 \mapsto d/c$.

Corollary 2.5. *Let* $x \in \mathbb{C} \setminus \mathbb{R}_{\geq 0}$, *and assume that*

$$\max_i(|a_i d/c|) < |t| < 1, \qquad |qcb_1 \cdots b_k| < |xa_1 \cdots a_{k+1}|.$$

Then the following identity holds:

$$\sum_{n=-\infty}^{\infty} \frac{(c)_n}{(d)_n} {}_{k+1}\varphi_k \left(\begin{matrix} a_1, \ldots, a_{k+1} \\ b_1, \ldots, b_k \end{matrix} ; q, xq^n \right) t^n$$

$$= \frac{(q, xt, q/xt, c, a_1, \ldots, a_{k+1}, b_1/t, \ldots, b_k/t; q)_\infty}{(t, x, q/x, d, a_1/t, \ldots, a_{k+1}/t, b_1, \ldots, b_k; q)_\infty}$$

$$\times {}_{k+2}\varphi_{k+1} \left(\begin{matrix} t, d/c, qt/b_1, \ldots, qt/b_k \\ qt/a_1, \ldots, qt/a_{k+1} \end{matrix} ; q, \frac{qcb_1 \cdots b_k}{xa_1 \cdots a_{k+1}} \right)$$

$$+ \frac{(q, d/c, a_1x, q/a_1x, ct/a_1, qa_1/ct, a_2, \ldots, a_{k+1}, b_1/a_1, \ldots, b_k/a_1; q)_\infty}{(x, q/x, q/c, d, a_1d/ct, t/a_1, a_2/a_1, \ldots, a_{k+1}/a_1, b_1, \ldots, b_k; q)_\infty}$$

$$\times {}_{k+2}\varphi_{k+1} \left(\begin{matrix} a_1, a_1d/ct, qa_1/b_1, \ldots, qa_1/b_k \\ qa_1/t, qa_1/a_2, \ldots, qa_1/a_{k+1} \end{matrix} ; q, \frac{qcb_1 \cdots b_k}{xa_1 \cdots a_{k+1}} \right)$$

$$+ \, \mathrm{idem}\,(a_1; a_2, \ldots, a_{k+1}).$$

Note that in the case $c = d$, the first ${}_{k+2}\varphi_{k+1}$ on the right-hand side reduces to 1 and the remaining terms to 0. Thus, we recover Lemma 2.2. We also remark that if we choose $k = 0$ in Corollary 2.5 and replace $x \mapsto a$, $a_1 \mapsto b/a$, we obtain the transformation formula

$$ {}_2\psi_2 \left(\begin{matrix} a, c \\ b, d \end{matrix} ; q, t \right) = \frac{(q, b/a, c, at, q/at; q)_\infty}{(q/a, b, d, t, b/at; q)_\infty} {}_2\varphi_1 \left(\begin{matrix} t, d/c \\ aqt/b \end{matrix} ; q, \frac{qc}{b} \right)$$

$$+ \frac{(q, q/b, d/c, act/b, qb/act; q)_\infty}{(q/a, q/c, d, at/b, bd/act; q)_\infty} {}_2\varphi_1 \left(\begin{matrix} b/a, bd/act \\ qb/at \end{matrix} ; q, \frac{qc}{b} \right).$$

This identity is also obtained by choosing $r = 2$, $c_1 = qa_2$, $c_2 = b_2$ in (Gasper and Rahman, 1990, (5.4.3)), and then applying (Gasper and Rahman, 1990, (III.1)) to both ${}_2\varphi_1$ series on the right-hand side.

References

Gasper, G. and Rahman, M. (1990). *Basic Hypergeometric Series.* Cambridge University Press, Cambridge.

Koelink, E. and Rosengren, H. (2002). Transmutation kernels for the little q-Jacobi function transform. *Rocky Mountain J. Math.*, 32:703–738.

Koelink, E. and Stokman, J. V. (2001). Fourier transforms on the quantum $su(1,1)$ group. *Publ. Res. Inst. Math. Sci.*, 37:621–715. With an appendix by M. Rahman.

Stokman, J. V. (2003). Askey-Wilson functions and quantum groups. Preprint.

THE HILBERT SPACE ASYMPTOTICS OF A CLASS OF ORTHONORMAL POLYNOMIALS ON A BOUNDED INTERVAL

S. N. M. Ruijsenaars
Centre for Mathematics and Computer Science
P.O. Box 94079
1090 GB Amsterdam
THE NETHERLANDS

Abstract We study the asymptotics of orthonormal polynomials $\{p_n(\cos x)\}_{n=0}^{\infty}$, associated with a certain class of weight functions $w(x) = 1/c(x)c(-x)$ on $[0, \pi]$. Our principal result is that the norm of the difference of $p_n(\cos x)$ and $D_n(x) \equiv c(x)e^{inx}+c(-x)e^{-inx}$ in $L^2([0,\pi],(2\pi)^{-1}w(x)dx)$ vanishes exponentially as $n \to \infty$. The decay rate is determined by analyticity properties of the c-function.

1. Introduction

Suppose $w(x)$ is a continuous function on $[0, \pi]$ that is positive on $(0, \pi)$. Then the functions

$$B_n(x) \equiv \mathcal{N}_n(e^{inx} + e^{-inx}), \quad \mathcal{N}_n \equiv \begin{cases} 1/2, & n = 0, \\ 1, & n \in \mathbb{N}^*, \end{cases} \tag{1.1}$$

form a total set in the Hilbert space

$$\mathcal{H} \equiv L^2\left([0,\pi],(2\pi)^{-1}w(x)dx\right). \tag{1.2}$$

By the Gram-Schmidt procedure, we can therefore construct an orthonormal base of the form

$$P_n(x) = \mu_n B_n(x) + \sum_{m<n} b_{nm}B_m(x), \quad \mu_n > 0, \quad b_{nm} \in \mathbb{R}, \quad n \in \mathbb{N}. \tag{1.3}$$

We may view $B_n(x)$ as a polynomial of degree n in $\cos x$ (the Tchebichev polynomial of the first kind). Doing so, it is clear that there are polynomials $p_n(v)$ such that

$$p_n(\cos x) = P_n(x). \tag{1.4}$$

As is well known (and easily verified), these polynomials satisfy a self-adjoint recurrence relation

$$a_n p_{n+1}(v) + b_n p_n(v) + a_{n-1} p_{n-1}(v) = 2v p_n(v), \tag{1.5}$$

with

$$a_{-1} \equiv 0, \quad a_n \in (0,2), \quad b_n \in (-2,2), \quad n \in \mathbb{N}. \tag{1.6}$$

In this paper we study the $n \to \infty$ asymptotics of $P_n(x), \mu_n, a_n$ and b_n. Our primary interest is in the $n \to \infty$ asymptotics of the functions $P_n(x)$ in the Hilbert space $\mathcal{H}$. We allow weight functions of the form

$$w(x) = 1/c(x)c(-x), \tag{1.7}$$

where $c(x)$ has certain analyticity properties specified in Section 2. Defining the dominant asymptotics function

$$D_n(x) \equiv c(x)e^{inx} + c(-x)e^{-inx}, \quad n \in \mathbb{N}, \tag{1.8}$$

our principal result is an exponential decay bound

$$\|P_n - D_n\| = O(\exp(-2nd)), \quad n \to \infty, \tag{1.9}$$

for weight functions $w(x)$ in the class $\mathcal{W}_e$ (2.18). Here, $\|\cdot\|$ denotes the norm derived from the scalar product of $\mathcal{H}$ (1.2), and the decay rate d is any positive number smaller than a number d_+ in $(0,\infty]$ that depends on the choice of c-function. (It is given by (2.2).)

We obtain no information on the pointwise decay of $|P_n(x) - D_n(x)|$ for $x \in (0,\pi)$, but we believe this is also governed by an $O(\exp(-2nd))$-bound. Our interest in obtaining a sharp L^2-bound was triggered by the possibility to use such a bound for an approximation argument. The application we have in mind concerns what may be viewed as 'relativistic' Lamé functions. To prove L^2-completeness of a suitable countable set of the latter, we need a sufficiently strong bound $\|P_n - D_n\| \leq \beta_n$ on orthonormal polynomials associated to certain weight functions in $\mathcal{W}_e$. Specifically, the sequence $\{\beta_n\}$ must be in l^2. It is however beyond our scope to elaborate on this; we return to this issue elsewhere (Ruijsenaars, 2003).

As it has turned out, the methods we have developed to prove the desired L^2-bounds can be generalized to the multi-variable case. The pertinent results have a striking application to the theory of quantum soliton systems, and have already been reported in our paper Ref. (Ruijsenaars, 2002). (For further generalizations, see work by van Diejen (van Diejen, 2003).)

In the one-variable case at issue here, the proofs are quite elementary and transparent. Furthermore, we can allow a larger class of weight functions than the one obtained by specializing Ref. (Ruijsenaars, 2002). It therefore seems useful to have an independent treatment of the one-variable case available.

At the outset, we should mention that the subclass $\mathcal{W}_{0,0}$ of $\mathcal{W}$ (cf. (2.15)) belongs to the huge class of weight functions for which Szegő has obtained uniform pointwise bounds on $|P_n(x) - D_n(x)|$, cf. Theorem 12.1.4 in his monograph (Szegő, 1975). The latter L^∞-bounds are however $O\left((\ln(n))^{-\lambda}\right)$ for some $\lambda > 0$, a decay that is far too slow to be in l^2. (To be sure, Szegő's bounds are presumably quite sharp for the class he handles.)

Another point of special note is that our class $\mathcal{W}_e$ of weight functions contains the weight function of the Askey-Wilson polynomials (Askey and Wilson, 1985; Gasper and Rahman, 1990). More precisely, setting

$$c_{AW}(x) \equiv \left(ae^{-ix};q\right)_\infty \left(be^{-ix};q\right)_\infty \left(ce^{-ix};q\right)_\infty \left(de^{-ix};q\right)_\infty / \left(e^{-2ix};q\right)_\infty, \tag{1.10}$$

the weight function

$$w_{AW}(x) \equiv 1/c_{AW}(x)c_{AW}(-x) \tag{1.11}$$

belongs to $\mathcal{W}_{1,1}$ (2.15) for

$$q, a, b, c, d \in (-1, 1). \tag{1.12}$$

We would like to emphasize that for this special case the exponential decay of $\|P_n - D_n\|$ is an obvious corollary of previous results by Ismail and Wilson (Ismail and Wilson, 1982; Ismail, 1986). Indeed, via generating function techniques they obtained an $L^\infty([0, \pi])$ exponential decay bound on $P_n - D_n$. It is an open question whether the generating functions of the much larger class of polynomials arising from $\mathcal{W}_e$ can again be exploited to obtain exponential decay in L^∞. (Note this would yield a stronger result than our L^2-decay, since we restrict attention to bounded weight functions.)

2. Main results

Our starting point consists of the space $\mathcal{A}$ of functions $f(z)$ that are analytic and zero-free in the closed unit disk, real-valued for $z \in [-1, 1]$, and normalized by

$$f(0) = 1. \tag{2.1}$$

Thus the convergence radii R_+ and R_- of the power series expansions of $f(z)$ and $1/f(z)$ are greater than 1, and the decay parameters

$$d_\pm \equiv \ln R_\pm \tag{2.2}$$

belong to $(0, \infty]$.

We now define a space $\mathcal{C}_r$ of 'reduced c-functions' $c_r(x), x \in \mathbb{R}$, by

$$\mathcal{C}_r \equiv \left\{c_r(x) = f\left(e^{-ix}\right) \mid f \in \mathcal{A}\right\}. \tag{2.3}$$

Thus any $c_r \in \mathcal{C}_r$ admits expansions of the form

$$c_r(x) = 1 + \sum_{k=1}^{\infty} \alpha_k^+ e^{-ikx}, \quad x \in \mathbb{R}, \quad \alpha_k^+ \in \mathbb{R}, \tag{2.4}$$

$$1/c_r(x) = 1 + \sum_{k=1}^{\infty} \alpha_k^- e^{-ikx}, \quad x \in \mathbb{R}, \quad \alpha_k^- \in \mathbb{R}, \tag{2.5}$$

$$\alpha_k^{\pm} = O\left(e^{-kd}\right), \quad d \in (0, d_{\pm}), \quad k \to \infty, \tag{2.6}$$

and we have

$$c_r(-x) = \overline{c_r(x)}, \quad x \in \mathbb{R}. \tag{2.7}$$

Next we introduce the u-function ('S-matrix')

$$u(x) \equiv c_r(x)/c_r(-x). \tag{2.8}$$

From (2.4)–(2.6) we readily deduce

$$u(x) = \sum_{l \in \mathbb{Z}} u_l e^{ilx}, \quad x \in \mathbb{R}, \quad u_l \in \mathbb{R}, \tag{2.9}$$

$$u_l = O\left(e^{-ld}\right), \quad d \in (0, d_-), \quad l \to \infty, \tag{2.10}$$

$$u_l = O\left(e^{ld}\right), \quad d \in (0, d_+), \quad l \to -\infty, \tag{2.11}$$

and from (2.7) we obtain

$$|u(x)| = 1, \quad x \in \mathbb{R}. \tag{2.12}$$

Finally, we define a reduced weight function

$$w_r(x) \equiv 1/c_r(x)c_r(-x). \tag{2.13}$$

Clearly, $w_r(x), x \in \mathbb{R}$, is a smooth, positive, 2π-periodic and even function.

We proceed to define classes of c-functions

$$\mathcal{C}_{M,N} \equiv \left\{c(x) = \left(1 - e^{-ix}\right)^{-M}\left(1 + e^{-ix}\right)^{-N} c_r(x) \mid c_r \in \mathcal{C}_r\right\}, \quad M, N \in \mathbb{N}, \tag{2.14}$$

and classes of w-functions

$$\mathcal{W}_{M,N} \equiv \{w(x) = 1/c(x)c(-x) \mid c(x) \in \mathcal{C}_{M,N}\}, \quad M, N \in \mathbb{N}, \tag{2.15}$$

$$\mathcal{W} \equiv \cup_{M,N\in\mathbb{N}} \mathcal{W}_{M,N}. \tag{2.16}$$

Thus we have

$$w(x) = \left[4\sin^2(x/2)\right]^M \left[4\cos^2(x/2)\right]^N w_r(x), \quad w \in \mathcal{W}_{M,N}. \tag{2.17}$$

(We point out that for the simplest case $c_r(x) = 1$, the weight functions (2.17) give rise to Jacobi polynomials (Szegő, 1975).) The class $\mathcal{W}_e$ for which all of our results hold true is now defined by

$$\mathcal{W}_e \equiv \mathcal{W}_{0,0} \cup \mathcal{W}_{1,0} \cup \mathcal{W}_{0,1} \cup \mathcal{W}_{1,1}. \tag{2.18}$$

Fixing $w \in \mathcal{W}_{M,N}$, we can write the function $D_n(x)$ (1.8) as

$$\begin{aligned} D_n(x) = {} & [2i\sin(x/2)]^{-M}[2\cos(x/2)]^{-N} c_r(x) \\ & \times \exp(ix[n + (M+N)/2]) + (x \to -x). \end{aligned} \tag{2.19}$$

By (2.17) this implies that $D_n(x)$ belongs to the Hilbert space $\mathcal{H}$ (1.2). We are now prepared for our first lemma.

Lemma 2.1. *Assume*

$$w \in \mathcal{W}_{M,N}, \quad M, N \in \mathbb{N}. \tag{2.20}$$

Then we have

$$(B_m, D_n) = \delta_{mn}, \quad m \le n, \tag{2.21}$$

$$(P_m, D_n) = \mu_n \delta_{mn}, \quad m \le n, \tag{2.22}$$

$$(D_n, D_n) = 1 + (-)^M u_{-2n-M-N}. \tag{2.23}$$

Proof. From the above definitions we have

$$\begin{aligned} (B_m, D_n) &= \frac{\mathcal{N}_m}{2\pi} \int_0^\pi dx \, \frac{1}{c(x)c(-x)} \left(e^{imx} + e^{-imx}\right) \left(c(x)e^{inx} + c(-x)e^{-inx}\right) \\ &= \frac{\mathcal{N}_m}{2\pi} \int_{-\pi}^\pi dx \, \frac{1}{c(-x)} \left(e^{i(n+m)x} + e^{i(n-m)x}\right) \\ &= \frac{\mathcal{N}_m}{2\pi} \int_{-\pi}^\pi dx \, \left[1 - e^{ix}\right]^M \left[1 + e^{ix}\right]^N \left(1 + \sum_{k=1}^\infty \alpha_k^- e^{ikx}\right) \\ &\quad \cdot \left(e^{i(n+m)x} + e^{i(n-m)x}\right). \end{aligned} \tag{2.24}$$

Since the sequence $\{\alpha_k^-\}$ has exponential decay, we may interchange the summation and the integration. For $m < n$ each of the resulting integrals has an integrand proportional to $\exp(ilx)$ with $l \in \mathbb{N}^*$, so they all vanish. For $m = n$ we pick up the constant term and obtain 1. Hence (2.21) follows.

Clearly, (2.22) follows from (2.21) and the definition (1.3) of P_n, so it remains to prove (2.23). To this end we calculate

$$\begin{aligned}(D_n, D_n) &= \frac{1}{2\pi}\int_0^\pi dx \left(\frac{c(x)}{c(-x)}\,e^{2inx} + 2 + \frac{c(-x)}{c(x)}\,e^{-2inx}\right) \\ &= 1 + \frac{1}{2\pi}\int_{-\pi}^{\pi} dx (-)^M e^{ix(M+N)} u(x) e^{2inx} \\ &= 1 + (-)^M u_{-2n-M-N},\end{aligned} \tag{2.25}$$

where we used (2.8)–(2.11). □

Notice that this lemma solely concerns equalities, as opposed to inequalities. As an obvious corollary, we obtain a further equality

$$(P_n - D_n, P_n - D_n) = 2\,(1 - \mu_n) + (-)^M u_{-2n-M-N}. \tag{2.26}$$

Using (2.11), the following inequality is also immediate from (2.23):

$$\|D_n\| = 1 + O(\exp(-2nd)), \quad d \in (0, d_+)\,, \quad n \to \infty. \tag{2.27}$$

From (2.22) and the Schwarz inequality we now get

$$\mu_n \le 1 + C\exp(-2nd), \quad C > 0, \quad d \in (0, d_+)\,, \ \forall n \in \mathbb{N}. \tag{2.28}$$

Next, we combine (2.26) and the upper bound (2.28) to deduce that we have an equivalence

$$\|P_n - D_n\| = O(\exp(-nd)) \ \Leftrightarrow \ \mu_n \ge 1 - C'\exp(-2nd), \ C' > 0, \ \forall n \in \mathbb{N}. \tag{2.29}$$

However, we will show in Section 3 that $\|P_n - D_n\|$ does not decay exponentially for $w \in \mathcal{W} \setminus \mathcal{W}_e$. We now pass to an auxiliary result that is still valid for all $w \in \mathcal{W}$.

To this end we introduce a truncated c_r-function

$$c_r^{(L)}(x) \equiv 1 + \sum_{l=1}^{L} \alpha_l^+ e^{-ilx}, \quad L \in \mathbb{N}^*, \tag{2.30}$$

and asymptotics function

$$D_n^{(L)}(x) \equiv [2i\sin(x/2)]^{-M}[2\cos(x/2)]^{-N}c_r^{(L)}(x) \\ \times \exp(ix[n+(M+N)/2]) + (x \to -x). \tag{2.31}$$

The following lemma shows that the remainder function

$$S_n^{(L)}(x) \equiv D_n(x) - D_n^{(L)}(x) \tag{2.32}$$

has exponential decay in L^2 sense as $L \to \infty$.

Lemma 2.2. *Assuming* (2.20) *and* $d \in (0, d_+)$, *we have*

$$\left\| S_n^{(L)} \right\| = O(\exp(-Ld)), \quad L \to \infty, \tag{2.33}$$

with the bound uniform for $n \in \mathbb{N}$.

Proof. By (2.17) and (2.30)–(2.32) we have

$$\begin{aligned} \left\| S_n^{(L)} \right\|^2 &= \frac{1}{2\pi} \int_0^\pi dx w_r(x) \left| \sum_{l=L+1}^{\infty} \alpha_l^+ \exp(ix[n+(M+N)/2-l]) \right. \\ &\quad \left. + (-)^M (x \to -x) \right|^2 \\ &\leq 2 \sup_{x\in[0,\pi]} w_r(x) \left(\sum_{l=L+1}^{\infty} |\alpha_l^+| \right)^2 . \end{aligned} \tag{2.34}$$

By virtue of boundedness of $w_r(x)$ and the estimates (2.6), the lemma follows. □

From (2.32)–(2.33) we deduce

$$\|P_n - D_n\| \leq \left\| P_n - D_n^{(2n-1)} \right\| + O(\exp(-2nd)). \tag{2.35}$$

To obtain exponential decay of $\|P_n - D_n\|$, therefore, we need only estimate $\left\| P_n - D_n^{(2n-1)} \right\|$. The significance of the choice $L = 2n-1$ is that it ensures that the only exponentials present in the expansion of $c_r^{(L)}(x)\exp(inx)$ are those occurring in the expansion of $P_n(x)$. For $M = N = 0$, it is therefore immediate that we have equalities

$$\begin{aligned} D_n^{(2n-1)}(x) &= B_n(x) + \sum_{m<n} c_{nm} B_m(x) \\ &= \mu_n^{-1} P_n(x) + \sum_{m<n} d_{nm} P_m(x), \quad n \in \mathbb{N}^*, \end{aligned} \tag{2.36}$$

for certain $c_{nm}, d_{nm} \in \mathbb{R}$. Now the functions on the rhs are manifestly bounded on $[0, \pi]$, whereas the functions on the lhs have a divergence as $x \to 0$ and/or as $x \to \pi$ whenever $M > 1$ and/or $N > 1$, cf. (2.31). Thus the restriction to $\mathcal{W}_e$ in the next lemma is essential.

Lemma 2.3. *Assume $w(x)$ belongs to $\mathcal{W}_e$ (2.18). Then the equalities (2.36) hold true.*

Proof. For $w \in \mathcal{W}_{0,0}$ we have already established (2.36). Consider next $w \in \mathcal{W}_{1,1}$. Then (2.30)–(2.31) yield

$$D_n^{(L)}(x) = \frac{\sin(n+1)x}{\sin x} + \sum_{l=1}^{L} \alpha_l^+ \frac{\sin(n+1-l)x}{\sin x}. \tag{2.37}$$

Since we have

$$\frac{\sin(k+1)x}{\sin x} = B_k(x) + \sum_{l<k} t_{kl} B_l(x), \tag{2.38}$$

we deduce (2.36).

Now let $w \in \mathcal{W}_{1,0}$. Then we obtain

$$D_n^{(L)}(x) = \frac{\sin(n+1/2)x}{\sin(x/2)} + \sum_{l=1}^{L} \alpha_l^+ \frac{\sin(n+1/2-l)x}{\sin(x/2)}, \tag{2.39}$$

and since

$$\frac{\sin(k+1/2)x}{\sin(x/2)} = B_k(x) + \sum_{l<k} s_{kl} B_l(x), \tag{2.40}$$

we obtain again (2.36). Finally, for $w \in \mathcal{W}_{0,1}$ we get

$$D_n^{(L)}(x) = \frac{\cos(n+1/2)x}{\cos(x/2)} + \sum_{l=1}^{L} \alpha_l^+ \frac{\cos(n+1/2-l)x}{\cos(x/2)}, \tag{2.41}$$

so (2.36) follows from

$$\frac{\cos(k+1/2)x}{\cos(x/2)} = B_k(x) + \sum_{l<k} r_{kl} B_l(x). \tag{2.42}$$

Hence the lemma is proved. □

We are now prepared to obtain the principal result of this paper.

Theorem 2.4. *Let $w \in \mathcal{W}_e$ and $d \in (0, d_+)$. Then we have the estimates*

$$\|P_n - D_n\| = O(\exp(-2nd)), \quad n \to \infty, \tag{2.43}$$

$$\mu_n - 1 = O(\exp(-2nd)), \quad n \to \infty. \tag{2.44}$$

Proof. By virtue of Lemma 2.3 we may invoke the expansions (2.36), which we rewrite as

$$D_n(x) = \mu_n^{-1} P_n(x) + \sum_{m<n} d_{nm} P_m(x) + S_n^{(2n-1)}(x), \tag{2.45}$$

cf. (2.32). Taking the inner product with $\mu_n P_n$ and using (2.22), we obtain

$$\mu_n^2 = 1 + \mu_n \left(P_n, S_n^{(2n-1)}\right). \tag{2.46}$$

By the Schwarz inequality and Lemma 2.2, we have

$$\left(P_n, S_n^{(2n-1)}\right) = O(\exp(-2nd)). \tag{2.47}$$

Now μ_n is positive, so in view of the upper bound (2.28) we may deduce

$$\mu_n^2 - 1 = O(\exp(-2nd)). \tag{2.48}$$

Hence the estimate (2.44) follows.

As a consequence of (2.44) and (2.26), we now obtain an $O(\exp(-nd))$-bound on $\|P_n - D_n\|$. To arive at the sharper bound (2.43), we introduce

$$d' \in (d, d_+). \tag{2.49}$$

Now we take the inner product of (2.45) with P_l for $l < n$, which yields due to (2.22)

$$0 = d_{nl} + \left(P_l, S_n^{(2n-1)}\right). \tag{2.50}$$

From the Schwarz inequality and (2.33) with $d \to d'$ we obtain

$$d_{nl} = O\left(\exp\left(-2nd'\right)\right). \tag{2.51}$$

Hence we have

$$\left\|D_n - \mu_n^{-1} P_n\right\| = O\left(n \exp\left(-2nd'\right)\right). \tag{2.52}$$

Telescoping and using (2.44) with $d \to d'$, we now get

$$\begin{aligned} \|P_n - D_n\| &\leq \left\|D_n - \mu_n^{-1} P_n\right\| + \left\|\left(1 - \mu_n^{-1}\right) P_n\right\| \\ &= O\left(n \exp\left(-2nd'\right)\right). \end{aligned} \tag{2.53}$$

Finally, recalling (2.49), we deduce (2.43). □

3. Further developments

In this section we derive some corollaries of the above results and consider the weight function class $\mathcal{W}_{M,N}$ with M and/or N greater than 1. First, we obtain the asymptotics of the recurrence coefficients a_n, b_n in (1.5).

Proposition 3.1. *Let $w \in \mathcal{W}_e$ and $d \in (0, d_+)$. Then we have*

$$a_n = 1 + O(\exp(-2nd)), \quad n \to \infty, \tag{3.1}$$

$$b_n = O(\exp(-2nd)), \quad n \to \infty. \tag{3.2}$$

Proof. From the relations (1.3)–(1.5) it is easily seen that

$$a_n = \mu_n / \mu_{n+1}, \quad n \in \mathbb{N}. \tag{3.3}$$

Therefore, (3.1) is clear from (2.44). To prove (3.2), we begin by noting that $D_n(x)$ (1.8) satisfies the recurrence

$$D_{n+1}(x) + D_{n-1}(x) = 2\cos(x) D_n(x). \tag{3.4}$$

Now the recurrence (1.5) entails

$$b_n = (P_n, 2vP_n), \tag{3.5}$$

where v denotes multiplication by $\cos x$. Telescoping, we obtain

$$b_n = (P_n - D_n, 2vP_n) + (D_n, 2v(P_n - D_n)) + (D_n, 2vD_n). \tag{3.6}$$

Using (3.4) and then (2.22), we get

$$(D_n, 2vD_n) = (D_n - P_n, D_{n+1}) + (D_n, D_{n-1} - P_{n-1}). \tag{3.7}$$

Finally, using the Schwarz inequality and the estimates (2.43), (2.27), we deduce (3.2). □

Next, we consider the special case where the function $f(z)$ defining $c_r(x)$ (recall (2.3)) is a polynomial.

Proposition 3.2. *Let $w \in \mathcal{W}_e$ and suppose that in the c_r-expansion (2.4) we have*

$$\alpha_k^+ = 0, \quad k > L, \quad L \in \mathbb{N}. \tag{3.8}$$

Then we have

$$P_n(x) = D_n(x), \quad 2n \geq L + 1. \tag{3.9}$$

Proof. From (2.30)–(2.32) we deduce

$$S_n^{(m)}(x) = 0, \quad m \geq L, \quad n \in \mathbb{N}. \tag{3.10}$$

Thus we have

$$S_n^{(2n-1)}(x) = 0, \quad 2n \geq L+1, \tag{3.11}$$

so that (2.45) yields

$$D_n(x) = \mu_n^{-1} P_n(x) + \sum_{m<n} d_{nm} P_m(x), \quad 2n \geq L+1. \tag{3.12}$$

Taking the inner product with $P_l, l \leq n$, we deduce from (2.22) that $d_{nl} = 0$ for $l < n$, and that $\mu_n = 1$ for $l = n$. Hence (3.9) follows. □

For w in our class $\mathcal{W}_{0,0}$, the result contained in this proposition can already be found in (Szegő, 1975), by looking rather hard in Section 12.4 ('Bernstein-Szegő polynomials'). In words, it says that, for $w \in \mathcal{W}_e$ corresponding to a polynomial $f(z)$ of degree L, the dominant asymptotics function $D_n(x)$ coincides with $P_n(x)$ for $2n \geq L+1$. As we have pointed out before, no such result can hold for $w \in \mathcal{W} \setminus \mathcal{W}_e$, since $P_n(x)$ is bounded on $(0, \pi)$, whereas in that case $D_n(x)$ is unbounded. Even so, in view of the results (2.21)–(2.29) (which are valid for all $w \in \mathcal{W}$), one might be inclined to guess that the sequence of numbers

$$c_n \equiv \|P_n - D_n\|, \quad n \in \mathbb{N}, \tag{3.13}$$

converges exponentially to 0 as $n \to \infty$. As we show next, this is not the case.

Proposition 3.3. *Let $w \in \mathcal{W} \setminus \mathcal{W}_e$. Then we have*

$$\sum_{n=0}^{\infty} |c_n|^2 = \infty. \tag{3.14}$$

Proof. To ease the exposition, we detail the proof for the case $w \in \mathcal{W}_{0,2}$ and then indicate how to proceed in general. We assume $\{c_n\} \in l^2$ so as to arrive at a contradiction.

Consider the identification map

$$\begin{aligned} I : \mathcal{H}_{0,2} \equiv L^2([0,\pi], w(x)dx) \to \mathcal{H}_{0,0} \equiv L^2\left([0,\pi], w_r(x)dx\right), \\ F(x) \mapsto 2\cos(x/2)F(x). \end{aligned} \tag{3.15}$$

Recalling (2.17), we see that I is an isometric isomorphism. Using obvious notation, we therefore have

$$c_n = \|IP_n - ID_n\|_{(0,0)} \,. \tag{3.16}$$

Next, recalling (2.19), we see that we may view $(ID_n)(x)$ as the dominant asymptotics function $\tilde{D}_{n+1}(x)$ corresponding to the polynomial

$\tilde{P}_{n+1}(x)$ in $\mathcal{H}_{0,0}$ that arises from $w_r(x)$. Doing so, it follows from the exponential decay of $\left\|\tilde{P}_{n+1} - \tilde{D}_{n+1}\right\|$ (which we proved in Section 2) and the assumption $\{c_n\} \in l^2$ that the sequence

$$\lambda_n \equiv \left\|IP_n - \tilde{P}_{n+1}\right\|_{(0,0)}, \quad n \in \mathbb{N}, \tag{3.17}$$

is in l^2.

We are now in the position to invoke a known completeness result, cf. Theorem A on p. 72 of Higgins' monograph (Higgins, 1977). Since I is unitary, the sequence $\{IP_n\}_{n=0}^{\infty}$ yields an orthonormal base in $\mathcal{H}_{0,0}$. Since the sequence $\{\lambda_n\}_{n=0}^{\infty}$ is in $l^2(\mathbb{N})$, the pertinent completeness result says that the sequence $\left\{\tilde{P}_{n+1}\right\}_{n=0}^{\infty}$ is complete in $\mathcal{H}_{0,0}$. This yields the desired contradiction, since $\tilde{P}_0$ is orthogonal to $\tilde{P}_1, \tilde{P}_2, \ldots$

Once this special case is understood, it will be clear how to proceed for any $\mathcal{W}_{M,N}$ with M and/or N greater than 1: one can identify $\mathcal{W}_{M,N}$ with one of the four spaces in $\mathcal{W}_e$ (2.18) (by reducing M and N modulo 2), and use (2.19) as in the special case $(M, N) = (0, 2)$ to arrive at a contradiction. □

Next, we note that thanks to (2.26) and (2.11), we have the equivalence

$$\lim_{n\to\infty} c_n = 0 \Leftrightarrow \lim_{n\to\infty} \mu_n = 1. \tag{3.18}$$

Consider now the weight function integral

$$I(w) \equiv \frac{1}{2\pi} \int_0^{\pi} w(x)dx. \tag{3.19}$$

Evidently, we have

$$P_0(x) = \mu_0 = I(w)^{-1/2}, \tag{3.20}$$

so using (3.3) we obtain

$$\mu_n = \mu_0 / \prod_{j=0}^{n-1} a_j, \quad n \in \mathbb{N}. \tag{3.21}$$

From this we readily infer a second equivalence

$$\lim_{n\to\infty} \mu_n = 1 \Leftrightarrow \lim_{n\to\infty} a_n = 1, \quad \lim_{n\to\infty} \prod_{j=0}^{n} a_j^{-2} = I(w). \tag{3.22}$$

At this point we insert a parenthetical remark. For the class $\mathcal{W}_e$ of weight functions, we have already shown that μ_n goes to 1. The class contains in particular the weight function w_{AW} (1.11), for the choice of parameters (1.12) yielding the Askey-Wilson integral $I(w_{AW})$ given by Eq. (6.1.1) in the monograph by Gasper and Rahman (Gasper and Rahman, 1990). Since the self-adjoint recurrence coefficients $a_{n,AW}$ are explicitly known, our results yield the explicit formula for $I(w_{AW})$ as a corollary. Quite likely, this relation between $I(w)$ and the a_j-product has been noticed before for the Askey-Wilson case, but we have not found this in the literature.

Of course, the normalization (2.1) of our weight functions is critical for the relations just pointed out. Indeed, when we switch from a given $w(x) \in \mathcal{W}$ to $\lambda w(x), \lambda > 0$, then we should multiply P_n by $\lambda^{-1/2}$ to retain unit norm. Thus the coefficients μ_n change to $\lambda^{-1/2}\mu_n$, whereas the recurrence coefficients a_n, b_n are invariant.

To elaborate on the normalization issue, let us fix $c_r \in \mathcal{C}_r$ and consider

$$c_R(x) \equiv \left(1 - e^{-ix}/R\right)^{-M} \left(1 + e^{-ix}/R\right)^{-N} c_r(x). \tag{3.23}$$

Clearly, we have

$$w_R(x) \equiv 1/c_R(x)c_R(-x) \in \mathcal{W}_{0,0}, \quad R > 1, \quad w_1(x) \in \mathcal{W}_{M,N}. \tag{3.24}$$

Now the polynomials $P_{n,R}(x)$ obviously converge to the polynomials $P_{n,1}(x)$ for $R \downarrow 1$, so the natural expectation is that the sequence $\mu_{n,1}$ still converges to 1 as $n \to \infty$. Unfortunately, there seems to be no short and simple way to control the pertinent interchange of limits.

In any case, the sequence $\mu_{n,1}$ does have limit 1 for $n \to \infty$. Equivalently, we have

$$\lim_{n\to\infty} \mu_n = 1, \quad \forall w \in \mathcal{W}, \tag{3.25}$$

so that we also have

$$\lim_{n\to\infty} c_n = 0, \quad \lim_{n\to\infty} a_n = 1, \quad \lim_{n\to\infty} \prod_{j=0}^{n} a_j^{-2} = I(w), \quad \forall w \in \mathcal{W}. \tag{3.26}$$

To substantiate this assertion, we invoke a limit theorem that can be found in Szegő's monograph, cf. Theorem 12.7.1 in (Szegő, 1975). The crux is that our normalization requirement (2.1) implies

$$\int_0^\pi \ln(w(x))dx = 0, \quad \forall w \in \mathcal{W}. \tag{3.27}$$

(To verify this for $w \in \mathcal{W}_{0,0}$, use the substitution $z = e^{-ix}$ and note that $f(z) \in \mathcal{A}$ has a one-valued analytic logarithm for $|z| \leq 1$ such that $\ln(f(0)) = 0$. To handle $w \in \mathcal{W}_{M,N}$, the above approximation (3.24) can be used.)

With the convergence question answered, it is natural to ask about the rate of convergence for $w \in \mathcal{W}_{M,N}$ with M or N greater than 1. In our last proposition we answer this for the simplest $w(x)$ in each class.

Proposition 3.4. *Assume $w \in \mathcal{W} \setminus \mathcal{W}_e$ corresponds to $c_r(x) = 1$. Then we have*

$$a_n = 1 + O\left(n^{-2}\right), \quad n \to \infty, \tag{3.28}$$

$$\mu_n = 1 + O\left(n^{-1}\right), \quad n \to \infty, \tag{3.29}$$

$$c_n = O\left(n^{-1/2}\right), \quad n \to \infty. \tag{3.30}$$

Proof. Inspecting (2.17), we see that the functions $P_n(x)$ are proportional to the Jacobi polynomials $P_n^{(M-1/2,N-1/2)}(\cos x)$. Their self-adjoint recurrence coefficients are given by

$$a_n^2 = \frac{(2n+2)(2n+2M+2N)(2n+2M+1)(2n+2N+1)}{(2n+M+N)(2n+M+N+1)^2(2n+M+N+2)}, \tag{3.31}$$

cf., e.g., (Koekoek and Swarttouw, 1994). Now a straightforward calculation yields

$$a_n^2 = 1 + C(M,N)n^{-2} + O\left(n^{-3}\right), \quad n \to \infty, \tag{3.32}$$

$$C(M,N) \equiv -[M(M-1) + N(N-1)]/2. \tag{3.33}$$

Hence (3.28) follows.

Since we have already seen that μ_n converges to 1, we may invoke (3.20)–(3.22), yielding

$$\mu_n = \prod_{j=n}^{\infty} a_j^2. \tag{3.34}$$

From this representation and (3.32) we readily deduce (3.29). Finally, (2.26) and (3.29) entail (3.30). □

It is natural to expect that the asymptotics (3.28)–(3.30) holds true for all $w \in \mathcal{W}_{M,N}$. After submission of this paper we learned that this is indeed the case, as follows from work by Kuijlaars, McLaughlin, van Assche and Vanlessen (Kuijlaars et al., 2003); moreover, we were informed that Kuijlaars has obtained results that imply in particular uniform exponential decay of $P_n(x) - D_n(x)$ on $[0, \pi]$ for $w \in \mathcal{W}_e$ (Kuijlaars, 2003). Both of these references make essential use of Riemann-Hilbert problem techniques (Deift, 1999).

Acknowledgments

We are indebted to M. Ismail for several illuminating discussions and comments. We would also like to thank W. van Assche for welcome information on pertinent literature. Finally, we acknowledge useful comments and suggestions from two referees, one of whom made us aware of the references (Kuijlaars et al., 2003) and (Kuijlaars, 2003).

References

Askey, R. A. and Wilson, J. A. (1985). Some basic hypergeometric orthogonal polynomials that generalize Jacobi polynomials. *Mem. Amer. Math. Soc.*, 54(319).

Deift, P. (1999). *Orthogonal polynomials and random matrices: A Riemann-Hilbert approach*, volume 3 of *Courant Lecture Notes in Mathematics*. New York University Courant Institute of Mathematical Sciences, New York, NY.

Gasper, G. and Rahman, M. (1990). *Basic Hypergeometric Series*, volume 35 of *Encyclopedia of Mathematics and its Applications*. Cambridge Univ. Press, Cambridge.

Higgins, J. R. (1977). *Completeness and basis properties of sets of special functions*, volume 72 of *Cambridge Tracts in Mathematics*. Cambridge Univ. Press, Cambridge.

Ismail, M. E. H. (1986). Asymptotics of the Askey-Wilson and q-Jacobi polynomials. *SIAM J. Math. Anal.*, 17:1475–1482.

Ismail, M. E. H. and Wilson, J. A. (1982). Asymptotic and generating relations for the q-Jacobi and ${}_4\varphi_3$ polynomials. *J. Approx. Theory*, 36:43–54.

Koekoek, R. and Swarttouw, R. F. (1994). The Askey-scheme of hypergeometric orthogonal polynomials and its q-analogue. report 94-05, Delft University of Technology.

Kuijlaars, A. B. J. (2003). Riemann-Hilbert analysis for orthogonal polynomials. In Koelink, E. and van Assche, W., editors, *Summer School on Orthogonal Polynomials and Special Functions, Leuven 2002*, Lecture Notes in Mathematics. Springer-Verlag. To appear.

Kuijlaars, A. B. J., McLaughlin, K. T.-R., van Assche, W., and Vanlessen, M. (2003). The Riemann-Hilbert approach to strong asymptotics for orthogonal polynomials on $[-1, 1]$. math.CA/0111252. Preprint.

Ruijsenaars, S. N. M. (2002). Factorized weight functions vs. factorized scattering. *Comm. Math. Phys.*, 228:467–494.

Ruijsenaars, S. N. M. (2003). Relativistic Lamé functions: Completeness vs. polynomial asymptotics. *Indag. Math. N.S.*, 11(3). Special issue dedicated to Tom Koornwinder.

Szegő, G. (1975). *Orthogonal polynomials*, volume XXIII of *Colloquium Publications*. American Mathematical Society, Providence, RI, fourth edition.

van Diejen, J. F. (2003). Asymptotic analysis of (partially) orthogonal polynomials associated with root systems. *Intern. Math. Res. Notices*, (7):387–410.

ABEL-ROTHE TYPE GENERALIZATIONS OF JACOBI'S TRIPLE PRODUCT IDENTITY

Michael Schlosser*
Institut für Mathematik der Universität Wien
Strudlhofgasse 4
A-1090 Wien
Austria
schlosse@ap.univie.ac.at

Abstract Using a simple classical method we derive bilateral series identities from terminating ones. In particular, we show how to deduce Ramanujan's ${}_1\psi_1$ summation from the q-Pfaff-Saalschütz summation. Further, we apply the same method to our previous q-Abel-Rothe summation to obtain, for the first time, Abel-Rothe type generalizations of Jacobi's triple product identity. We also give some results for multiple series.

Keywords: q-series, bilateral series, Jacobi's triple product identity, Ramanujan's ${}_1\psi_1$ summation, q-Rothe summation, q-Abel summation, Macdonald identities, A_r series, $U(n)$ series.

1. Introduction

Jacobi's (Jacobi, 1829) triple product identity,

$$\sum_{k=-\infty}^{\infty} q^{k^2} z^k = \prod_{j=0}^{\infty} \left(1-q^{2j+2}\right)\left(1+zq^{2j+1}\right)\left(1+z^{-1}q^{2j+1}\right), \qquad (1.1)$$

is one of the most famous and useful identities connecting number theory and analysis. Many grand moments in number theory rely on this result, such as the theorems on sums of squares (cf. (Gasper and Rahman, 1990, Sec. 8.11)), the Rogers-Ramanujan identities (cf. (Gasper and Rahman, 1990, Sec. 2.7)), or Euler's pentagonal number theorem (cf. (Bressoud, 1999, p. 51)). In addition to this identity, different extensions of it, including Ramanujan's (Hardy, 1940) ${}_1\psi_1$ summation formula

*The author was supported by an APART grant of the Austrian Academy of Sciences

(see (3.1)) and Bailey's (Bailey, 1936) very-well-poised ${}_6\psi_6$ summation formula, have served as effective tools for studies in number theory, combinatorics, and physics, see (Andrews, 1974).

In this paper, we derive new generalizations of Jacobi's triple product identity, in one variable and also in severable variables. Our new extensions look rather unusual. We classify these to be of "Abel-Rothe type," since they are derived from q-Abel-Rothe summations which we previously found in (Krattenthaler and Schlosser, 1999, Eq. (8.5)) and in (Schlosser, 1999, Th. 6.1). At the moment, we cannot tell if our new identities have interesting combinatorial or number-theoretic applications. Nevertheless, we believe that they are attractive by its own.

Our article is organized as follows. In Section 17.2, we review some basics in q-series. In addition to explaining some standard notation, we also briefly describe a well-known method employed in this article for obtaining a bilateral identity from a unilateral terminating identity, a method already utilized by Cauchy (Cauchy, 1843) in his second proof of Jacobi's triple product identity. In Section 17.3, we apply this classical method to derive Ramanujan's ${}_1\psi_1$ summation from the q-Pfaff-Saalschütz summation. According to our knowledge, this very simple proof of the ${}_1\psi_1$ summation has not been given explicitly before. In Section 17.4, we give two Abel-Rothe type generalizations of Jacobi's triple product identity, see Theorem 4.1 and Corollary 4.2. These are consequences of our q-Abel-Rothe summation from (Krattenthaler and Schlosser, 1999, Eq. (8.5)). In Section 17.5 we give multidimensional generalizations of our Abel-Rothe type identities, associated to the root system A_{r-1} (or equivalently, associated to the unitary group $U(r)$). As a direct consequence, we also give an Abel-Rothe type generalization of the Macdonald identities for the affine root system A_r. Finally, we establish the conditions of convergence of our multiple series in the Appendix.

Acknowledgments

I would like to thank George Gasper and Mourad Ismail for their comments on an earlier version of this article. Further, I am grateful to the anonymous referees for their very detailed comments. In particular, one of the referees asked me to find Abel-Rothe type extensions of the Macdonald identities. I thus included such an extension, see (5.6). Another referee insisted that I need to clarify the arguments I had used to prove the convergence of the multiple series in Theorems 5.2 and 5.3. This eventually lead me to find even slightly more general convergence conditions than I had originally stated (see Remark 0.4).

2. Some basics in q-series

First, we recall some standard notation for *q-series* and *basic hypergeometric series* (Gasper and Rahman, 1990). Let q be a (fixed) complex parameter (called the "base") with $0 < |q| < 1$. Then, for a complex parameter a, we define the *q-shifted factorial* by

$$(a)_\infty \equiv (a;q)_\infty := \prod_{j\geq 0} (1 - aq^j), \tag{2.1}$$

and

$$(a)_k \equiv (a;q)_k := \frac{(a;q)_\infty}{(aq^k;q)_\infty}, \tag{2.2}$$

where k is any integer. Since we work with the same base q throughout this article, we can readily omit writing out the base in the q-shifted factorials (writing $(a)_k$ instead of $(a;q)_k$, etc.) as this does not lead to any confusion. For brevity, we occasionally employ the condensed notation

$$(a_1, \ldots, a_m)_k := (a_1)_k \cdots (a_m)_k,$$

where k is an integer or infinity. Further, we utilize

$${}_s\varphi_{s-1}\left[\begin{matrix} a_1, a_2, \ldots, a_s \\ b_1, b_2, \ldots, b_{s-1} \end{matrix}; q, z\right] := \sum_{k=0}^{\infty} \frac{(a_1, a_2, \ldots, a_s; q)_k}{(q, b_1, \ldots, b_{s-1}; q)_k} z^k, \tag{2.3}$$

and

$${}_s\psi_s\left[\begin{matrix} a_1, a_2, \ldots, a_s \\ b_1, b_2, \ldots, b_s \end{matrix}; q, z\right] := \sum_{k=-\infty}^{\infty} \frac{(a_1, a_2, \ldots, a_s; q)_k}{(b_1, b_2, \ldots, b_s; q)_k} z^k, \tag{2.4}$$

to denote the *basic hypergeometric* ${}_s\varphi_{s-1}$ *series*, and the *bilateral basic hypergeometric* ${}_s\psi_s$ *series*, respectively.

A standard reference for basic hypergeometric series is Gasper and Rahman's text (Gasper and Rahman, 1990). Throughout this article, in our computations we make decent use of some elementary identities for q-shifted factorials, listed in (Gasper and Rahman, 1990, Appendix I).

We now turn our attention to identities. One of the simplest summations for basic hypergeometric series is the terminating q-binomial theorem,

$${}_1\varphi_0\left[\begin{matrix} q^{-n} \\ - \end{matrix}; q, z\right] = (zq^{-n})_n. \tag{2.5}$$

This can also be written (with $z \mapsto zq^n$) as

$$\sum_{k=0}^{n} \begin{bmatrix} n \\ k \end{bmatrix}_q (-1)^k q^{\binom{k}{2}} z^k = (z)_n, \tag{2.6}$$

where

$$\begin{bmatrix} n \\ k \end{bmatrix}_q := \frac{(q;q)_n}{(q;q)_k (q;q)_{n-k}} \tag{2.7}$$

denotes the *q-binomial coefficient.*

Cauchy's (Cauchy, 1843) second proof of Jacobi's triple product identity is very elegant and actually constitutes a useful method for obtaining bilateral series identities in general. It is worth looking closely at his proof: First he replaced in (2.6) n by $2n$ and then shifted the summation index $k \mapsto k+n$, which leads to

$$(z)_{2n} = \sum_{k=-n}^{n} \begin{bmatrix} 2n \\ n+k \end{bmatrix}_q (-1)^{n+k} q^{\binom{n+k}{2}} z^{n+k}. \tag{2.8}$$

Next, he replaced z by zq^{-n} and obtained after some elementary manipulations

$$(z, q/z)_n = \sum_{k=-n}^{n} \begin{bmatrix} 2n \\ n+k \end{bmatrix}_q (-1)^k q^{\binom{k}{2}} z^k. \tag{2.9}$$

Finally, after letting $n \to \infty$ he obtained

$$\sum_{k=-\infty}^{\infty} (-1)^k q^{\binom{k}{2}} z^k = (q, z, q/z)_\infty, \tag{2.10}$$

which is an equivalent form of Jacobi's triple product identity (1.1).

3. Ramanujan's ${}_1\psi_1$ summation

Hardy (Hardy, 1940, Eq. (12.12.2)) describes Ramanujan's ${}_1\psi_1$ summation (cf. (Gasper and Rahman, 1990, Appendix (II.29))),

$${}_1\psi_1 \begin{bmatrix} a \\ b \end{bmatrix}; q, z \end{bmatrix} = \frac{(q, b/a, az, q/az)_\infty}{(b, q/a, z, b/az)_\infty}, \tag{3.1}$$

where $|b/a| < |z| < 1$, as a "remarkable formula with many parameters." On the one hand, it bilaterally extends the nonterminating q-binomial theorem (which is the $b = q$ special case of (3.1)), on the other hand it also contains Jacobi's triple product identity as a special case. Namely, if in (3.1) we replace z by z/a, and then let $a \to \infty$ and $b \to 0$, we immediately obtain (2.10). Another important special case of (3.1) is obtained when $b = aq$, which is a bilateral q-series summation due to Kronecker, see Weil (Weil, 1976, pp. 70–71).

Ramanujan (who very rarely gave any proofs) did not provide a proof for the above summation formula. It is interesting that Bailey's (Bailey,

1936, Eq. (4.7)) very-well-poised ${}_6\psi_6$ summation formula, although it contains more parameters than Ramanujan's ${}_1\psi_1$ summation, does *not* include the latter as a special case. Hahn (Hahn, 1949, $\kappa = 0$ in Eq. (4.7)) independently established (3.1) by considering a first order homogeneous q-difference equation. Hahn thus published the first proof of the ${}_1\psi_1$ summation. Not much later, M. Jackson (Jackson, 1950, Sec. 4) gave the first elementary proof of (3.1). Her proof derives the ${}_1\psi_1$ summation from the q-Gauß summation, by manipulation of series. A simple and elegant proof of the ${}_1\psi_1$ summation formula was given by Ismail (Ismail, 1977) who showed that the ${}_1\psi_1$ summation is an immediate consequence of the q-binomial theorem and analytic continuation.

We provide yet another simple proof of the ${}_1\psi_1$ summation formula (which seems to have been unnoticed so far) by deriving it from the terminating q-Pfaff-Saalschütz summation (cf. (Gasper and Rahman, 1990, Eq. (II.12))),

$$
{}_3\varphi_2\left[\begin{matrix} a, b, q^{-n} \\ c, abq^{1-n}/c \end{matrix}; q, q\right] = \frac{(c/a, c/b)_n}{(c, c/ab)_n}. \tag{3.2}
$$

First, in (3.2) we replace n by $2n$ and then shift the summation index by n such that the new sum runs from $-n$ to n:

$$
\begin{aligned}
\frac{(c/a, c/b)_{2n}}{(c, c/ab)_{2n}} &= \sum_{k=0}^{2n} \frac{(a, b, q^{-2n})_k}{(q, c, abq^{1-2n}/c)_k} q^k \\
&= \frac{(a, b, q^{-2n})_n}{(q, c, abq^{1-2n}/c)_n} q^n \sum_{k=-n}^{n} \frac{(aq^n, bq^n, q^{-n})_k}{(q^{1+n}, cq^n, abq^{1-n}/c)_k} q^k.
\end{aligned}
$$

Next, we replace a by aq^{-n}, and we replace c by cq^{-n}.

$$
\begin{aligned}
&\sum_{k=-n}^{n} \frac{(a, bq^n, q^{-n})_k}{(q^{1+n}, c, abq^{1-n}/c)_k} q^k \\
&= \frac{(c/a, cq^{-n}/b)_{2n}\,(q, cq^{-n}, abq^{1-2n}/c)_n}{(cq^{-n}, c/ab)_{2n}\,(aq^{-n}, b, q^{-2n})_n} q^{-n} = \frac{(c/a)_{2n}(c/b, bq/c, q, q)_n}{(q)_{2n}(c, q/a, b, c/ab)_n}.
\end{aligned}
$$

Now, we may let $n \to \infty$ (assuming $|c/ab| < 1$ and $|b| < 1$) while appealing to Tannery's theorem (Bromwich, 1949) for being allowed to interchange limit and summation. This gives

$$
\sum_{k=-\infty}^{\infty} \frac{(a)_k}{(c)_k} \left(\frac{c}{ab}\right)^k = \frac{(c/a, c/b, bq/c, q)_\infty}{(c, q/a, b, c/ab)_\infty},
$$

where $|c/a| < |c/ab| < 1$. Finally, replacing b by c/az and then c by b gives (3.1).

Remark 3.1. *The elementary method we use in the above derivation (exactly the same method already utilized by Cauchy) has also been exploited by Bailey (Bailey, 1936, Secs. 3 and 6), (Bailey, 1950) (see also Slater (Slater, 1966, Sec. 6.2)). For instance, in (Bailey, 1950) Bailey applies the method to Watson's transformation formula of a terminating very-well-poised* $_8\varphi_7$ *into a multiple of a balanced* $_4\varphi_3$ *(Gasper and Rahman, 1990, Eq. (III.18)). As a result, he obtains a transformation for a* $_2\psi_2$ *series, see also Gasper and Rahman (Gasper and Rahman, 1990, Ex. 5.11).*

Remark 3.2. *We conjecture that any bilateral sum can be obtained from an appropriately chosen terminating identity by the above method (as a limit, without using analytic continuation). However, it is already not known whether Bailey's (Bailey, 1936, Eq. (4.7))* $_6\psi_6$ *summation formula (cf. (Gasper and Rahman, 1990, Eq. (II.33))) follows from such an identity.*

4. Abel-Rothe type generalizations of Jacobi's triple product identity

We apply the method of bilateralization[1] we just utilized now to the following q-Abel-Rothe summation (Krattenthaler and Schlosser, 1999, Eq. (8.5)),

$$\begin{aligned}(c)_n = \sum_{k=0}^{n} \begin{bmatrix} n \\ k \end{bmatrix}_q (1-a-b)\left(aq^{1-k}+bq\right)_{k-1} \\ \times \left(c\left(a+bq^k\right)\right)_{n-k} (-1)^k q^{\binom{k}{2}} c^k.\end{aligned} \tag{4.1}$$

This summation is different from the q-Rothe summation found by Johnson (Johnson, 1996, Th. 4) which he derived by means of umbral calculus. It is also different from Jackson's (Jackson, 1910) q-Abel summation. Our summation in (4.1) was originally derived in (Krattenthaler and Schlosser, 1999) by extracting coefficients of a nonterminating q-Abel-Rothe type expansion formula (actually, the $n \to \infty$ case of (4.1)), which in turn was derived by inverse relations. However, it can also be derived directly by inverse relations (by combining the q-Chu-Vandermonde summation with a specific non-hypergeometric matrix inverse), see (Schlosser, 1999, Sec. 6).

In (Krattenthaler and Schlosser, 1999), (Schlosser, 1999) and (Schlosser, 2000), we referred to (4.1) as a q-Rothe summation to distinguish it from the q-Abel summation

$$1 = \sum_{k=0}^{n} \begin{bmatrix} n \\ k \end{bmatrix}_q (a+b)\left(a+bq^k\right)^{k-1}\left(a+bq^k\right)_{n-k}, \qquad (4.2)$$

that we derived in (Krattenthaler and Schlosser, 1999, Eq. (8.1)). It appears that (4.2) is different from any of the q-Abel summations from Jackson (Jackson, 1910) or Johnson (Johnson, 1996). However, it is equivalent to Bhatnagar and Milne's (Bhatnagar and Milne, 1997) q-Abel summation (by reversing the sum). Above we decided to call (4.1) a q-Abel-Rothe summation since it is also contains (4.2) as a special case. In fact, if in (4.1) we replace a and b by a/c and b/c, and then let $c \to 0$, we obtain after some algebra (4.2).

Our q-Abel-Rothe summation in (4.1) is indeed a q-extension of the Rothe summation: If we divide both sides by $(q)_n$, do the replacements $a \mapsto q^A - B$, $b \mapsto B$, $c \mapsto q^{-A-C}$, and then let $q \to 1$, we obtain Rothe's (Rothe, 1793) summation formula

$$\binom{A+C}{n} = \sum_{k=0}^{n} \frac{A}{A+Bk} \binom{A+Bk}{k} \binom{C-Bk}{n-k}. \qquad (4.3)$$

Rothe's identity is an elegant generalization of the well-known Chu-Vandermonde convolution formula, to which it reduces for $B=0$. Similarly, (4.1) reduces for $b=0$ to the q-Chu-Vandermonde summation listed in Appendix II, Eq. (II.6) of (Gasper and Rahman, 1990), and for $a=0$ to the q-Chu-Vandermonde summation in Appendix II, Eq. (II.7) of (Gasper and Rahman, 1990).

Furthermore, Eq. (4.2) (and thus also the more general (4.1)) is indeed a q-extension of the Abel summation: If in (4.2) we replace a and b, by $\frac{A}{(A+C)} + \frac{B}{(A+C)(1-q)}$ and $\frac{-B}{(A+C)(1-q)}$, respectively, and then let $q \to 1$, we obtain Abel's generalization (Abel, 1826) of the q-binomial theorem,

$$(A+C)^n = \sum_{k=0}^{n} \binom{n}{k} A(A+Bk)^{k-1}(C-Bk)^{n-k}. \qquad (4.4)$$

Some historical details concerning the Abel and Rothe summations can be found in Gould (Gould, 1956; Gould, 1957), and in Strehl (Strehl, 1992).

We now present our main result, an Abel-Rothe type generalization of (2.10):

Theorem 4.1. *Let a, b, and z be indeterminate. Then*

$$\frac{(q,z,q/z)_\infty}{(1-b)} = \sum_{k=-\infty}^{\infty} \left(aq^{1-k}+bq\right)_\infty \left(z\left(a+bq^k\right)\right)_\infty (-1)^k q^{\binom{k}{2}} z^k, \tag{4.5}$$

provided $\max(|az|,|b|)<1$.

Proof. In (4.1), we first replace n by $2n$, and then shift the summation index $k \mapsto k+n$. This gives

$$(c)_{2n} = \sum_{k=-n}^{n} \begin{bmatrix} 2n \\ n+k \end{bmatrix}_q (1-a-b)\left(aq^{1-n-k}+bq\right)_{k+n-1} \times \left(c\left(a+bq^{n+k}\right)\right)_{n-k} (-1)^{n+k} q^{\binom{n+k}{2}} c^{n+k}.$$

Next, we replace a and c by aq^n and cq^{-n}. After some elementary manipulations we obtain

$$\frac{(c,q/c)_n}{(1-aq^n-b)} = \sum_{k=-n}^{n} \begin{bmatrix} 2n \\ n+k \end{bmatrix}_q \left(aq^{1-k}+bq\right)_{k+n-1} \times \left(c\left(a+bq^k\right)\right)_{n-k} (-1)^k q^{\binom{k}{2}} c^k.$$

Finally, replacing c by z and (assuming $|az|,|b|<1$) letting $n\to\infty$ while appealing to Tannery's theorem, we formally arrive at (4.5). However, it remains to establish the conditions of convergence.

Since

$$\left(aq^{1-k}+bq\right)_\infty = (-1)^k q^{-\binom{k}{2}} (a+bq^k)^k \left(1/\left(a+bq^k\right)\right)_k \left(aq+bq^{1+k}\right)_\infty,$$

it is easy to find that if $|az|<1$ then the positive part of the sum, i.e., $\sum_{k\ge 0}$, converges. Similarly, for the negative part of the sum, i.e., $\sum_{k<0}$, we use

$$\left(z\left(a+bq^k\right)\right)_\infty = (-1)^k q^{-\binom{k}{2}} z^{-k} \left(aq^{-k}+b\right)^{-k} \times \left(q/z\left(aq^{-k}+b\right)\right)_{-k} \left(z\left(aq^{-k}+b\right)\right)_\infty,$$

and determine that we need $|b|<1$ for absolute convergence. □

By reversing the sum in (4.5), we easily deduce the following:

Corollary 4.2. *Let a, b, and z be indeterminate. Then*

$$\frac{(q, zq, 1/z)_\infty}{(1-az)} = \sum_{k=-\infty}^{\infty} \left(aq^{-k}+b\right)_\infty \left(zq\left(a+bq^k\right)\right)_\infty (-1)^k q^{\binom{k+1}{2}} z^k, \tag{4.6}$$

provided $\max(|az|, |b|) < 1$.

Proof. In (4.5), we first replace k by $-k$, and then simultaneously replace a, b and z, by bz, az and $1/z$, respectively. □

We have given analytical convergence conditions for the identities (4.5) and (4.6). However, we would like to point out that these identities also hold when regarded as identities for *formal power series* over q.

It is obvious that both Theorem 4.1 and Corollary 4.2 reduce to Jacobi's triple product identity (2.10) when $a = b = 0$.

If we extract coefficients of z^n on both sides of (4.5), using (2.10) on the left and

$$(x)_\infty = \sum_{j\geq 0} \frac{(-1)^j q^{\binom{j}{2}}}{(q)_j} x^j \tag{4.7}$$

(cf. (Gasper and Rahman, 1990, Eq. (II.2))) on the right hand side, divide both sides by $(-1)^n q^{\binom{n}{2}}$ and replace a by aq^{-n}, we obtain

$$\frac{1}{1-b} = \sum_{j=0}^{\infty} \frac{\left(aq^j+b\right)^j}{(q)_j} \left(aq^{1+j}+bq\right)_\infty \tag{4.8}$$

(which is valid for $|b| < 1$), which is, modulo substitution of variables, our q-Abel-type expansion in (Schlosser, 1999, Eq. (3.4)). For a multivariable extension of (4.8), see (5.5). If we now replace a and b by $-BZ$ and $(1-q^A+B)Z$ and then let $q \to 1^-$, while using $\lim_{q\to 1^-}((1-q)Z)_\infty = e^{-Z}$, we recover Lambert's (Lambert, 1758) formula

$$\frac{e^{AZ}}{1-BZ} = \sum_{j=0}^{\infty} \frac{(A+Bj)^j}{j!} Z^j e^{-BZj}, \tag{4.9}$$

which is valid for $\left|BZe^{1-BZ}\right| < 1$. Note that in (4.9), Z is a redundant parameter. However, the advantage of writing (4.9) in this form is that here we have an identity of power series in the variable Z (having in mind the expansion of the geometric series and of the exponential function).

In (Krattenthaler and Schlosser, 1999), (Schlosser, 1999), and (Schlosser, 2000) we erroneously attributed (4.9) and some related expansions to Euler (Euler, 1779), but which are actually due to Lambert (Lambert,

1758). Nevertheless, Euler's article on Lambert's identities is significant and is often cited in the literature as sole reference for these identities (see e.g., Pólya and Szegő (Pólya and Szegő, 1925, pp. 301–302)).

5. Multidimensional generalizations

Here we extend Theorem 4.1 and Corollary 4.2 to multiple series associated to the root systems of type A, or equivalently, associated to the unitary groups. Multiple series, associated to root systems, or to Lie groups, have been investigated by various authors. Many different types of such series exist in the literature. For some results on the special type of series that are considered in this section, see, e.g., (Bhatnagar and Milne, 1997), (Bhatnagar and Schlosser, 1998), (Milne, 1997), (Milne and Schlosser, 2002), (Rosengren, 2003), (Schlosser, 1997), (Schlosser, 1999), and (Schlosser, 2000).

In the following, we consider r-dimensional series, where r is a positive integer. For brevity, we employ the notation $|\mathbf{k}| := k_1 + \cdots + k_r$.

If we apply the method of bilateralization to the multidimensional q-Abel-Rothe summations that were derived in (Krattenthaler and Schlosser, 1999) (see Theorems 8.2 and 8.3 therein), the multiple q-Abel-Rothe summations in Theorems 6.7 and 6.9 of (Schlosser, 1999), or Theorems 3.7 and 3.8 of (Schlosser, 2000), the resulting series do not converge for higher dimensions. The only multidimensional q-Abel-Rothe summations we are aware of that converge when bilateralized are Theorem 6.11 of (Schlosser, 1999), and (the slightly more general) Theorem 3.9 of (Schlosser, 2000). Both these theorems were derived by applying multidimensional inverse relations, in particular by combining different higher-dimensional q-Chu-Vandermonde summations with specific multidimensional non-hypergeometric matrix inverses.

For the sake of simplicity, we consider here only the multilateral identities arising from Theorem 6.11 of (Schlosser, 1999), a multiple q-Abel-Rothe summation associated to the root system A_{r-1}:

Theorem 5.1. *Let a, b, c, and $x_1, \dots, x_r$ be indeterminate, and let $n_1, \dots, n_r$ be nonnegative integers. Then there holds*

$$(c)_{|\mathbf{n}|} = \sum_{\substack{0 \le k_i \le n_i \\ i=1,\dots,r}} \left(\prod_{i,j=1}^{r} \left[\frac{\left(\frac{x_i}{x_j}q\right)_{n_i}}{\left(\frac{x_i}{x_j}q\right)_{k_i} \left(\frac{x_i}{x_j}q^{1+k_i-k_j}\right)_{n_i-k_i}} \right] \right.$$
$$\left. \times (1-a-b)\left(aq^{1-|\mathbf{k}|}+bq\right)_{|\mathbf{k}|-1} \left(c(a+bq^{|\mathbf{k}|})\right)_{|\mathbf{n}|-|\mathbf{k}|} (-1)^{|\mathbf{k}|} q^{\binom{|\mathbf{k}|}{2}} c^{|\mathbf{k}|}. \right). \tag{5.1}$$

Our multilateral extension of (4.5) is as follows:

Theorem 5.2. *Let a, b, z, and $x_1, \dots, x_r$ be indeterminate. Then*

$$\frac{(z, q/z)_\infty}{(1-b)} \prod_{i,j=1}^{r} \left(\frac{x_i}{x_j}q\right)_\infty$$
$$= \sum_{k_1,\dots,k_r=-\infty}^{\infty} \left(\prod_{1 \le i < j \le r} \left(\frac{x_i q^{k_i} - x_j q^{k_j}}{x_i - x_j} \right) \left(aq^{1-|\mathbf{k}|}+bq\right)_\infty \right.$$
$$\left. \times \left(z(a+bq^{|\mathbf{k}|})\right)_\infty (-1)^{r|\mathbf{k}|} q^{r\sum_{i=1}^{r}\binom{k_i}{2}} z^{|\mathbf{k}|} \prod_{i=1}^{r} x_i^{rk_i-|\mathbf{k}|} \right), \tag{5.2}$$

provided $\max(|az|, |b|) < 1$.

Proof. The proof is very similar to the one-dimensional case. In (5.1), we replace n_i by $2n_i$, for $i = 1, \dots, r$, and then shift all the summation indices $k_i \mapsto k_i + n_i$. This gives

$$(c)_{2|\mathbf{n}|} = \sum_{\substack{-n_i \le k_i \le n_i \\ i=1,\dots,r}} \left(\prod_{i,j=1}^{r} \left[\frac{\left(\frac{x_i}{x_j}q\right)_{2n_i}}{\left(\frac{x_i}{x_j}q\right)_{n_i+k_i} \left(\frac{x_i}{x_j}q^{1+n_i-n_j+k_i-k_j}\right)_{n_i-k_i}} \right] \right.$$
$$\times (1-a-b)\left(aq^{1-|\mathbf{n}|-|\mathbf{k}|}+bq\right)_{|\mathbf{n}|+|\mathbf{k}|-1}$$
$$\left. \times \left(c(a+bq^{|\mathbf{n}|+|\mathbf{k}|})\right)_{|\mathbf{n}|-|\mathbf{k}|} (-1)^{|\mathbf{n}|+|\mathbf{k}|} q^{\binom{|\mathbf{n}|+|\mathbf{k}|}{2}} c^{|\mathbf{n}|+|\mathbf{k}|} \right).$$

We replace a, c and x_i, by $aq^{|\mathbf{n}|}$, $cq^{-|\mathbf{n}|}$ and $x_iq^{-n_i}$, $i = 1, \dots, r$, respectively. After some elementary manipulations we obtain

$$\frac{(c, q/c)_{|\mathbf{n}|}}{(1 - aq^{|\mathbf{n}|} - b)} = \sum_{\substack{-n_i \le k_i \le n_i \\ i=1,\dots,r}} \left(\prod_{i,j=1}^{r} \left[\frac{\left(\frac{x_i}{x_j} q\right)_{n_i+n_j}}{\left(\frac{x_i}{x_j} q\right)_{n_j+k_i} \left(\frac{x_i}{x_j} q^{1+k_i-k_j}\right)_{n_i-k_i}} \right] \right.$$
$$\left. \times \left(aq^{1-|\mathbf{k}|} + bq\right)_{|\mathbf{n}|+|\mathbf{k}|-1} \left(c(a + bq^{|\mathbf{k}|})\right)_{|\mathbf{n}|-|\mathbf{k}|} (-1)^{|\mathbf{k}|} q^{\binom{|\mathbf{k}|}{2}} c^{|\mathbf{k}|} \right).$$

Next, we replace c by z and let $n_i \to \infty$, for $i = 1, \dots, r$ (assuming $|az| < 1$ and $|b| < 1$), while appealing to Tannery's theorem. Finally, we apply the simple identity

$$\prod_{i,j=1}^{r} \left(\frac{x_i}{x_j} q\right)_{k_i-k_j} = \prod_{1 \le i < j \le r} \left(\frac{x_i}{x_j} q\right)_{k_i-k_j} \left(\frac{x_j}{x_i} q\right)_{k_j-k_i}$$
$$= (-1)^{(r-1)|\mathbf{k}|} q^{-\binom{|\mathbf{k}|}{2} + r \sum_{i=1}^{r} \binom{k_i}{2}} \prod_{i=1}^{r} x_i^{rk_i - |\mathbf{k}|} \prod_{1 \le i < j \le r} \left(\frac{x_i q^{k_i} - x_j q^{k_j}}{x_i - x_j} \right) \tag{5.3}$$

and arrive at (5.2). For establishing the conditions of convergence of the series, see the Appendix. □

Next, we provide the following multilateral generalization of (4.6).

Theorem 5.3. *Let a, b, z, and $x_1, \dots, x_r$ be indeterminate. Then*

$$\frac{(zq, 1/z)_\infty}{(1 - az)} \prod_{i,j=1}^{r} \left(\frac{x_i}{x_j} q\right)_\infty$$
$$= \sum_{k_1,\dots,k_r=-\infty}^{\infty} \left(\prod_{1 \le i < j \le r} \left(\frac{x_i q^{k_i} - x_j q^{k_j}}{x_i - x_j} \right) \left(aq^{-|\mathbf{k}|} + b\right)_\infty \right.$$
$$\left. \times \left(zq(a + bq^{|\mathbf{k}|})\right)_\infty (-1)^{r|\mathbf{k}|} q^{|\mathbf{k}| + r \sum_{i=1}^{r} \binom{k_i}{2}} z^{|\mathbf{k}|} \prod_{i=1}^{r} x_i^{rk_i - |\mathbf{k}|} \right), \tag{5.4}$$

provided $\max(|az|, |b|) < 1$.

Proof. In (5.2), we first replace k_i by $-k_i$, for $i = 1, \dots, r$, and then simultaneously replace a, b, z and x_i, by bz, az, $1/z$ and $1/x_i$, for $i = 1, \dots, r$, respectively. □

We have given analytical convergence conditions for the identities (5.2) and (5.4). However, as we already observed in Section 17.4 when dealing with the respective one variable cases, these identities also hold when regarded as identities for *formal power series* over q.

We complete this section with an Abel-Rothe type generalization of the Macdonald identities for the affine root system A_r, as a direct consequence of Theorem 5.2.

If we multiply both sides of (5.2) by $\prod_{1\le i<j\le r}(1-x_i/x_j)$, extract the coefficient of z^M, using the Jacobi triple product identity (2.10) on the left hand side and (4.7) on the right hand side, and divide the resulting identity by $(-1)^M q^{\binom{M}{2}}$, we obtain

$$\frac{1}{1-b}(q)_\infty^{r-1}\prod_{1\le i<j\le r}\left(\frac{x_i}{x_j},\frac{x_j}{x_i}q\right)_\infty$$
$$=\sum_{\substack{k_1,\dots,k_r=-\infty\\|\mathbf{k}|\le M}}^{\infty}\left(\prod_{1\le i<j\le r}\left(1-\frac{x_i}{x_j}q^{k_i-k_j}\right)\left(aq^{1-|\mathbf{k}|}+bq\right)_\infty\frac{\left(a+bq^{|\mathbf{k}|}\right)^{M-|\mathbf{k}|}}{(q)_{M-|\mathbf{k}|}}\right.$$
$$\left.\times(-1)^{(r-1)|\mathbf{k}|}q^{-M|\mathbf{k}|+\binom{|\mathbf{k}|+1}{2}+\sum_{i=1}^{r}r\binom{k_i}{2}+(i-1)k_i}\prod_{i=1}^{r}x_i^{rk_i-|\mathbf{k}|}\right).\quad(5.5)$$

We use the Vandermonde determinant expansion (0.2) and a little bit of algebra and obtain

$$\frac{1}{1-b}(q)_\infty^{r-1}\prod_{1\le i<j\le r}\left(\frac{x_i}{x_j},\frac{x_j}{x_i}q\right)_\infty$$
$$=\sum_{\sigma\in\mathcal{S}_r}\operatorname{sgn}(\sigma)\prod_{i=1}^{r}x_i^{\sigma(i)-i}\sum_{\substack{k_1,\dots,k_r=-\infty\\|\mathbf{k}|\le M}}^{\infty}\left(\left(aq^{1-|\mathbf{k}|}+bq\right)_\infty\frac{\left(a+bq^{|\mathbf{k}|}\right)^{M-|\mathbf{k}|}}{(q)_{M-|\mathbf{k}|}}\right.$$
$$\left.\times(-1)^{(r-1)|\mathbf{k}|}q^{-M|\mathbf{k}|+\binom{|\mathbf{k}|+1}{2}+\sum_{i=1}^{r}r\binom{k_i}{2}+(\sigma(i)-1)k_i}\prod_{i=1}^{r}x_i^{rk_i-|\mathbf{k}|}\right).\quad(5.6)$$

If $a=0$ and $b=0$, the terms of the sum in (5.6) are zero unless $|\mathbf{k}|=M$. Specializing this further by setting $M=0$ gives an identity which has been shown to be equivalent to the Macdonald identities for the affine root system A_r, see Milne (Milne, 1985). Regarding this, we may consider the identity (5.6) as an Abel–Rothe type generalization of the Macdonald identities for the affine root system A_r.

Concluding, we want to point out that we could have given an even more general multidimensional Abel-Rothe type generalization of Jacobi's triple product identity than Theorem 5.2, by multilateralizing Theorem 3.9 of (Schlosser, 2000) instead of Theorem 6.11 of (Schlosser, 1999) as above. However, we feel that, because of the more complicated factors being involved, the result would be not as elegant as Theorem 5.2 which is sufficiently illustrative. We therefore decided to refrain from giving this more general identity.

Appendix: Convergence of multiple series

Here we prove the conditions of convergence of our multiple series identity in Theorem 5.2 (and thus also of Theorem 5.3).

We determine the condition for absolute convergence of the multilateral series in (5.2) by splitting the entire sum $\sum_{k_1,\dots,k_r=-\infty}^{\infty}$ into two sums, $\sum_{|\mathbf{k}|\geq 0}$ and $\sum_{|\mathbf{k}|<0}$, and show the absolute convergence for each of these separately.

We first consider the sum $\sum_{|\mathbf{k}|\geq 0}$. We obtain that for $|\mathbf{k}| \geq 0$ the sum in (5.2) converges absolutely provided

$$\sum_{\substack{k_1,\dots,k_r=-\infty\\|\mathbf{k}|\geq 0}}^{\infty} \Bigg| \left(a+bq^{|\mathbf{k}|}\right)^{|\mathbf{k}|} z^{|\mathbf{k}|} q^{-\binom{|\mathbf{k}|}{2}+r\sum_{i=1}^{r}\binom{k_i}{2}} \\ \times \prod_{i=1}^{r} x_i^{rk_i-|\mathbf{k}|} \prod_{1\leq i<j\leq r} \left(x_i q^{k_i} - x_j q^{k_j}\right) \Bigg| < \infty. \tag{0.1}$$

We use the Vandermonde determinant expansion

$$\prod_{1\leq i<j\leq r} \left(x_i q^{k_i} - x_j q^{k_j}\right) = \sum_{\sigma\in\mathcal{S}_r} \operatorname{sgn}(\sigma) \prod_{i=1}^{r} x_i^{r-\sigma(i)} q^{(r-\sigma(i))k_i}, \tag{0.2}$$

where $\mathcal{S}_r$ denotes the symmetric group of order r, interchange summations in (0.1) and obtain $r!$ multiple sums each corresponding to a permutation $\sigma \in \mathcal{S}_r$. Thus for $|\mathbf{k}| \geq 0$ the series in (5.2) converges provided

$$\sum_{\substack{k_1,\dots,k_r=-\infty\\|\mathbf{k}|\geq 0}}^{\infty} \Bigg| \left(a+bq^{|\mathbf{k}|}\right)^{|\mathbf{k}|} z^{|\mathbf{k}|} q^{-\binom{|\mathbf{k}|}{2}+r\sum_{i=1}^{r}\binom{k_i}{2}} \prod_{i=1}^{r} q^{(r-\sigma(i))k_i} x_i^{rk_i-|\mathbf{k}|} \Bigg| < \infty, \tag{0.3}$$

for any $\sigma \in \mathcal{S}_r$.

The next step is crucial and typically applies in the theory of multidimensional basic hypergeometric series over the root system A_{r-1} for a class of series. (For instance, it applies to several of the multilateral summations and transformations in (Milne and Schlosser, 2002).) In the summand of (0.3), we have

$$-\binom{|\mathbf{k}|}{2} + r\sum_{i=1}^{r}\binom{k_i}{2} = -\frac{(r-1)}{2}|\mathbf{k}| + \frac{1}{2}\sum_{1\leq i<j\leq r}(k_i-k_j)^2 \tag{0.4}$$

appearing in the exponent of q. Since $|\mathbf{k}| \geq 0$, we can assume, without loss of generality, that in particular $k_r \geq 0$. (At least one summation index is nonnegative—choose it to be the r-th, by relabelling, if necessary.)

In order to exploit the quadratic powers of q in the sum (which contribute to convergence), we make the substitutions

$$k_i \mapsto \sum_{i \leq l \leq r} m_l, \qquad \text{for } i = 1, \ldots, r.$$

Under these substitutions $|\mathbf{k}|$ becomes $\sum_{l=1}^{r} l m_l$, while for $i < j$, $k_i - k_j$ becomes $\sum_{i \leq l < j} m_l$. We now use (0.4) and replace $q^{-\binom{|\mathbf{k}|}{2} + r \sum_{i=1}^{r} \binom{k_i}{2}}$ by

$$q^{-\frac{(r-1)}{2} \sum_{l=1}^{r} l m_l + \frac{1}{2} \sum_{l=1}^{r-1} m_l^2}$$

(we left out some quadratic powers), for comparison with a dominating multiple series, and obtain that for $|\mathbf{k}| \geq 0$ and $k_r \geq 0$ the series in (5.2) converges provided

$$\sum_{\substack{m_1, \ldots, m_r = -\infty \\ m_r, \sum_{l=1}^{r} l m_l \geq 0}}^{\infty} \left| \left(a + b q^{\sum_{l=1}^{r} l m_l}\right)^{\sum_{l=1}^{r} l m_l} z^{\sum_{l=1}^{r} l m_l} q^{-\frac{(r-1)}{2} \sum_{l=1}^{r} l m_l + \frac{1}{2} \sum_{l=1}^{r-1} m_l^2} \right.$$
$$\left. \times \prod_{i=1}^{r} q^{(r-\sigma(i)) \sum_{i \leq l \leq r} m_l} x_i^{r \sum_{i \leq l \leq r} m_l - \sum_{1 \leq l \leq r} l m_l} \right| < \infty,$$

for any $\sigma \in \mathcal{S}_r$. The above series converges absolutely if

$$\sum_{\substack{m_1, \ldots, m_r = -\infty \\ m_r, \sum_{l=1}^{r} l m_l \geq 0}}^{\infty} \left| \left(a z q^{-\frac{(r-1)}{2}} \prod_{i=1}^{r} x_i^{-1}\right)^{\sum_{l=1}^{r} l m_l} q^{\frac{1}{2} \sum_{l=1}^{r-1} m_l^2} \right.$$
$$\left. \times \prod_{i=1}^{r} q^{(r-\sigma(i)) \sum_{i \leq l \leq r} m_l} x_i^{r \sum_{i \leq l \leq r} m_l} \right| < \infty. \tag{0.5}$$

Now, the series in (0.5) is dominated by

$$\prod_{l=1}^{r-1} \sum_{m_l = -\infty}^{\infty} \left| \left(a z q^{-\frac{r-1}{2}} \prod_{i=1}^{r} x_i^{-1}\right)^{l m_l} q^{\frac{1}{2} m_l^2} q^{\sum_{1 \leq i \leq l} (r-\sigma(i)) m_l} \prod_{1 \leq i \leq l} x_i^{r m_l} \right|$$
$$\times \sum_{m_r = 0}^{\infty} \left| \left(a z q^{-\frac{r-1}{2}} \prod_{i=1}^{r} x_i^{-1}\right)^{r m_r} q^{\sum_{1 \leq i \leq r} (r-\sigma(i)) m_r} \prod_{1 \leq i \leq r} x_i^{r m_r} \right|. \tag{0.6}$$

We deduce by d'Alembert's ratio test that the product of the first $r-1$ series converge everywhere due to the quadratic powers of q (since $|q| < 1$). Further, by the same test, the r-th series converges whenever

$$\left| \left(a z q^{-\frac{r-1}{2}} \prod_{i=1}^{r} x_i^{-1}\right)^r q^{\sum_{1 \leq i \leq r} (r-\sigma(i))} \prod_{i=1}^{r} x_i^r \right| = |az|^r < 1,$$

or equivalently, whenever $|az| < 1$.

The absolute convergence of the sum $\sum_{|\mathbf{k}|<0}$ is established in a similar manner. In this case we obtain that for $|\mathbf{k}| < 0$ the sum in (5.2) converges absolutely provided

$$\sum_{\substack{k_1,\dots,k_r=-\infty \\ |\mathbf{k}|<0}}^{\infty} \Bigg| \left(aq^{-|\mathbf{k}|}+b\right)^{-|\mathbf{k}|} q^{-\binom{|\mathbf{k}|}{2}+r\sum_{i=1}^{r}\binom{k_i}{2}} \times \prod_{i=1}^{r} x_i^{rk_i-|\mathbf{k}|} \prod_{1\le i<j\le r} \left(x_i q^{k_i} - x_j q^{k_j}\right) \Bigg| < \infty. \tag{0.7}$$

The further analysis is as follows. We use (0.2) and (0.4) and assume that for $|\mathbf{k}| < 0$, without loss of generality, $k_r < 0$. In a very similar analysis to above one easily finds the condition $|b| < 1$ for absolute convergence.

Remark 0.4. *In an earlier version of this article we had given a smaller region of convergence for the series in* (5.2). *In particular, instead of*

$$\max(|az|, |b|) < 1$$

we had given the condition

$$|az| < \left|q^{\frac{r-1}{2}} x_j^{-r} \prod_{i=1}^{r} x_i\right| < |q^{r-1}b^{-1}|, \qquad \textit{for } j = 1,\dots,r. \tag{0.8}$$

To see that this gives a smaller region of convergence (for $r > 1$), assume we would have instead $\max(|az|, |b|) \geq 1$. *Now take the "product" of the whole relation* (0.8) *over all $j = 1,\dots,r$. This gives $|az|^r < |q|^{\binom{r}{2}} < |q^{r(r-1)}b^{-r}|$, or equivalently, after taking r-th roots, $|az| < |q|^{\frac{r-1}{2}} < |q^{r-1}b^{-1}|$, which apparently contradicts* $\max(|az|, |b|) \geq 1$ *since $r \geq 1$.*

The same argument which leads to the convergence condition in Theorem 5.2 *(in contrast to the condition* (0.8)*) can be used to improve some analogous results given in the literature. In particular, this concerns the papers (Milne, 1997), (Milne and Schlosser, 2002), (Schlosser, 1999), (Schlosser, 2000) (and possibly others).*

Notes

1. The Merriam-Webster Online dictionary (`http://www.m-w.com/cgi-bin/dictionary`) gives for the entry '**-ize**': "... **1 a** (1): cause to be ... (2): cause to be formed into ... **2 a:** become ... **usage** The suffix *-ize* has been productive in English since the time of Thomas Nashe (1567–1601), who claimed credit for introducing it into English to remedy the surplus of monosyllabic words. Almost any noun or adjective can be made into a verb by adding *-ize* <hospital*ize*> <familiar*ize*>; many technical terms are coined this way <oxid*ize*> as well as verbs of ethnic derivation <American*ize*> and verbs derived from proper names <bowdler*ize*> <mesmer*ize*>. Nashe noted in 1591 that his coinages in *-ize* were being complained about, and to this day new words in *-ize* <final*ize*> <priorit*ize*> are sure to draw critical fire."

References

Abel, N. H. (1826). Beweis eines Ausdrucks, von welchem die Binomial-Formel ein einzelner Fall ist. *J. Reine Angew. Math.*, 1:159–160.

Andrews, G. E. (1974). Applications of basic hypergeometric functions. *SIAM Rev.*, 16:441–484.

Bailey, W. N. (1936). Series of hypergeometric type which are infinite in both directions. *Quart. J. Math. Oxford Series*, 7:105–115.

Bailey, W. N. (1950). On the basic bilateral hypergeometric series ${}_2\psi_2$. *Quart. J. Math. Oxford Series 2*, 1:194–198.

Bhatnagar, G. and Milne, S. C. (1997). Generalized bibasic hypergeometric series and their $u(n)$ extensions. *Adv. Math.*, 131:188–252.

Bhatnagar, G. and Schlosser, M. (1998). C_n and D_n very-well-poised ${}_{10}\varphi_9$ transformations. *Constr. Approx.*, 14:531–567.

Bressoud, D. M. (1999). *Proofs and Confirmations.* Cambridge University Press, Cambridge.

Bromwich, T. J. (1949). *An Introduction to the Theory of Infinite Series.* Macmillan, London, second edition.

Cauchy, A.-L. (1843). Mémoire sur les fonctions dont plusieurs valeurs .... *C.R. Acad. Sci. Paris*, 17:523. Reprinted in Oeuvres de Cauchy, Ser. 1, vol. 8, Gauthier-Villars, Paris (1893), 42–50.

Euler, L. (1779). De serie Lambertiana plurimisque eius insignibus proprietatibus. *Acta Acad. Sci. Petro.* II (1783), 29–51; reprinted in Opera Omnia, Ser. I, vol. 6, Teubner, Leibzig (1921), 350–369.

Gasper, G. and Rahman, M. (1990). *Basic Hypergeometric Series*, volume 35 of *Encyclopedia of Mathematics and its Applications.* Cambridge University Press, Cambridge.

Gould, H. W. (1956). Some generalizations of Vandermonde's convolution. *Amer. Math. Monthly*, 63:84–91.

Gould, H. W. (1957). Final analysis of Vandermonde's convolution. *Amer. Math. Monthly*, 64:409–415.

Hahn, W. (1949). Beiträge zur Theorie der Heineschen Reihen. Die hypergeometrische q-Differenzengleichung. Das q-Analogon der Laplace-Transformation. *Math. Nachr.*, 2:340–379.

Hardy, G. H. (1940). *Ramanujan.* Cambridge University Press, Cambridge. Reprinted by: Chelsea, New York, 1978.

Ismail, M. E. H. (1977). A simple proof of Ramanujan's ${}_1\psi_1$ sum. *Proc. Amer. Math. Soc.*, 63:185–186.

Jackson, F. H. (1910). A q-generalization of Abel's series. *Rend. Circ. Math. Palermo*, 29:340–346.

Jackson, M. (1950). On Lerch's transcendant and the basic bilateral hypergeometric series ${}_2\psi_2$. *J. London Math. Soc.*, 25:189–196.

Jacobi, C. G. J. (1829). *Fundamenta Nova Theoriae Functionum Ellipticarum.* Unknown. Reprinted in Jacobi's Gesammelte Werke, vol. 1, (Reimer, Berlin, 1881–1891), pp. 49–239; reprinted by Chelsea (New York, 1969); now distributed by the Amer. Math. Soc., Providence, RI.

Johnson, W. P. (1996). q-Extensions of identities of Abel-Rothe type. *Discrete Math.*, 159:161–177.

Krattenthaler, C. and Schlosser, M. (1999). A new multidimensional matrix inverse with applications to multiple q-series. *Discrete Math.*, 204:249–279.

Lambert, J. H. (1758). Observationes variae in mathesin puram. *Acta Helvetica*, 3:128–168.

Milne, S. C. (1985). An elementary proof of the Macdonald identities for $A_{\ell^{(1)}}$. *Adv. Math.*, 57:34–70.

Milne, S. C. (1997). Balanced ${}_3\varphi_2$ summation theorems for $U(n)$ basic hypergeometric series. *Adv. Math.*, 131:93–187.

Milne, S. C. and Schlosser, M. (2002). A new A_n extension of Ramanujan's ${}_1\psi_1$ summation with applications to multilateral A_n series. *Rocky Mount. J. Math.*, 32(2):759–792.

Pólya, G. and Szegő, G. (1925). *Aufgaben und Lehrsätze aus der Analysis, Band I.* Julius Springer, Berlin.

Rosengren, H. (2003). Reduction formulae for Karlsson-Minton type hypergeometric functions. *Constr. Approx.* To appear.

Rothe, H. A. (1793). Formulae de serierum reversione demonstratio universalis signis localibus combinatorico-analyticorum vicariis exhibita. Leipzig.

Schlosser, M. (1997). Multidimensional matrix inversions and A_r and D_r basic hypergeometric series. *Ramanujan J.*, 1:243–274.

Schlosser, M. (1999). Some new applications of matrix inversions in A_r. *Ramanujan J.*, 3:405–461.

Schlosser, M. (2000). A new multidimensional matrix inversion in A_r. In *q-series from a contemporary perspective (South Hadley, MA, 1998)*, volume 254 of *Contemp. Math.*, pages 413–432. Amer. Math. Soc., Providence, RI.

Slater, L. J. (1966). *Generalized Hypergeometric Functions.* Cambridge Univ. Press, London/New York.

Strehl, V. (1992). Identities of Rothe-Abel-Schläfli-Hurwitz-type. *Discrete Math.*, 99:321–340.

Weil, A. (1976). *Elliptic functions according to Eisenstein and Kronecker.* Springer-Verlag, Berlin. Ergebnisse der Mathematik und ihrer Grenzgebiete, Band 88.

SUMMABLE SUMS OF HYPERGEOMETRIC SERIES*

D. Stanton
School of Mathematics
University of Minnesota
Minneapolis, MN 55455
stanton@math.umn.edu

Abstract New expansions for certain ${}_2F_1$'s as a sum of r higher order hypergeometric series are given. When specialized to the binomial theorem, these r hypergeometric series sum. The results represent cubic and higher order transformations, and only Vandermonde's theorem is necessary for the elementary proof. Some q-analogues are also given.

1. Introduction

A hypergeometric series may always be written as a sum of two other hypergeometric series by splitting the series into its even and odd terms, for example

$$(1-x)^{-\alpha} = \sum_{k=0}^{\infty} \frac{(\alpha)_{2k}}{(2k)!} x^{2k} + \sum_{k=0}^{\infty} \frac{(\alpha)_{2k+1}}{(2k+1)!} x^{2k+1}.$$

One may also write a series as a sum of r series by reorganizing the terms modulo r.

In this paper a variation of this idea is considered, where the terms modulo r differ by a power of a linear function. We give in §2 four examples of this phenomena, Theorems 2.2, 2.5, 2.8, and 2.11, which may be considered as rth-degree transformations. Special cases of these expansions give new expansions for $(1-x)^{-\alpha}$: four cubic expansions are explicitly given as Corollaries 2.3, 2.6, 2.9, and 2.12 in §2. q-analogues may also be given, we state two such in §4: Theorems 2.2q and 2.5q. Thus these results are in the same spirit as those of Mizan Rahman in (Rahman, 1989; Rahman, 1993; Rahman, 1997), and his work with George

*Partially supported by NSF grant DMS 0203282.

Gasper (Gasper and Rahman, 1990; Gasper and Rahman, 1990b), although these are at a much lower level. They are motivated by Corollary 4.4 for $r = 2$, which was the key lemma in (Prellberg and Stanton, 2003).

2. Main Results

In this section we state and prove the main results of this paper, Theorems 2.2, 2.5, 2.8, and 2.11.

One may expand a formal power series $F(x)$ in x as a formal power series in $y = x(1-x)^{-1/r}$

$$F(x) = \sum_{k=0}^{\infty} a_k \frac{x^k}{(1-x)^{k/r}}.$$

The coefficients a_k may be found from the Lagrange inversion formula. If r is a positive integer, then this series may be rewritten as

$$F(x) = \sum_{i=0}^{r-1} \sum_{t=0}^{\infty} b_t^{(i)} \frac{x^{rt+i}}{(1-x)^{t+i/r}}. \tag{2.1}$$

We shall consider variations of (2.1) where the denominator exponent is either t or $t+1$ instead of $t + i/r$. We shall find expansions for the function

$$\sum_{k=0}^{\infty} \frac{(\alpha)_k}{(\beta)_k} x^k = {}_2F_1\left(\begin{matrix} \alpha, & 1 \\ & \beta \end{matrix} \,\middle|\, x \right).$$

First we consider what happens if all of the denominator exponents in (2.1) are t.

Proposition 2.1. *Let $F(x) = \sum_{s=0}^{\infty} F_s x^s$ be a formal power series in x, and let $r \geq 2$ be an integer. If*

$$F(x) = \sum_{i=0}^{r-1} \sum_{t=0}^{\infty} b_t^{(i)} \frac{x^{rt+i}}{(1-x)^t}$$

then

$$\sum_{s=0}^{t} \binom{t}{s} (-1)^s F_{rt+i-s} = \begin{cases} b_t^{(i)} & \text{if } 1 \leq i \leq r-1, \\ b_t^{(0)} - b_{t-1}^{(r-1)} & \text{if } i = 0. \end{cases}$$

Proof. Considering the coefficient of x^{rt+i}, $0 \leq i \leq r-1$, in

$$F(x)(1-x)^t = \sum_{j=0}^{r-1} \sum_{s=0}^{t} b_s^{(j)} x^{rs+j} (1-x)^{t-s} + \sum_{j=0}^{r-1} \sum_{s=t+1}^{\infty} b_s^{(j)} x^{rs+j} (1-x)^{t-s}$$

we find

$$b_t^{(i)} = \sum_{s=0}^{t} \binom{t}{s} (-1)^s F_{rt+i-s}, \quad 1 \le i \le r-1.$$

For $i = 0$ there are two terms ($j = 0$, $s = t$, $(b_t^{(0)})$ and $j = r-1$, $s = t-1$, $(b_{t-1}^{(r-1)})$) which contribute to the coefficient of x^{rt}. □

Theorem 2.2. *For any integer $r \ge 2$ we have*

$$\sum_{k=0}^{\infty} \frac{(\alpha)_k}{(\beta)_k} x^k = \frac{\beta - 1}{\beta - \alpha - 1}$$

$$-\frac{\alpha}{\beta-\alpha-1}\ {}_{r+1}F_r\left(\begin{matrix} \frac{\alpha+1}{r-1}, & \dots, & \frac{\alpha+r-1}{r-1}, & \beta-\alpha-1, & 1 \\ \frac{\beta}{r}, & \dots, & \frac{\beta+r-1}{r} & & \end{matrix}\middle| \frac{(r-1)^{r-1}x^r}{r^r(x-1)}\right)$$

$$+\sum_{i=1}^{r-1} \frac{(\alpha)_i x^i}{(\beta)_i}\ {}_{r+1}F_r\left(\begin{matrix} \frac{\alpha+i}{r-1}, & \dots, & \frac{\alpha+i+r-2}{r-1}, & \beta-\alpha, & 1 \\ \frac{\beta+i}{r}, & \dots, & \frac{\beta+i+r-1}{r} & & \end{matrix}\middle| \frac{(r-1)^{r-1}x^r}{r^r(x-1)}\right).$$

Proof. In Proposition 2.1 take

$$F_s = \frac{(\alpha)_s}{(\beta)_s}$$

and use the Vandermonde evaluation

$$\sum_{s=0}^{t} \binom{t}{s} (-1)^s F_{A-s} = \frac{(\alpha)_A (\beta-\alpha)_t}{(\beta)_A (1-A-\alpha)_t}. \tag{2.2}$$

□

When $\beta = 1$ in Theorem 2.2, the left side sums to $(1-x)^{-\alpha}$ by the binomial theorem. We explicitly state the $r = 3$ case as a corollary.

Corollary 2.3.

$$(1-x)^{-\alpha} = {}_3F_2\left(\begin{matrix} -\alpha, & (\alpha+1)/2, & (\alpha+2)/2 \\ & 1/3, & 2/3 \end{matrix}\middle| \frac{-4x^3}{27(1-x)}\right) +$$

$$\alpha x\ {}_3F_2\left(\begin{matrix} 1-\alpha, & (\alpha+1)/2, & (\alpha+2)/2 \\ & 2/3, & 4/3 \end{matrix}\middle| \frac{-4x^3}{27(1-x)}\right) +$$

$$\frac{\alpha(\alpha+1)x^2}{2}\ {}_3F_2\left(\begin{matrix} 1-\alpha, & (\alpha+2)/2, & (\alpha+3)/2 \\ & 4/3, & 5/3 \end{matrix}\middle| \frac{-4x^3}{27(1-x)}\right).$$

Next we consider a variation of Theorem 2.2 in which the denominator exponents for $b_t^{(i)}$ are $t+1$ for $1 \le i \le r-1$ and t for $b_t^{(0)}$.

Proposition 2.4. *Let* $F(x) = \sum_{s=0}^{\infty} F_s x^s$ *be a formal power series in* x, *and let* r *be a positive integer. If*

$$F(x) = \sum_{t=0}^{\infty} b_t^{(0)} \frac{x^{rt}}{(1-x)^t} + \sum_{i=1}^{r-1} \sum_{t=0}^{\infty} b_t^{(i)} \frac{x^{rt+i}}{(1-x)^{t+1}}$$

then

$$b_t^{(0)} = \sum_{s=0}^{t} \binom{t}{s} (-1)^s F_{rt-s}, \tag{2.3}$$

$$b_t^{(1)} = \sum_{s=0}^{t} \binom{t}{s} (-1)^s F_{rt+1-s}, \tag{2.4}$$

$$b_t^{(i)} = \sum_{s=0}^{t} \binom{t}{s} (-1)^s (F_{rt+i-s} - F_{rt+i-1-s}), \quad 2 \le i \le r-1. \tag{2.5}$$

Proof. Let's again find the coefficient of x^{rt+i}, $0 \le i \le r-1$, in

$$\begin{aligned} F(x)(1-x)^t &= \sum_{s=0}^{t} b_s^{(0)} x^{rs} (1-x)^{t-s} + \sum_{s=t+1}^{\infty} b_s^{(0)} x^{rs} (1-x)^{t-s} \\ &+ \sum_{j=1}^{r-1} \sum_{s=0}^{t-1} b_s^{(j)} x^{rs+j} (1-x)^{t-s-1} + \sum_{j=1}^{r-1} \sum_{s=t}^{\infty} b_s^{(j)} x^{rs+j} (1-x)^{t-s-1}. \end{aligned}$$

If $i = 0$ or 1, only the $s = t$ term contributes, and we obtain (1) and (2). For $2 \le i \le r-1$, the term $s = t$ contributes for $1 \le j \le i$ and we see that

$$\sum_{j=1}^{i} b_t^{(j)} = \sum_{s=0}^{t} \binom{t}{s} (-1)^s F_{rs+i-s},$$

which implies (3). □

Theorem 2.5. *For any integer $r \geq 2$ we have*

$$\sum_{k=0}^{\infty} \frac{(\alpha)_k}{(\beta)_k} x^k = {}_{r+1}F_r \left(\begin{matrix} \frac{\alpha}{r-1}, & \ldots, & \frac{\alpha+r-2}{r-1}, & \beta-\alpha, & 1 \\ \frac{\beta}{r}, & \ldots, & \frac{\beta+r-1}{r} & & \end{matrix} \middle| \frac{(r-1)^{r-1}x^r}{r^r(x-1)} \right) +$$

$$\frac{\alpha x}{\beta(1-x)} \, {}_{r+1}F_r \left(\begin{matrix} \frac{\alpha+1}{r-1}, & \ldots, & \frac{\alpha+r-1}{r-1}, & \beta-\alpha, & 1 \\ \frac{\beta+1}{r}, & \ldots, & \frac{\beta+r}{r} & & \end{matrix} \middle| \frac{(r-1)^{r-1}x^r}{r^r(x-1)} \right) +$$

$$\sum_{i=2}^{r-1} \frac{(\alpha)_{i-1}(\alpha-\beta)x^i}{(\beta)_i(1-x)}$$

$$\times \, {}_{r+1}F_r \left(\begin{matrix} \frac{\alpha+i-1}{r-1}, & \ldots, & \frac{\alpha+i+r-3}{r-1}, & \beta-\alpha+1, & 1 \\ \frac{\beta+i}{r}, & \ldots, & \frac{\beta+i+r-1}{r} & & \end{matrix} \middle| \frac{(r-1)^{r-1}x^r}{r^r(x-1)} \right).$$

Proof. In Proposition 2.4 take

$$F_s = \frac{(\alpha)_s}{(\beta)_s}$$

and use (2.2). □

The $\beta = 1$ case is the next corollary.

Corollary 2.6.

$$(1-x)^{-\alpha} = {}_3F_2 \left(\begin{matrix} 1-\alpha, & \alpha/2, & (\alpha+1)/2 \\ & 1/3, & 2/3 \end{matrix} \middle| \frac{-4x^3}{27(1-x)} \right)$$

$$+ \frac{\alpha x}{1-x} {}_3F_2 \left(\begin{matrix} 1-\alpha, & (\alpha+1)/2, & (\alpha+2)/2 \\ & 2/3, & 4/3 \end{matrix} \middle| \frac{-4x^3}{27(1-x)} \right)$$

$$+ \frac{\alpha(\alpha-1)x^2}{2(1-x)} {}_3F_2 \left(\begin{matrix} 2-\alpha, & (\alpha+1)/2, & (\alpha+2)/2 \\ & 4/3, & 5/3 \end{matrix} \middle| \frac{-4x^3}{27(1-x)} \right).$$

Proposition 2.4 may be generalized by taking the denominator exponents for $b_t^{(i)}$ to be $t+1$ for $v+1 \leq i \leq r-1$, and the exponents for $b_t^{(i)}$ to be t, $0 \leq i \leq v$, for some $0 \leq v \leq r-2$.

Proposition 2.7. *Let $F(x) = \sum_{s=0}^{\infty} F_s x^s$ be a formal power series in x, and let $0 \leq v \leq r-2$ be non-negative integers. If*

$$F(x) = \sum_{i=0}^{v} \sum_{t=0}^{\infty} b_t^{(i)} \frac{x^{rt+i}}{(1-x)^t} + \sum_{i=v+1}^{r-1} \sum_{t=0}^{\infty} b_t^{(i)} \frac{x^{rt+i}}{(1-x)^{t+1}}$$

then

$$b_t^{(i)} = \sum_{s=0}^{t} \binom{t}{s} (-1)^s F_{rt+i-s}, \quad 0 \leq i \leq v+1$$

$$b_t^{(i)} = \sum_{s=0}^{t} \binom{t}{s} (-1)^s (F_{rt+i-s} - F_{rt+i-1-s}), \quad v+2 \leq i \leq r-1.$$

Theorem 2.5 is the $v = 0$ case of Theorem 2.8.

Theorem 2.8. *For integers $0 \leq v \leq r-2$ we have*

$$\sum_{k=0}^{\infty} \frac{(\alpha)_k}{(\beta)_k} x^k = {}_{r+1}F_r \left(\begin{matrix} \frac{\alpha}{r-1}, & \ldots, & \frac{\alpha+r-2}{r-1}, & \beta-\alpha, & 1 \\ \frac{\beta}{r}, & \ldots, & \frac{\beta+r-1}{r} \end{matrix} \middle| \frac{(r-1)^{r-1}x^r}{r^r(x-1)} \right) +$$

$$\sum_{i=1}^{v} \frac{(\alpha)_i x^i}{(\beta)_i} {}_{r+1}F_r \left(\begin{matrix} \frac{\alpha+i}{r-1}, & \ldots, & \frac{\alpha+i+r-2}{r-1}, & \beta-\alpha, & 1 \\ \frac{\beta+i}{r}, & \ldots, & \frac{\beta+i+r-1}{r} \end{matrix} \middle| \frac{(r-1)^{r-1}x^r}{r^r(x-1)} \right) +$$

$$\frac{(\alpha)_{v+1} x^{v+1}}{(\beta)_{v+1}(1-x)}$$

$$\times {}_{r+1}F_r \left(\begin{matrix} \frac{\alpha+v+1}{r-1}, & \ldots, & \frac{\alpha+v+r-1}{r-1}, & \beta-\alpha, & 1 \\ \frac{\beta+v+1}{r}, & \ldots, & \frac{\beta+v+r}{r} \end{matrix} \middle| \frac{(r-1)^{r-1}x^r}{r^r(x-1)} \right) +$$

$$\sum_{i=v+2}^{r-1} \frac{(\alpha)_{i-1}(\alpha-\beta)x^i}{(\beta)_i(1-x)}$$

$$\times {}_{r+1}F_r \left(\begin{matrix} \frac{\alpha+i-1}{r-1}, & \ldots, & \frac{\alpha+i+r-3}{r-1}, & \beta-\alpha+1, & 1 \\ \frac{\beta+i}{r}, & \ldots, & \frac{\beta+i+r-1}{r} \end{matrix} \middle| \frac{(r-1)^{r-1}x^r}{r^r(x-1)} \right).$$

Corollary 2.9.

$$\begin{aligned} (1-x)^{-\alpha} = & \quad {}_3F_2 \left(\begin{matrix} 1-\alpha, & \alpha/2, & (\alpha+1)/2 \\ & 1/3, & 2/3 \end{matrix} \middle| \frac{-4x^3}{27(1-x)} \right) + \\ & \alpha x \; {}_3F_2 \left(\begin{matrix} 1-\alpha, & (\alpha+1)/2, & (\alpha+2)/2 \\ & 2/3, & 4/3 \end{matrix} \middle| \frac{-4x^3}{27(1-x)} \right) + \\ & \frac{\alpha(\alpha+1)x^2}{2(1-x)} {}_3F_2 \left(\begin{matrix} 1-\alpha, & (\alpha+2)/2, & (\alpha+3)/2 \\ & 4/3, & 5/3 \end{matrix} \middle| \frac{-4x^3}{27(1-x)} \right). \end{aligned}$$

A version of Proposition 2.7 exists for denominator exponents $t+1$ for $1 \leq i \leq v$, and t for $i = 0$ or $v+1 \leq i \leq r-1$, for some $1 \leq v \leq r-2$.

Proposition 2.10. *Let $F(x) = \sum_{s=0}^{\infty} F_s x^s$ be a formal power series in x, and let $1 \leq v \leq r-2$ be non-negative integers. If*

$$F(x) = \sum_{t=0}^{\infty} b_t^{(0)} \frac{x^{rt}}{(1-x)^t} + \sum_{i=1}^{v} \sum_{t=0}^{\infty} b_t^{(i)} \frac{x^{rt+i}}{(1-x)^{t+1}} + \sum_{i=v+1}^{r-1} \sum_{t=0}^{\infty} b_t^{(i)} \frac{x^{rt+i}}{(1-x)^t}$$

then

$$
\begin{aligned}
b_t^{(0)} - b_{t-1}^{(r-1)} &= \sum_{s=0}^{t} \binom{t}{s} (-1)^s F_{rt-s},\\
b_t^{(1)} &= \sum_{s=0}^{t} \binom{t}{s} (-1)^s F_{rt+1-s},\\
b_t^{(i)} &= \sum_{s=0}^{t} \binom{t}{s} (-1)^s (F_{rt+i-s} - F_{rt+i-1-s}), \quad 2 \le i \le v\\
b_t^{(i)} &= \sum_{s=0}^{t} \binom{t}{s} (-1)^s (F_{rt+i-s} - F_{rt+v-s}), \quad v+1 \le i \le r-1.
\end{aligned}
$$

Although the equations in Proposition 2.10 may be solved using (2.2), we give here only the $r = 3$, $v = 1$ version.

Theorem 2.11.

$$
\begin{aligned}
\sum_{k=0}^{\infty} \frac{(\alpha)_k}{(\beta)_k} x^k = \frac{1-\beta}{\alpha-1} + {}&\\
\frac{\alpha+\beta-2}{\alpha-1} {}_5F_4 &\left(\begin{matrix} \frac{\alpha-1}{2}, & \frac{\alpha}{2}, & \frac{\alpha+\beta+3}{5}, & \beta-\alpha, & 1 \\ & \frac{\beta}{3}, & \frac{\beta+1}{3}, & \frac{\beta+1}{3} & \frac{\alpha+\beta-2}{5} \end{matrix} \middle| \frac{-4x^3}{27(1-x)} \right) +\\
\frac{\alpha x}{\beta(1-x)} {}_4F_3 &\left(\begin{matrix} \frac{\alpha+1}{2}, & \frac{\alpha+2}{2}, & \beta-\alpha, & 1 \\ & \frac{\beta+1}{3}, & \frac{\beta+2}{3} & \frac{\beta+3}{3} \end{matrix} \middle| \frac{-4x^3}{27(1-x)} \right) +\\
\frac{\alpha(\alpha-\beta)x^2}{\beta(\beta+1)} {}_4F_3 &\left(\begin{matrix} \frac{\alpha+1}{2}, & \frac{\alpha+2}{2}, & \beta-\alpha+1, & 1 \\ & \frac{\beta+2}{3}, & \frac{\beta+3}{3}, & \frac{\beta+4}{3} \end{matrix} \middle| \frac{-4x^3}{27(1-x)} \right).
\end{aligned}
$$

Corollary 2.12.

$$
\begin{aligned}
(1-x)^{-\alpha} = \quad & {}_4F_3 \left(\begin{matrix} 1-\alpha, & \alpha/2, & (\alpha-1)/2, & (\alpha+4)/5 \\ & 1/3, & 2/3, & (\alpha-1)/5 \end{matrix} \middle| \frac{-4x^3}{27(1-x)} \right) +\\
\frac{\alpha x}{1-x} & {}_3F_2 \left(\begin{matrix} 1-\alpha, & (\alpha+1)/2, & (\alpha+2)/2 \\ & 2/3, & 4/3 \end{matrix} \middle| \frac{-4x^3}{27(1-x)} \right) +\\
\frac{\alpha(\alpha-1)x^2}{2} & {}_3F_2 \left(\begin{matrix} 2-\alpha, & (\alpha+1)/2, & (\alpha+2)/2 \\ & 4/3, & 5/3 \end{matrix} \middle| \frac{-4x^3}{27(1-x)} \right).
\end{aligned}
$$

3. Previous results

In (Gessel and Stanton, 1982, (5.12)) another expansion is given for the left side in Theorems 2.2, 2.5, 2.8, and 2.11,

$$
\sum_{k=0}^{\infty} \frac{(sb-a)_k}{(b+1)_k} x^k = (1-x)^a \sum_{k=0}^{\infty} \frac{b(-a-sk)_k}{k!(b+k)} (-x(1-x)^{s-1})^k. \tag{3.1}
$$

If $s = r$, a positive integer, then (3.1) becomes

$$\sum_{k=0}^{\infty} \frac{(rb-a)_k}{(b+1)_k} x^k = (1-x)^a \times {}_{r+1}F_r\left(\begin{matrix} \frac{a+1}{r}, & \ldots, & \frac{a+r}{r}, & b \\ \frac{a+1}{r-1}, & \ldots, & \frac{a+r-1}{r-1}, & b+1 \end{matrix} \,\middle|\, \frac{r^r x(1-x)^{r-1}}{(r-1)^{r-1}}\right) \quad (3.2)$$

while for $s = -r$ a negative integer we have

$$\sum_{k=0}^{\infty} \frac{(-rb-a)_k}{(b+1)_k} x^k = (1-x)^a \times {}_{r+2}F_{r+1}\left(\begin{matrix} \frac{-a}{r+1}, & \ldots, & \frac{-a+r}{r+1}, & b \\ \frac{-a}{r}, & \ldots, & \frac{-a+r-1}{r}, & b+1 \end{matrix} \,\middle|\, \frac{-(r+1)^{r+1}x}{r^r(1-x)^{r+1}}\right). \quad (3.3)$$

One may consider Theorems 2.2, 2.5, and 2.8, and Equations (3.2) and (3.3) as iterated higher order transformations.

4. q-analogues

It is a routine computation to find q-analogues of Propositions 2.1, 2.4, 2.7, and 2.10. The denominator terms of $(1-x)^t$ and $(1-x)^{t+1}$ are replaced by $(x;q)_t$ and $(x;q)_{t+1}$. What replaces (2.2) is the q-Vandermonde evaluation (Gasper and Rahman, 1990)

$$F_s = \frac{(a;q)_s}{(b;q)_s},$$

$$\sum_{s=0}^{t} \begin{bmatrix} t \\ s \end{bmatrix}_q (-1)^s q^{\binom{s}{2}} F_{A-s} = \frac{(a;q)_A (b/a;q)_t}{(b)_A (q^{1-A}/a;q)_t}. \quad (4.1)$$

We give the q-analogues of Theorems 2.2 and 2.5.

Theorem 4.1 (Theorem 2.2q). *For any integer $r \geq 2$,*

$$\sum_{k=0}^{\infty} \frac{(a;q)_k}{(b;q)_k} x^k = 1 + \sum_{t=0}^{\infty} \frac{(a;q)_{(r-1)t+1}(b/a;q)_{t-1}}{(b;q)_{rt}} (-a)^{t-1} q^{rt(t-1)-\binom{t}{2}} \frac{x^{rt}}{(x;q)_t} + \sum_{i=1}^{r-1}\sum_{t=0}^{\infty} \frac{(a;q)_{(r-1)t+i}(b/a;q)_t}{(b;q)_{rt+i}} (-a)^t q^{t(rt+i-1)-\binom{t}{2}} \frac{x^{rt+i}}{(x;q)_t}.$$

Theorem 4.2 (Theorem 2.5q). *For any integer $r \geq 2$,*

$$\sum_{k=0}^{\infty} \frac{(a;q)_k}{(b;q)_k} x^k = \sum_{t=0}^{\infty} \frac{(a;q)_{(r-1)t}(b/a;q)_t}{(b;q)_{rt}} (-a)^t q^{t(rt-1)-\binom{t}{2}} \frac{x^{rt}}{(x;q)_t} +$$
$$\sum_{t=0}^{\infty} \frac{(a;q)_{(r-1)t+1}(b/a;q)_t}{(b;q)_{rt+1}} (-a)^t q^{rt^2-\binom{t}{2}} \frac{x^{rt+1}}{(x;q)_{t+1}} +$$
$$\sum_{i=2}^{r-1} \sum_{t=0}^{\infty} \frac{(a;q)_{(r-1)t+i-1}(b/a;q)_{t+1}}{(b;q)_{rt+i}} (-a)^{t+1}$$
$$\times\, q^{(rt+i-1)(t+1)-\binom{t+1}{2}} \frac{x^{rt+i}}{(x;q)_{t+1}}.$$

The $a = q^N$, $b = q$ case of Theorem 2.5q was the key Lemma used in (Prellberg and Stanton, 2003) with $r = 2$. As corollaries we state the general $a = q^N$, $b = q$ case of Theorems 2.2q and 2.5q.

Corollary 4.3. *For any integer $r \geq 2$,*

$$\frac{1}{(x;q)_N} = 1 + \sum_{t=1}^{\infty} \begin{bmatrix} N+(r-1)t \\ rt \end{bmatrix}_q q^{rt(t-1)} \frac{x^{rt}}{(x;q)_t} +$$
$$\sum_{i=1}^{r-1} \sum_{t=0}^{\infty} \begin{bmatrix} N+(r-1)t+i-1 \\ rt+i \end{bmatrix}_q q^{t(rt+i)} \frac{x^{rt+i}}{(x;q)_t}.$$

Corollary 4.4. *For any integer $r \geq 2$,*

$$\frac{1}{(x;q)_N} = \sum_{t=0}^{\infty} \begin{bmatrix} N+(r-1)t-1 \\ rt \end{bmatrix}_q q^{rt^2} \frac{x^{rt}}{(x;q)_t} +$$
$$\sum_{t=0}^{\infty} \begin{bmatrix} N+(r-1)t \\ rt+1 \end{bmatrix}_q q^{t(rt+1)} \frac{x^{rt+1}}{(x;q)_{t+1}} +$$
$$\sum_{i=2}^{r-1} \sum_{t=0}^{\infty} \begin{bmatrix} N+(r-1)t+i-2 \\ rt+i \end{bmatrix}_q q^{(rt+i)(t+1)} \frac{x^{rt+i}}{(x;q)_{t+1}}.$$

References

Gasper, G. and Rahman, M. (1990a). *Basic Hypergeometric Series.* Cambridge University Press, Cambridge.

Gasper, G. and Rahman, M. (1990b). An indefinite bibasic summation formula and some quadratic, cubic and quartic sums. *Canad. J. Math.*, 42:1–27.

Gessel, I. and Stanton, D. (1982). Strange evaluations of hypergeometric series. *SIAM J. Math. Anal.*, 13:295–308.

Prellberg, T. and Stanton, D. (2003). Proof of a monotonicity conjecture. *J. Comb. Th. A*, 103:377–381.

Rahman, M. (1989). Some cubic summation formulas for basic hypergeometric series. *Utilitas Math.*, 36:161–172.

Rahman, M. (1993). Some quadratic and cubic summation formulas for basic hypergeometric series. *Canad. J. Math.*, 45:394–411.

Rahman, M. (1997). Some cubic summation and transformation formulas. *Ramanujan J.*, 1:299–308.

ASKEY-WILSON FUNCTIONS AND QUANTUM GROUPS

Jasper V. Stokman*
KdV Institute for Mathematics
Universiteit van Amsterdam
Plantage Muidergracht 24
1018 TV Amsterdam
THE NETHERLANDS
jstokman@science.uva.nl

Abstract Eigenfunctions of the Askey-Wilson second order q-difference operator for $0 < q < 1$ and $|q| = 1$ are constructed as formal matrix coefficients of the principal series representation of the quantized universal enveloping algebra $\mathcal{U}_q(\mathfrak{sl}(2,\mathbb{C}))$. The eigenfunctions are given in integral form. We show that for $0 < q < 1$ the resulting eigenfunction can be rewritten as a very-well-poised ${}_8\varphi_7$-series, and reduces for special parameter values to a natural elliptic analogue of the cosine kernel.

1. Introduction

The aim of this paper is to simplify the quantum group construction of explicit eigenfunctions of the second order Askey-Wilson q-difference operator, and to extend the results to the interesting and less well studied $|q| = 1$ case.

The approach is based on the known fact from (Koornwinder, 1993), (Noumi and Mimachi, 1992) and (Koelink, 1996) that the second order Askey-Wilson difference operator arises as radial part of the quantum Casimir element of $\mathcal{U}_q(\mathfrak{sl}(2,\mathbb{C}))$ when the radial part is computed with respect to Koornwinder's twisted primitive elements. Using this result, we construct nonpolynomial eigenfunctions of the Askey-Wilson second order difference operator as matrix coefficients of the principal series representation of $\mathcal{U}_q(\mathfrak{sl}(2,\mathbb{C}))$. The two cases $0 < q < 1$ and $|q| = 1$ will be treated seperately. The theory for $0 < q < 1$ is related to the

*Supported by the Royal Netherlands Academy of Arts and Sciences (KNAW).

noncompact quantum group $\mathcal{U}_q(\mathfrak{su}(1,1))$, while for $|q|=1$ it is related to the noncompact quantum group $\mathcal{U}_q(\mathfrak{sl}(2,\mathbb{R}))$.

This approach was considered in (Koelink and Stokman, 2001b) for $0<q<1$ using an explicit realization of the principal series representation on $l^2(\mathbb{Z})$. The resulting eigenfunction then appears as a non-symmetric Poisson type kernel involving nonterminating ${}_2\varphi_1$ series. With the help of a highly nontrivial summation formula, proved by Rahman in the appendix of (Koelink and Stokman, 2001b) (see (Koelink and Rosengren, 2002) for extensions), this eigenfunction was expressed as one of the explicit ${}_8\varphi_7$-solutions of the Askey-Wilson second order difference operator from (Ismail and Rahman, 1991). This eigenfunction was called the Askey-Wilson function in (Koelink and Stokman, 2001a), since it is a meromorphic continuation of the Askey-Wilson polynomial in its degree. In this paper we start by reproving this result, now using an explicit realization of the principal series representation of $\mathcal{U}_q(\mathfrak{sl}(2,\mathbb{C}))$ as difference operators acting on analytic functions on the complex plane Koornwinder's twisted primitive element then acts as a first order difference operator, hence eigenvectors are easily constructed (for the positive discrete series, this was observed in (Van der Jeugt and Jagannathan, 1998) and in (Rosengren, 2000)). The corresponding matrix coefficients lead to explicit integral representations for eigenfunctions of the Askey-Wilson second order difference operator. These matrix coefficients can be rewritten as the explicit ${}_8\varphi_7$-series representation of the Askey-Wilson function by a residue computation.

We also show that for a special choice of parameter values, the Askey-Wilson function reduces to an elliptic analogue of the cosine kernel. This is the analogue of the classical fact that the Jacobi function reduces to the cosine kernel for special parameter values, see e.g., (Koornwinder, 1984). We give two proofs, one proof uses an explicit expansion formula of the Askey-Wilson function in Askey-Wilson polynomials from (Stokman, 2002), the other proof uses Cherednik's Hecke algebra techniques from (Cherednik, 1997) and (Stokman, 2001).

In the second part of the paper we consider the quantum group techniques for $|q|=1$. In this case, the approach is similar to the construction of quantum analogues of Whittaker vectors and Whittaker functions from (Kharchev et al., 2002). The role of q-shifted factorials, or equivalently q-gamma functions, is now taken over by Ruijsenaars' (Ruijsenaars, 1997) hyperbolic gamma function. The hyperbolic gamma function is directly related to Barnes' double gamma function, as well as to Kurokawa's double sine function, see (Ruijsenaars, 1999) and references therein. The quantum group technique applied to this particular set-up leads to an eigenfunction of the Askey-Wilson second order difference operator for $|q| = 1$, given explicitly as an Euler type integral involving hyperbolic gamma functions.

The emphasis in this paper lies on exhibiting the similarities between the $0 < q < 1$ case and the $|q| = 1$ case as much as possible. Other approaches might very well lead to eigenfunctions for the Askey-Wilson second order difference operator for $|q| = 1$ which are "more optimal," in the sense that they satisfy *two* Askey-Wilson type difference equations in the geometric parameter, one with respect to base $q = \exp(2\pi i\tau)$, the other with respect to the "modular inverted" base $q = \exp(2\pi i/\tau)$, cf. (Kharchev et al., 2002) for q-Whittaker functions. Such eigenfunctions are expected to be realized as matrix coefficients of the *modular double* of the quantum group $\mathcal{U}_q(\mathfrak{sl}(2,\mathbb{C}))$ (a concept introduced by Faddeev in (Faddeev, 2000)), and are expected to be closely related to Ruijsenaars' (Ruijsenaars, 1999), (Ruijsenaars, 2001) R-function. The R-function is an eigenfunction of two Askey-Wilson type second order difference operators in the geometric parameter, which is explicitly given as a Barnes' type integral involving hyperbolic gamma functions. I hope to return to these considerations in a future paper.

Acknowledgments

It is a pleasure to dedicate this paper to Mizan Rahman. His important contributions to the theory of basic hypergeometric series and, more concretely, his kind help in the earlier stages of the research on the Askey-Wilson functions in (Koelink and Stokman, 2001b), have played, and still play, an important role in my research on Askey-Wilson functions.

2. Generalized gamma functions

In this section we discuss q-analogues of the gamma function for deformation parameter q in the regions $0 < |q| < 1$ and $|q| = 1$.

2.1 The q-gamma function for $0 < |q| < 1$.

Let τ be a fixed complex number in the upper half plane $\mathbb{H}$. The corresponding deformation parameter $q = q_\tau = \exp(2\pi i\tau)$ has modulus less than one. We write $q^u = \exp(2\pi i\tau u)$ for $u \in \mathbb{C}$.

Let $b, b_j \in \mathbb{C}$ and $n \in \mathbb{Z}_{\geq 0} \cup \{\infty\}$. The q-shifted factorial is defined by

$$(b;q)_n = \prod_{j=0}^{n-1}(1 - bq^j), \qquad (b_1, \ldots, b_m; q)_n = \prod_{j=1}^{m}(b_j;q)_n \,.$$

In q-analysis the function

$$x \mapsto \frac{(q;q)_\infty}{(q^x;q)_\infty}(1-q)^{1-x}$$

is known as the q-gamma function; see, e.g., (Gasper and Rahman, 1990). For our purposes, it is more convenient to work with the function

$$\Gamma_\tau(x) := \frac{q^{-\frac{x^2}{16}}}{\left(-q^{\frac{1}{2}(x+1)};q\right)_\infty}. \tag{2.1}$$

Observe that $\Gamma_\tau(x)$ is a zero-free meromorphic function with simple poles located at $-1+\tau^{-1}+2\mathbb{Z}_{\leq 0}+2\tau^{-1}\mathbb{Z}$. It furthermore satisfies the difference equation

$$\Gamma_\tau(x+2) = 2\cos(\pi(x+1)\tau/2)\Gamma_\tau(x). \tag{2.2}$$

Note furthermore that for $\tau \in i\,\mathbb{R}_{>0}$, i.e., $0<q<1$, the function $\Gamma_\tau(z)$ satisfies $\overline{\Gamma_\tau(x)} = \Gamma_\tau(\overline{x})$, where the bar stands for the complex conjugate.

Observe that the above defined q-gamma type functions are not τ^{-1}-periodic. It is probably for this reason that formulas in q-analysis are usually expressed in terms of q-shifted factorials instead of q-gamma functions. For our present purposes the expressions in terms of q-gamma type functions are convenient because it clarifies the similarities with the $|q|=1$ case.

2.2 The gamma function for $|q| = 1$.

In this subsection we take $\tau \in \mathbb{R}_{<0}$, whence $q = q_\tau = \exp(2\pi i\tau)$ satisfies $|q|=1$. As in the previous subsection, we write $q^u = \exp(2\pi i\tau u)$ for $u \in \mathbb{C}$.

It is easy to verify that the integral

$$\gamma_\tau(z) = \frac{1}{2i}\int_0^\infty \frac{dy}{y}\left(\frac{z}{y} - \frac{\sinh(\tau y z)}{\sinh(y)\sinh(\tau y)}\right) \tag{2.3}$$

converges absolutely in the strip $|\operatorname{Re}(z)| < 1-\tau^{-1}$. For $z \in \mathbb{C}$ in this strip we set

$$G_\tau(z) = \exp(i\gamma_\tau(z)). \tag{2.4}$$

Ruijsenaars' (Ruijsenaars, 1997, Sect. 3) hyperbolic gamma function $G(z) = G(a_+,a_-;z)$ with $a_+,a_- > 0$ is related to G_τ by

$$G_\tau(z) = G(a_+,a_-;ia_-z/2), \qquad \tau = -a_-/a_+.$$

In the following proposition we recall some of Ruijsenaars' results (Ruijsenaars, 1997, Sect. 3) on the hyperbolic gamma function.

Proposition 2.1. **(i)** *The function $G_\tau(z)$ satisfies the difference equation*

$$G_\tau(z+2) = 2\cos(\pi(z+1)\tau/2)G_\tau(z).$$

In particular, $G_\tau(z)$ admits a meromorphic continuation to the complex plane $\mathbb{C}$, which we again denote by $G_\tau(z)$.

(ii) *The zeros of $G_\tau(z)$ are located at $1-\tau^{-1}+2\mathbb{Z}_{\geq 0}+2\tau^{-1}\mathbb{Z}_{\leq 0}$, and the poles of $G_\tau(z)$ are located at $-1+\tau^{-1}+2\mathbb{Z}_{\leq 0}+2\tau^{-1}\mathbb{Z}_{\geq 0}$.*

(iii) *$G_\tau(-z)G_\tau(z)=1$ and $G_\tau(z)=G_{\tau^{-1}}(-\tau z)$.*

(iv) *The function $\gamma_\tau(z)$ has an analytic continuation to the cut plane*

$$\mathbb{C}\setminus\left\{(-\infty,-1+\tau^{-1}]\cup[1-\tau^{-1},\infty)\right\},$$

which we again denote by $\gamma_\tau(z)$. Set $r=\max\left(1,-\tau^{-1}\right)$ and choose $\epsilon>0$, then

$$\left|\mp\gamma_\tau(z)+\frac{\pi\tau z^2}{8}-\frac{\pi}{24}\left(\tau+\tau^{-1}\right)\right| = \mathcal{O}\left(\exp\left((\epsilon-\pi/r)|\operatorname{Im}(z)|\right)\right),$$

for $\operatorname{Im}(z)\to\pm\infty$ uniformly for $\operatorname{Re}(z)$ in compacts of $\mathbb{R}$.

From the explicit expression for $G_\tau(z)$ with $|\operatorname{Re}(z)|<1-\tau^{-1}$ and the first order difference equation for G_τ, we have $\overline{G_\tau(z)}=G_\tau(\overline{z})$.

As Ruijsenaars verifies in (Ruijsenaars, 1999, Appendix A), the hyperbolic gamma function is a quotient of Barnes' double gamma function, and it essentially coincides with Kurokawa's double sine function.

The hyperbolic gamma function is the important building block for q-analysis with $|q|=1$. It was used in (Nishizawa, 2001) and (Nishizawa and Ueno, 2001) to construct for $|q|=1$ explicit integral solutions of the q-Bessel difference equation and of the q-hypergeometric difference equation. Ruijsenaars (Ruijsenaars, 1999) used hyperbolic gamma functions to construct an eigenfunction for the Askey-Wilson second order difference operator for $|q|=1$ as an explicit Barnes' type integral. In this paper we construct an eigenfunction of the Askey-Wilson second order difference operator for $|q|=1$ using representation theory of quantum groups.

3. The Askey-Wilson second order difference operator and quantum groups

Throughout this section we require that $\tau\in\mathbb{C}\setminus\frac{1}{2}\mathbb{Z}$ and we write $q=q_\tau=\exp(2\pi i\tau)$ for the corresponding deformation parameter. The condition on τ implies $q \neq \pm 1$. As usual, we write $q^u = \exp(2\pi i \tau u)$ for $u \in \mathbb{C}$.

Definition 3.1. *The quantum group $\mathcal{U}_q$ is the unital associative algebra over $\mathbb{C}$ generated by $K^{\pm 1}$, X^+ and X^-, subject to the relations*

$$KK^{-1} = K^{-1}K = 1,$$
$$KX^+ = qX^+K, \qquad KX^- = q^{-1}X^-K,$$
$$X^+X^- - X^-X^+ = \frac{K^2 - K^{-2}}{q - q^{-1}}.$$

It is well known that $\mathcal{U}_q$ has the structure of a Hopf-algebra, and as such it is a quantum deformation of the universal enveloping algebra of the simple Lie algebra $\mathfrak{sl}(2, \mathbb{C})$. The Hopf-algebra structure does not play a significant role in the present paper, so the definition of the Hopf-algebra structure is omitted here. I only want to stipulate that the upcoming definition of Koornwinder's twisted primitive element is motivated by its transformation behaviour under the action of the comultiplication of $\mathcal{U}_q$, see (Koornwinder, 1993) for details.

It is convenient to work with an extended version of $\mathcal{U}_q$, which we define as follows. Write $\mathcal{A} = \bigoplus_{x \in \mathbb{C}} \mathbb{C}\,\widehat{x}$ for the group algebra of the additive group $(\mathbb{C}, +)$. Denote $\mathrm{End}_{alg}(\mathcal{U}_q)$ for the unital algebra homomorphisms $\varphi : \mathcal{U}_q \to \mathcal{U}_q$. There exists an algebra homomorphism

$$\kappa : \mathcal{A} \to \underset{alg}{\mathrm{End}}(\mathcal{U}_q),$$

with $\kappa(\widehat{x}) = \kappa_x \in \mathrm{End}_{alg}(\mathcal{U}_q)$ for $x \in \mathbb{C}$ defined by

$$\kappa_x(K^{\pm 1}) = K^{\pm 1}, \qquad \kappa_x\left(X^{\pm}\right) = q^{\pm x}X^{\pm}.$$

Note that for $m \in \mathbb{Z}$,

$$\kappa_m(X) = K^m X K^{-m} \qquad \forall X \in \mathcal{U}_q,$$

so the automorphisms κ_x generalize the inner automorphisms $K^m(\,\cdot\,)K^{-m}$ of $\mathcal{U}_q$ ($m \in \mathbb{Z}$). The extended algebra $\mathcal{U}_q$ is now defined as follows.

Definition 3.2. *The unital, associative algebra $\mathcal{U}_q$ is the vector space $\mathcal{A} \otimes \mathcal{U}_q$ with multiplication defined by*

$$(\widehat{x} \otimes X)\,(\widehat{y} \otimes Y) = \widehat{(x+y)} \otimes \kappa_{-y}(X)Y, \qquad \forall\, x, y \in \mathbb{C}, \qquad \forall\, X, Y \in \mathcal{U}_q.$$

The unit element is $\widehat{0} \otimes 1$.

Observe that $\mathcal{A}$ and $\mathcal{U}_q$ embed as algebras in $\mathcal{U}_q$ by the formulas

$$a \mapsto a \otimes 1, \qquad X \mapsto \widehat{0} \otimes X$$

for $a \in \mathcal{A}$ and $X \in \mathcal{U}_q$. We will use these canonical embeddings to identify the algebras $\mathcal{U}_q$ and $\mathcal{A}$ with their images in $\mathcal{U}_q$. The commutation relations between $\mathcal{A}$ and $\mathcal{U}_q$ within $\mathcal{U}_q$ then become

$$\widehat{x}\, X = \widehat{x} \otimes X = \kappa_x(X)\, \widehat{x}, \qquad x \in \mathbb{C},\ X \in \mathcal{U}_q.$$

The *quantum Casimir element*, defined by

$$\Omega := X^+X^- + \frac{q^{-1}K^2 + qK^{-2} - 2}{(q-q^{-1})^2} \in \mathcal{U}_q, \tag{3.1}$$

is an algebraic generator of the center $\mathcal{Z}(\mathcal{U}_q)$ of $\mathcal{U}_q$. Note that Ω is also in the center $\mathcal{Z}\left(\mathcal{U}_q\right)$ of the extended algebra $\mathcal{U}_q$.

We now consider the explicit realization of $\mathcal{U}_q$ as difference operators. Such realizations are well known; see, e.g., (Van der Jeugt and Jagannathan, 1998), (Rosengren, 2000) and (Kharchev et al., 2002). Let $\mathcal{M}$ be the space of meromorphic functions on the complex plane $\mathbb{C}$. For any $\lambda \in \mathbb{C}$, the assignment

$$\begin{aligned}
\left(\pi_\lambda(X^\pm)f\right)(z) &= q^{\pm z}\left(\frac{q^{-\frac{1}{2}-i\lambda}f(z\mp 1) - q^{\frac{1}{2}+i\lambda}f(z\pm 1)}{q^{-1}-q}\right),\\
\left(\pi_\lambda(K^{\pm 1})f\right)(z) &= f(z\pm 1),\\
\left(\pi_\lambda(\widehat{x})f\right)(z) &= f(z+x), \qquad x \in \mathbb{C},
\end{aligned}$$

uniquely extends to a representation of $\mathcal{U}_q$ on $\mathcal{M}$. The quantum Casimir element Ω acts as

$$\pi_\lambda(\Omega) = \left(\frac{q^{i\lambda} - q^{-i\lambda}}{q-q^{-1}}\right)^2 \mathrm{Id}. \tag{3.2}$$

Observe furthermore that $\pi_\lambda(\widehat{m}) = \pi_\lambda(K^m)$ for all $m \in \mathbb{Z}$.

In the present paper the quantum group input to the theory of q-special functions is based on the explicit connection between the radial part of the quantum Casimir element Ω and the second order Askey-Wilson difference operator. Here the *Askey-Wilson second order difference operator* $\mathcal{D} = \mathcal{D}^{a,b,c,d}$, depending on four parameters (a,b,c,d) called the Askey-Wilson parameters, is defined by

$$(\mathcal{D}f)(x) = A(x)(f(x+2) - f(x)) + A(-x)(f(x-2) - f(x)), \tag{3.3}$$

with $A(x) = A(x; a, b, c, d)$ the explicit function

$$A(x) = \frac{(1-q^{a+x})(1-q^{b+x})(1-q^{c+x})(1-q^{d+x})}{(1-q^{2x})(1-q^{2+2x})},$$

cf. (Askey and Wilson, 1985). The radial part of Ω is computed with respect to elements $Y_\rho - \mu_\alpha(\rho)1 \in \mathcal{U}_q$ for $\alpha, \rho \in \mathbb{C}$, where Y_ρ is Koornwinder's (Koornwinder, 1993) *twisted primitive element*,

$$Y_\rho = q^{\frac{1}{2}}X^+K - q^{-\frac{1}{2}}X^-K + \left(\frac{q^{-\rho}+q^\rho}{q^{-1}-q}\right)(K^2-1) \tag{3.4}$$

and $\mu_\alpha(\rho)$ is the constant

$$\mu_\alpha(\rho) = \left(\frac{q^\rho(1-q^\alpha)+q^{-\rho}(1-q^{-\alpha})}{q-q^{-1}}\right). \tag{3.5}$$

Consider the five dimensional space

$$\mathcal{U}_q^1 = \operatorname*{span}_{\mathbb{C}}\{X^+, X^-, K, K^{-1}, 1\} \subset \mathcal{U}_q. \tag{3.6}$$

The radial part computation of Ω leads to the following result.

Proposition 3.3. *Let $\rho, \sigma, \alpha, \beta \in \mathbb{C}$. For all $x \in \mathbb{C}$,*

$$\widehat{x}\,\Omega K = \widehat{x}\,\Omega(x)\,K \bmod (Y_\rho - \mu_\alpha(\rho))\,\widehat{x}\,\mathcal{U}_q^1 + \widehat{x}\,\mathcal{U}_q^1\,(Y_\sigma - \mu_\beta(\sigma))$$

with $\Omega(x) = \Omega(x; \alpha, \rho, \beta, \sigma)$ given explicitly by

$$\Omega(x) = \frac{q^{\beta-1}}{(q-q^{-1})^2}\left\{B(x)K^2 + \left(C(x) + \left(1-q^{1-\beta}\right)^2\right)1 + D(x)K^{-2}\right\}$$

where

$$\begin{aligned} B(x) &= q^{-\beta}\frac{(1-q^{a+x})(1-q^{2-a+x})(1-q^{b+x})(1-q^{2-b+x})}{(1-q^{2x})(1-q^{2+2x})}, \\ C(x) &= -A(x; a, b, c, d) - A(-x; a, b, c, d), \\ D(x) &= q^{-\beta}\frac{(1-q^{c-x})(1-q^{2-c-x})(1-q^{d-x})(1-q^{2-d-x})}{(1-q^{-2x})(1-q^{2-2x})}. \end{aligned}$$

Here the Askey-Wilson parameters (a, b, c, d) are related to the parameters $\alpha, \beta, \rho, \sigma$ by

$$(a, b, c, d) = (1+\rho+\sigma, 1-\rho+\sigma, 1+\alpha+\rho-\beta-\sigma, 1-\alpha-\rho-\beta-\sigma). \tag{3.7}$$

Proof. The proof generalizes the radial part computation in (Koornwinder, 1993), where the case $\alpha = \beta = 0$ and $x \in \mathbb{Z}$ is considered (see also (Noumi and Mimachi, 1992) and (Koelink, 1996) for extensions to discrete values of α and β). In the present set-up the computation is a bit more complex, and we gather more precise information on the remainder. For the convenience of the reader, I have included the main steps of the proof as appendix. □

Unless specified otherwise, we assume that the Askey-Wilson parameters (a, b, c, d) are related to the four parameters $(\alpha, \rho, \beta, \sigma)$ by (3.7).

Proposition 3.3 allows us to identify specific eigenfunctions of $\pi_\lambda(\Omega)$ with eigenfunctions of the second order difference operator

$$(\mathcal{L}f)(x) = \left(\mathcal{L}^{\alpha,\rho,\beta,\sigma} f\right)(x) = B(x)f(x+2) + C(x)f(x) + D(x)f(x-2). \tag{3.8}$$

The second order difference operator $\mathcal{L}$ is gauge equivalent to the Askey-Wilson second order difference equation $\mathcal{D} = \mathcal{D}^{a,b,c,d}$, since $\mathcal{L} = \Delta \circ \mathcal{D} \circ \Delta^{-1}$ with $\Delta(x) = \Delta(x; a, b, c, d)$ any meromorphic function satisfying the difference equation

$$\begin{aligned}\Delta(x+2) &= \frac{\sin(\pi(c+x)\tau)\sin(\pi(d+x)\tau)}{\sin(\pi(2-a+x)\tau)\sin(\pi(2-b+x)\tau)}\Delta(x) \\ &= \frac{\left(1-q^{c+x}\right)\left(1-q^{d+x}\right)}{(1-q^{2-a+x})(1-q^{2-b+x})} q^{\beta}\,\Delta(x).\end{aligned} \tag{3.9}$$

To make use of the above radial part computation, we first need to construct explicit eigenfunctions of $\pi_\lambda(Y_\rho)$ with eigenvalue $\mu_\alpha(\rho)$. As we will see in the proof of the following proposition,the operator $\pi_\lambda(Y_\rho)$ is a first order difference operator, hence eigenfunctions of $\pi_\lambda(Y_\rho)$ admit the following simple characterization.

Proposition 3.4. *Let $\alpha, \rho \in \mathbb{C}$. A meromorphic function $f \in \mathcal{M}$ is an eigenfunction of $\pi_\lambda(Y_\rho)$ with eigenvalue $\mu_\alpha(\rho)$ if and only if*

$$f(z+2) = \frac{\sin(\pi(-i\lambda-\alpha-\rho+z)\tau)\sin(\pi(-i\lambda+\alpha+\rho+z)\tau)}{\sin(\pi(i\lambda-\rho+1+z)\tau)\sin(\pi(i\lambda+\rho+1+z)\tau)} f(z).$$

Proof. A direct computation shows that $\pi_\lambda(Y_\rho) \in \mathrm{End}(\mathcal{M})$ is the explicit first order difference operator

$$\begin{aligned}(\pi_\lambda(Y_\rho)f)(z) = \frac{q^{-\rho}}{q-q^{-1}}\Big\{&\left(q^{\rho-1-i\lambda-z}-1\right)\left(1-q^{\rho+1+i\lambda+z}\right)f(z+2) \\ &-\left(q^{\rho+i\lambda-z}-1\right)\left(1-q^{\rho-i\lambda+z}\right)f(z)\Big\},\end{aligned}$$

and, more generally,

$$\begin{aligned}&((\pi_\lambda(Y_\rho)-\mu_\alpha(\rho))f)(z)\\&=\frac{q^{-\rho}}{q-q^{-1}}\Big\{\left(q^{\rho-1-i\lambda-z}-1\right)\left(1-q^{\rho+1+i\lambda+z}\right)f(z+2)\\&\qquad-\left(q^{\alpha+\rho+i\lambda-z}-1\right)\left(1-q^{\alpha+\rho-i\lambda+z}\right)q^{-\alpha}f(z)\Big\}.\end{aligned}$$

The eigenvalue equation $\pi_\lambda(Y_\rho)f=\mu_\alpha(\rho)f$ is thus equivalent to the first order difference equation

$$f(z+2)=\frac{\left(1-q^{\alpha+\rho+i\lambda-z}\right)\left(1-q^{\alpha+\rho-i\lambda+z}\right)}{\left(1-q^{\rho-1-i\lambda-z}\right)\left(1-q^{\rho+1+i\lambda+z}\right)}q^{-\alpha}f(z).$$

Rewriting this formula yields the desired result. □

4. The Askey-Wilson function for $0<q<1$.

In this section we take $\tau\in i\,\mathbb{R}_{>0}$, so that $0<q=q_\tau=\exp(2\pi i\tau)<1$. The assignment

$$\left(K^{\pm1}\right)^*=K^{\pm1},\qquad \left(X^{\pm}\right)^*=-X^{\mp} \tag{4.1}$$

uniquely extends to a unital, anti-linear, anti-algebra involution on $\mathcal{U}_q$. This particular choice of $*$-structure corresponds classically to choosing the real form $\mathfrak{su}(1,1)$ of $\mathfrak{sl}(2,\mathbb{C})$.

The elements K^m ($m\in\mathbb{Z}$), the quantum Casimir element Ω (see (3.1)), and the special family Y_ρ ($\rho\in\mathbb{R}$) of Koornwinder's twisted primitive elements (3.4) are $*$-selfadjoint elements in $\mathcal{U}_q$. The eigenvalue $\mu_\alpha(\rho)$ (see (3.5)) is real for $\alpha,\rho\in\mathbb{R}$. We consider now an explicit $*$-unitary pairing for the representation π_λ.

Lemma 4.1. *Let $\lambda\in\mathbb{R}$. Suppose that $f,g\in\mathcal{M}$ are τ^{-1}-periodic and analytic on the strip $\{z\in\mathbb{C}\mid |\mathrm{Re}(z)|\le 1\}$. Then*

$$\langle\pi_\lambda(X)f,g\rangle=\langle f,\pi_\lambda(X^*)g\rangle,\qquad \forall X\in\mathcal{U}_q^1,$$

with the pairing $\langle\cdot,\cdot\rangle$ defined by

$$\langle f,g\rangle:=\int_0^1 f(y/\tau)\overline{g(y/\tau)}dy.$$

Proof. This is an easy verification for the basis elements $1,K^{\pm1},X^{\pm}$ of $\mathcal{U}_q^1$, using Cauchy's Theorem to shift contours. □

Remark 4.2. *This lemma can be applied recursively. Let $f, g \in \mathcal{M}$ be τ^{-1}-periodic and analytic on the strip $\{z \in \mathbb{C} \mid |\mathrm{Re}(z)| \leq k\}$ with $k \in \mathbb{Z}_{>0}$. For any $X = X_1 X_2 \cdots X_m \in \mathcal{U}_q$ with $m \leq k$ and $X_i \in \mathcal{U}_q^1$,*

$$\langle \pi_\lambda(X) f, g \rangle = \langle f, \pi_\lambda(X^*) g \rangle.$$

In particular, the subspace $\mathcal{O}_{\tau^{-1}}$ of entire, τ^{-1}-periodic functions is an $*$-unitary π_λ-invariant subspace of $\mathcal{M}$ with respect to the pairing $\langle \cdot, \cdot \rangle$. This subspace is the algebraic version of the principal series representation of the $*$-algebra $(\mathcal{U}_q, *)$.

To combine this lemma with the radial part computation of the quantum Casimir element (see Proposition 3.3), we need to construct meromorphic τ^{-1}-periodic eigenfunctions of $\pi_\lambda(Y_\rho)$ ($\rho \in \mathbb{R}$) which are analytic in a large enough strip around the imaginary axis. We claim that the meromorphic function $f_\lambda(z) = f_\lambda(z; \alpha, \rho)$ defined by

$$f_\lambda(z) = \frac{\Gamma_{2\tau}\left(-1 - \frac{1}{2\tau} + \alpha + \rho - i\lambda + z\right) \Gamma_{2\tau}\left(-\frac{1}{2\tau} + \rho - i\lambda - z\right)}{\Gamma_{2\tau}\left(1 - \frac{1}{2\tau} + \alpha + \rho + i\lambda - z\right) \Gamma_{2\tau}\left(-\frac{1}{2\tau} + \rho + i\lambda + z\right)} \tag{4.2}$$

meets these criteria for special values of the parameters. By Proposition 3.4 and the difference equation for Γ_τ it follows that $f_\lambda(z; \alpha, \rho)$ is an eigenfunction of $\pi_\lambda(Y_\rho)$ with eigenvalue $\mu_\alpha(\rho)$ for $\alpha, \rho, \lambda \in \mathbb{C}$. Writing $f_\lambda(z; \alpha, \rho)$ in terms of q-shifted factorials leads to the expression

$$f_\lambda(z; \alpha, \rho) = C \frac{\left(q^{2+\alpha+\rho+i\lambda-z}, q^{1+\rho+i\lambda+z}; q^2\right)_\infty}{\left(q^{\alpha+\rho-i\lambda+z}, q^{1+\rho-i\lambda-z}; q^2\right)_\infty} q^{-\frac{\alpha z}{2}} \tag{4.3}$$

for some nonzero constant C (independent of z), hence $f_\lambda(z)$ is τ^{-1}-periodic if and only if $\alpha \in 2\mathbb{Z}$. Furthermore, observe that the poles of $f_\lambda(z)$ are located at

$$i\lambda - \alpha - \rho + 2\mathbb{Z}_{\leq 0} + \mathbb{Z}\tau^{-1}, \qquad -i\lambda + \rho + 1 + 2\mathbb{Z}_{\geq 0} + \mathbb{Z}\tau^{-1}, \tag{4.4}$$

so $f_\lambda(z)$ is analytic on the strip $\{z \in \mathbb{C} \mid |\mathrm{Re}(z)| \leq \rho\}$ when $\alpha > 0$, $\rho \geq 0$ and $\lambda \in \mathbb{R}$.

Definition 4.3. *Let $\lambda \in \mathbb{R}$, $\alpha, \beta \in 2\mathbb{Z}_{>0}$ and $\rho, \sigma \in \mathbb{R}_{\geq 3}$. For $x \in \mathbb{C}$ with $|\mathrm{Re}(x)| \leq 2$ we define $\varphi_\lambda(x) = \varphi_\lambda(x; \alpha, \rho, \beta, \sigma)$ by*

$$\varphi_\lambda(x) := \langle \pi_\lambda(\widehat{x}K) f_\lambda(\cdot; \beta, \sigma), f_\lambda(\cdot; \alpha, \rho) \rangle.$$

Note that the matrix coefficient $\varphi_\lambda(x)$ is given explicitly by

$$\varphi_\lambda(x) = \int_0^1 \frac{\Gamma_{2\tau}\left(-1+\frac{1}{2\tau}+\alpha+\rho+i\lambda-\frac{y}{\tau}\right)\Gamma_{2\tau}\left(\frac{1}{2\tau}+\rho+i\lambda+\frac{y}{\tau}\right)}{\Gamma_{2\tau}\left(1+\frac{1}{2\tau}+\alpha+\rho-i\lambda+\frac{y}{\tau}\right)\Gamma_{2\tau}\left(\frac{1}{2\tau}+\rho-i\lambda-\frac{y}{\tau}\right)}\,dy$$
$$\times\frac{\Gamma_{2\tau}\left(-\frac{1}{2\tau}+\beta+\sigma-i\lambda+x+\frac{y}{\tau}\right)\Gamma_{2\tau}\left(-1-\frac{1}{2\tau}+\sigma-i\lambda-x-\frac{y}{\tau}\right)}{\Gamma_{2\tau}\left(-\frac{1}{2\tau}+\beta+\sigma+i\lambda-x-\frac{y}{\tau}\right)\Gamma_{2\tau}\left(1-\frac{1}{2\tau}+\sigma+i\lambda+x+\frac{y}{\tau}\right)}, \tag{4.5}$$

and that $\varphi_\lambda(x)$ is analytic on the strip $\{x \in \mathbb{C} \mid |\mathrm{Re}(x)| \le 2\}$. The quantum group interpretation of this explicit integral leads to the following result.

Theorem 4.4. *Let $\lambda \in \mathbb{R}$, $\alpha, \beta \in 2\mathbb{Z}_{>0}$ and $\rho, \sigma \in \mathbb{R}_{\ge 3}$. The matrix coefficient $\varphi_\lambda(x) = \varphi_\lambda(x;\alpha,\rho,\beta,\sigma)$ satisfies the second order difference equation*

$$(\mathcal{L}\varphi_\lambda)(x) = E(\lambda)\varphi_\lambda(x)$$

for generic $x \in i\mathbb{R}$, with the eigenvalue $E(\lambda) = E(\lambda;\beta)$ given by

$$E(\lambda) = -1 - q^{2-2\beta} + q^{1-\beta}\left(q^{2i\lambda}+q^{-2i\lambda}\right). \tag{4.6}$$

Proof. For the duration of the proof we use the shorthand notations $f(z) = f_\lambda(z;\beta,\sigma)$ and $g(z) = f_\lambda(z;\alpha,\rho)$. By the conditions on the parameters, Y_ρ is $*$-selfadjoint, $\mu_\alpha(\rho)$ is real and the meromorphic functions $f(z)$ and $g(z)$ are analytic on the strip $\{z \in \mathbb{C} \mid |\mathrm{Re}(z)| \le 3\}$. By Lemma 4.1 we thus obtain for any $X \in \mathcal{U}_q^1$ and $x \in i\,\mathbb{R}$,

$$\langle \pi_\lambda\left(\left(Y_\rho - \mu_\alpha(\rho)\right)\widehat{x}X\right)f, g\rangle = \langle \pi_\lambda\left(\widehat{x}X\right)f, \pi_\lambda\left(Y_\rho-\mu_\alpha(\rho)\right)g\rangle = 0,$$

and obviously also $\langle \pi_\lambda\left(\widehat{x}X\left(Y_\sigma - \mu_\beta(\sigma)\right)\right)f, g\rangle = 0$. We conclude from Proposition 3.3 that

$$\begin{aligned}\langle \pi_\lambda\left(\widehat{x}\,\Omega K\right)f, g\rangle &= \langle \pi_\lambda\left(\widehat{x}\,\Omega(x)K\right)f, g\rangle \\ &= \frac{q^{\beta-1}}{(q-q^{-1})^2}\left(\mathcal{L}\varphi_\lambda\right)(x) + \frac{q^{\beta-1}\left(1-q^{1-\beta}\right)^2}{(q-q^{-1})^2}\varphi_\lambda(x).\end{aligned}$$

On the other hand, (3.2) and Definition 4.3 implies that

$$\langle \pi_\lambda\left(\widehat{x}\,\Omega K\right)f, g\rangle = \left(\frac{q^{i\lambda}-q^{-i\lambda}}{q-q^{-1}}\right)^2\varphi_\lambda(x),$$

hence $\left(\mathcal{L}\varphi_\lambda\right)(x) = E(\lambda)\varphi_\lambda(x)$. □

For the comparison with the results for $|q| = 1$, see Section 5, it is convenient to note that the matrix coefficient $\varphi_\lambda(x)$ can be rewritten as

$$\varphi_\lambda(x) = C \int_0^1 \frac{\Gamma_{2\tau}\left(-1 - \frac{1}{2\tau} + \alpha + \rho + i\lambda - \frac{y}{\tau}\right)\Gamma_{2\tau}\left(-\frac{1}{2\tau} + \rho + i\lambda + \frac{y}{\tau}\right)}{\Gamma_{2\tau}\left(1 - \frac{1}{2\tau} + \alpha + \rho - i\lambda + \frac{y}{\tau}\right)\Gamma_{2\tau}\left(-\frac{1}{2\tau} + \rho - i\lambda - \frac{y}{\tau}\right)}$$
$$\times \frac{\Gamma_{2\tau}\left(\frac{1}{2\tau} + \beta + \sigma - i\lambda + x + \frac{y}{\tau}\right)\Gamma_{2\tau}\left(-1 + \frac{1}{2\tau} + \sigma - i\lambda - x - \frac{y}{\tau}\right)}{\Gamma_{2\tau}\left(-\frac{1}{2\tau} + \beta + \sigma + i\lambda - x - \frac{y}{\tau}\right)\Gamma_{2\tau}\left(1 - \frac{1}{2\tau} + \sigma + i\lambda + x + \frac{y}{\tau}\right)} dy \tag{4.7}$$

for some x-independent nonzero constant C. Formula (4.7) follows from (4.5) and the difference equation

$$\Gamma_{2\tau}\left(x + \frac{1}{2\tau}\right) = \exp\left(-\frac{\pi i x}{2}\right)\Gamma_{2\tau}\left(x - \frac{1}{2\tau}\right).$$

Corollary 4.5. *Let $\lambda \in \mathbb{R}$, $\alpha, \beta \in 2\mathbb{Z}_{>0}$ and $\rho, \sigma \in \mathbb{R}_{\geq 3}$. Let the Askey-Wilson parameters (a, b, c, d) be given by (3.7). The function $F_\lambda(x) = F_\lambda(x; a, b, c, d)$ defined by $F_\lambda(x) = \Delta(x)^{-1}\varphi_\lambda(x)$, with $\Delta(x) = \Delta(x; a, b, c, d)$ the τ^{-1}-periodic, meromorphic function*

$$\Delta(x) = \frac{\Gamma_{2\tau}\left(-1 + \frac{1}{2\tau} + a - x\right)\Gamma_{2\tau}\left(-1 + \frac{1}{2\tau} + b - x\right)}{\Gamma_{2\tau}\left(1 + \frac{1}{2\tau} - c - x\right)\Gamma_{2\tau}\left(1 + \frac{1}{2\tau} - d - x\right)}, \tag{4.8}$$

satisfies the Askey-Wilson second order difference equation

$$(\mathcal{D}F_\lambda)(x) = E(\lambda)F_\lambda(x)$$

for generic $x \in i\mathbb{R}$ and generic ρ and σ.

Proof. Note that $\Delta(x)$ can be rewritten as

$$\Delta(x) = C\frac{\left(q^{2-c-x}, q^{2-d-x}; q^2\right)_\infty}{\left(q^{a-x}, q^{b-x}; q^2\right)_\infty} q^{-\frac{\beta x}{2}} \tag{4.9}$$

for some nonzero (x-independent) constant C. Since $\beta \in 2\mathbb{Z}_{>0}$, the gauge factor $\Delta(x)$ is τ^{-1}-periodic. Furthermore, $\Delta(x)^{-1}$ is regular at $x \in \pm 2 + i\mathbb{R}$ and $x \in i\mathbb{R}$ under generic conditions on the parameters ρ and σ. The proof is completed by observing that $\Delta(x)$ satisfies the difference equation (3.9). □

We end this section by extending these results to continuous parameters α and β. Changing integration variable and substituting the expression for $\Gamma_{2\tau}$ in terms of q-shifted factorials, we can rewrite the matrix coefficient $\varphi_\lambda(x)$ (see (4.5)) as

$$\varphi_\lambda(x) = C\,\frac{q^{-\frac{\beta x}{2}}}{2\pi i}\int_{\mathbb{T}} \frac{\left(q^{1+\beta+\sigma+i\lambda-x}/z, q^{2+\sigma+i\lambda+x}z; q^2\right)_\infty}{\left(q^{\sigma-i\lambda-x}/z, q^{1+\beta+\sigma-i\lambda+x}z, ; q^2\right)_\infty} \times \frac{\left(q^{2+\alpha+\rho-i\lambda}z, q^{1+\rho-i\lambda}/z; q^2\right)_\infty}{\left(q^{1+\rho+i\lambda}z, q^{\alpha+\rho+i\lambda}/z; q^2\right)_\infty}\, z^{\frac{\alpha-\beta}{2}}\,\frac{dz}{z} \tag{4.10}$$

for some x-independent nonzero constant C, where $\mathbb{T} = \{z \in \mathbb{C} \mid |z| = 1\}$ is the positively oriented unit circle in the complex plane. To allow α, β to be continuous parameters, we need to get rid of the term $z^{\frac{\alpha-\beta}{2}}$ in the integrand. This can be achieved by rewriting $\varphi_\lambda(x)$ as

$$\varphi_\lambda(x) = C\,\frac{q^{-\frac{\beta x}{2}}}{2\pi i}\int_{\mathbb{T}} \frac{\left(q^{1+\beta+\sigma+i\lambda-x}/z, q^{1+\alpha+\rho-\beta-i\lambda}/z; q^2\right)_\infty}{\left(q^{1-\rho+i\lambda}z, q^{1+\rho+i\lambda}z; q^2\right)_\infty} \times \frac{\left(q^{1-\alpha-\rho+\beta+i\lambda}z, q^{2+\sigma+i\lambda+x}z, q^{2+\alpha+\rho-i\lambda}z; q^2\right)_\infty}{\left(q^{\alpha+\rho+i\lambda}/z, q^{\sigma-i\lambda-x}/z, q^{1+\beta+\sigma-i\lambda+x}z; q^2\right)_\infty}\,\frac{dz}{z}, \tag{4.11}$$

with C again some (different) irrelevant x-independent nonzero constant. The integral formula (4.11) follows from (4.10) by substitution of the identity

$$z^{\frac{\alpha-\beta}{2}}\left(q^{1+\rho-i\lambda}/z; q^2\right)_\infty = \left(-q^{i\lambda-\rho}\right)^{\frac{\beta-\alpha}{2}} q^{\frac{(\beta-\alpha)^2}{4}} \times \frac{\left(q^{1-\alpha-\rho+\beta+i\lambda}z, q^{1+\alpha+\rho-\beta-i\lambda}/z; q^2\right)_\infty}{\left(q^{1-\rho+i\lambda}z; q^2\right)_\infty},$$

which in turn is a direct consequence of the functional equation

$$\theta(q^{2k}z) = (-z)^{-k}q^{-k^2}\theta(z), \qquad k \in \mathbb{Z}$$

for the modified Jacobi theta function $\theta(z) = \left(qz, q/z; q^2\right)_\infty$. By (4.9), the eigenfunction $F_\lambda(x) = F_\lambda(x; a, b, c, d)$ of the Askey-Wilson second order difference equation $\mathcal{D}$ (see Corollary 4.5) is equal to

$$\mathcal{F}_\lambda(x) = \mathcal{F}_\lambda(x; a, b, c, d) := \frac{\left(q^{1+\rho+\sigma-x}, q^{1-\rho+\sigma-x}; q^2\right)_\infty}{\left(q^{1-\alpha-\rho+\beta+\sigma-x}, q^{1+\alpha+\rho+\beta+\sigma-x}; q^2\right)_\infty} \times \frac{1}{2\pi i}\int_{\mathbb{T}} \frac{\left(q^{1+\beta+\sigma+i\lambda-x}/z, q^{1+\alpha+\rho-\beta-i\lambda}/z; q^2\right)_\infty}{\left(q^{1-\rho+i\lambda}z, q^{1+\rho+i\lambda}z; q^2\right)_\infty} \times \frac{\left(q^{1-\alpha-\rho+\beta+i\lambda}z, q^{2+\sigma+i\lambda+x}z, q^{2+\alpha+\rho-i\lambda}z; q^2\right)_\infty}{\left(q^{\alpha+\rho+i\lambda}/z, q^{\sigma-i\lambda-x}/z, q^{1+\beta+\sigma-i\lambda+x}z; q^2\right)_\infty}\,\frac{dz}{z} \tag{4.12}$$

up to some nonzero x-independent multiplicative constant.

Define for the Askey-Wilson parameters (a, b, c, d) given by (3.7), dual Askey-Wilson parameters (a, b, c, d) by

$$(a, b, c, d) = (1 - \beta, 1 + \beta + 2\sigma, 1 + \alpha + 2\rho, 1 - \alpha).$$

This notion of dual Askey-Wilson parameters coincides with the notion of dual parameters as used in, e.g., (Koelink and Stokman, 2001a).

In the following theorem we express $\mathcal{F}_\lambda(x)$ in terms of basic hypergeometric series by shrinking the radius of the integration circle $\mathbb{T}$ to zero while picking up residues. Recall that the very-well-poised ${}_8\varphi_7$ series is defined by

$$\begin{aligned}{}_8W_7\,(u; b_1, b_2, b_3, b_4, b_5; q, z) \\ = \sum_{k=0}^{\infty} \frac{(1 - uq^{2k})\,(u, b_1, b_2, b_3, b_4, b_5; q)_k\, z^k}{(1 - u)\,(q, qu/b_1, qu/b_2, qu/b_3, qu/b_4, qu/b_5; q)_k}\end{aligned}$$

for $|z| < 1$, see (Gasper and Rahman, 1990).

Theorem 4.6. *Let $\lambda \in \mathbb{R}$, $\alpha, \beta \in \mathbb{R}_{>0}$ and $\rho, \sigma \in \mathbb{R}_{\geq 3}$. Under the parameter correspondence (3.7), the function $\mathcal{F}_\lambda(x) = \mathcal{F}_\lambda(x; a, b, c, d)$ given by (4.12) can be expressed in terms of basic hypergeometric series as*

$$\begin{aligned}\mathcal{F}_\lambda(x) &= C\, \frac{\left(q^{2+a-d+2i\lambda+x}, q^{2+a-d+2i\lambda-x}; q^2\right)_\infty}{\left(q^{2-d+x}, q^{2-d-x}; q^2\right)_\infty} \\ &\times {}_8W_7\left(q^{-2+a+b+c+2i\lambda}; q^{a+x}, q^{a-x}, q^{a+2i\lambda}, q^{b+2i\lambda}, q^{c+2i\lambda}; q^2, q^{2-d-2i\lambda}\right)\end{aligned} \tag{4.13}$$

with the (irrelevant) generically nonzero, x-independent constant C given by

$$C = \frac{\left(q^{2+2\sigma}, q^{2+\alpha+2\rho-\beta}, q^{2+\alpha+2\rho+\beta+2\sigma}, q^{1+\beta+2i\lambda}, q^{1+\alpha-2i\lambda}; q^2\right)_\infty}{\left(q^2, q^{1+\alpha+2i\lambda}, q^{1+\alpha+2\rho+2i\lambda}, q^{1+\beta+2\sigma-2i\lambda}, q^{3+\alpha+2\rho+2\sigma+2i\lambda}; q^2\right)_\infty}.$$

Furthermore, $\mathcal{F}_\lambda(x)$ is τ^{-1}-periodic and satisfies the Askey-Wilson difference equation $(\mathcal{D}\mathcal{F}_\lambda)(x) = E(\lambda)\mathcal{F}_\lambda(x)$ for generic $x \in i\,\mathbb{R}$.

Proof. The expression for $\mathcal{F}_\lambda$ follows by shrinking the radius of the integration contour $\mathbb{T}$ to zero while picking up residues. It is actually a special case of (Gasper and Rahman, 1990, Exerc. 4.4, p. 122), in which one should replace the base q by q^2 and the parameters (a, b, c, d, f, g, h, k)

by

$$\Big(q^{1-\beta-\sigma-i\lambda+x}, q^{1-\rho+i\lambda}, q^{\alpha+\rho+i\lambda}, q^{\sigma-i\lambda-x}, q^{2+\sigma+i\lambda+x}, \\ q^{1+\rho+i\lambda}, q^{1+\beta+\sigma-i\lambda+x}, q^{1-\beta-2i\lambda}\Big).$$

We have seen that the difference equation $(\mathcal{D}\mathcal{F}_\lambda)(x) = E(\lambda)\mathcal{F}_\lambda(x)$ for generic $x \in i\,\mathbb{R}$ is valid under the extra assumption $\alpha, \beta \in 2\mathbb{Z}_{>0}$. For $\alpha, \beta \in \mathbb{R}_{>0}$ this difference equation has been proved by Ismail and Rahman (Ismail and Rahman, 1991) using the explicit expression of $\mathcal{F}_\lambda(x)$ as very-well-poised ${}_8\varphi_7$ series. □

Using the explicit expressions of $\mathcal{F}_\lambda(x)$, the conditions on x, λ and the four parameters $\alpha, \rho, \beta, \sigma$ can be relaxed by meromorphic continuation. The resulting function $\mathcal{F}_\lambda$ is a meromorphic, τ^{-1}-periodic eigenfunction of the Askey-Wilson second order difference operator $\mathcal{D} = \mathcal{D}^{a,b,c,d}$ with eigenvalue $E(\lambda) = E(\lambda; \beta)$.

Remark 4.7. *Some special cases of the matrix coefficients φ_λ were explicitly expressed in terms of very-well-poised ${}_8\varphi_7$ series in (Koelink and Stokman, 2001b) using the realization of the representation π_λ on the representation space $l^2(\mathbb{Z})$. In this approach the basic hypergeometric series manipulations are much harder, since one needs a highly nontrivial evaluation of a non-symmetric Poisson type kernel involving nonterminating ${}_2\varphi_1$-series which is due to Rahman, see the appendix of (Koelink and Stokman, 2001b) and (Koelink and Rosengren, 2002).*

Remark 4.8. *The explicit ${}_8\varphi_7$ expression* (4.13) *of $\mathcal{F}_\lambda(x)$ and its analytic continuation was named the Askey-Wilson function in (Koelink and Stokman, 2001a). Suslov (Suslov, 1997), (Suslov, 2002) established Fourier-Bessel type orthogonality relations for the Askey-Wilson function. Koelink and the author (Koelink and Stokman, 2001b), (Koelink and Stokman, 2001a) defined a generalized Fourier transform involving the Askey-Wilson function as the integral kernel, and established its Plancherel and inversion formula. This transform, called the Askey-Wilson function transform, arises as Fourier transform on the noncompact quantum group* $\mathrm{SU}_q(1,1)$ *(see (Koelink and Stokman, 2001b)), and may thus be seen as a natural analogue of the Jacobi function transform.*

5. The expansion formula and the elliptic cosine kernel

In this section we still assume that $\tau \in i\,\mathbb{R}_{>0}$, so $0 < q = q_\tau = \exp(2\pi i\tau) < 1$. To keep contact with the conventions of the previous section, we keep working in base q^2.

First we recall the (normalized) Askey-Wilson polynomials (Askey and Wilson, 1985). The Askey-Wilson polynomials $E_m(x) = E_m(x; a, b, c, d)$ $(m \in \mathbb{Z}_{\geq 0})$ are defined by

$$E_m(x) = {}_4\varphi_3\left(\begin{matrix} q^{-2m}, q^{2m-2+a+b+c+d}, q^{a+x}, q^{a-x} \\ q^{a+b}, q^{a+c}, q^{a+d} \end{matrix}; q^2, q^2\right),$$

with

$${}_4\varphi_3\left(\begin{matrix} a_1, a_2, a_3, a_4 \\ b_1, b_2, b_3 \end{matrix}; q, z\right) = \sum_{k=0}^{\infty} \frac{(a_1, a_2, a_3, a_4; q)_k}{(q, b_1, b_2, b_3; q)_k} z^k, \qquad |z| < 1.$$

The Askey-Wilson polynomial $E_m(x)$ is a polynomial in $q^x + q^{-x}$ of degree m, normalized by $E_m(a) = 1$. They satisfy the orthogonality relations

$$\int_0^1 \frac{E_m(x/\tau)E_n(x/\tau)\,(q^{2x}, q^{-2x}; q^2)_\infty}{(q^{a+x}, q^{a-x}, q^{b+x}, q^{b-x}, q^{c+x}, q^{c-x}, q^{d+x}, q^{d-x}; q^2)_\infty}\, dx = 0,$$

for $m \neq n$, provided that $\mathrm{Re}(a), \mathrm{Re}(b), \mathrm{Re}(c), \mathrm{Re}(d) > 0$. Observe that the Askey-Wilson polynomial $E_m(x)$ is regular at the special choice

$$(a, b, c, d) = \left(0, \frac{1}{2\tau}, 1, 1 - \frac{1}{2\tau}\right) \tag{5.1}$$

of Askey-Wilson parameters. The above orthogonality relations also extend (by continuity) to the Askey-Wilson parameters (5.1), leading to

$$\int_0^1 E_m\left(x/\tau; 0, \frac{1}{2\tau}, 1, 1 - \frac{1}{2\tau}\right) E_n\left(x/\tau; 0, \frac{1}{2\tau}, 1, 1 - \frac{1}{2\tau}\right) dx = 0,$$

for $m \neq n$, hence we conclude that

$$E_m\left(x; 0, \frac{1}{2\tau}, 1, 1 - \frac{1}{2\tau}\right) = \cos(2\pi m\tau x), \qquad m \in \mathbb{Z}_{\geq 0}, \tag{5.2}$$

which is the usual cosine kernel from the Fourier theory on the unit circle, cf. (Askey and Wilson, 1985, (4.25)). In this section we derive an analogous result for the Askey-Wilson function.

In the previous section we have introduced the dual Askey-Wilson parameters associated to a, b, c and d. They can be alternatively expressed as

$$(a, b, c, d) = \Big(-1 + \frac{1}{2}(a + b + c + d), 1 + \frac{1}{2}(a + b - c - d),$$
$$1 + \frac{1}{2}(a - b + c - d), 1 + \frac{1}{2}(a - b - c + d)\Big).$$

Note that the special choice (5.1) of Askey-Wilson parameters is self dual, $(a,b,c,d) = (a,b,c,d)$. For our present purposes it is convenient to use yet another normalization of the Askey-Wilson function, namely

$$\begin{aligned}\mathfrak{E}^+(\mu,x) &= \mathfrak{E}^+(\mu,x;a,b,c,d)\\ &:= \frac{\left(q^{2+a-d+\mu+x}, q^{2+a-d+\mu-x}, q^{2-a-d}, q^{2+a-d}; q^2\right)_\infty}{\left(q^{a+b+c+\mu}, q^{2-d+\mu}, q^{2-d+x}, q^{2-d-x}; q^2\right)_\infty}\\ &\quad\times {}_8W_7\left(q^{-2+a+b+c+\mu}; q^{a+x}, q^{a-x}, q^{a+\mu}, q^{b+\mu}, q^{c+\mu}; q^2, q^{2-d-\mu}\right),\end{aligned}$$

for $\left|q^{2-d-\mu}\right| < 1$. For fixed λ, the eigenfunction $\mathcal{F}_\lambda(\cdot)$ of the Askey-Wilson second order difference operator $\mathcal{D}$ is a constant multiple of $\mathfrak{E}^+(2i\lambda,\cdot)$. The present normalization of the Askey-Wilson function is convenient due to the properties

$$\mathfrak{E}^+(\mu,x;a,b,c,d) = \mathfrak{E}^+(x,\mu;a,b,c,d) \tag{5.3}$$

and $\mathfrak{E}^+(-a,-a) = 1$. The property (5.3) is called *duality* and can be proved using a transformation formula for very-well-poised ${}_8\varphi_7$ series, see (Koelink and Stokman, 2001a) for details. Furthermore,

$$\mathfrak{E}^+(a+2m,x) = E_m(x), \qquad m \in \mathbb{Z}_{\geq 0}, \tag{5.4}$$

see, e.g., (Koelink and Stokman, 2001a, (3.5)), thus the Askey-Wilson function $\mathfrak{E}^+(\mu,x)$ provides a natural meromorphic continuation of the Askey-Wilson polynomial in its degree.

The meromorphic continuation of $\mathfrak{E}^+(\mu,x)$ in μ and x can be established by the integral representation of the Askey-Wilson function (see the previous section), or by the expression of the Askey-Wilson function as a sum of two balanced ${}_4\varphi_3$'s (see, e.g., (Koelink and Stokman, 2001a, (3.3))). For our present purposes, it is most convenient to consider the meromorphic continuation via the expansion formula of the Askey-Wilson function in Askey-Wilson polynomials, given by

$$\begin{aligned}\mathfrak{E}^+(\mu,x) &= \frac{\left(q^{2-a-d}, q^{2+a-d}, q^{b+c}, q^{2+b-d}, q^{2+c-d}; q^2\right)_\infty}{\left(q^{2+a+b+c-d}, q^{2-d+x}, q^{2-d-x}, q^{2-d+\mu}, q^{2-d-\mu}; q^2\right)_\infty}\\ &\quad\times \sum_{m=0}^{\infty} E_m(x;a,b,c,2-d)E_m(\mu;a,b,c,2-d)\\ &\quad\times \frac{\left(1-q^{4m+a+b+c-d}\right)\left(q^{a+b+c-d}, q^{a+b}, q^{a+c}; q^2\right)_m}{\left(1-q^{a+b+c-d}\right)\left(q^2, q^{2+b-d}, q^{2+c-d}; q^2\right)_m}(-1)^m q^{(1-a-d)m} q^{m^2},\end{aligned} \tag{5.5}$$

see (Stokman, 2002, Thm. 4.2). The sum converges absolutely and uniformly on compacta of $(\mu,x) \in \mathbb{C}\times\mathbb{C}$ due to the Gaussian q^{m^2}. The expansion formula (5.5) shows that the Askey-Wilson function $\mathfrak{E}^+(\mu, x)$ is well defined and regular at the special choice (5.1) of Askey-Wilson parameters. In fact, for this special choice of parameters, the Askey-Wilson function can be expressed in terms of the (renormalized) Jacobi theta function

$$\vartheta(x) = \left(-q^{1+x}, -q^{1-x}; q^2\right)_\infty$$

as follows.

Proposition 5.1. *We have the identity*

$$\mathfrak{E}^+\left(\mu, x; 0, \frac{1}{2\tau}, 1, 1 - \frac{1}{2\tau}\right) = \frac{(-q, -q; q^2)_\infty}{2} \left(\frac{\vartheta(\mu + x) + \vartheta(\mu - x)}{\vartheta(\mu)\vartheta(x)}\right). \tag{5.6}$$

Proof. To simplify notations, we write

$$\mathfrak{E}_0^+(\mu, x) = \mathfrak{E}^+\left(\mu, x; 0, \frac{1}{2\tau}, 1, 1 - \frac{1}{2\tau}\right) \tag{5.7}$$

for the duration of the proof.

We substitute the special choice (5.1) of Askey-Wilson parameters in the expansion formula for $\mathfrak{E}^+(\mu, x)$. By simple q-series manipulations and by (5.2), we obtain the explicit formula

$$\mathfrak{E}_0^+(\mu, x) = \frac{(-q, -q, -q, q, -q^2; q^2)_\infty}{(q^2; q^2)_\infty\, \vartheta(\mu)\vartheta(x)} \times \left(1 + 2\sum_{m=1}^{\infty} \cos(2\pi m\tau\mu)\cos(2\pi m\tau x)\, q^{m^2}\right).$$

Using the well known Jacobi triple product identity

$$\sum_{m=-\infty}^{\infty} q^{m^2+mx} = 1 + 2\sum_{m=1}^{\infty} \cos(2\pi m\tau x)\, q^{m^2} = \left(q^2; q^2\right)_\infty \vartheta(x),$$

and the elementary identity

$$\cos(2\pi m\tau\mu)\cos(2\pi m\tau x) = \frac{1}{2}\left(\cos(2\pi m\tau(\mu + x)) + \cos\left(2\pi m\tau(\mu - x)\right)\right),$$

we deduce that

$$\mathfrak{E}_0^+(\mu, x) = \frac{(-q, -q, -q, q, -q^2; q^2)_\infty}{2} \left(\frac{\vartheta(\mu + x) + \vartheta(\mu - x)}{\vartheta(\mu)\vartheta(x)}\right).$$

Simplifying the multiplicative constant yields the desired result. □

Remark 5.2. *The Jacobi theta function $\vartheta(x)$ is the natural τ^{-1}-periodic analogue of the Gaussian q^{-x^2}, since*

$$\sum_{k=-\infty}^{\infty} q^{-(x+k\tau^{-1})^2} = \sqrt{-2i\tau}\,(q^2;q^2)_\infty\,\vartheta(x)$$

by the Jacobi triple product identity and the Jacobi inversion formula. In fact, in (Stokman, 2002) and (Stokman, 2001) it is shown that the function $(q^{2-d+x}, q^{2-d-x}; q^2)_\infty$, which reduces to $\vartheta(x)$ for the Askey-Wilson parameters (5.1), *plays the role of the Gaussian in the Askey-Wilson theory. If one replaces the theta functions by Gaussians in the right hand side of* (5.6), *then we obtain up to a multiplicative constant*

$$\frac{1}{2}\left(\frac{q^{-(\mu+x)^2}+q^{-(\mu-x)^2}}{q^{-\mu^2-x^2}}\right) = \frac{1}{2}\left(q^{-2\mu x}+q^{2\mu x}\right) = \cos(4\pi\tau\mu x),$$

which is essentially the classical cosine kernel. Thus the right hand side of (5.6) *is an elliptic analogue of the cosine kernel.*

Remark 5.3. *Using the quasi-periodicity*

$$\vartheta(x+2) = q^{-1-x}\vartheta(x) \tag{5.8}$$

of the Jacobi theta function, we obtain as a consequence of (5.6),

$$\mathfrak{C}^+\left(2m, x; 0, \frac{1}{2\tau}, 1, 1-\frac{1}{2\tau}\right) = \cos(2\pi m\tau x), \qquad m \in \mathbb{Z}_{\geq 0},$$

which is in accordance with (5.2) *and* (5.4).

Remark 5.4. *The orthogonality relations for the Askey-Wilson polynomials with Askey-Wilson parameters* (5.1) *are equivalent to the L^2-theory of the classical Fourier tansform on the unit circle. On the other hand, the L^2-theory of the Askey-Wilson function transform, see (Koelink and Stokman, 2001a), does not reduce to the L^2-theory of the classical Fourier theory on the real line for the Askey-Wilson parameters* (5.1). *Instead one obtains a Fourier type transform with integral kernel given by the elliptic cosine function* (5.6). *For the corresponding L^2 theory, the transform is defined on a weighted L^2-space consisting of functions that are supported on a finite closed interval and an infinite, unbounded sequence of discrete mass points. This transform, as well as the general Askey-Wilson function transform, still has many properties in common with the classical Fourier transform on the real line, see e.g., (Koelink and Stokman, 2001a), (Stokman, 2002) and (Stokman, 2001).*

Cherednik's (Cherednik, 1997) Hecke algebra approach to q-special functions leads to a direct proof that the right hand side of the expansion formula (5.5) is an eigenfunction of the Askey-Wilson second order difference operator $\mathcal{D}$, see (Stokman, 2001). The expansion formula (5.5) may thus be seen as the explicit link between Cherednik's approach and Ismail's and Rahman's (Ismail and Rahman, 1991) construction of eigenfunctions of $\mathcal{D}$ in terms of very-well-poised ${}_8\varphi_7$ series. We end this section by sketching a proof of Proposition 5.1 using Cherednik's Hecke algebra approach.

The affine Hecke algebra techniques for Askey-Wilson polynomials are developed in full detail in (Noumi and Stokman, 2000), and for Askey-Wilson functions in (Stokman, 2001). We first recall one of the main results from (Stokman, 2001), specialized to the present rank one situation.

Define two difference-reflection operators by

$$\begin{aligned}\left(T_0^{c,d}f\right)(x) &= -q^{-2+c+d}f(x) + \frac{\left(1-q^{c-x}\right)\left(1-q^{d-x}\right)}{\left(1-q^{2-2x}\right)}\left(f(2-x)-f(x)\right),\\ \left(T_1^{a,b}f\right)(x) &= -q^{a+b}f(x) + \frac{\left(1-q^{a+x}\right)\left(1-q^{b+x}\right)}{\left(1-q^{2x}\right)}\left(f(-x)-f(x)\right).\end{aligned} \tag{5.9}$$

The connection with affine Hecke algebras follows from the fact that $T_0 = T_0^{c,d}$ and $T_1 = T_1^{a,b}$ satisfy Hecke type quadratic relations. These relations imply that the operators T_0 and T_1 are invertible. Consider the (invertible) operator

$$Y = Y^{a,b,c,d} := T_1^{a,b} \circ T_0^{c,d}. \tag{5.10}$$

Remark 5.5. *The operator $Y + Y^{-1}$, acting on even functions, is essentially the Askey-Wilson second order difference operator $\mathcal{D}$; see, e.g., (Noumi and Stokman, 2000, Prop. 5.8).*

Theorem 5.17 in (Stokman, 2001) states that for generic Askey-Wilson parameters (a,b,c,d), there exists a unique meromorphic function $\mathfrak{E}(\cdot,\cdot) = \mathfrak{E}(\cdot,\cdot;a,b,c,d)$ on $\mathbb{C}\times\mathbb{C}$ satisfying the following six conditions:

1. $\mathfrak{E}(\mu,x)$ is τ^{-1}-periodic in μ and x,
2. $(\mu,x) \mapsto \left(q^{2-d+\mu}, q^{2-d-\mu}, q^{2-d+x}, q^{2-d-x}; q^2\right)_\infty \mathfrak{E}(\mu,x)$ is analytic,
3. For fixed generic $\mu\in\mathbb{C}$, $\mathfrak{E}(\mu,\cdot)$ is an eigenfunction of $Y^{a,b,c,d}$ with eigenvalue $q^{a-\mu}$,

4. For fixed generic $x \in \mathbb{C}$, $\mathfrak{E}(\cdot, x)$ is an eigenfunction of $Y^{a,b,c,d}$ with eigenvalue q^{a-x},

5. $\left(T_1^{a,b}\mathfrak{E}(\mu,\cdot)\right)(x) = -q^{a+b}\left(T_1^{a,b}\mathfrak{E}(\cdot,x)\right)(\mu)$,

6. $\mathfrak{E}(-a,-a) = 1$.

The existence of a kernel $\mathfrak{E}$ satisfying the above six conditions is proved by explicitly constructing $\mathfrak{E}$ as series expansion in nonsymmetric analogues of the Askey-Wilson polynomials, see (Stokman, 2001, (6.6)). This expansion formula for $\mathfrak{E}$ is very similar to the expansion formula (5.5) of the Askey-Wilson function $\mathfrak{E}^+$ in Askey-Wilson polynomials. In fact, a comparison of the formulas leads to the explicit link

$$\mathfrak{E}^+(\mu,x) = \left(C_{a,b}^+\mathfrak{E}(\mu,\cdot)\right)(x), \qquad C_{a,b}^+ := \frac{1}{1-q^{a+b}}\left(1+T_1^{a,b}\right), \quad (5.11)$$

see (Stokman, 2001, Thm. 6.20). These results allow us to study the Askey-Wilson function $\mathfrak{E}^+$ using the characterizing conditions **1**–**6** for the underlying kernel $\mathfrak{E}$, instead of focussing on the explicit expression for $\mathfrak{E}^+$.

The kernel $\mathfrak{E}(\mu,x)$ is regular at the Askey-Wilson parameters (5.1). The resulting kernel

$$\mathfrak{E}_0(\mu,x) = \mathfrak{E}\left(\mu,x;0,\frac{1}{2\tau},1,1-\frac{1}{2\tau}\right),$$

is the unique meromorphic kernel satisfying the six conditions **1**–**6** for the special Askey-Wilson parameters (5.1). Observe that the operators T_0, T_1 and Y for the special Askey-Wilson parameters (5.1) reduce to

$$(T_0 f)(x) = f(2-x), \qquad (T_1 f)(x) = f(-x), \qquad (Yf)(x) = f(2+x),$$

hence $\mathfrak{E}_0$ is the unique meromorphic kernel satisfying the six conditions

1′. $\mathfrak{E}_0(\mu,x)$ is τ^{-1}-periodic in μ and x,

2′. $(\mu,x) \mapsto \vartheta(\mu)\vartheta(x)\mathfrak{E}_0(\mu,x)$ is analytic,

3′. $\mathfrak{E}_0(\mu,x+2) = q^{-\mu}\mathfrak{E}_0(\mu,x)$,

4′. $\mathfrak{E}_0(\mu+2,x) = q^{-x}\mathfrak{E}_0(\mu,x)$,

5′. $\mathfrak{E}_0(\mu,-x) = \mathfrak{E}_0(-\mu,x)$,

6′. $\mathfrak{E}_0(0,0) = 1$.

We conclude that

$$\mathfrak{E}_0(\mu, x) = \left(-q, -q; q^2\right)_\infty \frac{\vartheta(\mu + x)}{\vartheta(\mu)\vartheta(x)},$$

since the right hand side satisfies **1′**–**6′** due to the quasi-periodicity (5.8) of $\vartheta(x)$. Thus $\mathfrak{E}_0(\mu, x)$ is an elliptic analogue of the exponential kernel $\exp(-4\pi i\tau\mu x)$, cf. Remark 5.2. Using the notation (5.7), we conclude that

$$\mathfrak{E}_0^+(\mu, x) = \left(C^+_{0,\frac{1}{2\tau}} \mathfrak{E}_0(\mu, \cdot)\right)(x) = \frac{1}{2}\left(\mathfrak{E}_0(\mu, x) + \mathfrak{E}_0(\mu, -x)\right),$$

which is the desired formula (5.6).

6. The Askey-Wilson function for $|q| = 1$.

In this section we take $-\frac{1}{2} < \tau < 0$, so that $q = q_\tau = \exp(2\pi i\tau)$ has modulus one and $q \neq \pm 1$. The assignment

$$(K^{\pm 1})^\star = K^{\pm 1}, \qquad (X^\pm)^\star = -X^\pm \tag{6.1}$$

uniquely extends to a unital, anti-linear, anti-algebra involution on $\mathcal{U}_q$. This particular choice of $\star$-structure corresponds to the real form $\mathfrak{sl}(2, \mathbb{R})$ of $\mathfrak{sl}(2, \mathbb{C})$. The $\star$-unitary sesquilinear form for the representations π_λ ($\lambda \in \mathbb{R}$) of $\mathcal{U}_q$ (see Section 3) is

$$\langle f, g\rangle' = \int_{-i\infty}^{i\infty} f(z)\overline{g(\overline{z})}\, dz,$$

where f and g are meromorphic functions which are regular on a large enough strip around the imaginary axes and decay sufficiently fast at $\pm i\infty$. Koornwinder's twisted primitive element $iY_\rho \in \mathcal{U}_q$ is $\star$-selfadjoint for $\rho \in \mathbb{R}$. Thus in principle we are all set to extend the construction of eigenfunctions of the Askey-Wilson second order difference operator $\mathcal{D}$ to the $|q| = 1$ case by simply replacing the role of the q-gamma function $\Gamma_{2\tau}$ by $G_{2\tau}$. We need to be careful though due to the following differences with the $0 < q < 1$ case:

a. The analogue of the explicit eigenfunction of iY_ρ (cf. (4.2)), given now as quotient of hyperbolic gamma functions $G_{2\tau}$, has more singularities.

b. No τ^{-1}-periodicity conditions have to be imposed. Consequently, the parameters α and β do not need to be discretized for the $|q| = 1$ case.

c. We have to take the decay rates at $\pm i\infty$ of integrands into account.

One needs to be careful with the decay rate (see **c**) in reproving the crucial Theorem 4.4 because acting by $\pi_\lambda(X^\pm)$ worsens the asymptotics at $\pm i\infty$ (the factor $q^z = \exp(2\pi i\tau z)$ is $\mathcal{O}(\exp(-2\pi\tau \operatorname{Im}(z))$ as $\operatorname{Im}(z) \to \infty$). The decay rate can be improved by considering different eigenfunctions of iY_ρ, but then the location of the singularities turns out to cause problems.

To get around these problems, we generalize the techniques of Section 4 to a *nonunitary* set-up. More concretely, we replace $\langle\cdot,\cdot\rangle'$ by a bilinear form, given as a contour integral over a certain deformation of $i\mathbb{R}$. With such a bilinear form, the singularity problems and the asymptotic problems can be resolved simultaneously.

The proper replacement of the involution $\star$ is the unique unital, *linear*, anti-algebra involution $\circ$ on $\mathcal{U}_q$ satisfying

$$(K^{\pm 1})^\circ = K^{\mp 1}, \qquad (X^\pm)^\circ = -X^\pm.$$

For $f \in \mathcal{M}$ we write $S(f) \subset \mathbb{C}$ for the singular set of f.

Definition 6.1. *$f \in \mathcal{M}$ is said to have (exponential) growth rate $\epsilon \in \mathbb{R}$ at $\pm i\infty$ when the following two conditions are satisfied:*

1. *For some compact subset $K_f \subset \mathbb{R}$,*

$$S(f) \subset \{z \in \mathbb{C} \mid \operatorname{Im}(z) \in K_f\}.$$

2. *The function f satisfies*

$$|f(x+iy)| = \mathcal{O}\left(\exp(\epsilon|y|)\right), \qquad y \to \pm\infty,$$

uniformly for x in compacts of $\mathbb{R}$.

We call a contour $\mathcal{C}$ a *deformation of $i\mathbb{R}$* when $\mathcal{C}$ intersects the line

$$l_c = \{z \in \mathbb{C} \mid \operatorname{Im}(z) = c\}$$

in exactly one point $z_c(\mathcal{C})$ for all $c \in \mathbb{R}$, and $z_c(\mathcal{C}) = ic$ for $|c| \gg 0$. For $k \in \mathbb{Z}_{\geq 0}$ we define the strip of radius k around $\mathcal{C}$ by

$$\bigcup_{c \in \mathbb{R}} \{z \in l_c \mid |z - z_c(\mathcal{C})| \leq k\}.$$

For $k = 0$, the strip of radius zero around $\mathcal{C}$ is the contour $\mathcal{C}$ itself.

Lemma 6.2. *Suppose that $f, g \in \mathcal{M}$ have growth rates ϵ_f and ϵ_g at $\pm i\infty$, respectively. Suppose furthermore that $\epsilon_f + \epsilon_g < 2\pi\tau$ and that f*

and g are analytic on the strip of radius one around a given (oriented) deformation $\mathcal{C}$ of $i\,\mathbb{R}$. Then

$$(\pi_\lambda(X)f,g)_{\mathcal{C}} = (f,\pi_{-\lambda}(X^\circ)\,g)_{\mathcal{C}}, \qquad \forall X \in \mathcal{U}_q^1,$$

where

$$(f,g)_{\mathcal{C}} = \int_{\mathcal{C}} f(z)g(z)dz.$$

Proof. The proof follows by an elementary application of Cauchy's Theorem. □

Remark 6.3. *The condition on the growth rates in Lemma 6.2 may be weakened to $\epsilon_f + \epsilon_g < 0$ when X is taken from the subspace* $\mathrm{span}_{\mathbb{C}}\,\{1, K^{-1}, K\}$ *of $\mathcal{U}_q^1$.*

Koornwinder's twisted primitive element Y_ρ (see (3.4)) is not $\circ$-invariant,

$$Y_\rho^\circ = -q^{\frac{1}{2}}K^{-1}X^+ + q^{-\frac{1}{2}}K^{-1}X^- + \left(\frac{q^{-\rho}+q^{\rho}}{q^{-1}-q}\right)\left(K^{-2}-1\right).$$

On the other hand, a direct computation shows that $\pi_{-\lambda}\left(Y_\rho^\circ\right)$ is still a first order difference operator. This leads to the following analogue of Proposition 3.4.

Proposition 6.4. *Let $\alpha, \rho \in \mathbb{C}$. A meromorphic function $f \in \mathcal{M}$ is an eigenfunction of $\pi_{-\lambda}\left(Y_\rho^\circ\right)$ with eigenvalue $\mu_\alpha(\rho)$ if and only if*

$$f(z-2) = \frac{\sin(\pi(i\lambda+\alpha+\rho-z)\tau)\sin(\pi(-i\lambda+\alpha+\rho+z)\tau)}{\sin(\pi(i\lambda+\rho-1+z)\tau)\sin(\pi(-i\lambda+\rho+1-z)\tau)}\,f(z).$$

We define two meromorphic functions by

$$\begin{aligned} g_\lambda(z;\beta,\sigma) &= \frac{G_{2\tau}\left(\frac{1}{2\tau}-1+\beta+\sigma-i\lambda+z\right)G_{2\tau}\left(\frac{1}{2\tau}+\sigma-i\lambda-z\right)}{G_{2\tau}\left(-\frac{1}{2\tau}+1+\beta+\sigma+i\lambda-z\right)G_{2\tau}\left(-\frac{1}{2\tau}+\sigma+i\lambda+z\right)},\\ h_\lambda(z;\alpha,\rho) &= \frac{G_{2\tau}\left(-\frac{1}{2\tau}-1+\alpha+\rho+i\lambda-z\right)G_{2\tau}\left(-\frac{1}{2\tau}+\rho+i\lambda+z\right)}{G_{2\tau}\left(-\frac{1}{2\tau}+1+\alpha+\rho-i\lambda+z\right)G_{2\tau}\left(-\frac{1}{2\tau}+\rho-i\lambda-z\right)}. \end{aligned} \tag{6.2}$$

The difference equation for $G_{2\tau}$, Proposition 3.4 and Proposition 6.4 imply

$$\begin{aligned} \pi_\lambda\left(Y_\sigma\right)g_\lambda(\cdot;\beta,\sigma) &= \mu_\beta(\sigma)g_\lambda(\cdot;\beta,\sigma),\\ \pi_{-\lambda}\left(Y_\rho^\circ\right)h_\lambda(\cdot;\alpha,\rho) &= \mu_\alpha(\rho)h_\lambda(\cdot;\alpha,\rho). \end{aligned} \tag{6.3}$$

We want to construct now eigenfunctions of the gauged Askey-Wilson second order difference operator $\mathcal{L}$ which are of the form

$$\begin{aligned}\psi_\lambda(x;\alpha,\rho,\beta,\sigma) &= (\pi_\lambda(\widehat{x}K)\,g_\lambda(\cdot;\beta,\sigma),h_\lambda(\cdot;\alpha,\rho))_{\mathcal{C}} \\ &= \int_{\mathcal{C}} g_\lambda(1+x+z;\beta,\sigma)h_\lambda(z;\alpha,\rho)dz \end{aligned} \tag{6.4}$$

for a suitable deformation $\mathcal{C}$ of $i\,\mathbb{R}$. To make sense of this integral, we need to take the singularities and the asymptotic behaviour at $\pm i\infty$ of the integrand into account. The singularities can be located using the precise information on the zeros and poles of the hyperbolic gamma function G_τ, see Proposition 2.1**(ii)**. It follows that the singularities of $z \mapsto g_\lambda(1+x+z;\beta,\sigma)$ are contained in the union of the four half lines

$$\begin{aligned} &-1-\beta-\sigma+i\lambda-x+\mathbb{R}_{\leq 0}, &&-1+\beta+\sigma+i\lambda-x+\mathbb{R}_{\leq 0},\\ &-\sigma-i\lambda-x+\mathbb{R}_{\geq 0}, &&\sigma-i\lambda-x+\mathbb{R}_{\geq 0},\end{aligned}$$

and the singularities of $z \mapsto h_\lambda(z;\alpha,\rho)$ are contained in the union of the four half lines

$$\begin{aligned} &-1+\frac{1}{\tau}-\rho-i\lambda+\mathbb{R}_{\leq 0}, &&-1+\rho-i\lambda+\mathbb{R}_{\leq 0},\\ &-\alpha-\rho+i\lambda+\mathbb{R}_{\geq 0}, &&-\frac{1}{\tau}+\alpha+\rho+i\lambda+\mathbb{R}_{\geq 0}.\end{aligned}$$

We call the above eight half lines the *singular half lines* with respect to the given, fixed parameters $\tau, x, \lambda, \alpha, \rho, \beta, \sigma$. Each singular half line is contained in some horizontal line l_c for some $c \in \mathbb{R}$. For generic parameters $\alpha, \rho, \beta, \sigma$, the eight singular half lines lie on different horizontal lines. Under these generic assumptions, there exists a deformation $\mathcal{C}_x$ of $i\,\mathbb{R}$ which separates the four singular half lines with real part tending to $-\infty$ from the four singular half lines with real part tending to ∞. We take such contour $\mathcal{C} = \mathcal{C}_x$ in the definition (6.4) of ψ_λ. The resulting function $\psi_\lambda(x)$ is well defined and independent of the particular choice of the deformation $\mathcal{C}_x$ of $i\mathbb{R}$, since $g_\lambda(\cdot;\beta,\sigma)$ (respectively, $h_\lambda(\cdot;\alpha,\rho)$) has growth rate $\pi\left((1-2\,\mathrm{Im}(\lambda))\tau-1\right)$ (respectively $\pi\left(1+2\,\mathrm{Im}(\lambda)\right)\tau$) at $\pm i\infty$. This follows from the asymptotic behaviour of the hyperbolic gamma function G_τ, see Proposition 2.1**(iv)**. The resulting function $\psi_\lambda(x)$ is analytic in x.

Theorem 6.5. *For generic parameters $\alpha, \rho, \beta, \sigma$, the function $\psi_\lambda(\cdot) = \psi_\lambda(\cdot; \alpha, \rho, \beta, \sigma)$ defined by*

$$\psi_\lambda(x) = \int_{\mathcal{C}_x} g_\lambda(1 + x + z; \beta, \sigma) h_\lambda(z; \alpha, \rho)\, dz$$

$$= \int_{\mathcal{C}_x} \frac{G_{2\tau}\left(\frac{1}{2\tau} + \beta + \sigma - i\lambda + x + z\right) G_{2\tau}\left(\frac{1}{2\tau} - 1 + \sigma - i\lambda - x - z\right)}{G_{2\tau}\left(-\frac{1}{2\tau} + \beta + \sigma + i\lambda - x - z\right) G_{2\tau}\left(-\frac{1}{2\tau} + 1 + \sigma + i\lambda + x + z\right)}$$
$$\times \frac{G_{2\tau}\left(-\frac{1}{2\tau} - 1 + \alpha + \rho + i\lambda - z\right) G_{2\tau}\left(-\frac{1}{2\tau} + \rho + i\lambda + z\right)}{G_{2\tau}\left(-\frac{1}{2\tau} + 1 + \alpha + \rho - i\lambda + z\right) G_{2\tau}\left(-\frac{1}{2\tau} + \rho - i\lambda - z\right)}\, dz \tag{6.5}$$

is an eigenfunction of the gauged Askey-Wilson difference operator $\mathcal{L}^{\alpha,\rho,\beta,\sigma}$ with eigenvalue $E(\lambda; \beta)$.

Proof. We adjust the proof of Theorem 4.4 to the present set-up. We simplify notations by writing $g(z) = g_\lambda(z; \beta, \sigma)$ and $h(z) = h_\lambda(z; \alpha, \rho)$. Choose the deformation $\mathcal{C}_x$ of $i\,\mathbb{R}$ such that g and h are analytic on the strip of radius ≥ 4 around $\mathcal{C}_x$.

Observe that $\pi_\lambda(\widehat{x}\Omega K)\, g$ has the same growth rate at $\pm i\infty$ as g. By (3.2) and the definition of $\psi_\lambda(x)$, we conclude that the integral $(\pi_\lambda(\widehat{x}\Omega K)g, h)_{\mathcal{C}_x}$ converges absolutely and equals

$$\left(\frac{q^{i\lambda} - q^{-i\lambda}}{q - q^{-1}}\right)^2 \psi_\lambda(x).$$

On the other hand, the radial part computation of Ω with respect to Koornwinder's twisted primitive element yields

$$\widehat{x}\Omega K = \widehat{x}\Omega(x)K + \left((Y_\rho - \mu_\alpha(\rho))\, K^{-1}\right) X \widehat{x} K + \widehat{x} Z\, (Y_\sigma - \mu_\beta(\sigma)) \tag{6.6}$$

for certain elements $X, Z \in \mathcal{U}_q^1$, cf. Proposition 3.3. Substituting this algebraic identity in $(\pi_\lambda(\widehat{x}\Omega K)g, h)_{\mathcal{C}_x}$ and using $\pi_\lambda(Y_\sigma)\, g = \mu_\beta(\sigma) g$, we have

$$\begin{aligned}(\pi_\lambda(\widehat{x}\Omega K)g, h)_{\mathcal{C}_x} &= (\pi_\lambda(\widehat{x}\Omega(x)K)g, h)_{\mathcal{C}_x} \\ &\quad + \left(\pi_\lambda\left((Y_\rho - \mu_\alpha(\rho))\, K^{-1} X \widehat{x} K\right) g, h\right)_{\mathcal{C}_x},\end{aligned} \tag{6.7}$$

provided that both integrals on the right hand side of (6.7) converge absolutely.

For the second term on the right hand side of (6.7), observe that the sum of the growth rates of g and h at $\pm i\infty$ equals $(2\tau - 1)\pi$. Furthermore,

$$(2\tau - 1)\pi < 4\pi\tau$$

since $-\frac{1}{2} < \tau < 0$, and

$$\left(Y_\rho - \mu_\alpha(\rho)\right) K^{-1} \in \mathcal{U}_q^1.$$

Hence the integral

$$\left(\pi_\lambda \left(\left(Y_\rho - \mu_\alpha(\rho)\right) K^{-1} X \widehat{x} K\right) g, h\right)_{\mathcal{C}_x}$$

converges absolutely, and Lemma 6.2 shows that this integral equals

$$\left(\pi_\lambda \left(X \widehat{x} K\right) g, \pi_{-\lambda}(K) \pi_{-\lambda}\left(Y_\rho^\circ - \mu_\alpha(\rho)\right) h\right)_{\mathcal{C}_x} = 0,$$

where the last equality follows from the eigenvalue equation $\pi_{-\lambda}(Y_\rho^\circ)h = \mu_\alpha(\rho)h$.

For the first term on the right hand side of (6.7), observe that $\pi_\lambda\left(\widehat{x}\Omega(x)K\right) g$ has the same growth rate at $\pm i\infty$ as g. Hence the integral $\left(\pi_\lambda\left(\widehat{x}\Omega(x)K\right) g, h\right)_{\mathcal{C}_x}$ converges absolutely. By the definition of the gauged Askey-Wilson second order difference operator $\mathcal{L}$, this integral equals

$$\frac{q^{\beta-1}}{(q-q^{-1})^2}\left(\mathcal{L}\psi_\lambda\right)(x) + \frac{q^{\beta-1}\left(1-q^{1-\beta}\right)^2}{\left(q-q^{-1}\right)^2}\psi_\lambda(x).$$

Combining the results, we conclude that

$$\left(\pi_\lambda\left(\widehat{x}\Omega K\right) g, h\right)_{\mathcal{C}_x} = \left(\pi_\lambda\left(\widehat{x}\Omega(x)K\right) g, h\right)_{\mathcal{C}_x},$$

and for both sides we have obtained an explicit expression in terms of ψ_λ. The resulting identity for ψ_λ yields the desired difference equation $\mathcal{L}\psi_\lambda = E(\lambda)\psi_\lambda$. □

Remark 6.6. *The eigenfunction ψ_λ of $\mathcal{L}$ as given by the integral* (6.5) *looks very similar to the eigenfunction φ_λ of $\mathcal{L}$ (for the case $\tau \in i\,\mathbb{R}_{>0}$) as given by the integral* (4.7)*, since the integrand of φ_λ is essentially the integrand of ψ_λ with the hyperbolic gamma functions $G_{2\tau}$ replaced by q-gamma functions $\Gamma_{2\tau}$. Note though that the integration cycles are different.*

We can reformulate the difference equation for ψ_λ in terms of the Askey-Wilson second order difference operator $\mathcal{D} = \mathcal{D}^{a,b,c,d}$ (see (3.3)) using an appropriate gauge factor, cf. Corollary 4.5. Using the parameter correspondence (3.7), we can for instance choose the gauge factor $\delta(x) = \delta(x; \alpha, \rho, \beta, \sigma)$ by

$$\delta(x) = \frac{G_{2\tau}\left(-1+\frac{1}{2\tau}+a-x\right)G_{2\tau}\left(-1+\frac{1}{2\tau}+b-x\right)}{G_{2\tau}\left(1+\frac{1}{2\tau}-c-x\right)G_{2\tau}\left(1+\frac{1}{2\tau}-d-x\right)}. \tag{6.8}$$

Corollary 6.7. *For generic parameters $\alpha, \rho, \beta, \sigma$, the function $H_\lambda = H_\lambda(\cdot; a, b, c, d) \in \mathcal{M}$ defined by $H_\lambda(x) = \delta(x)^{-1}\psi_\lambda(x)$ satisfies the Askey-Wilson second order difference equation*

$$(\mathcal{D}H_\lambda)(x) = E(\lambda)H_\lambda(x),$$

where the Askey-Wilson parameters (a, b, c, d) are given by (3.7).

Appendix

In this appendix we sketch a proof of Proposition 3.3, following closely the arguments in (Koornwinder, 1993) for the special case $\alpha = \beta = 0$.

Fix $x \in \mathbb{C}$ and write $X \equiv X'$ for $X, X' \in \mathcal{U}_q$ if

$$X - X' \in (Y_\rho - \mu_\alpha(\rho))\, \widehat{x}\,\mathcal{U}_q^1 + \widehat{x}\,\mathcal{U}_q^1\,(Y_\sigma - \mu_\beta(\sigma)).$$

In order to simplify notations, I use the notations $\mu = \mu_\alpha(\rho)$ and $\nu = \mu_\beta(\sigma)$ for the duration of the proof. To reduce $\widehat{x}\,\Omega K = \widehat{x}\,K\Omega$ in the desired form, we need to concentrate on the part $\widehat{x}\,KX^+X^-$. Using

$$\begin{aligned}\widehat{x}\,KX^+X^- &= q^{\frac{1}{2}+x}\left(q^{\frac{1}{2}}X^+K - q^{-\frac{1}{2}}X^-K\right)\widehat{x}X^- \\ &\quad + q^{-\frac{1}{2}+x}X^-\widehat{x}\left(q^{-\frac{1}{2}}X^-K - q^{\frac{1}{2}}X^+K\right) + q^{2x}\widehat{x}KX^-X^+\end{aligned}$$

and the commutation relation between X^- and X^+ in $\mathcal{U}_q$, we obtain

$$\begin{aligned}(1-q^{2x})\,\widehat{x}\,KX^+X^- &\equiv q^{\frac{1}{2}+x}\left(\frac{q^\rho+q^{-\rho}}{q-q^{-1}}(K^2-1)+\mu\right)\widehat{x}\,X^- \\ -q^{-\frac{1}{2}+x}X^-\widehat{x}&\left(\frac{q^\sigma+q^{-\sigma}}{q-q^{-1}}(K^2-1)+\nu\right) + q^{2x}\left(\frac{K^{-1}-K^3}{q-q^{-1}}\right)\widehat{x},\end{aligned}$$

hence we need to focus now only on the reduction of $\widehat{x}X^-$ and $\widehat{x}K^2X^-$. Using

$$\begin{aligned}\widehat{x}\,X^- &= q^{\frac{1}{2}-x}\left(q^{-\frac{1}{2}}X^-K - q^{\frac{1}{2}}X^+K\right)K^{-1}\widehat{x} \\ &\quad + q^{\frac{3}{2}-2x}K^{-1}\widehat{x}\left(q^{\frac{1}{2}}X^+K - q^{-\frac{1}{2}}X^-K\right) + q^{1-2x}K^{-1}\widehat{x}\,X^-K,\end{aligned}$$

we obtain

$$\begin{aligned}(1-q^{2-2x})\,\widehat{x}\,X^- &\equiv -\,q^{\frac{1}{2}-x}\left(\frac{q^\rho+q^{-\rho}}{q-q^{-1}}(K^2-1)+\mu\right)K^{-1}\widehat{x} \\ &\quad + q^{\frac{3}{2}-2x}K^{-1}\widehat{x}\left(\frac{q^\sigma+q^{-\sigma}}{q-q^{-1}}(K^2-1)+\nu\right),\end{aligned}$$

and a similar computation yields

$$\begin{aligned}(1-q^{-2-2x})\,\widehat{x}\,K^2X^- &\equiv -\,q^{-\frac{3}{2}-x}\left(\frac{q^\rho+q^{-\rho}}{q-q^{-1}}(K^2-1)+\mu\right)K\widehat{x} \\ &\quad + q^{-\frac{5}{2}-2x}K\widehat{x}\left(\frac{q^\sigma+q^{-\sigma}}{q-q^{-1}}(K^2-1)+\nu\right).\end{aligned}$$

Collecting all these results we arrive for generic $x \in \mathbb{C}$ at a formula of the form

$$\widehat{x}\, \Omega K \equiv \widehat{x}\, \left(B(x)K^3 + C(x)K + D(x)K^{-1}\right)$$

for explicit rational expressions $B(x)$, $C(x)$ and $D(x)$. Using

$$q^\rho + q^{-\rho} - (q - q^{-1})\mu = q^{\rho+\alpha} + q^{-\rho-\alpha},$$
$$q^\sigma + q^{-\sigma} - \left(q - q^{-1}\right)\nu = q^{\sigma+\beta} + q^{-\sigma-\beta},$$

the rational functions $B(x)$, $C(x)$ and $D(x)$ are explicitly given by

$$B(x) = \frac{\left(q^{\frac{1}{2}+x}(q^\rho + q^{-\rho}) - q^{\frac{3}{2}+2x}(q^\sigma + q^{-\sigma})\right)\left(q^{-\frac{5}{2}-2x}(q^\sigma + q^{-\sigma}) - q^{-\frac{3}{2}-x}(q^\rho + q^{-\rho})\right)}{(q-q^{-1})^2\,(1-q^{2x})\,(1-q^{-2-2x})}$$
$$- \frac{q^{2x}}{(q-q^{-1})\,(1-q^{2x})} + \frac{q^{-1}}{(q-q^{-1})^2},$$

$$C(x) = -\frac{2}{(q-q^{-1})^2}$$
$$+ \frac{\left(q^{\frac{1}{2}+x}(q^\rho + q^{-\rho}) - q^{\frac{3}{2}+2x}(q^\sigma + q^{-\sigma})\right)\left(q^{-\frac{3}{2}-x}\left(q^{\rho+\alpha} + q^{-\rho-\alpha}\right) - q^{-\frac{5}{2}-2x}\left(q^{\sigma+\beta} + q^{-\sigma-\beta}\right)\right)}{(q-q^{-1})^2(1-q^{2x})(1-q^{-2-2x})}$$
$$+ \frac{\left(q^{-\frac{1}{2}+2x}\left(q^{\sigma+\beta} + q^{-\sigma-\beta}\right) - q^{\frac{1}{2}+x}\left(q^{\rho+\alpha} + q^{-\rho-\alpha}\right)\right)\left(q^{\frac{3}{2}-2x}\left(q^\sigma + q^{-\sigma}\right) - q^{\frac{1}{2}-x}\left(q^\rho + q^{-\rho}\right)\right)}{(q-q^{-1})^2\,(1-q^{2x})\,(1-q^{2-2x})}$$

and

$$D(x) = \frac{q^{2x}}{(q-q^{-1})(1-q^{2x})} + \frac{q}{(q-q^{-1})^2}$$
$$+ \left(q^{-\frac{1}{2}+2x}\left(q^{\sigma+\beta} + q^{-\sigma-\beta}\right) - q^{\frac{1}{2}+x}\left(q^{\rho+\alpha} + q^{-\rho-\alpha}\right)\right)$$
$$\times \frac{\left(q^{\frac{1}{2}-x}\left(q^{\rho+\alpha} + q^{-\rho-\alpha}\right) - q^{\frac{3}{2}-2x}\left(q^{\sigma+\beta} + q^{-\sigma-\beta}\right)\right)}{(q-q^{-1})^2(1-q^{2x})(1-q^{2-2x})}.$$

By a tedious computation, it can now be proven that

$$B(x) = \frac{q^{\beta-1}}{(q-q^{-1})^2} B(x),$$
$$C(x) = \frac{q^{\beta-1}}{(q-q^{-1})^2}(-A(x) - A(x^{-1}) + (1-q^{1-\beta})^2),$$
$$D(x) = \frac{q^{\beta-1}}{(q-q^{-1})^2} D(x),$$

with the parameters a, b, c, d as in (3.7).

References

Askey, R. and Wilson, J. A. (1985). Some basic hypergeometric polynomials that generalize Jacobi polynomials. *Mem. Amer. Math. Soc.*, 54(319).

Cherednik, I. (1997). Difference Macdonald-Mehta conjecture. *Internat. Math. Res. Notices*, 10:449–467.

Faddeev, L. (2000). Modular double of a quantum group. In *Conférence Moshé Flato 1999, Vol. I (Dijon)*, volume 21 of *Math Phys. Stud.*, pages 149–156. Kluwer Acad. Publ., Dordrecht.

Gasper, G. and Rahman, M. (1990). *Basic hypergeometric series*, volume 35 of *Encyclopedia of Mathematics and its Applications.* Cambridge University Press, Cambridge.

Ismail, M. and Rahman, M. (1991). The associated Askey-Wilson polynomials. *Trans. Amer. Math. Soc.*, 328:201–237.

Kharchev, S., Lebedev, D., and Semenov-Tian-Shansky, M. (2002). Unitary representations of $u_q(\mathfrak{sl}(2,\mathbb{R}))$, the modular double and the multiparticle q-deformed Toda chains. *Comm. Math. Phys.*, 225(3):573–609.

Koelink, E. (1996). Askey-Wilson polynomials and the quantum SU(2) group: survey and applications. *Acta Appl. Math.*, 44(3):295–352.

Koelink, E. and Rosengren, H. (2002). Transmutation kernels for the little q-Jacobi function transform. *Rocky Mountain J. Math.*, 32(2):703–738. Conference on Special Functions (Tempe, AZ, 2000).

Koelink, E. and Stokman, J. V. (2001a). The Askey-Wilson function transform. *Internat. Math. Res. Notices*, 22:1203–1227.

Koelink, E. and Stokman, J. V. (2001b). Fourier transforms on the quantum $\mathrm{SU}(1,1)$ group. *Publ. Res. Inst. Math. Sci.*, 37(4):621–715. With an appendix by M. Rahman.

Koornwinder, T. H. (1984). Jacobi functions and analysis on noncompact semisimple Lie groups. In *Special Functions: Group Theoretical Aspects and Applications*, Math. Appl., pages 1–85. Reidel, Dordrecht.

Koornwinder, T. H. (1993). Askey-Wilson polynomials as zonal spherical functions on the $su(2)$ quantum group. *SIAM J. Math. Anal.*, 24:795–813.

Nishizawa, M. (2001). q-special functions with $|q| = 1$ and their application to discrete integrable systems. *J. Phys. A*, 34(48):10639–10646. Symmetries and Integrability of Difference Equations (Tokyo, 2000).

Nishizawa, M. and Ueno, K. (2001). Integral solutions of hypergeometric q-difference systems with $|q| = 1$. In *Physics and Combinatorics 1999 (Nagoya)*, pages 273–286. World Sci. Publishing, River Edge, NJ.

Noumi, M. and Mimachi, K. (1992). Askey-Wilson polynomials as spherical functions on $\mathrm{SU}_q(2)$. In *Quantum Groups (Leningrad, 1990)*, volume 1510 of *Lecture Notes in Math.*, pages 98–103. Springer, Berlin.

Noumi, M. and Stokman, J. V. (2000). Askey-Wilson polynomials: an affine Hecke algebra approach. Preprint.

Rosengren, H. (2000). A new quantum algebraic interpretation of the Askey-Wilson polynomials. In *q-Series from a Contemporary Perspective (South Hadley, MA, 1998)*, volume 254 of *Contemp. Math.*, pages 371–394. Amer. Math. Soc., Providence, RI.

Ruijsenaars, S. N. M. (1997). First order analytic difference equations and integrable quantum systems. *J. Math. Phys.*, 38(2):1069–1146.

Ruijsenaars, S. N. M. (1999). A generalized hypergeometric function satisfying four analytic difference equations of Askey-Wilson type. *Comm. Math. Phys.*, 206(3):639–690.

Ruijsenaars, S. N. M. (2001). Special functions defined by analytic difference equations. In Bustoz, J., Ismail, M. E. H., and Suslov, S. K., editors, *Special Functions 2000: Current Perspective and Future Directions*, volume 30 of *NATO Science Series, II. Math., Phys. and Chem.*, pages 281–333. Kluwer Acad. Publ.

Stokman, J. V. (2001). Difference Fourier transforms for nonreduced root systems. *Selecta Math. (N.S.)*. To appear. Available electronically: math.QA/0111221.

Stokman, J. V. (2002). An expansion formula for the Askey-Wilson function. *J. Approx. Theory*, 114:308–342.

Suslov, S. K. (1997). Some orthogonal very well poised ${}_8\varphi_7$-functions. *J. Phys. A*, 30:5877–5885.

Suslov, S. K. (2002). Some orthogonal very-well-poised ${}_8\varphi_7$-functions that generalize Askey-Wilson polynomials. *Ramanujan J.*, 5(2):183–218.

Van der Jeugt, J. and Jagannathan, R. (1998). Realizations of su(1, 1) and $u_q(\mathrm{su}(1,1))$ and generating functions for orthogonal polynomials. *J. Math. Phys.*, 39(9):5062–5078.

AN ANALOG OF THE CAUCHY-HADAMARD FORMULA FOR EXPANSIONS IN q-POLYNOMIALS

Sergei K. Suslov
Department of Mathematics and Statistics
Arizona State University
Tempe, AZ 85287-1804
sks@asu.edu

Abstract We derive an analog of the Cauchy-Hadamard formula for certain polynomial expansions and consider some examples.

Keywords: Basic hypergeometric functions, q-orthogonal polynomials, continuous q-ultraspherical polynomials, continuous q-Hermite polynomials, the Askey-Wilson polynomials, the Chebyshev polynomials, the Jacobi polynomials, Taylor's series and its generalizations.

1. Main Result

A problem of fundamental interest in classical analysis is to study the representation of an analytic function as a series of polynomials; see, for example, (Boas and Buck, 1964), (Szegő, 1982, p. 147) and references therein. In this note we consider polynomial expansions of the form

$$\sum_{n=0}^{\infty} c_n\, p_n(x) . \tag{1.1}$$

These series are obviously infinite sums of analytic functions and by the Weierstrass theorem the uniform convergence guarantees analyticity of the limit function is some region of the complex x-plane. This raises an interesting question about the maximum domain of analyticity of these series. In many cases a complete solution can be given by the following analog of Cauchy-Hadamard's formula well-known for the power series (Ahlfors, 1979, pp. 38–39), (Dienes, 1957, pp. 75–76), (Markushevich, 1985, pp. 344–346).

Theorem 1.1. *Let E_ε be an ellipse in the complex x-plane*

$$\left(\frac{\operatorname{Re} x}{a_\varepsilon}\right)^2 + \left(\frac{\operatorname{Im} x}{b_\varepsilon}\right)^2 = 1 \tag{1.2}$$

with the semiaxes given by

$$a_\varepsilon = \frac{1}{2}\left(q^\varepsilon + q^{-\varepsilon}\right), \qquad b_\varepsilon = \frac{1}{2}\left|q^\varepsilon - q^{-\varepsilon}\right| \tag{1.3}$$

and the focal points at ± 1; let $\operatorname{Int} E_\varepsilon$ be the interior of the ellipse E_ε and $\operatorname{Ext} E_\varepsilon$ be the exterior of E_ε. Suppose that a sequence of polynomials $\{p_n(x)\}_{n=0}^\infty$, $\deg p_n(x) = n$, $n = 0, 1, 2, \ldots$, such that all zeros of $p_n(x)$ lie in the interval $(-1, 1)$, satisfies the following inequalities

$$C_1(\varepsilon)\, q^{-n\varepsilon} < |p_n(x)| < C_2(\varepsilon)\, q^{-n\varepsilon}, \tag{1.4}$$

where $C_1(\varepsilon) > 0$, on the ellipse E_ε with $0 < q < 1$, $\varepsilon > 0$ for all sufficiently large n. Then the series (1.1) converges absolutely for every x in the interior of the ellipse E_ε :

$$\left(\frac{\operatorname{Re} x}{a_\varepsilon}\right)^2 + \left(\frac{\operatorname{Im} x}{b_\varepsilon}\right)^2 < 1 \tag{1.5}$$

with

$$q^\varepsilon = \limsup_{n\to\infty} |c_n|^{1/n}. \tag{1.6}$$

If $0 < \varepsilon' < \varepsilon$, the convergence is uniform in the closure of the interior of the subellipse $E_{\varepsilon'}$. For every x in the interior of the ellipse E_ε the sum of the series (1.1) is an analytic function. For every x in the exterior of the ellipse E_ε the terms of the series (1.1) are unbounded, and the series is consequently divergent.

We shall call the ellipse E_ε given by (1.2)–(1.3) and (1.6) the *ellipse of convergence*; nothing is claimed about the convergence on the ellipse.

The proof of Theorem 1.1 here is similar to the proof of Theorem 6.12 in (Suslov, 2003b); see also Exercise 6.21.

Proof. Let E_ε be the ellipse of convergence given above. When $0 < \delta < \varepsilon$ and $q^\varepsilon < q^{\varepsilon-\delta}$, by the definition of limit superior there exists an $n_0 > 0$ such that

$$|c_n|^{1/n} < q^{\varepsilon-\delta}, \qquad \text{or} \qquad |c_n| < q^{(\varepsilon-\delta)n} \tag{1.7}$$

for all $n > n_0$. If $0 < \varepsilon' < \varepsilon$, the following estimate holds

$$|c_n\, p_n(x)| < C_2\left(\varepsilon'\right) q^{(\varepsilon-\delta-\varepsilon')n} \tag{1.8}$$

for all $n > n_0$ and every x inside the ellipse $E_{\varepsilon'}$ due to the maximum principle for the analytic functions and the hypotheses of the theorem. Thus, when $\varepsilon' < \varepsilon - \delta$, $0 < \delta < \varepsilon$, for large n the series (1.1) has a convergent geometric series as a majorant, and consequently converges absolutely. For every x on the subellipse $E_{\varepsilon'}$, $0 < \varepsilon' < \varepsilon$ and in its interior $\operatorname{Int} E_{\varepsilon'}$ the convergence is uniform by the M-test and, therefore, by the Weierstrass theorem the limit function is analytic inside E_ε.

If $\varepsilon' > \varepsilon > 0$, we choose δ, $0 < \delta < \varepsilon' - \varepsilon$ so that $q^{\varepsilon'} < q^{\varepsilon+\delta} < q^{\varepsilon}$. Since $q^{\varepsilon+\delta} < q^{\varepsilon}$ there are arbitrary large n such that

$$|c_n|^{1/n} > q^{\varepsilon+\delta}, \qquad \text{or} \qquad |c_n| > q^{(\varepsilon+\delta)n}. \tag{1.9}$$

Thus

$$|c_n\, p_n(x)| > C_1\left(\varepsilon'\right) q^{-(\varepsilon'-\varepsilon-\delta)n} > 0, \tag{1.10}$$

when $x \in E_{\varepsilon'} \subset \operatorname{Ext} E_\varepsilon$, for infinitely many n, and the terms are unbounded. This completes the proof of the theorem. □

Remark 1.2. *If under the hypotheses of Theorem* 1.1 *the set of polynomials* $\{p_n(x)\}_{n=0}^{\infty}$, *instead of satisfying the inequalities* (1.4), *is uniformly bounded on compacts*

$$|\, p_n(x)| < D = constant \tag{1.11}$$

for all sufficiently large n *and*

$$\sum_{n=0}^{\infty} |c_n| < \infty, \tag{1.12}$$

then the series (1.1) *is an entire function in the complex* x*-plane.*

Remark 1.3. *It follows from the second part of the proof of Theorem* 1.1 *that the imposed condition that all zeros of* $p_n(x)$ *lie in the interval* $(-1, 1)$ *can be replaced by weaker requirements that those zeros lie inside the ellipse of convergence and the lower bound in* (1.4) *holds only outside the ellipse of convergence. The Cauchy-Hadamard formula holds also in the case of Szegő's class of orthogonal polynomials (Szegő, 1975), Theorems* 12.7.3–12.7.4 *and (Szegő, 1982, p. 377), but our approach does not require orthogonality; see (Suslov, 2003a) for more details.*

Remark 1.4. *Theorem* 1.1 *holds also for a sequence of analytic functions* $\{f_n(x)\}$ *if the conditions* (1.4) *are satisfied. By the Root Test the theorem holds also under conditions* $\lim_{n\to\infty} C_1^{1/n}(\varepsilon) = \lim_{n\to\infty} C_2^{1/n}(\varepsilon) = 1$; *cf. (Davis, 1963, Lemma 4.4.2, p. 89).*

2. Some Expansions in Classical q-Orthogonal Polynomials

Let us consider the maximum domain of analyticity for several familiar expansions in q-orthogonal polynomials.

2.1 The continuous q-ultraspherical polynomials

These polynomials are defined by

$$C_n(\cos\theta;\beta\mid q)=\sum_{k=0}^{n}\frac{(\beta;q)_k\,(\beta;q)_{n-k}}{(q;q)_k\,(q;q)_{n-k}}\,e^{i(n-2k)\theta}, \tag{2.1}$$

where $x=\cos\theta$; see, for example, (Andrews et al., 1999) and (Gasper and Rahman, 1990); we shall assume throughout the paper that $-1<\beta<1$ and $0<q<1$. Let us establish first a convenient upper bound.

Lemma 2.1. *The following inequalities hold*

$$|C_n(x;\beta\mid q)|<\frac{(-|\beta|;q)_\infty^2}{(q;q)_\infty^2}\,q^{-n\varepsilon}\,\frac{1-q^{2(n+1)\varepsilon}}{1-q^{2\varepsilon}}<\frac{(-|\beta|;q)_\infty^2\,q^{-n\varepsilon}}{(1-q^{2\varepsilon})\,(q;q)_\infty^2} \tag{2.2}$$

for $\varepsilon>0$, $-1<\beta<1$ *and* $0<q<1$ *when* x *is inside or on the ellipse* E_ε *given by* (1.2)–(1.3).

Proof. As in the proof of Lemma 6.8 in (Suslov, 2003b), we rewrite the relation (2.1) in the form

$$C_n(x;\beta\mid q)=\sum_{k=0}^{n}\frac{(\beta;q)_k\,(\beta;q)_{n-k}}{(q;q)_k\,(q;q)_{n-k}}\,q^{(n-2k)z},\quad x=\frac{1}{2}\left(q^z+q^{-z}\right). \tag{2.3}$$

Let $z=\operatorname{Re}z+i\operatorname{Im}z$ with $\operatorname{Re}z=\varepsilon'$, $q^{i\operatorname{Im}z}=e^{i\theta}$, or $q^z=e^{i\theta}q^{\varepsilon'}$, where $0\le\varepsilon'\le\varepsilon$. Then

$$x=\frac{1}{2}\left(q^{\varepsilon'}+q^{-\varepsilon'}\right)\cos\theta+i\frac{1}{2}\left(q^{\varepsilon'}-q^{-\varepsilon'}\right)\sin\theta \tag{2.4}$$

and the image of the vertical line segment given by $\operatorname{Re}z=\varepsilon'$, $\operatorname{Im}z=\theta\log q$, $-\pi\le\theta\le\pi$ is the ellipse $E_{\varepsilon'}$ in the complex x-plane; see Figure 1.

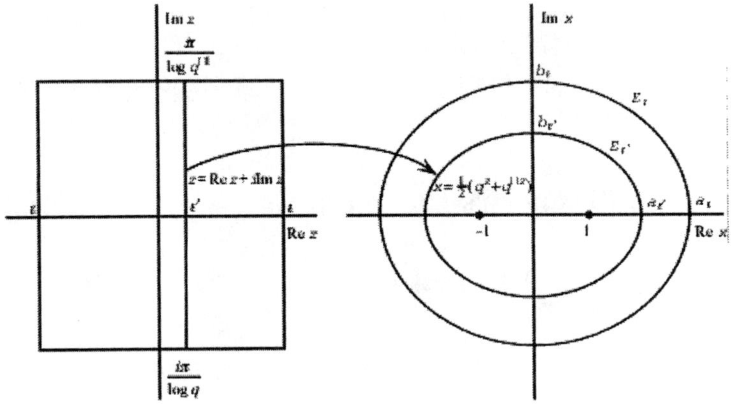

Figure 1. Mapping by $x = \left(q^z + q^{-z}\right)/2$.

The case $-\varepsilon \le -\varepsilon' \le 0$ corresponds to the opposite orientation of the ellipse $E_{\varepsilon'}$ and we will skip the details. Every x inside the largest ellipse E_ε belongs to some subellipse $E_{\varepsilon'}$ and, therefore, the following inequalities hold

$$\begin{aligned} |C_n(x;\beta \mid q)| &\le \sum_{k=0}^{n} \frac{(\beta;q)_k\,(\beta;q)_{n-k}}{(q;q)_k\,(q;q)_{n-k}}\, q^{(n-2k)\varepsilon'} \\ &= C_n\left(x\left(\varepsilon'\right);\beta \mid q\right) \le C_n\left(x(\varepsilon);\beta \mid q\right). \end{aligned} \tag{2.5}$$

One can also use the maximum modulus principle for analytic functions and the fact that the maxima of $|C_n(x;\beta \mid q)|$ on the ellipse E_ε occur at the points $\pm x = a_\varepsilon = x(\varepsilon)$. From (2.3) we obtain

$$\begin{aligned} C_n\left(x(\varepsilon);\beta \mid q\right) &= \sum_{k=0}^{n} \frac{(\beta;q)_k\,(\beta;q)_{n-k}}{(q;q)_k\,(q;q)_{n-k}}\, q^{(n-2k)\varepsilon} \\ < \frac{(-|\beta|;q)_\infty^2}{(q;q)_\infty^2} \sum_{k=0}^{n} q^{(n-2k)\varepsilon} &= \frac{(-|\beta|;q)_\infty^2}{(q;q)_\infty^2}\, q^{-n\varepsilon}\, \frac{1-q^{2(n+1)\varepsilon}}{1-q^{2\varepsilon}} \end{aligned} \tag{2.6}$$

because $(q;q)_\infty < (q;q)_k$ and $(\beta;q)_k < (-|\beta|;q)_\infty$ for $0<q<1$ and $-1<\beta<1$. As a result, for every x inside or on the ellipse E_ε with $\varepsilon > 0$ the following inequalities hold

$$|C_n(x;\beta \mid q)| < \frac{(-|\beta|;q)_\infty^2}{(q;q)_\infty^2}\, q^{-n\varepsilon}\, \frac{1-q^{2(n+1)\varepsilon}}{1-q^{2\varepsilon}} < \frac{(-|\beta|;q)_\infty^2\, q^{-n\varepsilon}}{\left(1-q^{2\varepsilon}\right)(q;q)_\infty^2}, \tag{2.7}$$

when $\varepsilon > 0$. This completes the proof of the lemma. □

We need to show that the hypotheses of Theorem 1.1 are satisfied for the continuous q-ultraspherical polynomials. This can be done with the help of the uniform asymptopic for these polynomials which one can establish with the help of the following result (Bromwich, 1965), (Ismail and Wilson, 1982).

Theorem 2.2 (Tannery's theorem). *Let*

$$S_n = a_0(n) + a_1(n) + \cdots + a_m(n) = \sum_{k=0}^{m} a_k(n); \tag{2.8}$$

$$\lim_{n\to\infty} a_k(n) = b_k, \qquad k \text{ being fixed}; \tag{2.9}$$

and

$$|a_k(n)| \le M_k, \qquad \sum_k M_k < \infty, \tag{2.10}$$

where M_k is independent of n. Then

$$\lim_{n\to\infty} S_n = \sum_{k=0}^{\infty} b_k, \tag{2.11}$$

it is being understood that $m = m(n) \to \infty$ as $n \to \infty$.

See (Bromwich, 1965) for the proof of this theorem; the result holds uniformly with respect to all parameters in compact sets if the convergence in (2.9) and bound in (2.10) hold uniformly.

In the case of the continuous q-ultraspherical polynomials Eq. (2.3) implies that

$$q^{nz} C_n(x; \beta \mid q) = \sum_{k=0}^{n} \frac{(\beta;q)_{n-k}\,(\beta;q)_k}{(q;q)_{n-k}\,(q;q)_k}\, q^{2kz}, \tag{2.12}$$

where

$$|a_k(n)| = \left| \frac{(\beta;q)_{n-k}\,(\beta;q)_k}{(q;q)_{n-k}\,(q;q)_k}\, q^{2kz} \right| < \frac{(-|\beta|;q)_\infty^2}{(q;q)_\infty^2}\, \left|q^{2z}\right|^k \tag{2.13}$$

with $-1 < \beta < 1$, $0 < q < 1$ and

$$\sum_{k=0}^{\infty} \left|q^{2z}\right|^k = \frac{1}{1 - |q^{2z}|}, \qquad |q^z| < 1. \tag{2.14}$$

Also

$$\lim_{n\to\infty} \frac{(\beta;q)_{n-k}\,(\beta;q)_k}{(q;q)_{n-k}\,(q;q)_k}\, q^{2kz} = \frac{(\beta;q)_\infty}{(q;q)_\infty} \frac{(\beta;q)_k}{(q;q)_k}\, q^{2kz}. \tag{2.15}$$

Thus by the Tannery theorem

$$\lim_{n\to\infty} q^{nz} C_n(x;\beta\mid q) = \frac{(\beta;q)_\infty}{(q;q)_\infty}\sum_{k=0}^{\infty}\frac{(\beta;q)_k}{(q;q)_k} q^{2kz} = \frac{(\beta;q)_\infty}{(q;q)_\infty}\frac{(\beta q^{2z};q)_\infty}{(q^{2z};q)_\infty}$$

due to the q-binomial theorem.

One can easily see that the resulting limiting relation

$$\lim_{n\to\infty} q^{nz} C_n(x;\beta\mid q) = \frac{(\beta,\beta q^{2z};q)_\infty}{(q,q^{2z};q)_\infty}, \qquad |q^z|<1 \tag{2.16}$$

holds uniformly on the ellipse E_ε with $x=(q^z+q^{-z})/2$ and $q^z=q^\varepsilon e^{i\theta}$, $\varepsilon>0$.

Indeed, the bound in (2.13) is uniform on E_ε and the convergence in (2.9) holds uniformly due to Cauchy's criteria. It is clear that

$$|a_k(m)-a_k(n)| = \frac{(\beta;q)_k}{(q;q)_k} q^{2\varepsilon k}\left|\frac{(\beta;q)_{m-k}}{(q;q)_{m-k}}-\frac{(\beta;q)_{n-k}}{(q;q)_{n-k}}\right|$$
$$\le \frac{(\beta;q)_k}{(q;q)_k} q^{2\varepsilon k}\left(\left|\frac{(\beta;q)_{m-k}}{(q;q)_{m-k}}-\frac{(\beta;q)_\infty}{(q;q)_\infty}\right|+\left|\frac{(\beta;q)_{m-k}}{(q;q)_{m-k}}-\frac{(\beta;q)_\infty}{(q;q)_\infty}\right|\right)$$

is arbitrary small on the ellipse for all sufficiently large m and n.

Now we can show that the conditions (1.4) of Theorem 1.1 are satisfied.

Lemma 2.3. *The following inequalities hold*

$$C_1(\varepsilon)\, q^{-n\varepsilon} < |C_n(x;\beta|q)| < C_2(\varepsilon)\, q^{-n\varepsilon} \tag{2.17}$$

for $\varepsilon>0$, $-1<\beta<1$ and $0<q<1$ for all sufficiently large n and every x on the ellipse E_ε given by (1.2)–(1.3).

Proof. Let $x(z)=(q^z+q^{-z})/2$ and $q^z=e^{i\theta}q^\varepsilon$ with $\varepsilon>0$, $-\pi\le\theta\le\pi$. In view of (2.16),

$$\lim_{n\to\infty}|q^{nz} C_n(x(z);\beta\mid q)| = \frac{(\beta;q)_\infty}{(q;q)_\infty}\left|\frac{(\beta q^{2\varepsilon}e^{2i\theta};q)_\infty}{(q^{2\varepsilon}e^{2i\theta};q)_\infty}\right|$$
$$= \frac{(\beta;q)_\infty}{(q;q)_\infty}\frac{(\beta q^{2\varepsilon}e^{2i\theta},\beta q^{2\varepsilon}e^{-2i\theta};q)_\infty^{1/2}}{(q^{2\varepsilon}e^{2i\theta},q^{2\varepsilon}e^{-2i\theta};q)_\infty^{1/2}} = L(\theta) \tag{2.18}$$

uniformly on the ellipse E_ε. This means that for a given $\varepsilon_0>0$ there is $n_0=n_0(\varepsilon_0)$ such that

$$L(\theta)-\varepsilon_0 < q^{n\varepsilon}|C_n(x;\beta\mid q)| < L(\theta)+\varepsilon_0 \tag{2.19}$$

for all $n \geq n_0$ and $x \in E_\varepsilon$. Introducing

$$L_1 = \min_{0 \leq \theta \leq \pi/2} L(\theta), \qquad L_2 = \max_{0 \leq \theta \leq \pi/2} L(\theta), \tag{2.20}$$

one can write

$$(L_1 - \varepsilon_0)\, q^{-n\varepsilon} < |C_n(x; \beta \mid q)| < (L_2 + \varepsilon_0)\, q^{-n\varepsilon} \tag{2.21}$$

for all $n \geq n_0$ and $x \in E_\varepsilon$. Choosing

$$C_1 = L_1 - \varepsilon_0 > 0, \qquad C_2 = L_2 + \varepsilon_0, \tag{2.22}$$

we arrive at inequalities (2.17) uniformly on the ellipse E_ε and the proof is complete. □

As the first example, let us consider convergence of the series in the Ismail and Zhang formula (Ismail and Zhang, 1994),

$$\begin{aligned} \mathcal{E}_q(x; i\omega) &= \frac{(q;q)_\infty \omega^{-\nu}}{(q^\nu;q)_\infty (-q\omega^2;q^2)_\infty} \\ &\times \sum_{n=0}^{\infty} i^n \left(1 - q^{\nu+n}\right) q^{n^2/4} J_{\nu+n}^{(2)}(2\omega;q)\, C_n(x;q^\nu \mid q), \end{aligned} \tag{2.23}$$

where $J_{\nu+n}^{(2)}(2\omega;q)$ is Jackson's q-Bessel function defined by

$$J_\nu^{(2)}(r;q) = \frac{\left(q^{\nu+1};q\right)_\infty}{(q;q)_\infty} \sum_{k=0}^{\infty} q^{(\nu+k)k} \frac{(-1)^k (r/2)^{\nu+2k}}{(q;q)_k \left(q^{\nu+1};q\right)_k} \tag{2.24}$$

with $\nu > -1$ and the basic exponential function on a q-quadratic grid is given by

$$\begin{aligned} \mathcal{E}_q(x;\alpha) &= \frac{\left(\alpha^2;q^2\right)_\infty}{(q\alpha^2;q^2)_\infty} \\ &\times \sum_{n=0}^{\infty} \frac{q^{n^2/4}, \alpha^n}{(q;q)_n} (-i)^n \left(-iq^{(1-n)/2} e^{i\theta}, -iq^{(1-n)/2} e^{-i\theta}; q\right)_n \end{aligned} \tag{2.25}$$

with $x = \cos\theta$ and $|\alpha| < 1$. This function was originally introduced by Ismail and Zhang (Ismail and Zhang, 1994) with different notation and normalization; see also (Atakishiyev and Suslov, 1992), (Suslov, 1997) for the corresponding solutions of a q-analog of the equation for harmonic motion; the above notation is due to Suslov (Suslov, 1997). Different proofs of the Ismail and Zhang formula were given in (Floreanini and

Vinet, 1995), (Ismail et al., 1999), (Ismail et al., 1996), and (Ismail and Stanton, 2000); see also (Suslov, 2003b), Section 4.5.

From (2.24) we obtain

$$\begin{aligned}\left|J_{\nu+n}^{(2)}(2\omega;q)\right| &< \frac{|\omega|^{n+\nu}}{(q;q)_\infty}\sum_{n=0}^{\infty} q^{k^2}\frac{\left(|\omega|^2 q^{\nu+n}\right)^k}{(q;q)_k} < \frac{|\omega|^{n+\nu}}{(q;q)_\infty^2}\sum_{n=0}^{\infty} q^{k^2}\left|\omega^2 q^\nu\right|^k \\ &< |\omega|^{n+\nu}\frac{(-q;q)_\infty}{(q;q)_\infty}\left(-q^{1+\nu}|\omega|^2, -q^{1-\nu}/|\omega|^2; q^2\right)_\infty \end{aligned} \tag{2.26}$$

by the Jacobi triple product identity (Andrews et al., 1999), (Gasper and Rahman, 1990). This upper bound is of an independent interest.

Introducing

$$c_n = i^n\left(1-q^{\nu+n}\right)q^{n^2/4}\,J_{\nu+n}^{(2)}(2\omega;q), \tag{2.27}$$

in view of (2.26) one gets

$$|c_n| < D\,q^{n^2/4}\,|\omega|^{n+\nu} \tag{2.28}$$

with

$$D = \frac{(-q;q)_\infty}{(q;q)_\infty}\left(-q^{1+\nu}|\omega|^2, -q^{1-\nu}/|\omega|^2; q^2\right)_\infty. \tag{2.29}$$

Therefore

$$|c_n|^{1/n} < D^{1/n}\,q^{n/4}\,|\omega|^{1+\nu/n} \to 0 \qquad \text{as} \quad n\to\infty \tag{2.30}$$

and

$$q^\varepsilon = \limsup_{n\to\infty}|c_n|^{1/n} = 0, \qquad \text{or} \quad \varepsilon=\infty. \tag{2.31}$$

The series in (2.23) is an entire function in the complex x-plane for all finite values of ω, thus providing an analytic continuation of the basic exponential function (2.25).

In a similar fashion, one can show that the ellipse of convergence of the series in Rogers' generating function for the continuous q-ultraspherical polynomials,

$$\sum_{n=0}^{\infty} C_n(\cos\theta;\beta\mid q)\,r^n = \frac{\left(\beta re^{i\theta}, \beta re^{-i\theta}; q\right)_\infty}{\left(re^{i\theta}, re^{-i\theta}; q\right)_\infty}, \qquad |r|<1 \tag{2.32}$$

see, for example, (Andrews et al., 1999) and (Gasper and Rahman, 1990), is the ellipse E_ε given by (1.2)–(1.3) with

$$q^\varepsilon = \limsup_{n\to\infty}\left(|r|^n\right)^{1/n} = |r|. \tag{2.33}$$

Observe that the first poles in the right side of (2.32) are located on the ellipse of convergence. In a similar manner, one can analyze the convergence of the series in the Poisson kernel for the continuous q-ultraspherical polynomials and in related bilinear generating functions (Gasper and Rahman, 1990).

2.2 The continuous q-Hermite polynomials

These polynomials are the special case of the continuous q-ultraspherical polynomials:

$$H_n(x \mid q) = (q;q)_n C_n(x; 0 \mid q). \tag{2.34}$$

They have two generating functions

$$\sum_{n=0}^{\infty} \frac{q^{n^2/4}}{(q;q)_n} \alpha^n H_n(x \mid q) = \left(q\alpha^2; q^2\right)_\infty \mathcal{E}_q(x;\alpha) \tag{2.35}$$

and

$$\sum_{n=0}^{\infty} \frac{r^n}{(q;q)_n} H_n(x \mid q) = \frac{1}{(re^{i\theta}, re^{-i\theta}; q)_\infty}, \qquad |r| < 1, \tag{2.36}$$

see, for example, (Suslov, 2003b) and (Gasper and Rahman, 1990), respectively.

By Lemma 2.1 and (2.34) the uniform upper bound is

$$\frac{|H(x \mid q)|}{(q;q)_n} < \frac{q^{-n\varepsilon}}{(1-q^{2\varepsilon})(q;q)_\infty^2} \tag{2.37}$$

and in the first case (2.35), the series defines an entire function in the complex x-plane for all finite values of α. Indeed,

$$|c_n|^{1/n} = C^{1/n} |\alpha| q^{n/4} \to 0 \qquad \text{as} \quad n \to \infty.$$

In the second case (2.36), by Lemma 2.3 and (2.34) the corresponding ellipse of convergence E_ε is given by (1.2)–(1.3) with

$$q^\varepsilon = \limsup_{n\to\infty} \left(|r|^n\right)^{1/n} = |r|.$$

The first poles in the right side of (2.36) occur on the ellipse of convergence.

Ismail and Stanton (Ismail and Stanton, 2000) have found the following expansion formula

$$\mathcal{E}_q(x;\alpha)\,\mathcal{E}_q(x;\beta) = \sum_{n=0}^{\infty} q^{n^2/4} \alpha^n H_n(x \mid q) \frac{\left(-\alpha\beta q^{(n+1)/2}; q\right)_\infty}{\left(q\alpha^2, q\beta^2; q^2\right)_\infty} \frac{\left(-q^{(1-n)/2}\beta/\alpha; q\right)_n}{(q;q)_n} \tag{2.38}$$

in the continuous q-Hermite polynomials. They call expansion (2.38) the "addition" theorem with respect to the parameter α because it becomes $\exp \alpha x \exp \beta x = \exp (\alpha + \beta) x$ in the limit $q \to 1^-$; see (Suslov, 1997), (Suslov, 2000) and (Suslov, 2003b, Chapter 3), for other addition theorems for the basic exponential functions.

One can easily verify with the help of (I.8) of (Gasper and Rahman, 1990) that

$$\begin{aligned} |c_{2k}|^{1/2k} &= q^{k/4} |\alpha\beta|^{1/2} \left| \left(-\alpha\beta q^{1/2+k}; q \right)_\infty \right|^{1/2k} \\ &\times \left| \frac{\left(-q^{1/2}\alpha/\beta, -q^{1/2}\beta/\alpha; q \right)_k}{(q;q)_{2k}} \right|^{1/2k} \to 0 \qquad \text{as } k \to \infty \end{aligned} \tag{2.39}$$

when $n = 2k$ and

$$\begin{aligned} |c_{2k+1}|^{1/(2k+1)} &= q^{k/4+(k+1)/4(2k+1)} |\alpha\beta|^{k/(2k+1)} \\ &\times \left| \left(-\alpha\beta q^{1+k}; q \right)_\infty \right|^{1/(2k+1)} \\ &\times \left| \frac{(\alpha+\beta) \left(-q\alpha/\beta, -q\beta/\alpha; q \right)_k}{(q;q)_{2k+1}} \right|^{1/(2k+1)} \to 0 \qquad \text{as } k \to \infty \end{aligned} \tag{2.40}$$

when $n = 2k+1$. Thus, the corresponding ellipse of convergence is E_∞ and the series represents an entire function.

2.3 The Askey–Wilson polynomials

These polynomials are given by

$$\begin{aligned} p_n(x) &= p_n(x; a, b, c, d) \\ &= a^{-n} (ab, ac, ad; q)_n \; {}_4\varphi_3 \left(\begin{matrix} q^{-n}, abcdq^{n-1}, ae^{i\theta}, ae^{-i\theta} \\ ab, ac, ad \end{matrix} ; q, q \right), \end{aligned} \tag{2.41}$$

where $x = \cos\theta$. They are the most general known classical orthogonal polynomials; see, for example, (Andrews and Askey, 1985), (Askey and Wilson, 1985), and (Gasper and Rahman, 1990).

The uniform asymptotic on the ellipse E_ε, namely,

$$\lim_{n\to\infty} q^{nz} p_n\left(\left(q^z + q^{-z}\right)/2; a, b, c, d \right) = \frac{(aq^z, bq^z, cq^z, dq^z; q)_\infty}{(q^{2z}; q)_\infty}, \tag{2.42}$$

when $\max(|a|, |b|, |c|, |d|) < 1$ and $|q^z| < 1$, was established in (Ismail and Wilson, 1982) with the help of the Tannery theorem; see also (Rahman, 1986) and (Gasper and Rahman, 1990) for another approach. The same consideration as in the proof of Lemma 2.3 shows that the conditions (1.4) of Theorem 1.1 are satisfied.

An example of the explicit expansion in the Askey-Wilson polynomials is the generating relation found by Ismail and Wilson (Ismail and Wilson, 1982):

$$\sum_{n=0}^{\infty} \frac{r^n}{(q, ab, cd; q)_n} p_n(x; a, b, c, d) = \frac{(ar, br, cr, dr; q)_\infty}{(ab, cd, re^{i\theta}, re^{-i\theta}; q)_\infty} \times {}_2\varphi_2\left(\begin{matrix} re^{i\theta}, re^{-i\theta} \\ ar, br \end{matrix}; q, ab\right) {}_2\varphi_2\left(\begin{matrix} re^{i\theta}, re^{-i\theta} \\ cr, dr \end{matrix}; q, cd\right), \tag{2.43}$$

where $|r| < 1$; we have used (A.3.5) of (Suslov, 2003b) in the right side. By Theorem 1 the corresponding ellipse of convergence is E_ε with $q^\varepsilon = |r|$; the first poles in the right occur on the ellipse.

3. Some Expansions in Classical Orthogonal Polynomials

Consider also several expansions in Chebyshev and Jacobi polynomials.

3.1 The Chebyshev polynomials

The polynomials $T_n(x)$ and $U_n(x)$ are usually defined as

$$T_n(\cos\theta) = \cos n\theta, \qquad U_n(\cos\theta) = \frac{\sin(n+1)\theta}{\sin\theta}. \tag{3.1}$$

They have several generating relations including

$$\sum_{n=0}^{\infty} T_n(x) r^n = \frac{1 - rx}{1 - 2rx + r^2}, \tag{3.2}$$

$$\sum_{n=0}^{\infty} U_n(x) r^n = \frac{1}{1 - 2rx + r^2}, \tag{3.3}$$

where $|r| < 1$,

$$\sum_{n=0}^{\infty} T_n(x) \frac{(1/2)_n}{n!} r^n = \frac{\sqrt{(1 - r + R)(1 + r + R)}}{2R}; \tag{3.4}$$

$$\sum_{n=0}^{\infty} U_n(x) \frac{(3/2)_n}{(n+1)!} r^n = \frac{2}{R\sqrt{(1 - r + R)(1 + r + R)}} \tag{3.5}$$

where $R = \sqrt{1 - 2rx + r^2}$, $|r| < 1$, and

$$\sum_{n=0}^{\infty} T_n(x) \frac{r^n}{n!} = e^{rx} \cosh\left(r\sqrt{x^2-1}\right), \tag{3.6}$$

$$\sum_{n=0}^{\infty} U_n(x) \frac{r^n}{(n+1)!} = e^{rx} \frac{\sinh\left(r\sqrt{x^2-1}\right)}{\sqrt{x^2-1}}, \tag{3.7}$$

see, for example, (Andrews et al., 1999), (Rainville, 1960).

From the definition of the Chebyshev polynomials one can establish the following lower and upper bounds on the ellipse E_ε with $\varepsilon > 0$ given by (1.2)–(1.3),

$$\frac{1}{2}\left(1 - q^{2\varepsilon}\right) q^{-n\varepsilon} < |T_n(x)| < \frac{1}{2}\left(1 + q^{2\varepsilon}\right) q^{-n\varepsilon}, \tag{3.8}$$

when $n = 1, 2, 3, \ldots$ and

$$\frac{1 - q^{2\varepsilon}}{1 + q^{2\varepsilon}} q^{-n\varepsilon} < |U_n(x)| < \frac{1 + q^{2\varepsilon}}{1 - q^{2\varepsilon}} q^{-n\varepsilon} \tag{3.9}$$

when $n = 0, 1, 2, \ldots$. Thus, by Theorem 1.1 the ellipse of convergence of the series in (3.2)–(3.5) is E_ε with $q^\varepsilon = |r| < 1$; the series in (3.6)–(3.7) represent entire functions.

3.2 The Jacobi polynomials

Expansions of analytic functions in series of Jacobi polynomials are discussed in (Askey, 1975), (Boas and Buck, 1964), (Erdélyi et al., 1953) and (Szegő, 1975). More details on the ellipse of convergence, asymptotics and inequalities can be found in (Antonov and Kholshevnikov, 1979), (Carlson, 1974b), (Carlson, 1974a) and references therein.

As an example, Jacobi's generating relation,

$$\sum_{n=0}^{\infty} P_n^{(\alpha,\beta)}(x)\, r^n = \frac{2^{\alpha+\beta}}{R\,(1 - r + R)^\alpha\,(1 + r + R)^\beta}, \tag{3.10}$$

where $R = \left(1 - 2xr + r^2\right)^{1/2}$, holds in the ellipse E_ε with $q^\varepsilon = |r| < 1$.

4. Some *q*-Taylor's Expansions

There is certain interest nowadays in expansions of polynomials and entire functions in the so-called q-Taylor series (Ismail and Stanton, 2002), (Ismail and Stanton, 2003a), and (Kac and Cheung, 2002); see also Exercise 3.8 of (Suslov, 2003b). We only consider here two examples related to the basic exponential function on a q-quadratic grid. The Taylor expansion of $\mathcal{E}_q(x;\alpha)$ with respect to α is

$$\begin{aligned}\mathcal{E}_q(x;\alpha) &= {}_2\varphi_1\left(\begin{matrix}-e^{2i\theta}, -e^{-2i\theta} \\ q\end{matrix}; q^2, q\alpha^2\right) \\ &+ \frac{2q^{1/4}}{1-q}\,\alpha\cos\theta\; {}_2\varphi_1\left(\begin{matrix}-qe^{2i\theta}, -qe^{-2i\theta} \\ q^3\end{matrix}; q^2, q\alpha^2\right) \\ &= \sum_{n=0}^{\infty} \frac{q^{n/2}\alpha^n}{(q;q)_n}\,\varphi_n(x;q),\end{aligned} \tag{4.1}$$

where by the definition

$$\begin{aligned}\varphi_{2k}(x;q) &= \left(-e^{2i\theta}, -e^{-2i\theta}; q^2\right)_k \\ &= 4x^2\prod_{p=1}^{k-1}\left(4x^2q^{2p} + \left(1-q^{2p}\right)^2\right)\end{aligned} \tag{4.2}$$

and

$$\begin{aligned}\varphi_{2k+1}(x;q) &= 2q^{-1/4}\cos\theta\left(-qe^{2i\theta}, -qe^{-2i\theta}; q^2\right)_k \\ &= 2q^{-1/4}\,x\prod_{p=0}^{k-1}\left(4x^2q^{2p+1} + \left(1-q^{2p+1}\right)^2\right).\end{aligned} \tag{4.3}$$

We assume that the empty products when $k = 0$ here are equal to 1. Formula (4.1) provides also an expansion of the basic exponential function with respect to the set of polynomials $\{\varphi_n(x;q)\}_{n=0}^{\infty}$. This basis has been used for the basic sine and cosine functions (Bustoz and Suslov, 1998) and for the basic exponential functions on a q-quadratic grid (Suslov, 2001)–(Suslov, 2002) and, recently, in a more general setting of the q-Taylor expansions (Ismail and Stanton, 2003a)–(Ismail and Stanton, 2003b).

One can easily establish the following upper bounds

$$\begin{aligned}|\varphi_{2k}(x;q)| &\le 4|x|^2\prod_{p=1}^{k-1}\left(4|x|^2q^{2p} + \left(1-q^{2p}\right)^2\right) \\ &< 4|x|^2\prod_{p=1}^{\infty}\left(4|x|^2q^{2p} + \left(1-q^{2p}\right)^2\right) \\ &= 4x^2(\varepsilon)\left(-q^{2\varepsilon}, -q^{-2\varepsilon}; q^2\right)_\infty\end{aligned} \tag{4.4}$$

and

$$
\begin{aligned}
|\varphi_{2k+1}(x;q)| &\le 2q^{-1/4}\,|x| \prod_{p=0}^{k-1} \left(4|x|^2 q^{2p+1} + \left(1-q^{2p+1}\right)^2\right) \\
&< 2q^{-1/4}\,|x| \prod_{p=0}^{\infty} \left(4|x|^2 q^{2p} + \left(1-q^{2p+1}\right)^2\right) \\
&= 2q^{-1/4} x(\varepsilon) \left(-q^{1+2\varepsilon}, -q^{1-2\varepsilon}; q^2\right)_\infty ,
\end{aligned}
\tag{4.5}
$$

where by the definition $|x| = x(\varepsilon) = \left(q^{\varepsilon} + q^{-\varepsilon}\right)/2$. These inequalities show that the polynomials $\{\varphi_n(x;q)\}_{n=0}^{\infty}$ are uniformly bounded with respect to n for any finite value of x. Thus, by Remark 1.2 the series (4.1) is an entire function in the complex x-plane when $q^{1/2}|\alpha| < 1$. The coefficients of this series coincide with those found by the q-Taylor series with respect to $\varphi_n(x;q)$. Indeed, in this case a formal q-Taylor formula has the form (Ismail and Stanton, 2003a)–(Ismail and Stanton, 2003b)

$$
f(x) = \sum_{n=0}^{\infty} f_n\, \varphi_n(x;q), \tag{4.6}
$$

where

$$
f_n = \frac{(1-q)^n}{2^n (q;q)_n} q^{n/4} \left(\mathcal{D}_q^n f(x)\right)\Big|_{x=0} \tag{4.7}
$$

and the operator $\mathcal{D}_q = \delta/\delta x$ is the standard first order Askey-Wilson divided difference operator

$$
\frac{\delta f(x(z))}{\delta x(z)} = \frac{f\left(x\left(z+1/2\right)\right) - f(x(z-1/2))}{x(z+1/2) - x(z-1/2)} \tag{4.8}
$$

with $x(z) = \left(q^z + q^{-z}\right)/2 = \cos\theta$, $q^z = e^{i\theta}$; see (Gasper and Rahman, 1990) and (Suslov, 2003b) for more details. Here we took also into account that the basic exponential function $f(x) = \mathcal{E}_q(x;\alpha)$ satisfies the difference equation

$$
\frac{\delta f}{\delta x} = \frac{2q^{1/4}\alpha}{1-q} f, \tag{4.9}
$$

which is a q-version of

$$
\frac{d}{dx} e^{dx} = \alpha\, e^{dx} \tag{4.10}
$$

on a q-quadratic grid.

In the recent papers (Ismail and Stanton, 2002)–(Ismail and Stanton, 2003a) Ismail and Stanton pointed out the following representations for the q-exponential function

$$\begin{aligned}\mathcal{E}_q(x;\alpha) &= \frac{\left(-\alpha;q^{1/2}\right)_\infty}{\left(q\alpha^2;q^2\right)_\infty}\,{}_2\varphi_1\left(\begin{matrix} q^{1/4}e^{i\theta},\, q^{1/4}e^{-i\theta} \\ -q^{1/2}\end{matrix};q^{1/2},-\alpha\right) \\ &= \frac{\left(\alpha;q^{1/2}\right)_\infty}{\left(q\alpha^2;q^2\right)_\infty}\,{}_2\varphi_1\left(\begin{matrix} -q^{1/4}e^{i\theta},\, -q^{1/4}e^{-i\theta} \\ -q^{1/2}\end{matrix};q^{1/2},\alpha\right),\end{aligned} \tag{4.11}$$

or

$$\begin{aligned}\mathcal{E}_q(x;\alpha) &= \frac{\left(-\alpha;q^{1/2}\right)_\infty}{\left(q\alpha^2;q^2\right)_\infty}\sum_{n=0}^{\infty}\frac{(-\alpha)^n}{(q;q)_n}\,\varphi_n(x;q) \\ &= \frac{\left(\alpha;q^{1/2}\right)_\infty}{\left(q\alpha^2;q^2\right)_\infty}\sum_{n=0}^{\infty}\frac{\alpha^n}{(q;q)_n}\,\varphi_n(-x;q),\end{aligned} \tag{4.12}$$

where $|\alpha| < 1$ and by the definition

$$\varphi_n(\cos\theta;q) = \left(q^{1/4}e^{i\theta}, q^{1/4}e^{-i\theta}; q^{1/2}\right)_n. \tag{4.13}$$

Two independent proofs of (4.11)–(4.12) are presented in (Suslov, 2003b); see Sections 2.3 and 3.4.3.

The uniform upper bound is

$$\begin{aligned}|\varphi_n(x;q)| &\le \left(-q^{1/4+\varepsilon}, -q^{1/4-\varepsilon}; q^{1/2}\right)_n \\ &< \left(-q^{1/4+\varepsilon}, -q^{1/4-\varepsilon}; q^{1/2}\right)_\infty\end{aligned} \tag{4.14}$$

and, therefore, the series in (4.12) is an entire function in the complex x-plane when $|\alpha| < 1$. The corresponding formal q-Taylor formula (Ismail and Stanton, 2002)–(Ismail and Stanton, 2003a)

$$f(x) = \sum_{k=0}^{\infty} f_n\,\varphi_n(x;q), \tag{4.15}$$

where

$$f_n = \frac{(q-1)^n}{2^n(q;q)_n}q^{-n/4}\left(\mathcal{D}_q^n f(x)\right)\Big|_{x=\eta}, \quad \eta = \left(q^{1/4}+q^{-1/4}\right)/2, \tag{4.16}$$

gives again the right coefficients in the expansions (4.11).

Acknowledgments

The author thanks Dick Askey, Joaquin Bustoz, George Gasper, and Dennis Stanton for valuable discussions; he also expresses his appreciation to the referees of this paper for their important comments.

References

Ahlfors, L. V. (1979). *Complex Analysis.* McGraw-Hill, New York, third edition.

Andrews, G. E. and Askey, R. A. (1985). Classical orthogonal polynomials. In *Orthogonal Polynomials and Applications (Bar-le-Duc, 1984)*, volume 1171 of *Lecture Notes in Math.*, pages 36–62. Springer-Verlag, Berlin.

Andrews, G. E., Askey, R. A., and Roy, R. (1999). *Special Functions.* Cambridge University Press, Cambridge.

Antonov, V. A. and Kholshevnikov, K. V. (1979). An estimate for Jacobi polynomials in the complex domain. *Vestnik Leningrad. Univ. Mat. Mekh. Astronom.*, (2):10–13, 118.

Askey, R. A. (1975). *Orthogonal Polynomials and Special Functions.* CBMS-NSF Regional Conferences Series in Applied Mathematics. Society for Industrial and Applied Mathematics, Philadelphia, PA.

Askey, R. A. and Wilson, J. A. (1985). Some basic hypergeometric orthogonal polynomials that generalize Jacobi polynomials. *Memoirs Amer. Math. Soc.*, 54(319).

Atakishiyev, N. M. and Suslov, S. K. (1992). Difference hypergeometric functions. In Gonchar, A. A. and Saff, E. B., editors, *Progress in Approximation Theory (Tampa, FL, 1990)*, volume 19 of *Computational Mathematics*, pages 1–35. Springer-Verlag, New York.

Boas, R. P. and Buck, R. C. (1964). *Polynomial Expansions of Analytic Functions.* Springer-Verlag, Berlin. Second printing, corrected. Ergebnisse der Mathematik und ihrer Grenzgebiete, N.F., Bd. 19.

Bromwich, T. J. (1965). *An Introduction to the Theory of Infinite Series.* Macmillan, New York. Second Edition, revised.

Bustoz, J. and Suslov, S. K. (1998). Basic analog of Fourier series on a q-quadratic grid. *Methods Appl. Anal.*, 5:1–38.

Carlson, B. C. (1974a). Expansions of analytic functions in Jacobi series. *SIAM J. Math. Anal.*, 5:586–596.

Carlson, B. C. (1974b). Inequalities for Jacobi polynomials and Dirichlet averages. *SIAM J. Math. Anal.*, 5:797–808.

Davis, P. J. (1963). *Interpolation and Approximation.* Blaisdell Publishing Co. Ginn and Co., New York-Toronto-London.

Dienes, P. (1957). *The Taylor Series: An Introduction to the Theory of Functions of a Complex Variable.* Dover, New York.

Erdélyi, A., Magnus, W., Oberhettinger, F., and Tricomi, F. G. (1953). *Higher Transcendental Functions, Vol. II.* McGraw-Hill, New York-Toronto-London. Based, in part, on notes left by Harry Bateman.

Floreanini, R. and Vinet, L. (1995). A model for the continuous q-ultraspherical polynomials. *J. Math. Phys.*, 36:3800–3813.

Gasper, G. and Rahman, M. (1990). *Basic Hypergeometric Series.* Cambridge University Press.

Ismail, M. E. H., Rahman, M., and Stanton, D. (1999). Quadratic q-exponentials and connection coefficient problems. *Proc. Amer. Math. Soc.*, 127(10):2931–2941.

Ismail, M. E. H., Rahman, M., and Zhang, R. (1996). Diagonalization of certain integral operators II. *J. Comp. Appl. Math.*, 68:163–196.

Ismail, M. E. H. and Stanton, D. (2000). Addition theorems for the q-exponential functions. In Ismail, M. E. H. and Stanton, D. W., editors, *q-Series from a Con-*

temporary Perspective, volume 254 of *Contemporary Mathematics*, pages 235–245. American Mathematical Society, Providence, RI.

Ismail, M. E. H. and Stanton, D. (2002). q-integral and moment representations for q-orthogonal polynomials. *Canad. J. Math.*, 54(4):709–735.

Ismail, M. E. H. and Stanton, D. (2003a). Applications of q-Taylor theorems. *J. Comp. Appl. Math.*, 153:259–272.

Ismail, M. E. H. and Stanton, D. (2003b). q-Taylor theorems, polynomial expansions, and interpolation of entire functions. *J. Approx. Theory*, 123(1):125–146.

Ismail, M. E. H. and Wilson, J. A. (1982). Asymptotic and generating relations for the q-Jacobi and ${}_4\varphi_3$ polynomials. *J. Approx. Theory*, 36:43–54.

Ismail, M. E. H. and Zhang, R. (1994). Diagonalization of certain integral operators. *Advances in Math.*, 109:1–33.

Kac, V. and Cheung, P. (2002). *Quantum Calculus.* Universitext. Springer-Verlag, New York.

Markushevich, A. I. (1985). *Theory of Functions of a Complex Variable*, volume I. Chelsea Publishing Company, New York, second edition.

Rahman, M. (1986). q-Wilson functions of the second kind. *SIAM J. Math. Anal.*, 17:1280–1286.

Rainville, E. D. (1960). *Special Functions.* The Macmillan Company, New York.

Suslov, S. K. (1997). "Addition" theorems for some q-exponential and q-trigonometric functions. *Methods Appl. Anal.*, 4:11–32.

Suslov, S. K. (2000). Another addition theorem for the q-exponential function. *J. Phys. A: Math. Gen.*, 33:L375–380.

Suslov, S. K. (2001). Basic exponential functions on a q-quadratic grid. In Bustoz, J., Ismail, M. E. H., and Suslov, S. K., editors, *Special Functions 2000: Current Perspective and Future Directions*, volume 30 of *NATO Science Series II: Mathematics, Physics and Chemistry*, pages 411–456. Kluwer Academic Publishers, Dordrecht-Boston-London.

Suslov, S. K. (2002). Some expansions in basic Fourier series and related topics. *J. Approx. Theory*, 115(2):289–353.

Suslov, S. K. (2003a). Expansion of analytic functions in q-orthogonal polynomials. Submitted.

Suslov, S. K. (2003b). *An Introduction to Basic Fourier Series*, volume 9 of *Developments in Mathematics.* Kluwer Academic Publishers, Dordrecht-Boston-London.

Szegő, G. (1975). *Orthogonal Polynomials*, volume XXIII of *Amer. Math. Soc. Colloq. Publ.* American Mathematical Society, Providence, RI, fourth edition.

Szegő, G. (1982). *Collected Papers. Vol. 1.* Contemporary Mathematicians. Birkhäuser, Boston, MA. 1915–1927, Edited by Richard Askey, Including commentaries and reviews by George Pólya, P. C. Rosenbloom, Askey, L. E. Payne, T. Kailath and Barry M. McCoy.

STRONG NONNEGATIVE LINEARIZATION OF ORTHOGONAL POLYNOMIALS

Ryszard Szwarc *
Institute of Mathematics
Wroclaw University
pl. Grunwaldzki 2/4
50–384 Wroclaw
POLAND

and

Institute of Mathematics
Polish Academy of Science
ul. Śniadeckich 8
00–950 Warszawa
POLAND

Abstract A stronger notion of nonnegative linearization of orthogonal polynomials is introduced. It requires that also the associated polynomials of any order have nonnegative linearization property. This turns out to be equivalent to a maximal principle of a discrete boundary value problem associated with orthogonal polynomials through the three term recurrence relation. The property is stable for certain perturbations of the recurrence relation. Criteria for the strong nonnegative linearization are derived. The range of parameters for the Jacobi polynomials satisfying this new property is determined.

Keywords: Orthogonal polynomials, recurrence relation, nonnegative linearization, discrete boundary value problem.

1. Introduction

One of the main problems in the theory of orthogonal polynomials is to determine whether the expansion of the product of two orthogonal

*This work was partially supported by KBN (Poland) under grant 5 P03A 034 20 and by European Commission via TMR network "Harmonic Analysis and Related Problems," RTN2-2001-00315.

polynomials in terms of these polynomials has nonnegative coefficients. We want to decide which orthogonal systems $\{p_n\}_{n=0}^{\infty}$ have the property

$$p_n(x)p_m(x) = \sum c(n,m,k)p_k(x)$$

with nonnegative coefficients $c(n,m,k)$ for every n, m and k.

Numerous classical orthogonal polynomials as well as their q-analogues satisfy nonnegative linearization property (Gasper, 1970a; Gasper, 1970b; Gasper, 1983), (Gasper and Rahman, 1990), (Ramis, 1992), (Rogers, 1894), (Szwarc, 1992b; Szwarc, 1995). There are many criteria for nonnegative linearization given in terms of the coefficients of the recurrence relation the orthogonal polynomials satisfy (Askey, 1970), (Młotkowski and Szwarc, 2001), (Szwarc, 1992a; Szwarc, 1992b; Szwarc, 2003), that can be applied to general orthogonal polynomials systems. These criteria are based on the connection between the linearization property and a certain discrete boundary value problem of hyperbolic type.

In this paper we are going to show that many polynomials systems satisfy even a stronger version of nonnegative linearization. Namely let $\{p_n\}_{n=0}^{\infty}$ be an orthogonal polynomial system. Let $\{p_n^{[l]}\}_{n=0}^{\infty}$ denote the associated polynomials of order l. We say that the polynomials $\{p_n\}_{n=0}^{\infty}$ satisfy *the strong nonnegative linearization property* if

$$p_n(x)p_m(x) = \sum c(n,m,k)p_k(x),$$
$$p_n^{[l]}(x)p_m^{[l]}(x) = \sum c_l(n,m,k)p_k^{[l]}(x).$$

with nonnegative coefficients $c(n,m,k)$ and $c_l(n,m,k)$ for any n, m, k and l.

The interesting feature of this property is the fact that it is equivalent to a maximum principle of the associated boundary value problem (see Theorem 2). Also this property is invariant for certain transformations of the recurrence relation (see Proposition 2), unlike the usual nonnegative linearization property.

In the last part of this work we are going to show that the Jacobi polynomials have the strong linearization property if and only if either $\alpha = \beta \geq -1/2$ or $\alpha > \beta > -1$ and $\alpha + \beta \geq 0$.

2. Strong nonnegative linearization

Let p_n denote a sequence of orthogonal polynomials, relative to a measure μ, satisfying the recurrence relation

$$xp_n = \gamma_n p_{n+1} + \beta_n p_n + \alpha_n p_{n-1}, \qquad n \geq 0, \tag{2.1}$$

where $\gamma_n, \alpha_{n+1} > 0$ and $\beta_n \in \mathbb{R}$. We use the convention that $p_0 = 1$ and $\alpha_0 = p_{-1} = 0$. For any nonnegative integer l let $p_n^{[l]}$ denote the sequence of polynomials satisfying

$$xp_n^{[l]} = \gamma_n p_{n+1}^{[l]} + \beta_n p_n^{[l]} + \alpha_n p_{n-1}^{[l]}, \qquad n \geq l+1, \tag{2.2}$$

$$p_0^{[l]} = p_1^{[l]} = \ldots = p_l^{[l]} = 0, \; p_{l+1}^{[l]} = \frac{1}{\gamma_l}. \tag{2.3}$$

For $n \geq l+1$ the polynomial $p_n^{[l]}$ is of degree $n-l-1$. The polynomials $p_n^{[l]}$ are called the *associated polynomial of order* $l+1$. These polynomials are orthogonal, as well. Let μ_l denote any orthogonality measure associated with $\{p_n^{[l]}\}_{n=l+1}^{\infty}$.

For $n \geq m \geq l+1 \geq 0$ consider the polynomials $p_n(x)p_m(x)$ and $p_n^{[l]}(x)p_m^{[l]}(x)$. We can express these products in terms of $p_k(x)$ or $p_k^{[l]}(x)$ to obtain the following.

$$p_n(x)p_m(x) = \sum_{k=0}^{\infty} c(n,m,k)p_k(x),$$

$$p_n^{[l]}(x)p_m^{[l]}(x) = \sum_{k=0}^{\infty} c_l(n,m,k)p_k^{[l]}(x).$$

The polynomial $p_n(x)p_m(x)$ has degree $n+m$ while $p_n^{[l]}(x)p_m^{[l]}(x)$ has degree $n+m-2l-2$. Hence the expansions have finite ranges and by the recurrence relation we obtain expansions of the form

$$p_n(x)p_m(x) = \sum_{k=|n-m|}^{n+m} c(n,m,k)p_k(x), \tag{2.4}$$

$$p_n^{[l]}(x)p_m^{[l]}(x) = \sum_{k=|n-m|+l+1}^{n+m-l-1} c_l(n,m,k)p_k^{[l]}(x). \tag{2.5}$$

Definition 2.1. *The system of orthogonal polynomials p_n satisfies* ***the strong nonnegative linearization property*** (SNLP) *if*

$$c(n,m,k) \geq 0, \tag{2.6}$$

$$c_l(n,m,k) \geq 0, \tag{2.7}$$

for any $n, m, k \geq 0$ and $l \geq 0$.

The form of recurrence relation used in (2.1) and (2.2) is suitable for applications. For technical reasons we will work with the renormalized polynomials P_n and $P_n^{[l]}$ defined as

$$P_n(x) = \frac{\gamma_0\gamma_1\cdots\gamma_{n-1}}{\alpha_1\alpha_2\dots\alpha_n}p_n(x), \quad n \geq 1,$$
$$P_n^{[l]}(x) = \frac{\gamma_l\gamma_{l+1}\cdots\gamma_{n-1}}{\alpha_{l+1}\alpha_{l+2}\cdots\alpha_n}p_n^{[l]}(x), \quad n \geq l+1.$$

Clearly the property of strong nonnegative linearization is equivalent for the systems $\{p_n\}_{n=0}^{\infty}$ and $\{P_n\}_{n=0}^{\infty}$, so we can work with the latter system from now on.

The polynomials P_n satisfy

$$xP_n = \alpha_{n+1}P_{n+1} + \beta_nP_n + \gamma_{n-1}P_{n-1}, \qquad n \geq 0, \tag{2.8}$$

where $\gamma_{-1} = 0$. On the other hand the polynomials $P_n^{[l]}$ satisfy

$$xP_n^{[l]} = \alpha_{n+1}P_{n+1}^{[l]} + \beta_nP_n^{[l]} + \gamma_{n-1}P_{n-1}^{[l]}, \qquad n \geq l+1. \tag{2.9}$$

Moreover by (2.4) and (2.5) we have

$$P_n(x)P_m(x) = \sum_{k=|n-m|}^{n+m} C(n,m,k)P_k(x), \tag{2.10}$$

$$P_n^{[l]}(x)P_m^{[l]}(x) = \sum_{k=|n-m|+l+1}^{n+m-l-1} C_l(n,m,k)P_k^{[l]}(x). \tag{2.11}$$

Let L denote a linear operator acting on sequences $a = \{a_n\}_{n=0}^{\infty}$ by the rule

$$(La)_n = \alpha_{n+1}a_{n+1} + \beta_na_n + \gamma_{n-1}a_{n-1}, \qquad n \geq 0. \tag{2.12}$$

For any real number x set

$$P(x) = \{P_n(x)\}_{n=0}^{\infty},$$
$$P^{[l]}(x) = \{P_n^{[l]}(x)\}_{n=0}^{\infty}.$$

Let δ_l denote the sequence whose terms are equal to zero except for the lth term which is equal to 1. The formulas (2.8), (2.9) and the fact that $P_{l+1}^{[l]} = \alpha_{l+1}^{-1}$ immediately imply that

$$LP(x) = xP(x), \tag{2.13}$$
$$LP^{[l]}(x) = xP^{[l]}(x) + \delta_l. \tag{2.14}$$

3. Hyperbolic boundary value problem and basic solutions

Let $u(n,m)$ be a matrix defined for $n \geq m \geq 0$. We introduce the operator H acting on the matrices by the rule

$$\begin{aligned}(Hu)(n,m) &= \alpha_{n+1}u(n+1,m) + \beta_n u(n,m) + \gamma_{n-1}u(n-1,m) \\ &\quad - \alpha_{m+1}u(n,m+1) - \beta_m u(n,m) - \gamma_{m-1}u(n,m-1),\end{aligned} \tag{3.1}$$

for $n > m \geq 0$. By (2.13), if we take $u(n,m) = P_n(x)P_m(x)$ for some x, then

$$(Hu)(n,m) = 0. \tag{3.2}$$

Similarly by (2.14), if we take $u(n,m) = P_n^{[l]}(x)P_m^{[l]}(x)$, then

$$(Hu)(n,m) = P_m^{[l]}(x)\delta_l(n) - P_n^{[l]}(x)\delta_l(m).$$

Assume $n > m$. Then $n = l$ implies $P_m^{[l]}(x) = 0$. Hence

$$(Hu)(n,m) = -P_n^{[l]}(x)\delta_l(m), \quad \text{for } n > m \geq 0. \tag{3.3}$$

Proposition 3.1. *Given a matrix* $v = \{v(n,m)\}_{n>m\geq 0}$ *and a sequence* $f = \{f(n)\}_{n\geq 0}$. *Let* $u = \{u(n,m)\}_{n\geq m\geq 0}$ *satisfy*

$$\begin{aligned} Hu(n,m) &= v(n,m), \quad \textit{for } n > m \geq 0, \\ u(n,0) &= f(n), \quad \textit{for } n \geq 0. \end{aligned}$$

Then

$$u(n,m) = -\sum_{k>l\geq 0} v(k,l)C_l(n,m,k) + \sum_{k\geq 0} f(k)C(n,m,k).$$

Proof. The formula (3.1) and the fact that $\alpha_m > 0$ imply that u is uniquely determined.

Let $u_k(n,m) = C(n,m,k)$. By (2.10) we have

$$u_k(n,m) = \left(\int_{\mathbb{R}} P_k^2(x)\,d\mu(x)\right)^{-1} \int_{\mathbb{R}} P_n(x)P_m(x)P_k(x)\,d\mu(x).$$

Therefore by (3.2) we obtain

$$\begin{aligned} (Hu_k)(n,m) &= 0, \quad \text{for } n > m \geq 0, \\ u_k(n,0) &= \delta_k(n), \quad \text{for } n \geq 0. \end{aligned}$$

For $k > l \geq 0$ let $u_{k,l}(n,m) = C_l(n,m,k)$. By (2.11) we have

$$u_{k,l}(n,m) = \left(\int_{\mathbb{R}} \left\{ P_k^{[l]}(x) \right\}^2 d\mu_l(x) \right)^{-1} \int_{\mathbb{R}} P_n^{[l]}(x) P_m^{[l]}(x) P_k^{[l]}(x)\, d\mu_l(x).$$

Thus by (3.3) we get

$$\begin{aligned} (Hu_{k,l})(n,m) &= -\delta_{(k,l)}(n,m), \ \text{for } n > m \geq 0, \\ u_{k,l}(n,0) &= 0, \ \text{for } n \geq 0. \end{aligned}$$

Hence the matrix

$$u(n,m) = -\sum_{k>l\geq 0} v(k,l) u_{k,l}(n,m) + \sum_{k\geq 0} f(k) u_k(n,m)$$

satisfies the assumptions of Proposition 1. By uniqueness we have $u = u$. □

Let H^* denote the adjoint operator to H with respect to the inner product of matrices

$$\langle u, v \rangle = \sum_{n>m\geq 0}^{\infty} u(n,m)\overline{v(n,m)}.$$

The explicit action of this operator is given by the following.

$$\begin{aligned} (H^*v)(n,m) =& \gamma_n v(n+1,m) + \beta_n v(n,m) + \alpha_n v(n-1,m) \\ & - \gamma_m v(n,m+1) - \beta_m v(n,m) - \alpha_m v(n,m-1). \end{aligned}$$

For each point (n,m) with $n \geq m \geq 0$, let $\Delta_{n,m}$ denote the set of lattice points located in the triangle with vertices in $(n-m+1,0)$, $(n+m-1,0)$ nad $(n,m-1)$, i.e.

$$\Delta_{n,m} = \{(i,j) \,|\, 0 \leq j \leq i, |n-i| < m-j\}.$$

The points of $\Delta_{n,m}$ are marked in the picture below with empty circles.

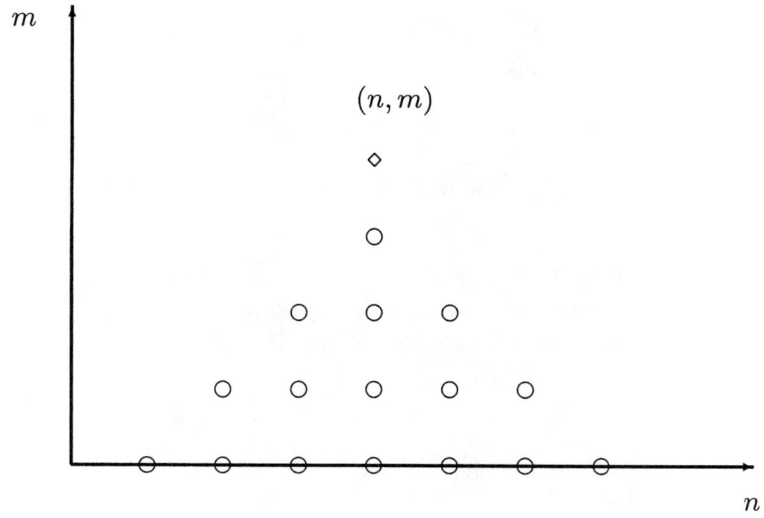

By (Szwarc, 2003, Theorem 1) nonnegative linearization is equivalent to the fact that for every (n, m) with $n \geq m \geq 0$ there exists a matrix v such that

$$\operatorname{supp} v \subset \Delta_{n,m}, \tag{3.4}$$

$$(H^*v)(n, m) < 0, \tag{3.5}$$

$$(H^*v)(i, j) \geq 0, \text{ for } (i, j) \neq (n, m). \tag{3.6}$$

Definition 3.2. *Any matrix v satisfying* (3.4) *and* (3.5) *will be called **a triangle function.***

Definition 3.3. *Let $v_{n,m}$ denote a matrix satisfying*

$$\operatorname{supp} v_{n,m} \subset \Delta_{n,m}, \tag{3.7}$$

$$(H^*v_{n,m})(n, m) = -1, \tag{3.8}$$

$$(H^*v_{n,m})(i, j) = 0, \text{ for } 0 < j < m \tag{3.9}$$

*The matrix $v_{n,m}$ will be called **the basic triangle function.***

The main result of this section relates the values of $v_{n,m}(k, l)$ to the coefficients $C_l(n, m, k)$.

Theorem 3.4. *For any $n \geq m \geq 0$ and $k > l \geq 0$ we have*

$$v_{n,m}(k, l) = C_l(n, m, k).$$

Moreover

$$H^*v_{n,m} = -\delta_{(n,m)} + \sum_{k=n-m}^{n+m} C(n, m, k)\delta_{(k,0)}.$$

Proof. Let $u(n,m) = P_n^{[l]}(x)P_m^{[l]}(x)$. We have $P_0^{[l]} = 0$, hence by (3.3), (3.8) and (3.9) we obtain

$$\begin{aligned} -P_n^{[l]}(x)P_m^{[l]}(x) &= -u(n,m) = \langle H^* v_{n,m}, u\rangle = \langle v_{n,m}, Hu\rangle \\ &= \sum_{k,j} v_{n,m}(k,j)(Hu)(k,j) = -\sum_k v_{n,m}(k,l)P_k^{[l]}(x). \end{aligned}$$

Thus by (2.11) we get $v_{n,m}(k,l) = C_l(n,m,k)$. The second part of the statement follows from (Szwarc, 2003, Lemma), but we will recapitulate the proof here for completeness. By (3.8) and (3.9) we have

$$H^* v_{n,m} = -\delta_{(n,m)} + \sum_k d_k \delta_{(k,0)}.$$

Let $u(n,m) = P_n(x)P_m(x)$. Since $Hu = 0$, we have

$$\begin{aligned} P_n(x)P_m(x) = u(n,m) &= -\langle H^* v_{n,m}, u\rangle + \sum_k d_k u(k,0) \\ &= \langle v_{n,m}, Hu\rangle + \sum_k d_k P_k(x) = \sum_k d_k P_k(x). \end{aligned}$$

Hence $d_k = C(n,m,k)$. □

4. Main results

The main result of this paper is the following.

Theorem 4.1. *Let p_n be a system of orthogonal polynomials satisfying the recurrence relation*

$$xp_n = \gamma_n p_{n+1} + \beta_n p_n + \alpha_n p_{n-1},$$

where $p_{-1} = 0$ and $p_0 = 1$. Then the following four conditions are equivalent.

(a) *The polynomials p_n satisfy the strong nonnegative linearization property.*

(b) *Let $u = \{u(n,m)\}_{n \ge m \ge 0}$ satisfy*

$$\begin{cases} (Hu)(n,m) \le 0, & \text{for } n > m \ge 0, \\ u(n,0) \ge 0. \end{cases}$$

Then $u(n,m) \ge 0$ for every $n \ge m \ge 0$.

(c) *For every $n \ge m \ge 0$ there exists a triangle function v, satisfying*

(i) $\operatorname{supp} v \subset \Delta_{n,m}$.

(ii) $(H^*v)(n,m) < 0$.

(iii) $(H^*v)(i,j) \geq 0$ *for* $(i,j) \neq (n,m)$.

(iv) $v \geq 0$.

(d) *The basic triangle functions* $v_{n,m}$ *(see* (3.7), (3.8), (3.9)) *satisfy*

(i) $(H^*v_{n,m})(i,0) \geq 0$.

(ii) $v_{n,m} \geq 0$.

Proof.
(b)⇒(a)
By the proof of Proposition 1 we have that if $u_k(n,m) = C(n,m,k)$ and $u_{k,l}(n,m) = C_l(n,m,k)$ then

$$(Hu_k)(n,m) = 0, \qquad (Hu_{k,l})(n,m) = -\delta_{(k,l)}(n,m)$$
$$u_k(n,0) = \delta_k(n), \qquad u_{k,l}(n,0) = 0.$$

for $n > m \geq 0$. Thus $C(n,m,k) \geq 0$ and $C_l(n,m,k) \geq 0$ for $n \geq m \geq 0$.
(a)⇒(d)
This follows immediately by Theorem 1.
(d)⇒(c)
This is clear by definition.
(c)⇒(b)
Let $u = \{u(n,m)\}_{n \geq m \geq 0}$ satisfy $(Hu)(n,m) \leq 0$, for $n > m \geq 0$ and $u(n,0) \geq 0$. We will show that $u(n,m) \geq 0$, by induction on m. Assume that $u(i,j) \geq 0$ for $j < m$. Let v be a triangle function satisfying the assumptions (c). Then

$$0 \geq \langle Hu, v \rangle = \langle u, H^*v \rangle = u(n,m)(H^*v)(n,m) + \sum_{\substack{i \geq j \geq 0 \\ j < m}} u(i,j)(H^*v)(i,j)$$

Therefore

$$-u(n,m)(H^*v)(n,m) \geq \sum_{\substack{i \geq j \geq 0 \\ j < m}} u(i,j)(H^*v)(i,j),$$

and the conclusion follows. □

Remark 4.2. *Theorem 2 should be juxtaposed with the following result which can be derived from (Szwarc, 2003, Theorem 1).*

Theorem 4.3. *Let p_n be a system of orthogonal polynomials satisfying the recurrence relation*

$$xp_n = \gamma_n p_{n+1} + \beta_n p_n + \alpha_n p_{n-1},$$

where $p_{-1} = 0$ and $p_0 = 1$. Then the following four conditions are equivalent.

(a) *The polynomials p_n satisfy nonnegative linearization property.*

(b) *Let $u = \{u(n,m)\}_{n\geq m\geq 0}$ satisfy*

$$\begin{cases} (Hu)(n,m) & = & 0, \text{ for } n > m \geq 0, \\ u(n,0) & \geq & 0. \end{cases}$$

Then $u(n,m) \geq 0$ for every $n \geq m \geq 0$.

(c) *For every $n \geq m \geq 0$ there exists a triangle function v, satisfying*

(i) $\operatorname{supp} v \subset \Delta_{n,m}$.

(ii) $(H^*v)(n,m) < 0$.

(iii) $(H^*v)(i,j) \geq 0$ *for* $(i,j) \neq (n,m)$.

(d) *The basic triangle functions $v_{n,m}$ (see (3.7), (3.8), (3.9)) satisfy*

(i) $(H^*v_{n,m})(i,0) \geq 0$.

One of the advantages of the strong nonnegative linearization property is its stability for a certain perturbation of the coefficients in the recurrence relation. Namely the following holds.

Proposition 4.4. *Assume orthogonal polynomial system $\{p_n\}_{n=0}^{\infty}$ satisfies* (SNLP). *Let ε_n be a nondecreasing sequence. Let q_n be a sequence of polynomials satisfying the perturbed recurrence relation*

$$xq_n = \gamma_n q_{n+1} + (\beta_n + \varepsilon_n) q_n + \alpha_n q_{n-1},$$

for $n \geq 0$. Then the system $\{q_n\}_{n=0}^{\infty}$ satisfies (SNLP).

Proof. We will make use of Theorem 4.1(c). Let H and H_ε denote the hyperbolic operators corresponding to the unperturbed and perturbed system, respectively. For any matrix $v(i,j)$ we have

$$(H_\varepsilon^* v)(i,j) = (H^*v)(i,j) + (\varepsilon_i - \varepsilon_j)v(i,j). \tag{4.1}$$

By assumptions for any $n \geq m \geq 0$, there exists a triangle function v satisfying the assumptions of Theorem 4.1(c) with respect to H. By

(4.1) the same matrix v satisfies these assumptions with respect to H_ε. Indeed, the assumptions (i) and (iv) do not depend on the perturbation. Since $v(n,m) = 0$ the assumption (ii) is not affected, as well. Concerning (iii), since $v \geq 0$ and ε_n is nondecreasing we have

$$(H_\varepsilon^* v)(i,j) \geq (H^* v)(i,j) \geq 0,$$

for $i \geq j \geq 0$ and $j < m$. Hence the perturbed system of polynomials satisfies (SNLP). □

5. Some necessary and sufficient conditions

We begin with the following generalization of Theorem 1 of (Szwarc, 1992a).

Theorem 5.1. *Let orthogonal polynomials $\{p_n\}_{n=0}^\infty$ satisfy (2.1). Let $\{c_n\}_{n=0}^\infty$ be a fixed sequence of positive numbers with $c_0 = 1$ and*

$$\alpha'_n = \frac{c_{n-1}}{c_n}\alpha_n, \qquad \gamma'_n = \frac{c_{n+1}}{c_n}\gamma_n, \qquad \text{for } n \geq 1.$$

Assume that

(i) $\beta_m \leq \beta_n$ *for* $m \leq n$.

(ii) $\alpha_m \leq \alpha'_n$ *for* $m < n$.

(iii) $\alpha_m + \gamma_m \leq \alpha'_n + \gamma'_n$ *for* $m < n$.

(iv) $\alpha_m \leq \gamma'_n$ *for* $m \leq n$.

Then the system $\{p_n\}_{n=0}^\infty$ satisfies the strong nonnegative linearization property.

Proof. It suffices to construct a suitable triangle function for every (n,m), with $n \geq m$, i.e., a matrix v satisfying the assumptions of Theorem 4.1(c). Fix (n,m). Define the matrix v according to the following.

$$v(i,j) = \begin{cases} c_i & (i,j) \in \Delta_{n,m},\ (n+m)-(i+j) \text{ odd} \\ 0 & \text{otherwise} \end{cases} \tag{5.1}$$

The points in the support of $v_{n,m}$ are marked by empty circles in the picture below.

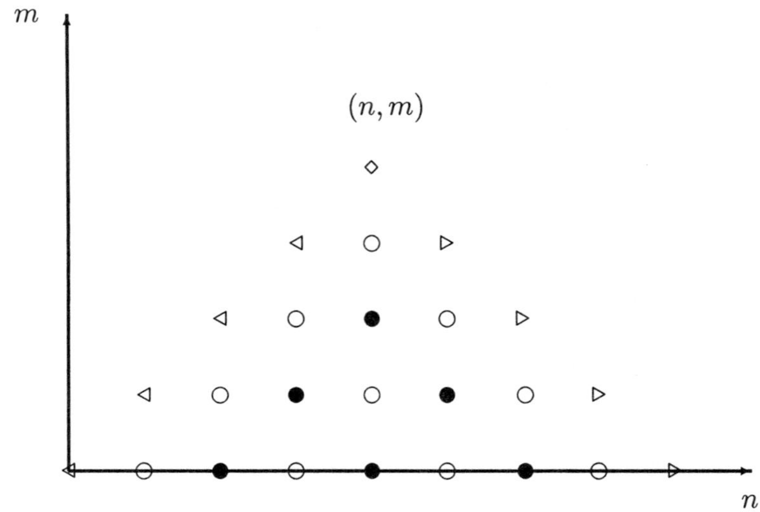

Then supp H^*v consists of the points marked by $\circ, \bullet, \triangleleft, \triangleright$ and $\diamond$. A straightforward computation gives

$$(H^*v)(i,j) = \begin{cases} -\alpha_m c_n & (i,j) = (n,m) \\ (\beta_i - \beta_j)c_i & (i,j) - \circ \\ \alpha_i c_{i-1} + \gamma_i c_{i+1} - \alpha_j c_i - \gamma_j c_i & (i,j) - \bullet \\ \alpha_i c_{i-1} - \alpha_j c_i & (i,j) - \triangleright \\ \gamma_i c_{i+1} - \alpha_j c_i & (i,j) - \triangleleft \end{cases}$$

Hence H^*v satisfies the assumptions of Theorem 4.1(c). □

Applying Theorem 5.1 to the sequences

$$c_n = 1 \quad \text{or} \quad c_n = \frac{\alpha_1 \alpha_2 \ldots \alpha_n}{\gamma_0 \gamma_1 \cdots \gamma_{n-1}}$$

gives the following.

Corollary 5.2. *Let orthogonal polynomials $\{p_n\}_{n=0}^{\infty}$ satisfy* (2.1). *If the sequences α_n, β_n, $\alpha_n + \gamma_n$ are nondecreasing and $\alpha_n \le \gamma_n$ for all n, then the system $\{p_n\}_{n=0}^{\infty}$ satisfies the strong nonnegative linearization property.*

Corollary 5.3. *Let orthogonal polynomials $\{p_n\}_{n=0}^{\infty}$ satisfy* (2.1). *Assume that*

(i) $\beta_m \le \beta_n$ *for* $m \le n$

(ii) $\alpha_m \le \gamma_n$ *for* $m \le n$

(iii) $\alpha_m + \gamma_m \le \alpha_{n-1} + \gamma_{n+1}$ *for* $m < n$

(iv) $\alpha_m \le \alpha_n$ *for* $m \le n$

Then the system $\{p_n\}_{n=0}^{\infty}$ *satisfies the strong nonnegative linearization property.*

Now we turn to necessary conditions for (SNLP).

Proposition 5.4. *Assume a system* $\{p_n\}_{n=0}^{\infty}$ *satisfies the strong nonnegative linearization property. Then the sequence* β_n *is nondecreasing.*

Proof. By (2.2) we can compute that for $n \ge 2$ we have

$$p_n^{[n-2]}(x) = \frac{1}{\gamma_{n-2}\gamma_{n-1}}(x - \beta_{n-1}).$$

But by (2.1) we have

$$(x - \beta_{n-1})p_n^{[n-2]} = \gamma_n p_{n+1}^{[n-2]} + (\beta_n - \beta_{n-1})p_n^{[n-2]} + \alpha_n p_{n-1}^{[n-2]}.$$

Thus $\beta_n \ge \beta_{n-1}$ for $n \ge 2$. On the other hand

$$p_1(x) = \frac{1}{\gamma_0}(x - \beta_0)$$

and

$$(x - \beta_0)p_1 = \gamma_1 p_2 + (\beta_1 - \beta_0)p_1 + \alpha_1 p_0.$$

Hence $\beta_1 \ge \beta_0$. □

6. Jacobi polynomials

The Jacobi polynomials $J_n^{(\alpha,\beta)}$ satisfy the recurrence relation

$$\begin{aligned} xJ_n^{(\alpha,\beta)} &= \frac{2(n+1)(n+\alpha+\beta+1)}{(2n+\alpha+\beta+1)(2n+\alpha+\beta+2)} J_{n+1}^{(\alpha,\beta)} \\ &+ \frac{\beta^2-\alpha^2}{(2n+\alpha+\beta)(2n+\alpha+\beta+2)} J_n^{(\alpha,\beta)} \\ &+ \frac{2(n+\alpha)(n+\beta)}{(2n+\alpha+\beta)(2n+\alpha+\beta+1)} J_{n-1}^{(\alpha,\beta)}. \end{aligned} \tag{6.1}$$

Theorem 6.1. *The Jacobi polynomials satisfy the strong nonnegative linearization property if and only if either* $\alpha > \beta > -1$ *and* $\alpha + \beta \ge 0$ *or* $\alpha = \beta \ge -\frac{1}{2}$.

Proof. Assume the Jacobi polynomials satisfy (SNLP). In particular they have nonnegative linearization property. By (Gasper, 1970a) we know that the condition $\alpha \geq \beta$ is necessary for nonnegative linearization to hold. Also if $\alpha = \beta$ then the condition $\alpha \geq -\frac{1}{2}$ is necessary (see (Askey, 1975)). Let $\alpha > \beta$. By Proposition 5.4 the sequence

$$\beta_n = \frac{\beta^2 - \alpha^2}{(2n+\alpha+\beta)(2n+\alpha+\beta+2)}$$

should be nondecreasing, which holds only if $\alpha + \beta \geq 0$. Hence the conditions on α and β are necessary for (SNLP).

Now we are going to show that the conditions on the parameters are also sufficient for (SNLP). Assume first that $\alpha = \beta \geq -1/2$. Let

$$R_n(x) = \frac{J_n^{(\alpha,\alpha)}(x)}{J_n^{(\alpha,\alpha)}(1)}.$$

Then by (Koekeok and Swarttouw, 1998, (1.8.1), (1.8.3)) the polynomials satisfy

$$xR_n(x) = \frac{n+2\alpha+1}{2n+2\alpha+1}R_{n+1}(x) + \frac{n}{2n+2\alpha+1}R_{n-1}(x).$$

Hence by Corollary 5.2 the polynomials satisfy (SNLP).

Assume now that $\alpha > \beta > -1$ and $\alpha + \beta \geq 0$. Let $p_n(x)$ denote the monic version of Jacobi polynomials, i.e., let

$$p_n(x) = \frac{1}{2^n}\binom{2n+\alpha+\beta}{n} J_n^{(\alpha,\beta)}(x).$$

By (Askey, 1970) the polynomials p_n satisfy the assumptions of Corollary 5.2 if $\alpha + \beta \geq 1$. Hence they satisfy (SNLP).

We have to consider the remaining case when $\alpha > \beta > -1$ and $0 \leq \alpha + \beta < 1$. By (6.1) we have

$$\alpha_n = \frac{2(n+\alpha)(n+\beta)}{(2n+\alpha+\beta)(2n+\alpha+\beta+1)}, \quad n > 0, \tag{6.2}$$

$$\beta_n = \frac{\beta^2 - \alpha^2}{(2n+\alpha+\beta)(2n+\alpha+\beta+2)}, \tag{6.3}$$

$$\gamma_n = \frac{2(n+1)(n+\alpha+\beta+1)}{(2n+\alpha+\beta+1)(2n+\alpha+\beta+2)}. \tag{6.4}$$

These numbers satisfy the assumptions of Corollary 5.3 for $\alpha \geq \beta$ and $0 \leq \alpha + \beta \leq 1$. Indeed, observe that for $n > 0$ we have

$$2\alpha_n - 1 = -\frac{(\alpha-\beta)^2}{2n+\alpha+\beta} + \frac{(\alpha-\beta)^2 - 1}{2n+\alpha+\beta+1},$$

$$2\gamma_{n-1} - 1 = -\frac{(\alpha+\beta)^2 - 1}{2n+\alpha+\beta-1} + \frac{(\alpha+\beta)^2}{2n+\alpha+\beta},$$

and

$$
\begin{aligned}
&2(\alpha_n+\gamma_n)-2\\
&=\frac{-4\alpha\beta}{(2n+\alpha+\beta+1)(2n+\alpha+\beta+2)}-\frac{2(\alpha-\beta)^2}{(2n+\alpha+\beta)(2n+\alpha+\beta+2)}\\
&=\frac{4\alpha\beta}{(2n+\alpha+\beta)(2n+\alpha+\beta+1)}-\frac{2(\alpha+\beta)^2}{(2n+\alpha+\beta)(2n+\alpha+\beta+2)}.
\end{aligned}
\tag{6.5}
$$

These calculations are valid only for $n>0$, because $\alpha_0=0$ does not coincide with (6.2). The formulas (6.2) and (6.4) show that α_n is nondecreasing and γ_n is nonincreasing when $\alpha+\beta\le 1$. Both sequences tend to $\frac{1}{2}$. This gives the conditions (ii) and (iv) of Corollary 5.3. The formula (6.5) shows that $\alpha_n+\gamma_n$ is nondecreasing for $n>0$, regardless the sign of $\alpha\beta$. This and the fact that α_n is nondecreasing imply

$$
\alpha_m+\gamma_m\le\alpha_{n-1}+\gamma_{n-1}\le\alpha_{n+1}+\gamma_{n-1},\quad 0<m<n-1.
$$

Thus the condition (iii) of Corollary 5.3 is satisfied for $0<m<n-1$. It remains to show the condition (iii) for $m=0$, i.e.

$$
\alpha_0+\gamma_0=\gamma_0=\frac{2}{2+\alpha+\beta}\le\alpha_{n+1}+\gamma_{n-1}.
$$

By (6.2) and (6.4) the above inequality is equivalent to the following.

$$
\begin{aligned}
&-\frac{(\alpha-\beta)^2}{2n+\alpha+\beta+2}+\frac{(\alpha-\beta)^2-1}{2n+\alpha+\beta+3}\\
&\qquad-\frac{(\alpha+\beta)^2-1}{2n+\alpha+\beta-1}+\frac{(\alpha+\beta)^2}{2n+\alpha+\beta}\ge-\frac{2(\alpha+\beta)}{2+\alpha+\beta}.
\end{aligned}
\tag{6.6}
$$

Observe that the left hand side of (6.6) is a decreasing function of $\alpha-\beta$. Therefore we can assume that $\alpha-\beta$ attains the maximal possible value, i.e., $\beta=-1$. Let $\beta=-1$ and $x=2n+\alpha+\beta+1$. Then $x\ge 2+\alpha+\beta+1\ge 3$. The left hand side of (6.6) can be now written as follows.

$$
\begin{aligned}
&-\frac{(\alpha+1)^2}{x+1}+\frac{(\alpha+1)^2-1}{x+2}-\frac{(\alpha-1)^2-1}{x-2}+\frac{(\alpha-1)^2-1}{x-1}\\
&=\frac{4}{(x-2)(x+2)}-\frac{(\alpha+1)^2}{(x+1)(x+2)}-\frac{(\alpha-1)^2}{(x-1)(x-2)}\\
&=\frac{4}{(x-2)(x+2)}-\frac{4}{(x+1)(x+2)}+\frac{4-(\alpha+1)^2}{(x+1)(x+2)}-\frac{(\alpha-1)^2}{(x-1)(x-2)}.
\end{aligned}
$$

The first two terms of the last expression give a positive contribution to the sum because $x > 2$. Hence it suffices to show that

$$\frac{4-(\alpha+1)^2}{(x+1)(x+2)} - \frac{(\alpha-1)^2}{(x-1)(x-2)} \geq -\frac{2(\alpha-1)}{\alpha+1}. \qquad (6.7)$$

Note that $\alpha - 1 \geq 0$ (as $\beta = -1$). Thus $\alpha + 1 \geq 2$ and $4 - (\alpha+1)^2 \leq 0$. Hence the left hand side of (6.7) is a nondecreasing function of x. Therefore we can verify (6.7) only for the smallest value of x, that is for $x = 2 + \alpha + \beta + 1 = 2 + \alpha$. Under substitution $x = 2 + \alpha$ the inequality (6.7) takes the form

$$\frac{4-(\alpha+1)^2}{(\alpha+3)(\alpha+4)} - \frac{(\alpha-1)^2}{(\alpha+1)\alpha} \geq -\frac{2(\alpha-1)}{\alpha+1}.$$

After simple transformations it reduces to

$$\frac{1}{\alpha+4} \leq \frac{1}{\alpha},$$

which is true because α is nonnegative. Summarizing, Corollary 5.3 yields that for $\alpha \geq \beta$ and $0 \leq \alpha + \beta \leq 1$ we get (SNLP). □

References

Askey, R. (1970). Linearization of the product of orthogonal polynomials. In Gunning, R., editor, *Problems in Analysis*, pages 223–228. Princeton University Press, Princeton, NJ.

Askey, R. (1975). *Orthogonal polynomials and special functions*, volume 21 of *Regional Conference Series in Applied Mathematics*. Society for Industrial and Applied Mathematics, Philadelphia, PA.

Gasper, G. (1970a). Linearization of the product of Jacobi polynomials, I. *Canad. J. Math.*, 22:171–175.

Gasper, G. (1970b). Linearization of the product of Jacobi polynomials, II. *Canad. J. Math.*, 22:582–593.

Gasper, G. (1983). A convolution structure and positivity of a generalized translation operator for the continuous q-Jacobi polynomials. In *Conference on harmonic analysis in honor of Antoni Zygmund, Vol. I, II (Chicago, Ill., 1981)*, Wadsworth Math. Ser., pages 44–59. Wadsworth, Belmont, CA.

Gasper, G. and Rahman, M. (1983). Nonnegative kernels in product formulas for q-racah polynomials I. *J. Math. Anal. Appl.*, 95:304–318.

Hylleraas, E. (1962). Linearization of products of Jacobi polynomials. *Math. Scand.*, 10:189–200.

Koekeok, R. and Swarttouw, R. F. (1998). The Askey-scheme of hypergeometric orthogonal polynomials and its q-analogue. Faculty of Technical Mathematics and Informatics 98–17, TU Delft.

Markett, C. (1994). Linearization of the product of symmetric orthogonal polynomials. *Constr. Approx.*, 10:317–338.

Młotkowski, W. and Szwarc, R. (2001). Nonnegative linearization for polynomials orthogonal with respect to discrete measures. *Constr. Approx.*, 17:413–429.

Rahman, M. (1981). The linearization of the product of continuous q-Jacobi polynomials. *Can. J. Math.*, 33:961–987.

Rogers, L. J. (1894). Second memoir on the expansion of certain infinite products. *Proc. London Math. Soc.*, 25:318–343.

Szwarc, R. (1992a). Orthogonal polynomials and a discrete boundary value problem, I. *SIAM J. Math. Anal.*, 23:959–964.

Szwarc, R. (1992b). Orthogonal polynomials and a discrete boundary value problem, II. *SIAM J. Math. Anal.*, 23:965–969.

Szwarc, R. (1995). Nonnegative linearization and q-ultraspherical polynomials. *Methods Appl. Anal.*, 2:399–407.

Szwarc, R. (2003). A necessary and sufficient condition for nonnegative linearizatiom of orthogonal polynomials. To appear.

REMARKS ON SOME BASIC HYPERGEOMETRIC SERIES

Changgui Zhang
Laboratoire AGAT (UMR - CNRS 8524)
UFR Math.
Université des Sciences et Technologies de Lille
Cité Scientifique
59655 Villeneuve d'Ascq cedex
FRANCE
zhang@agat.univ-lille1.fr

Abstract Many results in Mathematical Analysis seem to come from some "obvious" computations. For a few years, we have been interested in the analytic theory of linear q-difference equations. One of the problems we are working on is the analytical classification of q-difference equations. Recall that this problem was already considered by G. D. Birkhoff and some of his students ((Birkhoff, 1913), (Birkhoff and Güenther, 1941)). An important goal of these works is to be able to derive transcendental analytical invariants from the divergent power series solutions; that is, to be able to define a good concept of Stokes' multiplier for divergent q-series! Very recently, we noted ((Zhang, 2002), (Ramis et al., 2003)) that this problem can be treated in a satisfactory manner by a new summation theory of divergent power series through the use of Jacobian theta functions and some basic integral calculus. The purpose of the present article is to explain how much "obvious" this mechanism of summation may be if one practises some elementary calculations on q-series. It would be a very interesting question to understand (Di Vizio et al., 2003) whether Ramanujan's mysterious formulas are related to this transcendental invariant analysis...

The article contains four sections. In the first section, we explain how to use the theta function for giving a q-integral representation which remains valid for the sum function of every convergent power series. It is this integral representation which leads us to a new process of summation of divergent series. Some identities then follow on the convergent power series, in the spirit of a Stokes analysis: the convergence takes place only in spite of the Stokes phenomenon!

It is well known that every basic hypergeometric series ${}_2\varphi_0$ is formal limit of a family of ${}_2\varphi_1$'s. In (Zhang, 2002), we have applied the new summation method to this divergent power series ${}_2\varphi_0$. In the third section of our present article, we prove that each sum of ${}_2\varphi_0$ is exactly the limit of an associated family of ${}_2\varphi_1$ while the parameter tends to infinity following a q-spiral.

In the last section, we give a remark about the Euler's Γ function. The best known q-analog of Γ is certainly the Jackson's Γ_q (Askey, 1978), which satisfies a first order q-difference equation deduced from the fundamental equation of Γ. This q-difference equation has a formal Laurent series solution that is divergent everywhere in $\mathbb{C}$ (Zhang, 2001). Using a summation formula of Ramanujan for ${}_1\psi_1$ and the above-mentioned summation method, one gets a new q-analog of Γ which is meromorphic on $\mathbb{C}^*$.

Some basic *notations.* In the following, q denotes a real number in the open interval $(0,1)$ and some notations of the book (Gasper and Rahman, 1990) will be used. For example, if $a \in \mathbb{C}$, we set $(a;q)_0 = 1$,

$$(a;q)_n = \prod_{m=0}^{n-1} (1 - aq^m), \qquad \forall n \in \mathbb{N}^* \cup \{+\infty\};$$

for $a_1, \ldots, a_\ell \in \mathbb{C}$, we set:

$$(a_1, \ldots, a_\ell; q)_n = (a_1;q)_n \times \cdots \times (a_\ell;q)_n, \qquad \forall n \in \mathbb{N} \cup \{+\infty\}.$$

To each fixed $\lambda \in \mathbb{C}^*$, we associate its so-called q-spiral $[\lambda;q]$ by setting $\lambda q^{\mathbb{Z}} = [\lambda;q] = \{\lambda q^n : n \in \mathbb{Z}\}$. If λ, $\mu \in \mathbb{C}^*$, the following conditions are equivalent:

$$[\lambda;q] \cap [\mu;q] \neq \emptyset \qquad \Longleftrightarrow \qquad [\lambda;q] = [\mu;q] \qquad \Longleftrightarrow \qquad \lambda/\mu \in q^{\mathbb{Z}}.$$

1. How to get the sum of a power series by means of θ

Let us denote by $\theta_q(x)$ or more shortly $\theta(x)$ the theta function of Jacobi, given by the following series:

$$\theta(x) = \sum_{n\in\mathbb{Z}} q^{n(n-1)/2} x^n, \qquad x \in \mathbb{C}^*.$$

Jacobi's triple product formula says that:

$$\theta(x) = (q, -x, -q/x; q)_\infty;$$

from this, it follows that $\theta(x) = 0$ if and only if $x \in [-1; q]$. Recall also that θ verifies the fundamental equation $\theta(x) = x\theta(qx)$ or, more generally,

$$\theta\left(q^n x\right) = q^{-n(n-1)/2} x^{-n} \theta(x) \tag{1.1}$$

for any $n \in \mathbb{Z}$.

Consider any given power series $f = \sum_{n\geq 0} a_n x^n$ with complex coefficients and let λ be an arbitrary nonzero complex number. **Suppose the radius R of convergence of f is > 0.** It is obvious that the product $f(x)\theta(\lambda/x)$ defines an analytical function in the truncated disc $0 < |x| < R$. From direct computations, one obtains the following identity:

$$f(x)\theta\left(\frac{\lambda}{x}\right) = \sum_{n\in\mathbb{Z}} \varphi_f\left(\lambda q^n\right) q^{n(n-1)/2} \left(\frac{\lambda}{x}\right)^n, \tag{1.2}$$

where φ_f denotes the entire function, depending upon f, defined as follows:

$$\varphi_f(\xi) \stackrel{\text{def}}{=} \sum_{n\geq 0} a_n q^{n(n-1)/2} \xi^n. \tag{1.3}$$

It is useful to note that for any $B > 1/R$, there exists $C > 0$ such that

$$\forall \xi \in \mathbb{C}^*, \qquad |\varphi_f(\xi)| < C\theta(B|\xi|).$$

According to (Ramis, 1992), the function φ_f is said *to have a q-exponential growth of order (at most) one at infinity.*

Let $\mathbb{C}\{x\}$ be the ring of all power series that converge near $x = 0$ and $\mathbb{E}_{q;1}$ the set of all entire functions having at most a q-exponential growth of order one at infinity.

Proposition 1.1. *The map $f \mapsto \varphi_f$ given in (1.3) establishes a bijection between $\mathbb{C}\{x\}$ and $\mathbb{E}_{q;1}$.*

More precisely, if $\varphi = \varphi_f$, $f \in \mathbb{C}\{x\}$ and if $R > 0$ is the radius of convergence of f, then the following assertions hold.

1. *For any $0 < r < R$, let $C_{0,r}^+$ be the counterclockwise-oriented circle centered at the origin and of radius r, then*

$$\varphi(\xi) = \frac{1}{2\pi i} \int_{C_{0,r}^+} f(x)\theta\left(\frac{\xi}{x}\right)\frac{dx}{x}, \qquad \xi \in \mathbb{C}^*. \tag{1.4}$$

2. *For any $\lambda \in \mathbb{C}^*$, the following q-integral*

$$\frac{1}{1-q}\int_0^{\lambda\infty} \frac{\varphi(\xi)}{\theta(\frac{\xi}{x})}\frac{d_q\xi}{\xi} \stackrel{\text{def}}{=} \sum_{n\in\mathbb{Z}} \frac{\varphi(\lambda q^n)}{\theta(\frac{\lambda q^n}{x})} \tag{1.5}$$

defines an analytical function which is equal to the sum of the convergent power series f in the open disc $0 < |x| < R$ minus the points of the set $(-\lambda q^{\mathbb{Z}})$ (i.e., the q-spiral $[-\lambda; q]$).

Accordingly, one can obtain the sum of a convergent power series by the following process:

$$f \longrightarrow \varphi = \varphi_f \longrightarrow \frac{1}{1-q} \int_0^{\lambda\infty} \frac{\varphi(\xi)}{\theta\left(\frac{\xi}{x}\right)} \frac{d_q\xi}{\xi} \; (= \mathcal{S}f). \tag{1.6}$$

Proof. Applying Cauchy's formula to φ_f gives the growth of its coefficients, from which we deduce that the map $f \mapsto \varphi_f$ is surjective (see (Ramis, 1992)). It is obvious that this map is also injective. Note that the q-integral representation of 1.1 is a reformulation of the identity (1.2), while the formula (1.4) follows directly from Cauchy. □

According to (Zhang, 2002) and (Ramis and Zhang, 2002), we shall denote by $\hat{\mathcal{B}}_{q;1}f$ the power series φ_f of (1.3) and by $\mathcal{L}_{q;1}^{[\lambda;q]}\varphi$ the q-integral of (1.5). It is important to remark that the convergence of $\mathcal{L}_{q;1}^{[\lambda;q]}\varphi$ depends only on the asymptotic behaviour of the function $\varphi(\xi)$ as ξ goes to ∞ along the q-spiral $[\lambda; q]$. So, let $\mathbb{H}_{q;1}^{[\lambda;q]}$ be the set of analytic function germs at the origin that can be analytically continued to a function having a q-exponential growth of order one at infinity in a neighborhood of $[\lambda; q]$. Here we call *neighborhood of* $[\lambda; q]$ any domain $V \subset \mathbb{C}$ such that there exists a neighborhood U of λ in $\mathbb{C}^*$ for which the inclusion $\{\xi q^n : \xi \in U, n \in \mathbb{Z}\} \subset V$ holds. It is essential to notice that $\mathbb{E}_{q;1} \subset \mathbb{H}_{q;1}^{[\lambda;q]}$ for every $\lambda \in \mathbb{C}^*$, the inclusion being strict. Therefore, the process (1.6) allows us to **sum** not only the convergent power series, but also any power series $\hat{f}$ that can be transformed by $\hat{f} \mapsto \hat{\mathcal{B}}_{q;1}\hat{f}$ to be an element of $\mathbb{H}_{q;1}^{[\lambda;q]}$. By definition, these power series $\hat{f}$ are called $[\lambda; q]$-*summable, of sum* $\mathcal{L}_{q;1}^{[\lambda;q} \circ \hat{\mathcal{B}}_{q;1}\hat{f}$. Here one uses the notation "$\hat{f}$" instead of "f", because the series under consideration is not necessarily convergent: one will have to distinguish a power series from "its sum(s)".

In (Zhang, 2002), it is shown that the summation process $\hat{f} \rightarrow \mathcal{L}_{q;1}^{[\lambda;q]} \circ \hat{\mathcal{B}}_{q;1}\hat{f}$ can be applied to every formal power series solution of any q-difference equation if this equation is, in some sense, generically singular. For example, all basic hypergeometric series ${}_2\varphi_0(a, b; —; q, x)$ are summable by this method. For more details, see (Zhang, 2002) and (Ramis and Zhang, 2002).

2. Identities characterizing the convergence of a power series

Let $\varphi = \varphi_f$, $f \in \mathbb{C}\{x\}$ (i.e., f is a convergent power series) and suppose R is the radius of convergence of f. Since $\theta(-1) = 0$, putting $\lambda = \xi$ and $x = -\xi$ in (1.2) gives the following formula:

$$\sum_{n\in\mathbb{Z}}(-1)^n q^{n(n-1)/2}\varphi\left(q^n\xi\right) = 0, \tag{2.1}$$

which is valid for any $\xi \in \mathbb{C}$ such that $|\xi| < R$.

The equality (2.1) can be viewed as an identity characterizing the convergence of the power series f such that $\varphi = \varphi_f$. Indeed, let φ be any analytical function known in a domain U of the form $U = \{\xi \in \mathbb{C} : |\xi| < 1\} \cup \{\xi \in \mathbb{C}^* : -\alpha < \arg x < \alpha\}$, where $\alpha \in (0, \pi)$. Suppose that φ has a q-exponential growth of order at most one at infinity. For each $x \in U$ such that $-\alpha < \arg x < \alpha$ and $|x| < R$ (for a suitable fixed real $R > 0$), we write

$$\Delta\varphi(\xi) = \sum_{n\in\mathbb{Z}}(-1)^n q^{n(n-1)/2}\varphi(q^n\xi).$$

Theorem 2.1. *Let $V_{\alpha;R} = \{\xi \in \mathbb{C}^* : -\alpha < \arg x < \alpha, |\xi| < R\}$. The function $\Delta\varphi$ is identically equal to zero on the sector $V_{\alpha;R}$ if and only if there exists a convergent power series f (i.e., $f \in \mathbb{C}\{x\}$) such that $\varphi = \varphi_f$.*

Proof. The "if" part has been explained at the beginning of this section. We shall prove the "only if" part.

For each $\delta \in (-\alpha, \alpha)$, we note

$$D_R^\delta = \{x \in \mathbb{C}^* : |x| < R, \delta - \pi < \arg x < \delta + \pi\}$$

and we set, if $x \in D_R^\delta$:

$$f^\delta(x) = \frac{1}{\ln(1/q)}\int_0^{e^{i\delta}\infty}\frac{\varphi(\xi)}{\theta\left(\frac{\xi}{x}\right)}\frac{d\xi}{\xi}. \tag{2.2}$$

Since φ is holomorphic in U, the functions f^δ can be glued into an analytic function on the sector

$$V = \left\{x \in \tilde{\mathbb{C}}^* : |x| < R, -\alpha - \pi < \arg x < \alpha + \pi\right\}$$

of the Riemann surface of the logarithm function. Let f be this function just constructed on V; by the theorem of residues, we get:

$$f\left(xe^{\pi i}\right) - f\left(xe^{-\pi i}\right) = \frac{2\pi i}{\ln(1/q)}\Delta\varphi(\xi)$$

for all $x \in V_{\alpha,R}$. By assumption, $\Delta\varphi(\xi) = 0$ on $V_{\alpha;R}$; hence, it follows that f can be identified to an analytical function in the truncated disc $0 < |x| < R$. We write again f for the latter function. By means of direct estimations for (2.2), one can check that f is bounded in a neighborhood of zero. Therefore, by Riemann's Removable Singularities Theorem the function f is holomorphic at the origin, i.e., f is the sum function of a convergent power series in the disc $|x| < R$.

It only remains to verify that the Taylor expansion of φ at zero coincides with the power series φ_f. To do this, one can use the following formula:

$$\frac{1}{\ln(1/q)} \int_0^{e^{i\delta}\infty} \frac{\xi^n}{\theta\left(\frac{\xi}{x}\right)} \frac{d\xi}{\xi} = q^{-n(n-1)/2}\xi^n;$$

for more details, see (Zhang, 2000). □

Now let's go back again to the formula (1.2) and let be given $\lambda \in \mathbb{C}^*$, $\mu \in \mathbb{C}^*$; one gets:

$$\theta\left(\frac{\mu}{x}\right) \sum_{n\in\mathbb{Z}} \varphi\left(\lambda q^n\right) q^{n(n-1)/2} \left(\frac{\lambda}{x}\right)^n = \theta\left(\frac{\lambda}{x}\right) \sum_{n\in\mathbb{Z}} \varphi\left(\mu q^n\right) q^{n(n-1)/2} \left(\frac{\mu}{x}\right)^n. \tag{2.3}$$

Recall that for any analytic function g given in an open disc $0 < |x| < r$, if $\sum_{n\in\mathbb{Z}} a_n x^n$ is its Laurent series expansion, then the formula (1.2) can be extended in the following way:

$$\theta\left(\frac{\lambda}{x}\right) g(x) = \sum_{m\in\mathbb{Z}} \sum_{n\in\mathbb{Z}} a_n q^{n(n-1)/2} \left(\lambda q^m\right)^n q^{m(m-1)/2} \left(\frac{\lambda}{x}\right)^m. \tag{2.4}$$

At the same time, the equality (2.3) takes the following form:

$$\sum_{m\in\mathbb{Z}} \sum_{n\in\mathbb{Z}} \varphi\left(\lambda q^n\right) q^{n^2} \left(\frac{\lambda}{\mu q^m}\right)^n q^{m(m-1)/2} \left(\frac{\mu}{x}\right)^m$$
$$= \sum_{m\in\mathbb{Z}} \sum_{n\in\mathbb{Z}} \varphi\left(\mu q^n\right) q^{n^2} \left(\frac{\mu}{\lambda q^m}\right)^n q^{m(m-1)/2} \left(\frac{\lambda}{x}\right)^m;$$

this implies: $\forall m \in \mathbb{Z}$,

$$\mu^m \sum_{n\in\mathbb{Z}} \varphi(\lambda q^n) \left(\frac{\lambda}{\mu}\right)^n q^{n(n-m)} = \lambda^m \sum_{n\in\mathbb{Z}} \varphi(\mu q^n) \left(\frac{\mu}{\lambda}\right)^n q^{n(n-m)}|.$$

So, $\forall \lambda \in \mathbb{C}^*$, $\forall \mu \in \mathbb{C}^*$, $\forall \epsilon \in \{0, 1/2\}$:

$$\sum_{n\in\epsilon+\mathbb{Z}} \varphi(\lambda q^n) \left(\frac{\lambda}{\mu}\right)^n q^{n^2} = \sum_{n\in\epsilon+\mathbb{Z}} \varphi(\mu q^n) \left(\frac{\mu}{\lambda}\right)^n q^{n^2}. \tag{2.5}$$

The identity (2.5) is also a formula that characterizes the convergence of the power series f such that $\varphi = \varphi_f$ $(= \hat{\mathcal{B}}_{q;1} f)$. Indeed, consider any function $\varphi \in \mathbb{H}_{q;1}^{[\lambda;q]}$; it may be noticed that $\varphi \in \mathbb{H}_{q;1}^{[\mu;q]}$ for all $\mu \in \mathbb{C}^*$ close enough to λ. The following result is essentially related to the PRINCIPLE that the $[\lambda; q]$-summation is a totally discontinuous mapping with respect to λ unless the power series to sum has in fact a radius of convergence > 0!

Theorem 2.2. *Let $\lambda \in \mathbb{C}^*$ and consider a function $\varphi \in \mathbb{H}_{q;1}^{[\lambda;q]}$. We have $\varphi \in \mathbb{E}_{q;1}$ if and only if there exists $\mu \in \mathbb{C}^*$ such that $[\lambda; q] \neq [\mu; q]$, $\varphi \in \mathbb{H}_{q;1}^{[\mu;q]}$ and that (2.5) holds for $\epsilon = 0$ and $1/2$.*

Proof. The identity (2.5) holds if, and only if, according to (2.3)–(2.4), we have the following:

$$\mathcal{L}_{q;1}^{[\lambda;q]} \varphi = \mathcal{L}_{q;1}^{[\mu;q]} \varphi.$$

Remember that each integral $\mathcal{L}_{q;1}^{[\lambda;q]} \varphi$ defines an analytic function in the truncated disc $0 < |x| < R$ minus all points of the q-spiral $[-\lambda; q]$. Hence, neither $\mathcal{L}_{q;1}^{[\lambda;q]} \varphi$ nor $\mathcal{L}_{q;1}^{[\mu;q]} \varphi$ has singularity in the disc $0 < |x| < R$. The proof may be completed by the Removable Singularities Theorem and the fact that $\mathcal{L}_{q;1}^{[\lambda;q]} \varphi$ is asymptotic to the power series $\hat{f}$ such that $\varphi = \hat{\mathcal{B}}_{q;1} \hat{f}$; see (Zhang, 2002) and (Ramis and Zhang, 2002). □

It is obvious that the formulas (2.1), (2.5) can be extended to a very larger class of functions that may possess a singularity at zero; cf. (2.4). On the other hand, if we only restricted to the class of rational functions $\varphi = P/Q$ $(P, Q \in \mathbb{C}[\xi])$, the equalities (2.1) and (2.5) would give criteria on the divisibility of P by Q: what unexpected criteria (**how to make them effective** ?)!

3. Confluence of ${}_2\varphi_1$ to ${}_2\varphi_0$ along a q-spiral

For any a, b, $c \in \mathbb{C}$ such that $c \notin q^{-\mathbb{N}}$, we write ${}_2\varphi_1(a, b; c; q; x)$, as in the book (Gasper and Rahman, 1990), the following Heine's series:

$$ {}_2\varphi_1(a, b; c; q; x) = \sum_{n \geq 0} \frac{(a; q)_n (b; q)_n}{(q; q)_n (c; q)_n} x^n. $$

If $ab \neq 0$ and $a/b \notin q^{\mathbb{Z}}$, the sum of the power series ${}_2\varphi_1(a, b; c; q, x)$ can be analytically continued in the cut plane $\mathbb{C} \setminus [1, +\infty)$ by the following formula, due to G. N. Watson:

$$ \begin{aligned} {}_2\varphi_1(a, b; c; q, x) = {} & \frac{(b, c/a; q)_\infty}{(c, b/a; q)_\infty} \frac{\theta(-ax)}{\theta(-x)} \, {}_2\varphi_1\left(a, aq/c; aq/b; q, \frac{cq}{abx}\right) \\ & + \frac{(a, c/b; q)_\infty}{(c, a/b; q)_\infty} \frac{\theta(-bx)}{\theta(-x)} \, {}_2\varphi_1\left(b, bq/c; bq/a; q, \frac{cq}{abx}\right). \end{aligned} \tag{3.1} $$

Note that if $c \to \infty$, the series ${}_2\varphi_1(a, b; c; q, ct)$ converges termwise to ${}_2\varphi_0(a, b; —; q, t)$, where

$$ {}_2\varphi_0(a, b; —; q, t) = \sum_{n \geq 0} \frac{(a, b; q)_n}{(q; q)_n} q^{-n(n-1)/2} (-t)^n. \tag{3.2} $$

Unless a or $b \in q^{-\mathbb{N}}$, the last power series is divergent for all $t \neq 0$ and, according to Theorem 2.2.1 (Zhang, 2002), it is $[\lambda; q]$-summable for all $\lambda \in \mathbb{C}^* \setminus (-q^{\mathbb{Z}})$. More precisely, let ${}_2f_0(a, b; \lambda, q, t)$ be the sum of ${}_2\varphi_0(a, b; —; q, t)$ corresponding to the "path" $[\lambda; q]$. By Theorem 2.2.1 (Zhang, 2002), one has the following formula, which is similar to the above-mentioned one (3.1):

$$ \begin{aligned} {}_2f_0(a, b; \lambda, q, t) = {} & \frac{(b; q)_\infty}{(b/a; q)_\infty} \frac{\theta(a\lambda)}{\theta(\lambda)} \frac{\theta(qat/\lambda)}{\theta(qt/\lambda)} \, {}_2\varphi_1\left(a, 0; \frac{aq}{b}; q, \frac{q}{abt}\right) \\ & + \frac{(a; q)_\infty}{(a/b; q)_\infty} \frac{\theta(b\lambda)}{\theta(\lambda)} \frac{\theta(qbt/\lambda)}{\theta(qt/\lambda)} \, {}_2\varphi_1\left(b, 0; \frac{bq}{a}; q, \frac{q}{abt}\right). \end{aligned} \tag{3.3} $$

Theorem 3.1. *Let $\lambda \in \mathbb{C}^* \setminus \left(-q^{\mathbb{Z}}\right)$. For all $t \in \mathbb{C}^* \setminus \left(-\lambda q^{\mathbb{Z}}\right)$, one has:*

$$ \lim_{\substack{n \in \mathbb{N} \\ n \to \infty}} {}_2\varphi_1\left(a, b; -\frac{q^{-n}}{\lambda}; q, -\frac{q^{-n}}{\lambda} t\right) = {}_2f_0(a, b; \lambda, q, t). \tag{3.4} $$

Remark that this Theorem has a "fast" proof: one verifies that the right side of (3.1) converges, as $c = -\frac{q^{-n}}{\lambda}$, $n \in \mathbb{N}$, $x = ct$ and $n \to +\infty$, to the right side of (3.3); to do this, the following lemma is helpful.

Lemma 3.2. *Let λ, $a \in \mathbb{C}^*$ such that $\lambda \notin [-1;q]$. One has the following:*

$$\lim_{\substack{n\in\mathbb{N}\\ n\to\infty}} \frac{(-a\lambda q^{-n};q)_\infty}{(-\lambda q^{-n};q)_\infty} a^{-n} = \frac{\theta(a\lambda)}{\theta(\lambda)}. \tag{3.5}$$

Therefore, if α, β, γ and δ are four complex numbers such that $\alpha\beta = \delta\gamma \neq 0$, $\delta \notin [-1;q]$, $\gamma \notin [-1;q]$, then the following limit holds:

$$\lim_{\substack{n\in\mathbb{N}\\ n\to\infty}} \frac{(-\alpha q^{-n}, -\beta q^{-n})_\infty}{(-\gamma q^{-n}, -\delta q^{-n};q)_\infty} = \frac{\theta(\alpha)\theta(\beta)}{\theta(\gamma)\theta(\delta)}.$$

Proof. It suffices to notice that, for any $n \in \mathbb{N}$ one has:

$$(-\lambda q^{-n};q)_\infty = (-\lambda;q)_\infty \left(-\frac{1}{\lambda};q\right)_n \lambda^n q^{-n(n+1)/2}.$$

□

Now, we shall give a "direct" proof of Theorem 3.1: it is helpful to understand in which way the formula (1.4) goes to its limit form after the confluence along a q-spiral.

Proof of Theorem 3.1. Let $n \in \mathbb{N}$; thanks to the formula (1.2), one has:

$$\theta\left(\frac{\lambda}{t}\right) {}_2\varphi_1\left(a, b; -\frac{q^{-n}}{\lambda}; q, -\frac{q^{-n}}{\lambda}t\right) = \sum_{m\in\mathbb{Z}} \varphi_{f_n}(q^m\lambda)\, q^{m(m-1)/2}\left(\frac{\lambda}{t}\right)^m,$$

where

$$\begin{aligned}\varphi_{f_n}(q^m\lambda) &= {}_2\varphi_2\left(a, b; -\frac{q^{-n}}{\lambda}, 0; q, q^{-n+m}\right)\\ &= \frac{1}{2\pi i}\int_{C^+_{0;r}} {}_2\varphi_1\left(a, b; -\frac{q^{-n}}{\lambda}; q, x\right)\theta\left(-\frac{q^{-n+m}}{x}\right)\frac{dx}{x}.\end{aligned} \tag{3.6}$$

Next, one expands the function ${}_2\varphi_1\left(a, b; -\frac{q^{-n}}{\lambda}; q, x\right)$ at infinity by means of the formula (3.1). Hence, the integral of (3.6) gives the following:

$$\begin{aligned}&\varphi_{f_n}(q^m\lambda)\\ &= \frac{(b, -q^{-n}/(a\lambda);q)_\infty}{(-q^{-n}/\lambda, b/a;q)_\infty} a^{1-m+n}\, {}_2\varphi_2\left(a, -a\lambda q^{n+1}; \frac{aq}{b}, 0; q, -\frac{q^{2-m}}{b}\right)\\ &+ \frac{(a, -q^{-n}/b\lambda);q)_\infty}{(-q^{-n}/\lambda, a/b;q)_\infty} b^{1-m+n}\, {}_2\varphi_2\left(b, -b\lambda q^{n+1}; \frac{bq}{a}, 0; q, -\frac{q^{2-m}}{a}\right).\end{aligned}$$

which, together with the formula (3.5), leads to the conclusion of Theorem 3.1. □

4. q-analogs of Γ and summation of a basic bilateral series

The Jackson's q-Gamma function

$$\Gamma_q(z) = \frac{(q;q)_\infty}{(q^z;q)_\infty}(1-q)^{1-z}$$

satisfies the functional equation

$$G(z+1) = \frac{q^z-1}{q-1}G(z). \tag{4.1}$$

Letting $x = q^z$, this equation leads us to the following q-difference equation:

$$(q-1)y(qx) = (x-1)y(x). \tag{4.2}$$

Put $y = \sum\limits_{n\in\mathbb{Z}} \alpha_n x^n$ in (4.2); by checking the corresponding coefficients, one gets:

$$\alpha_n = \frac{1}{1-(1-q)q^n}\,a_{n-1}, \qquad \forall n \in \mathbb{Z}.$$

It follows that, if $1-q \notin q^{\mathbb{Z}}$, then:

$$\alpha_n = \frac{\alpha_0}{((1-q)q;q)_n} \qquad \left(\stackrel{\text{def}}{=} \frac{\left((1-q)q^{1+n};q\right)_\infty}{((1-q)q;q)_\infty}\alpha_0\right). \tag{4.3}$$

In the rest of this paper, we suppose that $1-q \notin q^{\mathbb{Z}}$. It is immediate to observe that if $1-q \in q^{\mathbb{Z}}$, any formal solution will be convergent.

Let's denote by $\hat{g}$ the Laurent series corresponding to $\alpha_0 = 1$:

$$\hat{g}(x) = \sum_{n\in\mathbb{Z}} \frac{x^n}{((1-q)q;q)_n}.$$

If $n \to -\infty$, then $\left((1-q)q^{1+n};q\right)_\infty = O\left((q-1)^{-n}q^{-n(n+1)/2}\right)$; it follows that the polar part of $\hat{g}(x)$ is divergent in $\mathbb{C}$ everywhere. Recall that the summation method described in (1.6) is valid for power series. Now one extends this method to the Laurent series $\hat{g}$ in the following way (see (2.4)):

$$\begin{aligned}\hat{g}(x) &\overset{\hat{\mathcal{B}}_{q;1}}{\Longrightarrow} {}_0\psi_1(-;(1-q)q;q,-\xi) = \sum_{n\in\mathbb{Z}} \frac{q^{n(n-1)/2}}{((1-q)q;q)_n}\xi^n \\ &\overset{\mathcal{L}_{q;1}^{[\lambda;q]}}{\Longrightarrow} \mathcal{L}_{q;1}^{[\lambda;q]} \circ \hat{\mathcal{B}}_{q;1}\hat{g}(x) \stackrel{\text{def}}{=} \mathcal{L}_{q;1}^{[\lambda;q]}\,{}_0\psi_1(-;(1-q)q;q,-\xi)(x),\end{aligned} \tag{4.4}$$

where $\lambda \in \mathbb{C}^* \setminus \big((q-1)q^{\mathbb{Z}}\big)$.

Lemma 4.1. *For all $\beta \notin q^{-\mathbb{N}}$, one has:*

$$ {}_0\psi_1(-;\beta;q,x) = \frac{\left(q,x,\frac{q}{x};q\right)_\infty}{\left(\beta,\frac{\beta}{x};q\right)_\infty}. $$

Proof. It suffices to use Ramanujan's summation formula for ${}_1\psi_1$ (*cf* (Gasper and Rahman, 1990), page 126, (5.2.1)), noticing also the fact that

$$ {}_0\psi_1(-;\beta;q,x) = \lim_{\alpha\to\infty} {}_1\psi_1\left(\alpha;\beta;q,\frac{x}{\alpha}\right). $$

□

In particular, the following identity holds:

$$ \hat{\mathcal{B}}_{q;1}\hat{g}(\xi) = \frac{\theta(\xi)}{\theta\left(\frac{\xi}{1-q}\right)} \frac{(q;q)_\infty}{((1-q)q;q)_\infty} \left(-\frac{\xi}{1-q};q\right)_\infty. $$

If $n \in \mathbb{Z}$ and $\lambda \in \mathbb{C}^* \setminus (q-1)q^{\mathbb{Z}}$, one gets, from the formula (1.1):

$$ \hat{\mathcal{B}}_{q;1}\hat{g}\left(\lambda q^n\right) = \frac{\left(q,-\frac{\lambda}{1-q};q\right)_\infty}{((1-q)q;q)_\infty} \frac{\theta(\lambda)}{\theta\left(\frac{\lambda}{1-q}\right)} \frac{1}{(1-q)^n\left(-\frac{\lambda}{1-q};q\right)_n}; $$

from this and Lemma 4.1 one deduces that

$$ \begin{aligned} &\mathcal{L}_{q;1}^{[\lambda;q]} \circ \hat{\mathcal{B}}_{q;1}\hat{g}(x) \\ &= \frac{1}{\left((1-q)q,-\frac{(1-q)q}{\lambda};q\right)_\infty} \frac{\theta(\lambda)}{\theta\left(\frac{\lambda}{x}\right)} {}_0\psi_1\left(-;-\frac{\lambda}{1-q};q,-\frac{\lambda}{(1-q)x}\right) \\ &= \frac{(q;q)_\infty}{((1-q)q,x;q)_\infty} \frac{\theta(\lambda)\theta\left(\frac{\lambda}{(1-q)x}\right)}{\theta\left(\frac{\lambda}{x}\right)\theta\left(\frac{\lambda}{1-q}\right)}. \end{aligned} $$

If $x = q$, one verifies without difficulties that

$$ \mathcal{L}_{q;1}^{[\lambda;q]} \circ \hat{\mathcal{B}}_{q;1}\hat{g}(q) = \frac{1}{(1-q;q)_\infty}; $$

hence we are ready to state the following result.

Theorem 4.2. *Consider the formal solution* $\hat{\Gamma}(q;z) = (1-q;q)_\infty \hat{g}(q^z)$ *of the equation* (4.1). *One has*

$$\hat{\Gamma}(q;z) = (1-q)\sum_{n\in\mathbb{Z}} ((1-q)q^n;q)_\infty \, q^{nz}$$

and the series $\hat{\Gamma}(q;z)$ *is* $[\lambda;q]$*-summable in the variable* q^z *for all* $\lambda \in \mathbb{C}\setminus(q-1)q^{\mathbb{Z}}$*, with sum:*

$$\mathcal{L}^{[\lambda;q]}_{q;1} \circ \hat{\mathcal{B}}_{q;1}\hat{\Gamma}(q;z) = (1-q)\,\frac{(q;q)_\infty}{(q^z;q)_\infty}\,\frac{\theta(\lambda)\theta\left(\frac{\lambda}{(1-q)}q^{-z}\right)}{\theta\left(\frac{\lambda}{1-q}\right)\theta\left(\lambda q^{-z}\right)}.$$

When q *tends to* 1*, the following limit holds:*

$$\lim_{q\to 1} \mathcal{L}^{[\lambda;q]}_{q;1} \circ \hat{\mathcal{B}}_{q;1}\hat{\Gamma}(q;z) = \Gamma(z)$$

for all $z \in \mathbb{C}\setminus\{0,-1,-2,\dots\}$*, the convergence being uniform on every compact subset of* $\mathbb{C}\setminus\{0,-1,-2,\dots\}$.

Proof. It only remains to check the limit, which can be deduced from the following formulas (see (Askey, 1978)):

$$\lim_{q\to 1^-} \frac{(q;q)_\infty}{(q^a;q)_\infty}(1-q)^{1-a} = \Gamma(a), \qquad \lim_{q\to 1^-} \frac{\theta(q^a x)}{\theta(q^b x)} = x^{b-a},$$

$$\lim_{q\to 1^-} \frac{\theta((1-q)q^a x)}{\theta((1-q)q^b x)}(1-q)^{a-b} = x^{b-a}.$$

□

Acknowledgments

The author would like to thank Jacques Sauloy for his constructive suggestions.

References

Askey, R. (1978). The q-gamma and q-beta functions. *Appl. Anal.*, 8:125–141.

Birkhoff, G. D. (1913). The generalized Riemann problem for linear differential equations and the allied problems for linear difference and q-difference equations. *Proc. Am. Acad.*, 49:521–568.

Birkhoff, G. D. and Güenther, P. E. (1941). Note on a canonical form from the linear q-difference system. *Proc. Nat. Acad. of Sci.*, 27:218–222.

Di Vizio, L., Ramis, J.-P., Sauloy, J., and Zhang, C. (2003). Equations aux q-différences. *Gazette des mathématiciens (SMF)*, 96:20–49.

Gasper, G. and Rahman, M. (1990). *Basic Hypergeometric Series.* Encycl. Math. Appl. Cambridge Univ. Press, Cambridge.

Ramis, J.-P. (1992). About the growth of entire functions solutions of linear algebraic q-difference equations. *Annales de la Fac. de Toulouse Math. (6)*, I(1):53–94.

Ramis, J.-P., Sauloy, J., and Zhang, C. (2003). Classification analytique locale des équations aux q-différences. In preparation.

Ramis, J.-P. and Zhang, C. (2002). Développement asymptotique q-Gevrey et fonction thêta de Jacobi. *C. R. Acad. Sci. Paris*, 335(11):899–902.

Zhang, C. (2000). Transformations de q-Borel-Laplace au moyen de la fonction thêta de Jacobi. *C. R. Acad. Sci. Paris Sér. I Math.*, 331:31–34.

Zhang, C. (2001). Sur la fonction q-Gamma de Jackson. *Aequationes Math.*, 62:60–78.

Zhang, C. (2002). Une sommation discrète pour des équations aux q-différences linéaires et à coefficients analytiques: théorie générale et exemples. In Braaksma, B. L. J., Immink, G. K., der Put, M. V., and Top, J., editors, *Proceedings of the Conference on Differential Equations and the Stokes Phenomenon (Groningen, 28–30 May 2001)*, pages 309–329. World Scientific.